CHANCE AND CH[...]
CARDPACK AND CH[...]

I0043495

AN INTRODUCTION TO PROBABILITY IN PRACTICE BY VISUAL AIDS

VOLUME I

by

LANCELOT HOGBEN M.A. (Cantab.) D.Sc. (Lond.) F.R.S.

PROFESSOR OF MEDICAL STATISTICS IN THE UNIVERSITY OF BIRMINGHAM

We do ill to exalt the powers of the human mind,
when we should rather seek out its proper helps
FRANCIS BACON

One time seeing is worth a thousand times hearing
Old Chinese proverb

LONDON
MAX PARRISH & CO LIMITED
1950

MAX PARRISH AND CO LIMITED
ADPRINT HOUSE RATHBONE PLACE LONDON W I
IN ASSOCIATION WITH
CHANTICLEER PRESS INC
41 EAST 50TH STREET NEW YORK 22 N.Y.
ADPRINT LIMITED LONDON

CHARTS DRAWN
BY
GLADYS HAINES
FROM DESIGNS BY THE AUTHOR

PRINTED IN GREAT BRITAIN AT
THE UNIVERSITY PRESS
ABERDEEN

AUTHOR'S FOREWORD

UNLESS higher education is to assume an increasingly authoritarian aspect, increasing demand for statistical instruction by an ever-widening range of specialties is a challenge inviting more diligent concern than it has so far provoked. That less than 1 per cent. of research workers clearly apprehend the rationale of statistical techniques they commonly invoke would be less disturbing if the logicians of science were themselves unanimous on matters of anything but trivial importance, or if those who resort to prescribed procedures had alternative means of checking their claims. Neither the one nor the other is true. With or without a knowledge of its theoretical rationale, the laboratory worker can call to his aid with a clear conscience a device such as the glass electrode or the spectrometer, because he can, and in practice usually does, calibrate a new physical method by recourse to a known standard; but there is no known standard by which we can assess the long-run results of acting on the assumption that the famous postulate of Thomas Bayes leads more or less often to correct judgments about the world we live in. None the less, its implications are by no means trivial ; and statisticians of good repute are of diverse persuasions concerning its credentials.

One therefore hopes that an attempt to present the elements of statistical theory by exploiting a new educational technique will commend itself to the sympathetic consideration of statisticians who are also interested in education. If the execution of the undertaking falls short of what more expert use of visual aids may accomplish, it is the author's hope that a novel approach to the logical assumptions inherent in the symbolic treatment of statistical methods will stimulate others to undertake with greater success what he has failed to achieve. Since it is the fashion to decry that part of statistical theory which rightly pertains to large samples, it is necessary to explain that the second volume in preparation for the press is an attempt to present the basis of more modern procedures without leaving the solid ground of statistical models to clarify the nature of the null hypothesis implicit in the mathematical postulates. Such is the method of this volume, illustrated at the outset (p. 96) by Fisher's illuminating tea-cup parable.

Many useful manuals set forth standard methods of statistical analysis for the student who is prepared to take the formulæ on trust. A few authoritative treatises deal with the rationale of statistical techniques for the benefit of the reader who can employ mathematical operations for which only trained mathematicians have the requisite facility. This book is neither for the one nor for the other. Vol. I has little to say about statistical techniques which have lately come into widespread use among research workers content to apply them without an understanding of their credentials. Though avoiding issues of rigour which the author is not competent to arbitrate upon, it deals only with statistical methods for which it can offer the reader a rationale *en rapport* with the rules of algebra and differential calculus nowadays included in the higher school leaving syllabus, if supplemented with a few less accessible theorems set forth in the introductory chapter. Thus it is for those who prefer to use traditional methods of which they understand the implications—and therefore the limitations—rather than to rely on more refined devices whose theoretical postulates are open to controversy or beyond their powers of appraisal.

It is a pleasure to acknowledge the patience with which Miss Gladys Haines collaborated with me in the lay-out of the visual material. Should the method of presentation appeal to other teachers, some may be interested to know that the illustrations are obtainable from the publishers as a 2-colour complete set of 84 wall charts.

TABLE OF CONTENTS

TABLE OF CONTENTS

*

TO

ENID CHARLES

CHAPTER 1

FIGURATE NUMBERS AND FUNDAMENTAL APPROXIMATIONS

Introductory Remarks

THE student who is not an expert mathematician experiences difficulties of two sorts in appraising the validity and scope of statistical techniques. Some arise because the type of mathematics on which the latter rely calls for prolonged exercise and considerable aptitude ; but difficulties of this sort do not constitute an insuperable obstacle to an understanding of the basic principles of probability. Discouragement of another kind arises less because essential manipulations are intrinsically difficult than because they are unfamiliar. Indeed, it is well-nigh impossible to recapture the thought of those who laid the foundations of the theory of chance in the seventeenth and eighteenth centuries, unless one has some knowledge of relatively elementary algebraic devices without a niche in the college or higher school text-books of to-day. The historical reason for this lacuna is not far to seek. Triumphs of mathematical skill in the domain of physical measurements have encouraged increasing preoccupation with the infinitesimal calculus and with such parts of algebra as lie on the road to it. In short, the continuum is the keynote of contemporary courses from which the beginner may carry away the conviction that algorithms now mastered before the teen ages circumscribe any practical use pertaining to the properties of *whole numbers* as such.

This is unfortunate, because the theory of statistics * is always about what we can *count* individually ; and only sometimes about what we can also *measure*. It is important to be explicit about this feature of statistical analysis at the outset, the more so because it is not uncommon ·to see methods devised for treatment of small samples specified by metrical characteristics extended beyond their legitimate terms of reference or the presumptive intentions of their authors to the treatment of small samples specified by all-or-none attributes. Numbers which specify individuals or items of an assemblage are necessarily whole numbers like the number of sheep in a field. There may be 50 sheep, 51 sheep, or 50 sheep plus a certain amount of mutton ; but there cannot be 50·987 sheep in any field.

For this reason, we here start, as Pascal started, with some preliminary theorems not commonly dealt with in modern text-books of elementary scope though by no means recondite on that account. Our first chapter contains some material which will not be new to many readers who may find more novel substance in later ones, and much that may well be new to some readers sufficiently familiar with the elements of differential calculus to follow subsequent arguments which invoke its use. In either case, the reader will be wise to skim through its pages at the outset to get a bird's-eye view of its contents before devoting detailed study to sections which break new ground. The first three sections (1.01-1.03) call for comment because they introduce symbols employed consistently throughout the rest of the book. Others, notably 1.05 and 1.06, introduce essential algebraic theorems which will be new to readers whose knowledge of algebra does not embrace the study of finite differences and the elementary properties of hyper-

---

* i.e. *Sampling* theory in contradistinction to *frequency* theory of aggregates of molecules, genes, etc. (See remarks on page 99.)

geometric series. The latter half of 1.02 and the greater part of 1.08 are not essential to what follows. The same is true of pp. 72-74 in Chapter 2. The justification for their inclusion in the text is that they provide peculiar opportunities for exercise in algebraic manipulations of a type rarely invoked in elementary text-books but useful for the student of mathematical probability.

For the reader not familiar with more advanced modern text-books the following explanatory remarks about convenient symbols may be necessary. If we think of zero as an arbitrary point in a scale or sequence with the domain of negative and positive integers respectively to the left and right in accordance with the Cartesian convention, $-3 < 0 < +2$ or $+5 > 0 > -4$, so that in general $+a > -a$ and $-b < +b$. When we wish to signify that x is *numerically* greater than y regardless of sign we write $x > |y|$ ("x greater than *mod* y") or $|y| < x$. For x *not less than* y (i.e. x equal to or greater than y) we write $x \geqslant y$ in preference to $x \nless y$. Similarly, for x *not greater than* y (i.e. x equal to or less than y) we write $x \leqslant y$ in preference to $x \ngtr y$. One other symbol is of great importance, since so many statistical formulæ are convenient approximations precise enough for practical needs. In contradistinction to $=$, our symbol $\simeq$ signifies *approximately equal to*.

It is reasonable to assume that some readers with sufficient background to benefit from any uses this book may have will have forgotten some of the algebra, more especially differential calculus, necessary for the exposition of principles set forth in later chapters. Accordingly, the writer has taken the precaution to intercalate sections to provide opportunities of revision for readers who have not ready access to text-books of mathematics. The exercises are designed, where possible, to anticipate subsequent themes ; and an asterisk signifies that a result will be of use at a later stage. The insertion of certain lemmas as exercises on a relevant class of operations is intentional to dispense with the need for digressions which distract attention from the main issue ; and the reader will be well advised to pay attention to them, as they arise.

1.01   Symbolism of Continued Addition and Multiplication

In the mathematical analysis of choice and chance we frequently have recourse to expressions involving continued sums or products. It is customary to represent the operation of summation briefly by use of the Greek capital s (*sigma*) thus :

$$a_0 + a_1 + a_2 + \ldots + a_{n-1} + a_n = \sum_{x=0}^{x=n} a_x.$$

In operations involving summation, it is important to remember the following relation :

$$\sum_{x=a}^{x=b} a_x = \sum_{x=1}^{x=b} a_x - \sum_{x=1}^{x=(a-1)} a_x \qquad . \qquad . \qquad . \qquad . \qquad . \quad \text{(i)}$$

By analogy, we may use the Greek capital p (*pi*) to indicate briefly a product of terms referable to a common pattern as thus :

$$a_0 \cdot a_1 \cdot a_2 \ldots a_{n-1} \cdot a_n = \prod_{x=0}^{x=n} \cdot a_x.$$

If k is a constant, we have

$$\sum_{x=0}^{x=n} k a_x = k \sum_{x=0}^{x=n} a_x \; ; \quad \prod_{x=0}^{x=n} k a_x = k^{n+1} \prod_{x=0}^{x=n} a_x.$$

For the class of continued products represented by $n!$ or $\underline{|n}$ the corresponding formula would be

$$n! = n(n-1)(n-2) \ldots 3 \cdot 2 \cdot 1 = \prod_{x=0}^{x=(n-1)} (n-x).$$

For such products of which successive factors differ by unity, there is a more economical notation of *factorial powers*, based on analogy with the meaning of whole number exponents, as shown below :

n	$= n^1$	n	$= n^{(1)}$
$n \cdot n$	$= n^2$	$n(n-1)$	$= n^{(2)}$
$n \cdot n \cdot n$	$= n^3$	$n(n-1)(n-2)$	$= n^{(3)}$
$n \cdot n \cdot n \cdot n$	$= n^4$	$n(n-1)(n-2)(n-3)$	$= n^{(4)}$
$n \cdot n \cdot n \cdot n \ldots r$ factors $= n^r$		$n(n-1)(n-2)(n-3) \ldots r$ factors $= n^{(r)}$	

The last of r factors in such a product is $(n - r + 1)$. Accordingly

$$n^{(r)} = n(n-1)(n-2)(n-3) \ldots (n-r+2)(n-r+1) . \qquad . \qquad . \text{ (ii)}$$

By definition

$$n^{(n)} = n(n-1)(n-2)(n-3) \ldots n \text{ factors} = n! \qquad . \qquad . \text{ (iii)}$$

If $x > r$, by definition

$$(x-r)! = (x-r)(x-r-1)(x-r-2) \ldots 3 . 2 . 1$$
$$x^{(r)} = x(x-1)(x-2) \ldots (x-r+1)$$
$$\therefore x^{(r)}(x-r)! = x(x-1)(x-2) \ldots (x-r+1)(x-r)(x-r-1) \ldots 3 . 2 . 1$$
$$= x!$$
$$\therefore x^{(r)} = \frac{x!}{(x-r)!} . \qquad . \qquad . \qquad . \qquad . \qquad . \qquad . \qquad . \text{ (iv)}$$

Thus

$$7^{(4)} = 7 . 6 . 5 . 4 = (7 . 6 . 5 . 4 . 3 . 2 . 1) \div (3 . 2 . 1).$$

Another useful formula involving factorial powers is deducible by reversing the order of the factors of a product :

$$n(n+1)(n+2) \ldots r \text{ factors} = n(n+1)(n+2) \ldots (n+r-2)(n+r-1).$$

If we reverse the order, we have

$$(n+r-1)(n+r-2) \ldots (n+2)(n+1)n$$
$$= (n+r-1)\overline{(n+r-1-1)}\overline{(n+r-1-2)} \ldots \overline{(n+r-1-r+1)}$$
$$= (n+r-1)(n+r-2) \ldots r \text{ factors}.$$
$$\therefore n(n+1)(n+2) \ldots r \text{ factors} = (n+r-1)^{(r)} \qquad . \qquad . \text{ (v)}$$

Several important properties of factorial powers follow from the following relation which holds good if $n > m$:

$$x^{(n)} = x(x-1)(x-2) \ldots (x-m+2)(x-m+1)(x-m)(x-m-1) \ldots$$
$$(x-n+2)(x-n+1)$$
$$= [x(x-1) \ldots (x-m+1)][\overline{(x-m)}\overline{(x-m-1)} \ldots$$
$$\overline{(x-m-\overline{n-m}+2)}\overline{(x-m-\overline{n-m}+1)}]$$
$$= x^{(m)} . (x-m)^{(n-m)}.$$

Subject to the same condition, the laws of composition of factorial indices follow at once from this, *viz.* :

$$x^{(m)} . x^{(n)} = [x^{(m)}]^2(x-m)^{(n-m)} \quad and \quad \frac{x^{(n)}}{x^{(m)}} = (x-m)^{(n-m)} \qquad . \qquad . \text{ (vi)}$$

If we put $m = 0$ in either of the last two expressions, we get

$$x^{(n)} = x^{(0)} \cdot x^{(n)}$$
$$\therefore x^{(0)} = 1 \qquad . \qquad . \qquad . \qquad . \qquad . \qquad . \qquad \text{(vii)}$$

Also from (iii) and (iv)

$$\frac{n^{(n)}}{n!} = 1 = \frac{n!}{n!(n-n)!} = \frac{1}{0!}$$
$$\therefore 0! = 1 \qquad . \qquad . \qquad . \qquad . \qquad . \qquad . \qquad \text{(viii)}$$

It is useful to *memorise* formulæ (i)-(viii) above.

EXERCISE 1.01

1.★ Show that
$$\prod_{x=0}^{x=n} ab_x = a^{n+1} \prod_{x=0}^{x=n} b_x.$$

2. Make an 11 × 11 table exhibiting the values of $a^{(r)}$ for all values of a from 0 to 10 inclusive and all values of r from 0 to 10 inclusive.

3. Make a table of factorial $n!$ from 0 to 12 inclusive.

4. Find the numerical values of :
$$\frac{7^{(4)}}{3^{(2)}} ; \quad \frac{10^{(5)}}{5^{(5)}} ; \quad \frac{4^{(3)}}{9^{(6)}} ; \quad \frac{6^{(4)}}{10^{(2)}}.$$

5.★ Investigate the meaning of $x^{(-n)}$ by recourse to equation (vi) above.

6.★ By recourse to (vi) and (vii), show that the reciprocal of the factorial of a negative number is zero.

7.★ If $u_x = s^{(x)} \div f^{(x)}$, show that
$$u_{x+1} = \frac{s^{(x+1)}}{f^{(x+1)}} = \frac{(s-x)}{(f-x)} u_x.$$

8.★ In the same way, show that
$$u_{x-1} = \frac{(f-x+1)}{(s-x+1)} \cdot u_x.$$

9.★ Show that
$$(kn)! = k^{kn} \prod_{r=0}^{r=kn-1} \left(n - \frac{r}{k} \right).$$

10.★ Establish the following identities :

(a) $\quad (ab)^{(r)} = a^r b^r \prod_{m=0}^{m=r-1} \left(1 - \frac{m}{ab} \right).$

(b) $\quad \dfrac{(2n)!}{(n!)^2} = \dfrac{(2n)^{(n)}}{n!}.$

11.★ Use the results of Ex. 10 to show that

(a) $\quad \dfrac{2.4.6.8 \ldots 2n}{1.3.5.7 \ldots (2n-1)} = W = \dfrac{(n! \, 2^n)^2}{(2n)!}.$

$$(b) \quad W = \left(\frac{n!}{n^n}\right)^2 \div \prod_{r=0}^{r=2n-1}\left(1 - \frac{r}{2n}\right).$$

12.* Show that

$$\log n! = n \cdot \log n + \sum_{r=0}^{r=n-1} \log\left(1 - \frac{r}{n}\right).$$

1.02. FIGURATE SERIES

Certain series of numbers have a history which goes back far into the past. They intrigued the Pythagorean brotherhoods of antiquity. They attracted much attention from the Hindu pioneers of algebra and their Moorish successors. Pascal, the father of mathematical probability, wrote a treatise on *Figurate* series, and we may appropriately follow his example by examining their properties as a prelude to a modern treatment of choice and chance.

The figurate series which are our chief concern in the context of modern statistics have this in common. We can generate terms of any such series from its predecessor in accordance with the rule for generating the natural numbers from a succession of units—the parent of all such series—as shown below :

Units :   1   1   1   1   1 . . .
Natural numbers : 1   (1 + 1)   (1 + 1 + 1)   (1 + 1 + 1 + 1)   (1 + 1 + 1 + 1 + 1) . . .
           = 1   = 2   = 3   = 4   = 5 . . .

The first of the succeeding series generated in accordance with this plan is the *triangular numbers.*

Natural numbers :   1   2   3   4   5 . . .
Triangular numbers : 1   (1 + 2)   (1 + 2 + 3)   (1 + 2 + 3 + 4)   (1 + 2 + 3 + 4 + 5) . . .
          1   = 3   = 6   = 10   = 15 . . .

The next series of the same family is the *tetrahedral* numbers :

1   (1 + 3)   (1 + 3 + 6)   (1 + 3 + 6 + 10)   (1 + 3 + 6 + 10 + 15)
1   = 4   = 10   = 20   = 35

By analogous summation we may build up super-solid numbers of 4, 5, . . . etc., dimensions, e.g. :

1   (1 + 4)   (1 + 4 + 10)   (1 + 4 + 10 + 20)   (1 + 4 + 10 + 20 + 35)
1   = 5   = 15   = 35   = 70
1   (1 + 5)   (1 + 5 + 15)   (1 + 5 + 15 + 35)   (1 + 5 + 15 + 35 + 70)
1   = 6   = 21   = 56   = 126

In accordance with the visual representation of the first four of these series we specify each by its dimension *d*, as follows :

Representation	d			Series		
Point	0	1	1	1	1	1 . . .
Line	1	1	2	3	4	5 . . .
Triangle	2	1	3	6	10	15 . . .
Tetrahedron	3	1	4	10	20	35 . . .
Super-solid	4	1	5	15	35	70 . . .
,,	5	1	6	21	56	126 . . .
,,	6	1	7	28	84	210 . . .

ROW
(r)

	$^0F_1=$ 1	$^0F_2=$ 1	$^0F_3=$ 1	$^0F_4=$ 1	$^0F_5=$ 1	$^0F_6=$ 1
O						
I	$^1F_1=$ 1	$^1F_2=$ 2	$^1F_3=$ 3	$^1F_4=$ 4	$^1F_5=$ 5	$^1F_6=$ 6
2	$^2F_1=$ 1	$^2F_2=$ 3	$^2F_3=$ 6	$^2F_4=$ 10	$^2F_5=$ 15	$^2F_6=$ 21
3	$^3F_1=$ 1	$^3F_2=$ 4	$^3F_3=$ 10	$^3F_4=$ 20	$^3F_5=$ 35	$^3F_6=$ 56
4	$^4F_1$ 1	$^4F_2$ 5	$^4F_3$ 15	$^4F_4$ 35	$^4F_5$ 70	$^4F_6$ 126
5	$^5F_1$ 1	$^5F_2$ 6	$^5F_3$ 21	$^5F_4$ 56	$^5F_5$ 126	$^5F_6$ 252
	1	2	3	4	5	6 COLUMN (c)

FIG. 1. The first five Figurate Numbers Series generated by successive Addition from a Sequence of Units.

The term of *rank* 1 of all such series is itself unity. The *one-dimensional* figurate of the same class is the arithmetic progression whose common difference is 1. The two-dimensional is the series whose terms are sums of this A.P. We shall here denote a number of rank n and dimension d in series of this family by the symbol ^dF_n in accordance with the following scheme:

$$^0F_1 = 1 \quad ^0F_2 = 1 \quad ^0F_3 = 1 \quad ^0F_4 = 1 \ldots$$
$$^1F_1 = 1 \quad ^1F_2 = 2 \quad ^1F_3 = 3 \quad ^1F_4 = 4 \ldots$$
$$^2F_1 = 1 \quad ^2F_2 = 3 \quad ^2F_3 = 6 \quad ^2F_4 = 10 \ldots$$
$$^3F_1 = 1 \quad ^3F_2 = 4 \quad ^3F_3 = 10 \quad ^3F_4 = 20 \ldots$$
$$^4F_1 = 1 \quad ^4F_2 = 5 \quad \text{etc., etc.}$$

From the law of generation exhibited in preceding tables it is evident that

$$^1F_n = \sum_{x=1}^{x=n} {}^0F_x ; \quad ^2F_n = \sum_{x=1}^{x=n} {}^1F_x ; \quad ^3F_n = \sum_{x=1}^{x=n} {}^2F_x ,$$

and in general, for *positive integral* values of d and n,

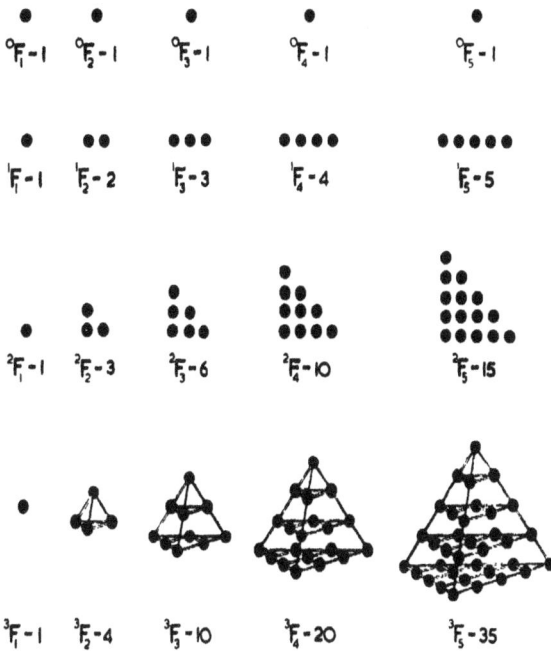

FIG. 2. Visual Representation of the first four Figurate Series of Fig. 1 as *Points* (*zero* dimensions), *Lines* (*one* dimensional figures), *Triangles* (*two* dimensions) and *Tetrahedra* (*three* dimensions).

$$^{d+1}F_n = \sum_{x=1}^{x=n} {}^{d}F_x \qquad \qquad \text{(i)}$$

An expression for the triangular (^2F_n) numbers of rank n follows from the familiar formula for the sum (S_n) of an A.P. $a_1, a_2, a_3, \ldots a_n$, *viz.* $S_n = \frac{1}{2}n(a_1 + a_n)$. The A.P. is in this case the natural number series $1, 2, 3, \ldots n$, so that $a_1 = 1$ and $a_n = n$:

$$\therefore \ S_n = \tfrac{1}{2}n(1 + n) = \tfrac{1}{2}(n + 1)(\overline{n + 1} - 1) = \tfrac{1}{2}(n + 1)^{(2)}.$$
$$\therefore \ ^2F_n = \frac{(n + 1)^{(2)}}{2} = \frac{(\overline{n + 2} - 1)^{(2)}}{2!}.$$

Fig. 3 shows a visual representation of the genesis of the formula, and Fig. 4 shows the genesis of the corresponding formula for the tetrahedral numbers ($1, 4, 10, 20 \ldots$) for which it is easy to derive an expression by trial and error, *viz.* :

$$^3F_n = \frac{(n + 2)(n + 1)n}{\cdot \ 6} = \frac{(\overline{n + 3} - 1)^{(3)}}{3!}.$$

Formulæ for the series of higher dimensions can be severally obtained by the method of 1.05 below, but the generalisation of the pattern common to the above is justifiable by induction. For the first four series we have

$^0F_n = 1 = \overline{(n+0-1)}^{(0)} \div 0!$ in accordance with (viii) of 1.01.

$^1F_n = n = \overline{(n+1-1)}^{(1)} \div 1!.$

$^2F_n = \overline{(n+2-1)}^{(2)} \div 2!.$

$^3F_n = \overline{(n+3-1)}^{(3)} \div 3!.$

In general, we have

$$^dF_n = \frac{(n+d-1)^{(d)}}{d!} \qquad\qquad (ii)$$

In accordance with (iv) in 1.01 we may also write this as

$$^dF_n = \frac{(n+d-1)!}{d!\,(n-1)!} \qquad\qquad (iii)$$

Thus, the 4th term of the series 1, 7, 28 . . . etc., in agreement with the foregoing table is

$$^6F_4 = \frac{(6+4-1)!}{6!\,(4-1)!} = \frac{9!}{6!\,3!} = \frac{9.8.7}{3.2.1} = 84.$$

From (i) it follows that

$$^{d+1}F_1 = {}^dF_1 = \frac{(d+1-1)!}{d!\,(1-1)!}$$
$$= \frac{d!}{d!\,0!}$$
$$= \frac{(d+1)!}{(d+1)!\,0!}$$
$$= \frac{\overline{(d+1}+1-1)!}{(d+1)!\,(1-1)!};$$

$$^{d+1}F_2 = {}^dF_1 + {}^dF_2 = \frac{(d+1)!}{(d+1)!\,0!} + \frac{(d+2-1)!}{d!(2-1)!}$$
$$= \frac{(d+1)!}{(d+1)!}(d+2)$$
$$= \frac{\overline{(d+1}+2-1)!}{(d+1)!\,(2-1)!};$$

$$^{d+1}F_3 = {}^dF_1 + {}^dF_2 + {}^dF_3 = \frac{(d+2)!}{(d+1)!} + \frac{(d+3-1)!}{d!\,(3-1)!}$$
$$= \frac{(d+2)!}{(d+1)!}\left\{1+\frac{d+1}{2}\right\}$$
$$= \frac{(d+3)!}{(d+1)!\,2!}$$
$$= \frac{\overline{(d+1}+3-1)!}{(d+1)!\,(3-1)!};$$

and, in general, $\quad {}^{d+1}F_n = \frac{\overline{(d+1}+n-1)!}{(d+1)!\,(n-1)!}.$

The form of this expression is identical with (iii) above, and we know that (iii) is true when $d = 2$. Hence it is true when $d = 3$. If true when $d = 3$, it is also true when $d = 4$, and so on.[*]

In Fig. 1, identical rows and columns succeed one another *in the same order*. If we label the rows and columns of figurate series of successive dimensions, so that figurates of row r are r-dimensional and the column c specifies the rank, we see (Fig. 11) that

$$^2F_4 = {}^3F_3 = {}^{4-1}F_{2+1},$$
$$^4F_6 = {}^5F_5 = {}^{6-1}F_{4+1},$$
$$^5F_3 = {}^2F_6 = {}^{3-1}F_{5+1}.$$

In general, therefore,

$$^rF_c = {}^{c-1}F_{r+1}.$$

This is consistent with (ii) and (iii), since we then have

$$^{c-1}F_{r+1} = \frac{\overline{(r+1+c-1}-1)^{(c-1)}}{(c-1)!} = \frac{(r+c-1)^{(c-1)}}{(c-1)!}$$
$$= \frac{(r+c-1)!}{(r+c-1-c-1)!\,(c-1)!}$$
$$= \frac{(c+r-1)!}{r!\,(c-1)!}$$

By means of (i) and (ii) or (iii), it is possible to sum many series of a type we commonly meet in the theory of choice. We may combine (i) and (ii) in the formula :

$$\sum_{x=1}^{x=n} \frac{(x+d-1)^{(d)}}{d!} = \frac{(n+d)^{(d+1)}}{(d+1)!} \qquad . \qquad . \qquad . \qquad . \qquad \text{(iv)}$$

Example.—We may use (iv) to find the sum of the squares, cubes, etc., of the natural numbers, by proceeding as follows :

$$^2F_n = \frac{n(n+1)}{2} = \frac{n^2}{2} + \frac{n}{2}.$$
$$\therefore \frac{n^2}{2} = {}^2F_n - \frac{n}{2}.$$
$$\therefore \Sigma n^2 = 2\Sigma\,{}^2F_n - \Sigma n = 2 \cdot {}^3F_n - {}^2F_n.$$

In accordance with (i) in 1.01 we may generalise the limits of summation in (iv) as follows when b and a are both *positive* :

$$\sum_{x=a}^{x=b} \frac{(x+d-1)^{(d)}}{d!} = \sum_{x=1}^{x=b} \frac{(x+d-1)^{(d)}}{d!} - \sum_{x=1}^{x=(a-1)} \frac{(x+d-1)^{(d)}}{d!}$$
$$= \frac{(b+d)^{(d+1)}}{(d+1)!} - \frac{(a+d-1)^{(d+1)}}{(d+1)!} \qquad . \qquad . \qquad . \qquad \text{(v)}$$

The required summation may take a form which demands redefinition of the limits, e.g. :

$$\sum_{x=a}^{x=b} {}^dF_{x-c} = \sum_{x=a}^{x=b} \frac{(x-c+d-1)^{(d)}}{d!}.$$

[*] For an alternative proof of (ii) see the method of *Example* 1 in 1.10.

2

When it is necessary to sum a series in this form, we proceed thus : Put $y = (x - c)$, so that $x = (y + c)$. When $x = b$, $y = (b - c)$ and when $x = a$, $y = (a - c)$:

$$\therefore \sum_{x=a}^{x=b} {}^{d}F_{x-c} = \sum_{y=(a-c)}^{y=(b-c)} {}^{d}F_{y}$$

$$= \sum_{y=1}^{y=(b-c)} {}^{d}F_{y} - \sum_{y=1}^{y=(a-c-1)} {}^{d}F_{y}$$

$$= \frac{(b-c+d)^{(d+1)}}{(d+1)!} - \frac{(a-c+d-1)^{(d+1)}}{(d+1)!} \qquad . \qquad . \qquad . \qquad \text{(vi)}$$

In this context, it is important to note a peculiarity of the triangular family of series defined by (ii). In accordance with (v) in 1.01 we may put

$$^{d}F_{n} = \frac{n(n+1)(n+2) \dots (n+d-1)}{d!},$$

$$\therefore \ {}^{d}F_{0} = 0, \quad \text{when} \quad d > 0,$$

$$\therefore \sum_{x=1}^{x=n} {}^{d}F_{x} = \sum_{x=0}^{x=n} {}^{d}F_{x} = {}^{d+1}F_{n} \qquad . \qquad . \qquad . \qquad . \qquad . \qquad . \qquad \text{(vii)}$$

When $d = 0$, this relation does not hold good. For ${}^{0}F_{0} = 1$.

★★★★ The operation defined by (i)-(iii) and (vii) is valid only if n is positive. By applying the formula ${}^{2}F_{n} = \frac{1}{2}n(n+1)$, we can extend the series $0, 1, 3, 6, 10, \dots$ into the negative domain ; but the addition rule which relates this series to ${}^{1}F_{n} = n$ is no longer valid. For the series ${}^{1}F_{n} \dots {}^{4}F_{n}$ we may calculate results of the application of the appropriate formulæ thus :

${}^{1}F_{n}$	-6	-5	-4	-3	-2	-1	0	1	2	3	4	5
${}^{2}F_{n}$	15	10	6	3	1	0	0	1	3	6	10	15
${}^{3}F_{n}$	-20	-10	-4	-1	0	0	0	1	4	10	20	35
${}^{4}F_{n}$	15	5	1	0	0	0	0	1	5	15	35	70

The rule of summation in the negative range is deducible from the above when $d > 0$, viz. :

$$\sum_{x=0}^{x=-n} {}^{d}F_{x} = (-1)^{d} . {}^{d+1}F_{n-d+1} \qquad . \qquad . \qquad . \qquad . \qquad \text{(viii)}$$

$$\sum_{x=-m}^{x=+n} {}^{d}F_{x} = {}^{d+1}F_{n} + (-1)^{d} . {}^{d+1}F_{m-d+1} \qquad . \qquad . \qquad \text{(ix)}$$

Thus

$$\sum_{x=-6}^{x=5} {}^{3}F_{x} = {}^{4}F_{5} + (-1)^{3} {}^{4}F_{4} = 70 - 35 = 35$$

$$= (-20 - 10 - 4 - 1 + 0 + 1 + 4 + 10 + 20 + 35).$$

NOTE (for *exercise only*).—The one-dimensional series of the family discussed above is an A.P. whose common difference is 1. From A.P.s (Fig. 5) of which the term of rank 1 is unity we may build up analogous families defined by the common difference. From the series of *odd* (numbers (common difference 2) we can generate series in accordance with the foregoing principle Fig. 6) as follows :

★★★★ Omit on first reading.

Row	1	3	5	7	9 ...
		$(1+3)$	$(1+3+5)$	$(1+3+5+7)$	$(1+3+5+7+9)$ ...
Square	1	$=4$	$=9$	$=16$	$=25$ ...
		$(1+4)$	$(1+4+9)$	$(1+4+9+16)$	$(1+4+9+16+25)$ ...
Pyramid	1	$=5$	$=14$	$=30$	$=55$ ...

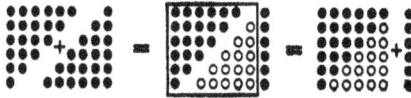

$$2\,{}^{2}F_{n} \;=\; n^{2}+n$$

$${}^{2}F_{n} \;=\; \frac{n^{2}+n}{2} \;=\; \frac{(n+1)^{(2)}}{2}$$

Fig. 3. The build-up of a Formula for the Triangular Number Series of Fig. 2.

$$1 + 3 + 6 + 10 + 15 + 21$$
$$= {}^{2}F_{1} + {}^{2}F_{2} + {}^{2}F_{3} + {}^{2}F_{4} + {}^{2}F_{5} + {}^{2}F_{6}$$
$$= {}^{3}F_{6}$$

$$(6+1).21 = (6+1)\,{}^{2}F_{6}$$

$$(6+1)\,{}^{2}F_{6} + {}^{2}F_{6} - {}^{3}F_{6}$$

$$= (6+2)\quad {}^{2}F_{6} - {}^{3}F_{6}$$

$$= 2\,({}^{2}F_{1} + {}^{2}F_{2} + {}^{2}F_{3} + {}^{2}F_{4} + {}^{2}F_{5} + {}^{2}F_{6})$$
$$= 2\,{}^{3}F_{6}$$

$$(6+2)\,{}^{2}F_{6} - {}^{3}F_{6} = 2\,{}^{3}F_{6}$$

$$\therefore 3.\,{}^{3}F_{6} = (6+2)\,{}^{2}F_{6}$$

$$\therefore {}^{3}F_{6} = \frac{(6+2)(6+1)6}{3.\,2.}$$

$${}^{3}F_{n} = \frac{(n+2)(n+1)n}{3.\,2.\,1}$$

Fig. 4. The build-up of a Formula for the *Tetrahedral* Numbers of Fig. 2.

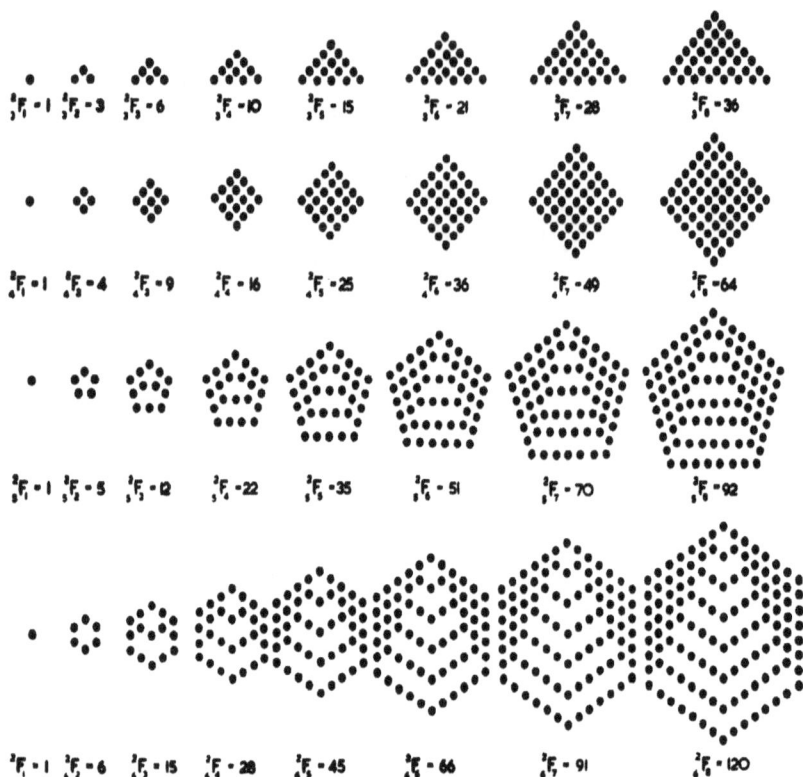

FIG. 5. Two-dimensional Figurates representing the sum of Arithmetic Progressions whose first Term is Unity. The Common Differences are (in successive rows) 1, 2, 3, 4.

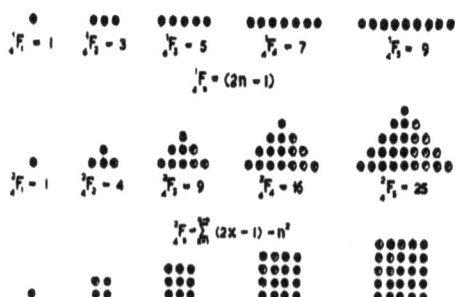

FIG. 6. Finding a Formula for the sum of the first n Odd Numbers by the Figurate Method.

From the series whose common difference is 3 we may generate the pentagonal numbers, 6-faced pyramids and 4-dimensional super-solids :

Row	1	4	7	10	13	16 ...
Pentagon	1	5	12	22	35	51 ...
6-Face pyramid	1	6	18	40	75	126 ...
Supersolid	1	7	25	65	140	266 ...

$${}_3^2F_n - 1 + 2(n-1) + {}_3^2F_{n-2}$$
$$-1 + (3-1)(n-1) + (3-2){}_3^2F_{n-2}$$

$${}_4^2F_n - 1 + 3(n-1) + 2{}_3^2F_{n-2}$$
$$-1 + (4-1)(n-1) + (4-2){}_3^2F_{n-2}$$

$${}_5^2F_n - 1 + (5-1)(n-1) + (5-2){}_3^2F_{n-2} \quad {}_6^2F_n - 1 + (6-1)(n-1) + (6-2){}_3^2F_{n-2}$$

$${}_s^2F_n - 1 + (s-1)(n-1) + (s-2){}_3^2F_{n-2}$$

$$-1 + (s-1)(n-1) + \frac{(s-2)(n-2)(n-1)}{2}$$

$$-\frac{n}{2}(s-2)(n-1+2)$$

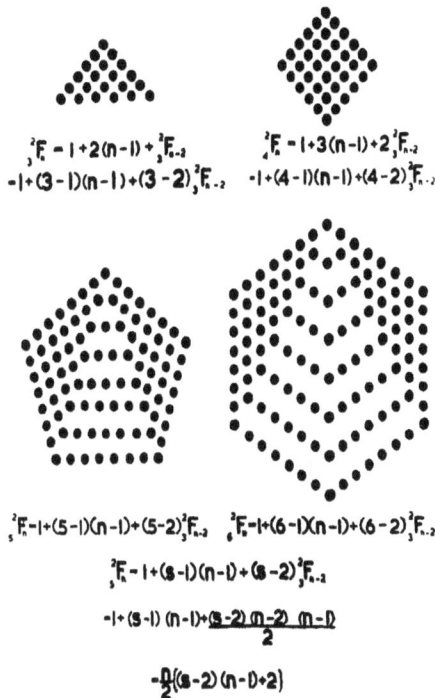

FIG. 7. Finding a Formula for the Figurate Numbers of Fig. 5.

In families of this pattern the relation between the common difference (c) of the parent 1-dimensional A.P. and the number (s) of *sides of the* 2-*dimensional* figurate representation is given by $c = (s-2)$ or $s = (c+2)$. If we represent the term of rank n and dimension d of any such series by the general symbol ${}_s^dF_n$, the appropriate symbol for the series whose generative A.P. is the natural numbers is ${}_3^dF_n$. The latter are the only ones we shall make use of in what follows ; but it is a helpful *proficiency* exercise to explore the properties of others. With the help of Fig. 7, the student should be able to establish by induction and to check the more general formula of which (ii) above is a particular case :

$$ {}_s^dF_n = \frac{(n+d-2)^{(d-1)}}{d!} \cdot (\overline{s-2} \cdot \overline{n-1} \cdot + d). $$

When $s = 3$, this reduces to (ii), *viz.* :

$$ {}_3^dF_n = \frac{(n+d-2)^{(d-1)}}{d!} \cdot (n-1+d) $$

$$ = \frac{(n+d-1)^{(d)}}{d!}. $$

The general property of all such series is given by

$$ \sum_{x=1}^{x=n} {}_s^dF_x = {}_s^{d+1}F_n = \frac{(n+d-1)^{(d)}}{(d+1)!} \cdot (\overline{s-2} \cdot \overline{n-1} \cdot + d + 1). $$

The families of plane figurate numbers represented by ${}_s^2F_n$ and of solid figures represented by ${}_s^3F_n$ do not exhaust number series we can represent by regular patterns of the sort shown in Figs. 1 and 5. Another family is the *central polygonals* (Figs. 8, 9 and 10) whose formula is

$$1 + s(n-1) + s \cdot {}_s^2F_{n-2} = {}_s^2P_n = 1 + s \cdot {}_s^2F_{n-1}.$$

The octagonals ($s = 8$) of this type are the squares of the odd numbers

$$1, 9, 25, 49, 81, \ldots$$

The corresponding 3-dimensional series (pyramid of 9 faces) is

$$\begin{aligned}
{}_8^3P_n &= 1, \quad 1+9, \quad 1+9+25, \quad 1+9+25+49, \ldots \\
&= 1, \quad\ 10, \qquad 35, \qquad\quad 84, \ldots
\end{aligned}$$

The summation formula is

$$\begin{aligned}
{}_s^3P_n = \sum_1^n {}_s^2P_x &= n + s\sum_{x=1}^{x=n} {}_s^2F_{x-1} \\
&= n + s\sum_{y=0}^{y=n-1} {}_s^2F_y \\
&= n + \frac{s(n+1)^{(3)}}{3!} \\
&= \frac{n}{3!}(sn^2 - s + 6).
\end{aligned}$$

Check : ${}_8^3P_4 = \frac{4}{6}(8.16 - 8 + 6) = 84$ (*as above*).

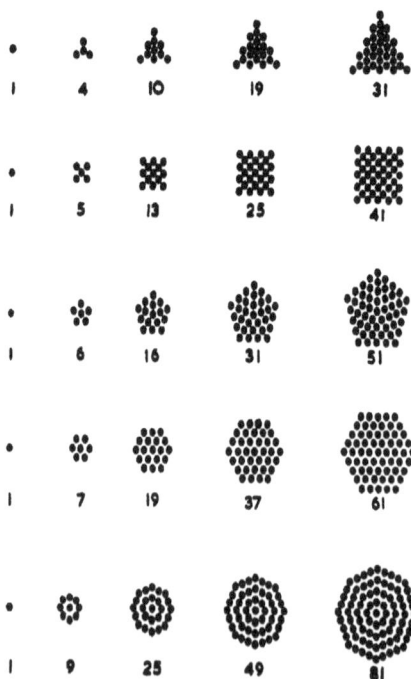

FIG. 8. The Central Polygonal Series.

The visual representations of figurate numbers of this type in 2 dimensions are reducible to rows and triangles (Figs. 8-9), and those of 3-dimensional series to rows, triangles and tetrahedra. Hence the algebraic expressions are deducible from the formulæ for triangular and tetrahedral numbers. Such expressions are also obtainable by recourse to the general method dealt with in 4.03 below. The accompanying charts exhibit the visual approach which made possible exploration of the properties of a large class of number series before the introduction of *symbolic* algebra.★★★★

EXERCISE 1.02

1. Write down the first 12 triangular numbers, the tetrahedral numbers and the 4th-dimensional numbers of the same class.

2. Find the numerical values of :

$$\tfrac{1}{3}F_{21}\,;\ \ \tfrac{2}{3}F_{10}\,;\ \ \tfrac{3}{3}F_{7}\,;\ \ \tfrac{4}{3}F_{6}\,;\ \ \tfrac{5}{3}F_{6}\,;\ \ \tfrac{6}{3}F_{5}.$$

3. Write down the first 10 members of the series

$$\tfrac{2}{4}F_{n}\,;\ \ \tfrac{2}{5}F_{n}\,;\ \ \tfrac{2}{6}F_{n}\,;\ \ \tfrac{2}{7}F_{n}\,;\ \ \tfrac{2}{8}F_{n}\,;$$
$$\tfrac{3}{4}F_{n}\,;\ \ \tfrac{3}{5}F_{n}\,;\ \ \tfrac{3}{6}F_{n}\,;\ \ \tfrac{3}{7}F_{n}\,;\ \ \tfrac{3}{8}F_{n}.$$

4.★ Show that

$$\tfrac{3}{3}F_{n} = \tfrac{1}{2}\sum_{x=1}^{x=n} x^{2} + \tfrac{1}{2}\left(\tfrac{2}{3}F_{n}\right).$$

5.★ Show that

$$\tfrac{4}{3}F_{n} - \tfrac{1}{3}\left(\tfrac{2}{3}F_{n}\right) = \tfrac{1}{6}\sum_{x=0}^{x=n} x^{3} + \tfrac{1}{2}\sum_{x=0}^{x=n} x^{2}\,;\quad \tfrac{5}{3}F_{n} - \tfrac{1}{4}\left(\tfrac{2}{3}F_{n}\right) = \tfrac{1}{24}\sum_{x=0}^{x=n} x^{4} + \tfrac{1}{4}\sum_{x=0}^{x=n} x^{3} + \tfrac{11}{24}\sum_{x=0}^{x=n} x^{2}.$$

6.★ By (vii) and (ix) show that if $k \geqslant 2$

$$\sum_{x=1}^{x=n} x^{(k)} = \frac{(n+1)^{(k+1)}}{k+1} \quad (n \geqslant k).$$

7.★ Obtain an expression involving one unknown (n) alone for

$$\sum_{x=0}^{x=n} x^{2}\,;\ \ \sum_{x=0}^{x=n} x^{3}\,;\ \ \sum_{x=0}^{x=n} x^{4}\,;\ \ \sum_{x=0}^{x=n} x^{5}.$$

8. Find a formula for

$$\sum_{x=1}^{x=n-1} x^{2}\,;\ \ \sum_{x=1}^{x=n-1} x^{3}\,;\ \ \sum_{x=1}^{x=n-1} x^{4}.$$

9. Find the numerical values of

$$\tfrac{2}{8}P_{7}\,;\ \ \tfrac{3}{5}P_{4}\,;\ \ \tfrac{4}{6}P_{3}\,;\ \ \tfrac{3}{7}P_{6}.$$

10. Show that

$$\sum_{x=0}^{x=n} (2n+1)^{2} = 1 + 8 \cdot \tfrac{3}{3}F_{n}.$$

$$^2_3P_n - 1 + 3(n-1) + 3 \, ^2_3F_{n-2}$$

$$^2_5P_n - 1 + 5(n-1) + 5 \, ^2_3F_{n-2}$$

$$^2_7P_n - 1 + 7(n-1) + 7 \, ^2_3F_{n-2}$$

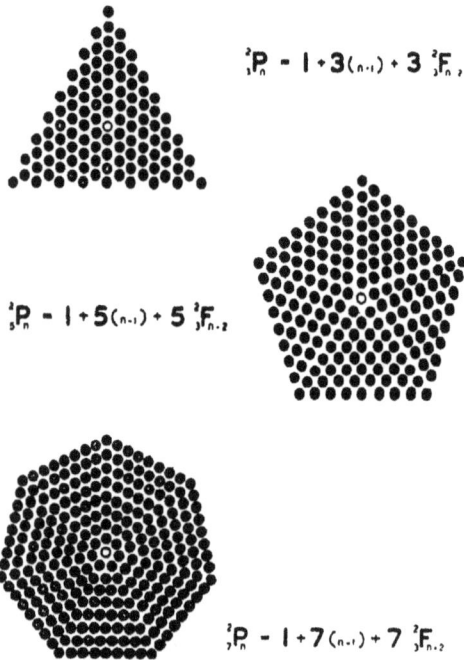

FIG. 9. One way of finding a Formula for the Figurate Numbers of Fig. 8.

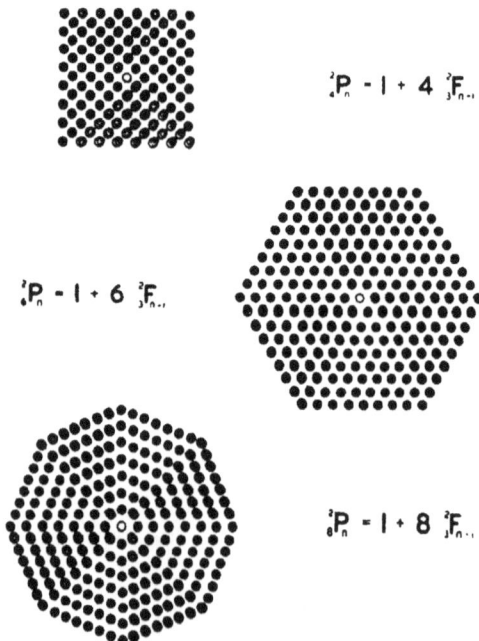

$$^2_4P_n - 1 + 4 \, ^2_3F_{n-1}$$

$$^2_6P_n - 1 + 6 \, ^2_3F_{n-1}$$

$$^2_8P_n - 1 + 8 \, ^2_3F_{n-1}$$

FIG. 10. Another way of finding a Formula for the Figurate Numbers of Fig. 8.

11. Find an expression for

$$\text{(a) } {}_3^d F_{n+1}; \quad \text{(b) } {}_3^d F_{n-x+1}.$$

12. Find the numerical values of

$$\text{(a) } \sum_{x=5}^{x=10} {}_3^6 F_x; \quad \text{(b) } \sum_{x=5}^{x=12} {}_3^5 F_x; \quad \text{(c) } \sum_{x=3}^{x=11} {}_3^7 F_x; \quad \text{(d) } \sum_{x=4}^{x=10} {}_4^3 F_x; \quad \text{(e) } \sum_{x=6}^{x=12} {}_5^4 F_x; \quad \text{(f) } \sum_{x=7}^{x=11} {}_6^3 F_x.$$

13. Evaluate

$$\text{(i) } \sum_{x=3}^{x=8} {}_8^3 P_x; \quad \text{(ii) } \sum_{x=2}^{x=7} {}_5^2 P_x; \quad \text{(iii) } \sum_{x=3}^{x=10} {}_4^3 P_x.$$

14.* Devise a formula for

$$\text{(a) } \sum_{x=a}^{x=b} {}_3^d F_{n-2x+1}; \quad \text{(b) } \sum_{x=c}^{x=b} {}_4^d F_{n+3x-2}.$$

15. Repeat No. 5 of Ex. 1.01 with due regard to the meaning of ${}_s^0 F_n$ when $s > 3$.

1.03 PASCAL'S TRIANGLE

We may set out the series generated successively by the units, natural numbers and triangular numbers of 1.02 in successive columns, as in Fig. 11. In such an arrangement there is a *row* of terms corresponding to those of each column. By sliding successive columns downwards through 1, 2, 3, . . . etc., rows we get the arrangement (Fig. 12) known as Pascal's triangle, though it dates at least from the time of Omar Khayyám. In this arrangement the rth row has $(r + 1)$ terms whose rank we label from 0 to c, so that the cth column starts at the end of the cth row.

We shall label a Pascal number $r_{(c)}$, as the number in row r of column rank c. By reference to Figs. 11 and 12 we see that

$$2_{(0)} = 1 = {}^0F_3 = {}^0F_{2-0+1} = 1 = {}^2F_1 = {}^{2-0}F_{0+1}$$
$$2_{(1)} = 2 = {}^1F_2 = {}^1F_{2-1+1} = 2 = {}^1F_2 = {}^{2-1}F_{1+1}$$
$$2_{(2)} = 1 = {}^2F_1 = {}^2F_{2-2+1} = 1 = {}^0F_3 = {}^{2-2}F_{2+1}$$
$$3_{(0)} = 1 = {}^0F_4 = {}^0F_{3-0+1} = 1 = {}^3F_1 = {}^{3-0}F_{0+1}$$
$$3_{(1)} = 3 = {}^1F_3 = {}^1F_{3-1+1} = 3 = {}^2F_2 = {}^{3-1}F_{1+1}$$
$$3_{(2)} = 3 = {}^2F_2 = {}^2F_{3-2+1} = 3 = {}^1F_3 = {}^{3-2}F_{2+1}$$
$$3_{(3)} = 1 = {}^3F_1 = {}^3F_{3-3+1} = 1 = {}^0F_4 = {}^{3-3}F_{3+1}$$

In general, the rule is

$$r_{(c)} = {}^cF_{r-c+1} \quad or \quad r_{(c)} = {}^{r-c}F_{c+1}.$$

From (ii) in 1.02

$$\begin{aligned}
{}^cF_{r-c+1} &= \frac{\overline{(r-c+1+c-1)}^{(c)}}{c!} \\
&= \frac{r^{(c)}}{c!}.
\end{aligned}$$

$${}^{r}F_{c} = {}^{c-1}F_{r+1}$$

ROW (r) \ COLUMN (c)	1	2	3	4	5	6
0	${}^0F_1$ 1 ${}^0F_1$	${}^0F_2$ 1 ${}^1F_1$	${}^0F_3$ 1 ${}^2F_1$	${}^0F_4$ 1 ${}^3F_1$	${}^0F_5$ 1 ${}^4F_1$	${}^0F_6$ 1 ${}^5F_1$
1	${}^1F_1$ 1 ${}^0F_2$	${}^1F_2$ 2 ${}^1F_2$	${}^1F_3$ 3 ${}^2F_2$	${}^1F_4$ 4 ${}^3F_2$	${}^1F_5$ 5 ${}^4F_2$	${}^1F_6$ 6 ${}^5F_2$
2	${}^2F_1$ 1 ${}^0F_3$	${}^2F_2$ 3 ${}^1F_3$	${}^2F_3$ 6 ${}^2F_3$	${}^2F_4$ 10 ${}^3F_3$	${}^2F_5$ 15 ${}^4F_3$	${}^2F_6$ 21 ${}^5F_3$
3	${}^3F_1$ 1 ${}^0F_4$	${}^3F_2$ 4 ${}^1F_4$	${}^3F_3$ 10 ${}^2F_4$	${}^3F_4$ 20 ${}^3F_4$	${}^3F_5$ 35 ${}^4F_4$	${}^3F_6$ 56 ${}^5F_4$
4	${}^4F_1$ 1 ${}^0F_5$	${}^4F_2$ 5 ${}^1F_5$	${}^4F_3$ 15 ${}^2F_5$	${}^4F_4$ 35 ${}^3F_5$	${}^4F_5$ 70 ${}^4F_5$	${}^4F_6$ 126 ${}^5F_5$
5	${}^5F_1$ 1 ${}^0F_6$	${}^5F_2$ 6 ${}^1F_6$	${}^5F_3$ 21 ${}^2F_6$	${}^5F_4$ 56 ${}^3F_6$	${}^5F_5$ 126 ${}^4F_6$	${}^5F_6$ 252 ${}^5F_6$

FIG. 11. Two ways of symbolising the Figurate Numbers of Fig. 1.

From (iv) in 1.01

$$\frac{r^{(c)}}{c!} = \frac{r!}{c!(r-c)!}$$

$$\therefore r_{(c)} = \frac{r!}{c!(r-c)!} \quad or \quad \frac{r^{(c)}}{c!} \quad . \quad . \quad . \quad . \quad . \quad . \quad (i)$$

By (viii) and (iii) in 1.01 we also have

$$r_{(0)} = 1 = r_{(r)} \quad . \quad . \quad . \quad . \quad . \quad . \quad (ii)$$

We may arrive at (i) by another route, since

$$\begin{aligned} r_{(c)} &= {}^{r-c}F_{c+1} \\ &= \frac{(\overline{c+1} + \overline{r-c} - 1)^{(r-c)}}{(r-c)!} \\ &= \frac{r^{(r-c)}}{(r-c)!}; \end{aligned}$$

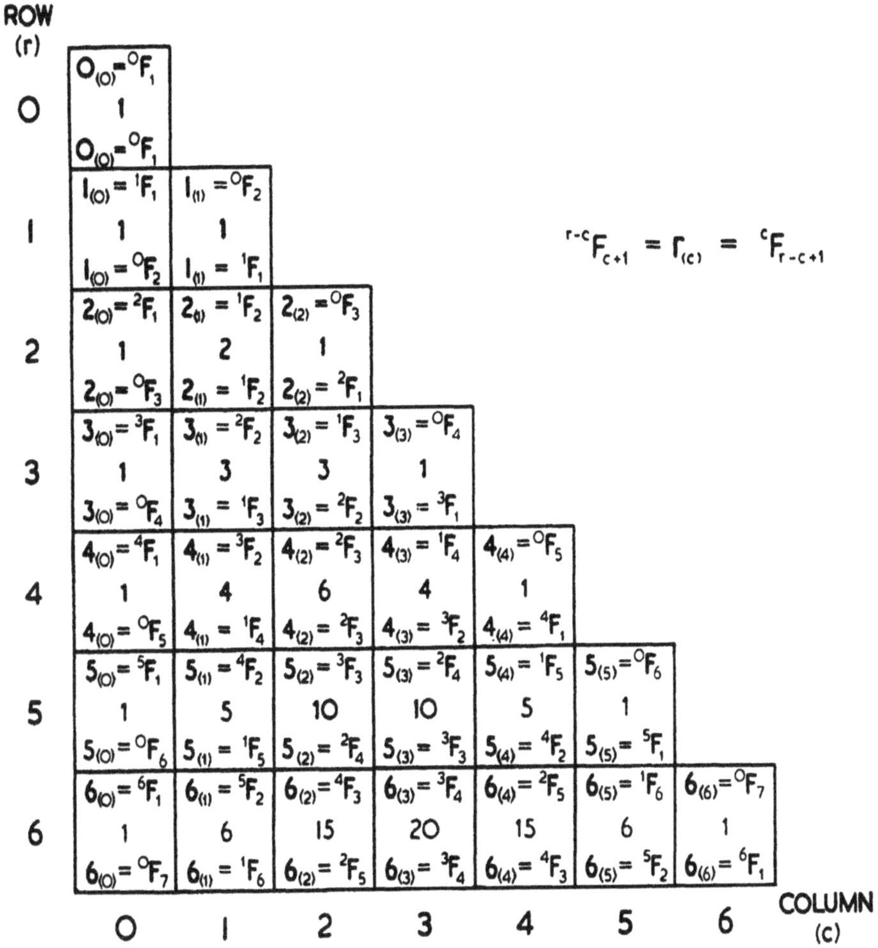
ROW (r)
0
$0_{(0)} = {}^{0}F_1$
1
$0_{(0)} = {}^{0}F_1$
1
$1_{(0)} = {}^{1}F_1$ | $1_{(1)} = {}^{0}F_2$
1 | 1
$1_{(0)} = {}^{0}F_2$ | $1_{(1)} = {}^{1}F_1$
2
$2_{(0)} = {}^{2}F_1$ | $2_{(1)} = {}^{1}F_2$ | $2_{(2)} = {}^{0}F_3$
1 | 2 | 1
$2_{(0)} = {}^{0}F_3$ | $2_{(1)} = {}^{1}F_2$ | $2_{(2)} = {}^{2}F_1$
3
$3_{(0)} = {}^{3}F_1$ | $3_{(1)} = {}^{2}F_2$ | $3_{(2)} = {}^{1}F_3$ | $3_{(3)} = {}^{0}F_4$
1 | 3 | 3 | 1
$3_{(0)} = {}^{0}F_4$ | $3_{(1)} = {}^{1}F_3$ | $3_{(2)} = {}^{2}F_2$ | $3_{(3)} = {}^{3}F_1$
4
$4_{(0)} = {}^{4}F_1$ | $4_{(1)} = {}^{3}F_2$ | $4_{(2)} = {}^{2}F_3$ | $4_{(3)} = {}^{1}F_4$ | $4_{(4)} = {}^{0}F_5$
1 | 4 | 6 | 4 | 1
$4_{(0)} = {}^{0}F_5$ | $4_{(1)} = {}^{1}F_4$ | $4_{(2)} = {}^{2}F_3$ | $4_{(3)} = {}^{3}F_2$ | $4_{(4)} = {}^{4}F_1$
5
$5_{(0)} = {}^{5}F_1$ | $5_{(1)} = {}^{4}F_2$ | $5_{(2)} = {}^{3}F_3$ | $5_{(3)} = {}^{2}F_4$ | $5_{(4)} = {}^{1}F_5$ | $5_{(5)} = {}^{0}F_6$
1 | 5 | 10 | 10 | 5 | 1
$5_{(0)} = {}^{0}F_6$ | $5_{(1)} = {}^{1}F_5$ | $5_{(2)} = {}^{2}F_4$ | $5_{(3)} = {}^{3}F_3$ | $5_{(4)} = {}^{4}F_2$ | $5_{(5)} = {}^{5}F_1$
6
$6_{(0)} = {}^{6}F_1$ | $6_{(1)} = {}^{5}F_2$ | $6_{(2)} = {}^{4}F_3$ | $6_{(3)} = {}^{3}F_4$ | $6_{(4)} = {}^{2}F_5$ | $6_{(5)} = {}^{1}F_6$ | $6_{(6)} = {}^{0}F_7$
1 | 6 | 15 | 20 | 15 | 6 | 1
$6_{(0)} = {}^{0}F_7$ | $6_{(1)} = {}^{1}F_6$ | $6_{(2)} = {}^{2}F_5$ | $6_{(3)} = {}^{3}F_4$ | $6_{(4)} = {}^{4}F_3$ | $6_{(5)} = {}^{5}F_2$ | $6_{(6)} = {}^{6}F_1$
0 1 2 3 4 5 6
COLUMN (c)
${}^{r-c}F_{c+1} = r_{(c)} = {}^{c}F_{r-c+1}$

FIG. 12. The derivation of the Pascal Triangle and of the Formula for Pascal Numbers from the Figurate Numbers of Fig. 11.

$$\therefore\ r_{(c)} = \frac{r!}{(r - \overline{r - c})!\,(r-c)!}$$

$$= \frac{r!}{c!(r-c)!}.$$

EXERCISE 1.03

1. Make a table of values of $r_{(x)}$ for $r = 0$ to $r = 10$.

2. Write down the series defined by

$$12_{(x)}\,;\quad 16_{(x)}\,;\quad 25_{(x)}.$$

3. By direct multiplication find $(a + b)^r$ for all values of r from 0 to 10 inclusive.

4. Tabulate the results of 3 and identify the nature of the numerical coefficients by recourse to the results of 1.

5. Indicate the pattern of the general term $A_x . a^m b^l$ by appropriate symbols for the numerical coefficient A_x and the powers l, m in the expansion of Ex. 3 above.

1.04 THE BINOMIAL THEOREM

Pascal's numbers provide the clue to a general rule for expanding an expression of the form $(b + a)^r$ as a *power series*, i.e. in ascending and descending powers of a and b. By direct multiplication we obtain the following results :

$$(b + a)^0 = 1$$
$$(b + a)^1 = b + a$$
$$(b + a)^2 = b^2 + 2ab + a^2$$
$$(b + a)^3 = b^3 + 3b^2a + 3ba^2 + a^3$$
$$(b + a)^4 = b^4 + 4b^3a + 6b^2a^2 + 4ba^3 + a^4$$
$$(b + a)^5 = b^5 + 5b^4a + 10b^3a^2 + 10b^2a^3 + 5ba^4 + a^5,$$
$$\text{etc., etc.}$$

By inspection, we recognise the coefficients of the above as Pascal's numbers ; and it is convenient to lay out the results in a schema like that of Pascal's triangle, *viz.* :

Power (= rank of row)

0	$1 . b^0 a^0$					
1	$1 . b^1 a^0$	$1 . b^0 a^1$				
2	$1 . b^2 a^0$	$2 . b^1 a^1$	$1 . b^0 a^2$			
3	$1 . b^3 a^0$	$3 . b^2 a^1$	$3 . b^1 a^2$	$1 . b^0 a^3$		
4	$1 . b^4 a^0$	$4 . b^3 a^1$	$6 . b^2 a^2$	$4 . b^1 a^3$	$1 . b^0 a^4$	
5	$1 . b^5 a^0$	$5 . b^4 a^1$	$10 . b^3 a^2$	$10 . b^2 a^3$	$5 . b^1 a^4$	$1 . b^0 a^5$
Term	0	1	2	3	4	5

(= rank of column)

It is now easy to recognise the pattern of the series by the *general* term of $(b + a)^r$, *viz.* :

$$r_{(x)} . b^{r-x} . a^x = \frac{r!}{(r - x)! \, x!} b^{r-x} a^x \qquad . \qquad . \qquad . \qquad . \qquad . \quad \text{(i)}$$

To show that this expression for the term involving the xth power of a in the expansion of $(b + a)^r$ is valid for *all* positive integral values of r, we assume that it is true for a particular value of r as shown above. It then suffices to show that we get the correct expression for the term involving a^x in the expansion of $(b + a)^{r+1}$, if we substitute $(r + 1)$ for r in (i) above, i.e. $(r + 1)_{(x)} . a^x b^{r-x+1}$.

We first note that

$$(b + a)^{r+1} = (b + a)(b + a)^r.$$

The term involving a^x in the expansion of $(b + a)^{r+1}$ is therefore the sum of b times the term involving a^x and a times the term involving a^{x-1} in the expansion of $(b + a)^r$. If we denote by A_x the coefficient of the general term of $(b + a)^{r+1}$, we may therefore write

$$A_x \cdot b^{r-x+1} a^x = b \cdot r_{(x)} b^{r-x} a^x + a \cdot r_{(x-1)} b^{r-x+1} a^{x-1}.$$

$$\therefore A_x = r_{(x)} + r_{(x-1)}$$

$$= \frac{r!}{x!\,(r-x)!} + \frac{r!}{(x-1)!\,(r-x+1)!}$$

$$= \frac{r!\,(r-x+1)}{x!\,(r-x+1)!} + \frac{r!\,x}{x!\,(r-x+1)!}$$

$$= \frac{r!}{x!\,(r-x+1)!}\overline{(r-x+1+x)}$$

$$= \frac{r!\,(r+1)}{x!\,(r+1-x)!}$$

$$= \frac{(r+1)!}{x!\,(r+1-x)!}.$$

Hence the general term of $(b+a)^{r+1}$ is

$$\frac{(r+1)!}{x!\,(r+1-x)!}\, b^{r+1-x}\, a^x.$$

This is the expression we get by substituting $(r+1)$ for r in (i) above. The complete expansion is most easy to memorise in the form

$$(b+a)^r = \frac{r^{(0)}}{0!}b^r a^0 + \frac{r^{(1)}}{1!}b^{r-1}a^1 + \frac{r^{(2)}}{2!}b^{r-2}a^2 + \frac{r^{(3)}}{3!}b^{r-3}a^3 + \ldots + \frac{r^{(r)}}{r!}b^{r-r}a^r \qquad \text{(ii)}$$

Alternatively, and more briefly :

$$(b+a)^r = \sum_{x=0}^{x=r} \frac{r!}{x!\,(r-x)!}\, b^{r-x}\, a^x.$$

From (i) it follows that

$$2^r = (1+1)^r = 1 + r + \frac{r^{(2)}}{2!} + \frac{r^{(3)}}{3!} + \ldots + \frac{r^{(r)}}{r!}$$

$$\therefore\ 2^r = \sum_{x=0}^{x=r} r_{(x)} \qquad . \qquad . \qquad . \qquad . \qquad . \qquad . \qquad . \qquad \text{(iii)}$$

$$\therefore\ 2^r - 1 = \sum_{x=1}^{x=r} r_{(x)} \quad \text{and} \quad 2(2^{r-1}-1) = \sum_{x=1}^{x=r-1} r_{(x)}$$

EXERCISE 1.04

1. Write out the numerical values of the expansion $(\tfrac{1}{2}+\tfrac{1}{2})^r$ and $(\tfrac{1}{4}+\tfrac{3}{4})^r$ for all values of r from 0 to 12 inclusive.

2. Write out the numerical values of

$$(\tfrac{1}{3}-\tfrac{2}{3})^8;\quad (\tfrac{1}{2}+\tfrac{1}{2})^{16};\quad (\tfrac{1}{2}+\tfrac{1}{2})^{25};\quad (\tfrac{1}{4}+\tfrac{3}{4})^{16}.$$

3. Find the sum of the first four terms and the three mid-terms of

$$(\tfrac{1}{4}+\tfrac{3}{4})^{12}.$$

4. Expand $(0\cdot35 + 0\cdot65)^8$ and $(0\cdot65 + 0\cdot35)^{12}$.

5. Find the rth term of

$(x + a)^{16}$ for $r = 10$; $(x - a)^{12}$ for $r = 7$;

$(1 + 2a)^{15}$ for $r = 5$; $(3x - 1)^{20}$ for $r = 10$.

6. Find the middle (or two middle) terms of the expansions

$$\left(3p - \frac{q}{3}\right)^9; \quad (13u + 5v)^6; \quad \left(\frac{x}{2} - \frac{y}{5}\right)^{10}; \quad \left(\frac{3m}{5} + \frac{2n}{3}\right)^{15}.$$

7. Find the coefficient of

x^8 in $(x + a)^{20}$; a^5 in $(x - a)^{10}$; u^{18} in $(u + v)^{25}$; t^{15} in $(s + t)^{24}$.

8. Calculate by binomial summation

$$(1 \cdot 01)^6; \quad (1 \cdot 3)^8; \quad (1 \cdot 05)^{10}; \quad (1 \cdot 04)^9.$$

9. Evaluate by the same method correct to 3 decimal places

$$(5 \cdot 9)^6; \quad (3 \cdot 95)^7; \quad (6 \cdot 997)^4; \quad (4 \cdot 89)^5.$$

Hint—$3 \cdot 95 = (4 - 0 \cdot 05)$.

10.* If y_x, y_{x+1} and y_{x-1} respectively stand for the xth,* $(x + 1)$th and $(x - 1)$th terms of the expansion of the binomial $(q + p)^r$, show that

$$y_{x+1} = \frac{p}{q} \cdot \frac{r - x}{x + 1} \cdot y_x.$$

$$y_{x-1} = \frac{q}{p} \cdot \frac{x}{r - x + 1} \cdot y_x.$$

11.* In the symbolism of No. 5 in Ex. 1.03, what is the mean value of (a) the xth term and the $(x + 1)$th; (b) the $(x - 1)$th and the xth term ?

12.* What is the difference between (a) the xth and $(x + 1)$th; (b) the xth and $(x - 1)$th term; and what is the proportionate error involved in equating the two differences ?

13.* What is the difference between the mean values of (a) the xth and $(x + 1)$th term; (b) the xth and $(x - 1)$th term of $(q + p)^r$?

14. On the assumption that the binomial theorem is valid for positive fractional indices, investigate the proportionate error involved in the following approximations for $x = 0 \cdot 5$, $x = 0 \cdot 1$, $x = 0 \cdot 05$ and $x = 0 \cdot 001$:

(a) $(1 + x)^{\frac{1}{2}} \simeq 1 + \frac{1}{2}x$.

(b). $(1 + x)^{\frac{1}{2}} \simeq 1 + \frac{1}{2}x + \frac{1}{2}(-\frac{1}{2})\frac{x^2}{2} = 1 + \frac{1}{2}x - \frac{1}{8}x^2$.

(c) $\sqrt[3]{1 - x} \simeq 1 - \frac{x}{3}$.

15. On the assumption that the binomial theorem is valid for fractional indices write down the first six terms of

(a) $\sqrt{1 + x}$;   (b) $\sqrt{1 + x^2}$;   (c) $\sqrt{1 - x}$;   (d) $\sqrt{1 - x^2}$.

* By the xth term we here signify the term involving p^x. We label the *initial* term of the binomial as the 0th (*see* p. 31).

16. Compare the answers to 14 and 15 when $x = 0.5$ by extraction of the square roots of the expressions involved or by logs.

17. On the assumption that the binomial theorem is valid for negative as well as positive indices, fractional or integral, show that

$$\frac{1}{\sqrt{1 - x^2}} = 1 + \frac{1}{2} \cdot x^2 + \frac{1.3}{2.4} \cdot x^4 + \frac{1.3.5}{2.4.6} \cdot x^6 \ldots$$

18. What is the general term of the expression in 12 ?

19. Find an infinite series for $\sqrt{2}$ by expanding $(1 + 1)^{\frac{1}{2}}$. Compare the result with the expansion of $(\frac{3}{2} + \frac{1}{2})^{\frac{1}{2}}$.

20. Find an infinite series for $\sqrt{\frac{1}{2}}$ by expanding $(1 - 0.5)^{\frac{1}{2}}$, and ascertain whether it is convergent.

1.05 The Vanishing Triangle

We shall now explore an operation (Δ) suggested by the build-up of figurate series as a means of deriving an expression for a *discrete* function of x, if reducible to a power series such as the general formula of the tetrahedral numbers

$$\frac{(x + 2)^{(3)}}{3!} = \frac{x^3}{6} + \frac{x^2}{2} + \frac{x}{3}.$$

The use of the term *discrete* in this context calls for explanation. In everyday life, we use numbers in two ways, for *measurements* which are necessarily approximate, and for *enumeration* which is necessarily exact, if correct. Corresponding to these two ways of applying numbers to the real world are two classes of functions, both of which are illustrable by reference to the statement that 2^x is a function of x.

(*a*) If the independent variable x can increase only by steps of one at a time, so that x must be a whole number, the expression 2^x specifies successive terms of the series which we may write alternatively as

1	2	4	8	16 $\ldots$
2^0	2^1	2^2	2^3	2^4 $\ldots$

In what follows we shall represent a function of this sort by the symbol y_x for the *dependent* variable. In this case $y_x = 2^x$ and x is the *rank* of the term of the series so defined. We may exhibit the relation thus :

$y_x = 2^0$	2^1	2^2	2^3	$2^4 \ldots$
$x = 0$	1	2	3	4 $\ldots$

Note.—Elementary text-books often label the *initial* term of a series as the term of *rank* 1. This entails a change of the form of the function. If we put $y_1 = 1 = 2^0 = 2^{1-1}$, the *general term* of the series defining the form of the function y_x is 2^{x-1}. How we represent the sum of the first n terms of a series depends on what convention we use to label the rank of a term. If we label the initial term as the term of rank 0, the first n terms are the terms from y_0 to y_{n-1} and there are $(n + 1)$ terms in the range y_0 to y_n inclusive. If we label the initial term as the term of rank 1, the first n terms are the terms from y_1 to y_n inclusive. In the y_0 symbolism used in what follows, the general (xth) term of the G.P. whose common ratio is r is $y_0 r^x$ and that of the

A.P. whose common difference is d is $y_0 + xd$. In the alternative symbolism the corresponding formulæ are $y_1 r^{x-1}$ and $y_1 + (x - 1)d$. The corresponding operations of summation require appropriate changes of the limits, e.g. :

$$\sum_{x=1}^{x=n} 2^{x-1} = \sum_{x=0}^{x=n-1} 2^x.$$

(b) If a function of x may assume a value corresponding to *any* value of x fractional or integral, rational or irrational, real or imaginary, we usually denote it as $f(x) = y$. Thus $f(x) \doteq 2^x$ may mean $1 \cdot 414 \ldots$ when $x = \frac{1}{2}$; but $y_x = 2^x$ has no meaning corresponding to $x = \frac{1}{2}$. Since $y = f(x)$ has values corresponding to any value of x, we can represent it by a graph on which there is a point y corresponding to any such value.

We cannot correctly represent the discrete function y_x as a graph, since its corresponding track is a series of points which lie equally spaced in the x-dimension of the Cartesian grid. Strictly speaking, the convention of visualising it as the staircase figure called a *histogram* is open to the same objection. The reason for doing so will appear later, when we seek a method for approximate summation of discrete functions.

When dealing with discrete functions with respect to which x is the rank of y_x, the smallest increment of x itself is $\Delta x = 1$. The corresponding increment of y_x when x increases to $x + 1 = (x + \Delta x)$ is

$$y_{x+1} - y_x = \Delta y_x.$$

For example, $\Delta 2^x = 2^{x+1} - 2^x = 2^x(2 - 1) = 2^x$, and $\Delta 3^x = 3^{x+1} - 3^x = 3^x(3 - 1) = 2 . 3^x$ Thus the increment of y_x per unit increment of x, i.e. the *rate of change* of y_x, is

$$\Delta y_x \div \Delta x = \Delta y_x.$$

If a function $f(x)$ is not *discrete* in this sense, possible increments of x have no smallest limit, and the rate of change of y with respect to x is the limiting ratio called its first differential coefficient denoted by the symbol

$$\frac{dy}{dx} = f'(x).$$

To derive Δy_x the increment of the discrete function y_x when x increases to $(x + 1)$ we subtract y_x from y_{x+1} in accordance with the following procedure with respect to the series of tetrahedral numbers defined by

$$y_x = \frac{(x + 2)^{(3)}}{3!}.$$

Since x stands for the rank of the term y_x in the above

$$y_0 = \frac{(0 + 2)(0 + 1)(0 + 0)}{3!} = 0.$$

We may therefore set out the first six terms of the series as in the first two lines below :

y_0	y_1	y_2	y_3	y_4	y_5
0	1	4	10	20	35
$(1 - 0)$	$(4 - 1)$	$(10 - 4)$	$(20 - 10)$	$(35 - 20)$	
$= 1$	$= 3$	$= 6$	$= 10$	$= 15$	
Δy_0	Δy_1	Δy_2	Δy_3	Δy_4	

We may apply this procedure successively as follows :

0		1		4		10		20		35
	1		3		6		10		15	
		2		3		4		5		
			1		1		1			
				0		0				
					0					

From the laws of formation of figurate series successive application of the operation denoted by Δ necessarily generates a succession of zeros. Successive rows of the *vanishing triangle*, as we may call this arrangement, reveal in reverse order the process of successive summation which generates the original series. The student should be able to satisfy himself or herself that the number of successive Δ operations which intervene before a row of zeros is the dimension (d) of the series, i.e. the highest power of x in the expanded form of y_x.

When performing the Δ operation successively, it is useful to label successive differences appropriately by an exponent which shows how many times we have to repeat the operation of subtraction. Thus

$$\Delta^2 y_x = \Delta y_{x+1} - \Delta y_x = (y_{x+2} - y_{x+1}) - (y_{x+1} - y_x)$$
$$= y_{x+2} - 2y_{x+1} + y_x.$$
$$\Delta^3 y_x = (\Delta y_{x+2} - \Delta y_{x+1}) - (\Delta y_{x+1} - \Delta y_x)$$
$$= [(y_{x+3} - y_{x+2}) - (y_{x+2} - y_{x+1})] - [(y_{x+2} - y_{x+1}) - (y_{x+1} - y_x)]$$
$$= y_{x+3} - 3y_{x+2} + 3y_{x+1} - y_x.$$

In conformity with this symbolism we may write Δy_x as $\Delta^1 y_x$ just as $x = x^1$; and $\Delta^0 y_x = y_x$ signifies that we do *not* perform on y_x the Δ operation, i.e. *subtraction from its successor*.

In general,

$$\Delta^{m+1} y_x = \Delta^m y_{x+1} - \Delta^m y_x,$$
$$\therefore \ \Delta^m y_{x+1} = \Delta^m y_x + \Delta^{m+1} y_x,$$

or

$$\Delta^m y_x = \Delta^m y_{x-1} + \Delta^{m+1} y_{x-1}.$$

By definition :

$$y_{x+1} = y_x + \Delta y_x.$$

We may write this in the form

$$y_{x+1} = (1 + \Delta) y_x.$$

The operator $(1 + \Delta)$ obeys the fundamental laws of arithmetic, as the following example illustrates :

$$y_{x+2} = y_{x+1} + \Delta y_{x+1} = (1 + \Delta) y_{x+1}$$
$$= (y_x + \Delta y_x) + (\Delta y_x + \Delta^2 y_x)$$
$$= y_x + 2\Delta y_x + \Delta^2 y_x$$
$$= (1 + 2\Delta + \Delta^2) y_x$$
$$= (1 + \Delta)^2 y_x.$$

Hence, if $x = 0$,

$$y_2 = (1 + \Delta)^2 y_0,$$

and, in general,

$$y_x = (1 + \Delta)^x y_0$$
$$= y_0 + x\Delta y_0 + \frac{x^{(2)}}{2!} \Delta^2 y_0 + \frac{x^{(3)}}{3!} \Delta^3 y_0 \ . \ . \ . \ \text{etc.} \ . \qquad . \qquad . \qquad . \quad \text{(i)}$$

3

This expression is a very powerful instrument for detecting the law of formation of a numerical series, as illustrated below. The accompanying table which refers to the tetrahedral numbers makes the connotation of all the symbols explicit, and exposes the arithmetical rationale of the formula itself.

Rank	Term	1st Difference	2nd Difference	3rd Difference	4th Difference
0	$y_0 = 0$				
		$\Delta^1 y_0 = 1$			
1	$y_1 = 1$		$\Delta^2 y_0 = 2$		
		$\Delta^1 y_1 = 3$		$\Delta^3 y_0 = 1$	
2	$y_2 = 4$		$\Delta^2 y_1 = 3$		$\Delta^4 y_0 = 0$
		$\Delta^1 y_2 = 6$		$\Delta^3 y_1 = 1$	
3	$y_3 = 10$		$\Delta^2 y_2 = 4$		$\Delta^4 y_1 = 0$
		$\Delta^1 y_3 = 10$		$\Delta^3 y_2 = 1$	
4	$y_4 = 20$		$\Delta^2 y_3 = 5$		
		$\Delta^1 y_4 = 15$			
5	$y_5 = 35$				

By reference to the foregoing table, it is possible to exhibit a very useful relation connecting any value of y_x with the terms (y_0, $\Delta^1 y_0$, $\Delta^2 y_0$, etc.) in the uppermost diagonal. To make it as explicit as possible, the following sequence shows the appropriate numerical values involved when the series defined by y_x is the tetrahedral numbers, but the symbolic relations are, of course, valid in their own right :

$$
\begin{aligned}
y_5 &= y_4 + \Delta^1 y_4 \\
&= 20 + 15 \\
&= (y_3 + \Delta^1 y_3) + (\Delta^1 y_3 + \Delta^2 y_3) \\
&= y_3 + 2\Delta^1 y_3 + \Delta^2 y_3 \\
&= 10 + 2(10) + 5 \\
&= (y_2 + \Delta^1 y_2) + 2(\Delta^1 y_2 + \Delta^2 y_2) + (\Delta^2 y_2 + \Delta^3 y_2) \\
&= y_2 + 3\Delta^1 y_2 + 3\Delta^2 y_2 + \Delta^3 y_2 \\
&= 4 + 3(6) + 3(4) + 1 \\
&= (y_1 + \Delta^1 y_1) + 3(\Delta^1 y_1 + \Delta^2 y_1) + 3(\Delta^2 y_1 + \Delta^3 y_1) + (\Delta^3 y_1 + \Delta^4 y_1) \\
&= y_1 + 4\Delta^1 y_1 + 6\Delta^2 y_1 + 4\Delta^3 y_1 + \Delta^4 y_1 \\
&= 1 + 4(3) + 6(3) + 4(1) + 0 \\
&= (y_0 + \Delta^1 y_0) + 4(\Delta^1 y_0 + \Delta^2 y_0) + 6(\Delta^2 y_0 + \Delta^3 y_0) + 4(\Delta^3 y_0 + \Delta^4 y_0) + (\Delta^4 y_0 + \Delta^5 y_0) \\
\therefore\ \underline{y_5 = y_0 + 5\Delta^1 y_0 + 10\Delta^2 y_0 + 10\Delta^3 y_0 + 5\Delta^4 y_0 + \Delta^5 y_0} \\
&= 0 + 5(1) + 10(2) + 10(1) + 5(0) + 0 = 35
\end{aligned}
$$

The expression underlined illustrates the general rule (i) stated above :

$$
\begin{aligned}
y_x &= y_0 + x_{(1)}\Delta^1 y_0 + x_{(2)}\Delta^2 y_0 + x_{(3)}\Delta^3 y_0 \ldots \\
&= \sum_{r=0}^{r=x} x_{(r)} \Delta^r y_0 \quad or \quad \sum_{r=0}^{r=x} \frac{x^{(r)}}{r!} \Delta^r y_0 \ . \qquad\qquad (ii)
\end{aligned}
$$

By applying (i) or (ii) we can derive a formula for any series of numbers which lead to a vanishing triangle. As an illustration of its use, let us consider the series which represents the sum of the cubes of the first x natural numbers, viz. :

Rank (x)—

0	1	2	3	4	5
0	(0^3+1^3)	$(0^3+1^3+2^3)$	$(0^3+1^3+2^3+3^3)$	$(0^3+1^3+2^3+3^3+4^3)$	$(0^3+1^3+2^3+3^3+4^3+5^3)$
y_x 0	1	9	36	100	225

The appropriate vanishing triangle is

$$
\begin{array}{ccccccccccccc}
0 & & 1 & & 9 & & 36 & & 100 & & 225 \\
& 1 & & 8 & & 27 & & 64 & & 125 & \\
& & 7 & & 19 & & 37 & & 61 & & \\
& & & 12 & & 18 & & 24 & & & \\
& & & & 6 & & 6 & & & & \\
& & & & & 0 & & & & &
\end{array}
$$

Whence we find

$$y_0 = 0 \ ; \ \Delta^1 y_0 = 1 \ ; \ \Delta^2 y_0 = 7 \ ; \ \Delta^3 y_0 = 12 \ ; \ \Delta^4 y_0 = 6 \ ; \ \Delta^5 y_0 = 0.$$

By substitution in (i) we now have

$$
\begin{aligned}
y_x &= 0 + x_{(1)} \cdot 1 + x_{(2)} \cdot 7 + x_{(3)} \cdot 12 + x_{(4)} \cdot 6 \\
&= x + \frac{7x(x-1)}{2} + \frac{12x(x-1)(x-2)}{6} + \frac{6x(x-1)(x-2)(x-3)}{24} \\
&= \tfrac{1}{4}[4x + 14x(x-1) + 8x(x-1)(x-2) + x(x-1)(x-2)(x-3)] \\
&= \frac{x^2(x+1)^2}{4}.
\end{aligned}
$$

Note.—This result checks itself, since it is the square of $\tfrac{1}{2}x(x+1)$, the general formula for the xth triangular number (p. 16). Successive values of y_x are in fact the squares of the triangular numbers, *viz.* : $0^2, 1^2, 3^2, 6^2, 10^2, 15^2. \ldots$

The relation defined by (i) is due to Gregory, a seventeenth century Scots mathematician who communicated it to Newton. It suggested the analogous formula discovered later by another Scots mathematician, Colin Maclaurin. If we denote the value $y = f(x)$ assumes when $x = 0$ by $f(0)$, the value $dy/dx = f^1(x)$ assumes when $x = 0$ by $f^1(0)$, the value d^2y/dx^2 assumes when $x = 0$ by $f^2(0)$ and so on :

$$y = f(0) + xf^1(0) + \frac{x^2}{2!}f^2(0) + \frac{x^3}{3!}f^3(0) \qquad \cdot \quad \cdot \quad \cdot \quad \cdot \quad \text{(iii)}$$

In Maclaurin's series simple powers of x replace factorial exponents of the coefficients in the Gregory-Newton formula. It is an interesting fact that many expressions involving the operator Δ are analogous to corresponding expressions involving the operator d/dx, if we replace simple powers by factorial indices. Consider, for example, the expression

$$
\begin{aligned}
y_x &= x^{(r)} \\
&= x(x-1)(x-2) \ldots (x-r+2)(x-r+1) \\
&= (x-r+1)x^{(r-1)}, \\
y_{x+1} &= (x+1)^{(r)} \\
&= (x+1)x(x-1) \ldots (x-r+2) \\
&= (x+1)x^{(r-1)}, \\
\therefore \ y_{x+1} - y_x &= \Delta y_x = (x+1)x^{(r-1)} - (x-r+1)x^{(r-1)}, \\
\therefore \ \Delta y_x &= rx^{(r-1)} \qquad \cdot \quad \cdot \quad \cdot \quad \cdot \quad \cdot \quad \cdot \quad \cdot \quad \cdot \quad \text{(iv)}
\end{aligned}
$$

This recalls the familiar expression for the differential coefficient of $y = x^r$, viz. :

$$\frac{dy}{dx} = rx^{r-1}.$$

When $y_x = (m + x)^{(r)}$ the same procedure shows that

$$\Delta y_x = r(m + x)^{(r-1)} \qquad \cdots \qquad \cdots \qquad (v)$$
$$\therefore \Delta^2 y_x = r(r - 1)(m + x)^{(r-2)}$$
$$= r^{(2)}(m + x)^{(r-2)},$$

and, in general,

$$\Delta^n y_x = r^{(n)}(m + x)^{(r-n)} \qquad \cdots \qquad \cdots \qquad (vi)$$

We shall use this result below.

It is sometimes convenient to express the xth term of a series in terms of the term of unit rank. We then have

$$u_2 = u_1 + \Delta u_1 = (1 + \Delta)u_1,$$
$$u_3 = u_2 + \Delta u_2 = (1 + \Delta)u_2 = (1 + \Delta)^2 u_1,$$

and, in general,

$$u_x = (1 + \Delta)^{x-1} u_1$$
$$= u_1 + (x - 1)\Delta u_1 + \frac{(x - 1)(x - 2)}{2!}\Delta^2 u_1 + \frac{(x - 1)(x - 2)(x - 3)}{3!}\Delta^3 u_1, \text{ etc. } . \quad (vii)$$

EXERCISE 1.05

1. Find formulæ for the following series :

(a) 0, 1, 3, 6, 10, 15.

(b) 0, 2, 6, 12, 20, 30.

(c) 0, 2, 8, 18, 32, 50.

(d) 0, 1, 8, 27, 64, 125.

(e) 0, 5, 24, 57, 104, 165.

(f) 0, 1, 12, 33, 64, 105.

(g) 0, 1, 6, 18, 40, 75.

(h) 1, 1, 5, 13, 25, 41.

(j) 4, 1, 12, 37, 76, 129.

(k) 0, 1, 4, 10, 20, 35.

2.* Find formulæ for the sums of the squares, cubes, 4th powers and 5th powers of the natural numbers and check by reference to Nos. 4–6 of Ex. 1.02.

3. Add an additional term to each of the following series :

(a) 0, 5, 20, 45, 80, 125.

(b) 0, 9, 28, 57, 96, 145.

(c) 0, 1, 4, 9, 16, 25.

(d) 8, 13, 18, 23, 28.

(e) 1, 6, 18, 40, 75.

(f) 0, 1, 12, 33, 64, 105.

(g) 1, 12, 37, 76, 129.

(h) 2, 14, 36, 68, 110.

(j) 0, 1, 7, 25, 65, 140.

(k) 0, 1, 8, 21, 40, 65.

4. Write down the values of

$$\Delta x^{(3)} ; \quad \Delta x^{(5)} ; \quad \Delta x^{(6)} ; \quad \Delta x^{(-4)}.$$

5. Find a formula for Δa^x and show that $\Delta 2^x = 2^x$.

6. Show that

$$2^x = 1 + x + \frac{x^{(2)}}{2!} + \frac{x^{(3)}}{3!} + \frac{x^{(4)}}{4!}, \text{ etc.,}$$

(a) by expanding $(1 + 1)^x$;   (b) by the operation $(1 + \Delta)^x 2^0$.

7. Show that

$$\Delta \log x = \log\left(1 + \frac{1}{x}\right).$$

8. By recourse to the formulæ for $\sin(A+B)$ and $\cos(A+B)$ find expressions for $\underline{\Delta \sin x}$ and $\underline{\Delta \cos x}$.

9. Find expressions for

(a) $\Delta 3x(1-x^2)$.   (f) Δx^n.

(b) $\Delta(a+bx)$.   (g) $\Delta(x+a)^{-3}$.

(c) $\Delta(a+bx^{(r)})$.   (h) $\Delta \dfrac{a+b^x}{x}$.

(d) Δx^3.   (j) $\Delta \dfrac{ax+b^x}{1+x}$.

(e) $\Delta \log(a+bx)$.   (k) $\Delta \tan x$.

1.06 THE BINOMIAL EXPANSION FOR FACTORIAL POWERS *

If a and b are both *integers* we may expand $(b+a)^{(r)}$ in a form analogous to the power series for $(b+a)^r$ of which (ii) in 1.04 defines the general term. We can show that this is true for particular *positive integral* values of r by direct multiplication. Thus, if $r=3$,

$$b^{(3)} + 3b^{(2)}a + 3b^{(1)}a^{(2)} + a^{(3)} = b(b-1)(b-2) + 3ab(b-1) + 3ba(a-1) + a(a-1)(a-2)$$
$$= (b^3 - 3b^2 + 2b) + (3ab^2 - 3ab) + (3a^2 b - 3ab) + (a^3 - 3a^2 + 2a)$$
$$= (b^3 + 3b^2 a + 3ba^2 + a^3) - 3(b^2 + 2ba + a^2) + 2(b+a)$$
$$= (b+a)^3 - 3(b+a)^2 + 2(b+a)$$
$$= (b+a)[(b+a)^2 - 3(b+a) + 2]$$
$$= (b+a)(b+a-1)(b+a-2)$$
$$= (b+a)^{(3)}.$$

The general term of the expansion $(b+a)^{(r)}$ is :

$$\frac{r!}{x!(r-x)!}b^{(r-x)}a^{(x)} \qquad . \qquad . \qquad . \qquad . \qquad . \qquad . \qquad (i)$$

A proof of the theorem for positive integral values of r and for integral values of b and a is obtainable from the *Gregory-Newton* expansion (i) of 1.05. Consider the series whose general term is

$$y_a = (b+a)^{(r)}.$$

Successive terms of this series obtained by putting $a = 0, 1, 2$, etc., are

$$b^{(r)}, \quad (b+1)^{(r)}, \quad (b+2)^{(r)}, \ldots \text{ etc.}$$

By (v) and (vi) in 1.05

$$\Delta y_a = r(b+a)^{(r-1)},$$
$$\Delta^2 y_a = r(r-1)(b+a)^{(r-2)} = r^{(2)}(b+a)^{(r-2)}.$$

$$. \qquad . \qquad . \qquad . \qquad . \qquad . \qquad . \qquad . \qquad .$$

$$\Delta^x y_a = r^{(x)}(b+a)^{(r-x)}.$$

Hence we have

$$\Delta^x y_0 = r^{(x)} b^{(r-x)} ; \quad y_0 = b^{(r)}.$$

* *Vandermonde's* Theorem.

In accordance with (ii) in 1.05

$$y_a = y_0 + a\Delta y_0 + \frac{a^{(2)}}{2!}\Delta^2 y_0 \ldots \text{ etc.,}$$

$$\therefore\ y_a = b^{(r)} + rb^{(r-1)}a + \frac{r^{(2)}}{2!}b^{(r-2)}a^{(2)} \ldots \text{ etc.}$$

Thus the general term of the expansion $y_a = (b + a)^{(r)}$ is as given by (i) above :

$$\frac{r^{(x)}}{x!}b^{(r-x)}a^{(x)} = \frac{r!}{x!(r-x)!}b^{(r-x)}a^{(x)}.$$

That is to say,

$$(b + a)^{(r)} = \sum_{x=0}^{x=r} \frac{r!}{x!(r-x)!}b^{(r-x)}a^{(x)} \qquad . \qquad . \qquad . \qquad . \text{ (ii)}$$

An important peculiarity of the expansion $(b + a)^{(r)}$ arises when r is greater than b and/or a. By definition $x^{(x)} = x(x-1)(x-2) \ldots 3.2.1$; and $x^{(x+1)} = x(x-1)(x-2) \ldots 3.2.1.0 = 0$. In general $x^{(r)} = 0$, if $r > x$. Hence $(b + a)^{(r)}$ vanishes if $r > (a + b)$ and all terms in the power series corresponding to $(b + a)^{(r)}$ vanish if the index (x) of a exceeds a, or that of b exceeds b. Consider the expansion

$$(2 + 3)^{(5)} = 5^{(5)} = 5!$$

By (i) above

$$(2 + 3)^{(5)} = 2^{(5)} + 5 . 2^{(4)}3 + 10 . 2^{(3)}3^{(2)} + 10 . 2^{(2)}3^{(3)} + 5 . 2 . 3^{(4)} + 3^{(5)}.$$

In every term except one, namely $10 . 2^{(2)}3^{(3)}$, the index of 2 exceeds 2 or the index of 3 exceeds 3. Accordingly, all terms vanish except

$$10 . 2^{(2)} . 3^{(3)} = 10 . 2 . 1 . 3 . 2 . 1 = 5 . 4 . 3 . 2 . 1 = 5!$$

The number of vanishing terms in a hypergeometric series, i.e. in a series involving factorial powers, is of some interest in connection with the calculus of choice, and is therefore worthy of examination. We shall use the symbol ${}^2S_r$ for the number of terms in the power series of the expansion $(b + a)^r$ and ${}^2S_{(r)}$ for the number of *residual* terms, i.e. terms which do not vanish, in that of $(b + a)^{(r)}$. We assume that b, a, and r are whole numbers. Since we label the terms of the binomial from 0 to r inclusive, ${}^2S_r = (r + 1)$.

Since $(b + a)^{(0)} = 1 = (b + a)^0$ and $(b + a)^{(1)} = (b + a) = (b + a)^1$,

$${}^2S_{(0)} = {}^2S_0 \text{ and } {}^2S_{(1)} = {}^2S_1.$$

For values of r exceeding 0 and 1, we have to distinguish four cases :

Case 1. When r exceeds neither b nor a ($r < b$ or a), no terms of $(b + a)^{(r)}$ vanish, so ${}^2S_{(r)} = {}^2S_r = (r + 1)$.

Case 2. When r exceeds b only ($b < r < a$) every term vanishes if the exponent of b exceeds b. Thus the residual terms are those involving $b^{(0)} \ldots b^{(b)}$ inclusive. Hence ${}^2S_{(r)} = (1 + b)$. *Mutatis mutandis* the same is true if r exceeds a only.

Case 3. When r exceeds both b and a but not their sum ($r > b, r > a, r < b + a$), we can put $r = (b + a - c)$, in which $c > 0*$ and $< r$ so that

$$(b + a)^{(r)} = (r + c)^{(r)}.$$

---

* Since $(a + b - c) = r$ and $r > a$, $(r + b - c) > r$, and since $r > b$, so that $(2r - c)$ is also greater than r. Hence $(r - c) > 0$, so that $r > c$.

For the reason stated with respect to the preceding case, this expression has $(1 + c)$ residual terms, i.e. ${}^2S_{(r)} = 1 + c$,

$$\therefore \ {}^2S_{(r)} = 1 + b + a - r.$$

If $r = (b + a)$ so that $c = 0$, there is only one residual term, viz.: $r_{(a)}b^{(b)}a^{(a)}$ as illustrated by the expansion $(2 + 3)^{(5)}$ above.

Case 4. When r exceeds the sum of b and a $(r > b + a)$, the expression reduces to zero, i.e. ${}^2S_{(r)} = 0$.

EXERCISE 1.06

1. Expand $(2 + 3)^{(4)}$; $(1 + 3)^{(2)}$; $(3 + 5)^{(3)}$.

2. Expand $\dfrac{(13 + 39)^{(4)}}{(52)^{(4)}}$ and $\dfrac{(26 + 26)^{(4)}}{(52)^{(4)}}$.

3.* If y_x and y_{x+1} are respectively the xth and $(x + 1)$th terms of the expansion $(s + f)^{(r)}$, show that
$$y_{x+1} = \frac{(s - x)(r - x)}{(x + 1)(f - r + x + 1)} \cdot y_x.$$

4.* Find a similar expression for y_{x-1}.

5.* Find an expression for Δy_x and for Δy_{x-1}, also the mean value of the two differences.

6. Expand $(4 + 6)^{(7)}$, and cite the numerical value of each term.

7. What terms vanish in the expansion of $(3 + 5)^{(8)}$?

8. How many residual terms are there in the expansion of $(3 + 5)^{(7)}$?

9. Determine the number of residual and vanishing terms in
$$(2 + 4)^{(5)}; \ (4 + 5)^{(8)}; \ (4 + 6)^{(10)}.$$

10.* If b, c, r are all positive integers, show that
$$(b + c)^{(r)} = r! \, [b_{(r)} + cb_{(r-1)} + c_{(2)}b_{(r-2)} \cdots c_{(r-1)}b + c_{(r)}],$$
and hence that
$$\frac{(b + c)!}{r! \, (r - b - c)!} = \sum_{x=0}^{x=r} c_{(x)}b_{(r-x)}.$$

1.07 THE MULTINOMIAL THEOREM

The binomial theorem exposes the law for deriving a power series corresponding to the repeated multiplication of an expression involving the sum of two *basic terms*, viz.: b and a in $(b + a)^r$. To expand an expression which contains three basic terms we may proceed as follows in conformity with (ii) of 1.04 :

$$(c + b + a)^r = (c + \overline{b + a})^r$$
$$= c^r + r \cdot c^{r-1}(b + a) + \frac{r^{(2)}}{2!}c^{r-2}(b + a)^2 + \frac{r^{(3)}}{3!}c^{r-3}(b + a)^3 + \ldots + (b + a)^r.$$

The general term of this binomial is

$$\frac{r^{(x)}}{x!}c^{r-x}(b+a)^x \qquad . \qquad . \qquad . \qquad . \qquad . \qquad \text{(i)}$$

If we expand the factor within the brackets we get

$$b^x + xb^{x-1}a + \frac{x^{(2)}}{2!}b^{x-2}a^2 + \frac{x^{(3)}}{3!}b^{x-3}a^3 \ldots + a^x.$$

The general term of this, i.e. the pattern to which all terms conform, is

$$\frac{x^{(m)}}{m!}b^{x-m}a^m \qquad . \qquad . \qquad . \qquad . \qquad . \qquad \text{(ii)}$$

The general pattern to which the terms of the completely expanded trinomial conform is obtainable by substitution of (ii) and (i), viz. :

$$\frac{r^{(x)}}{x!}c^{r-x} \cdot \frac{x^{(m)}}{m!}b^{x-m}a^m = \frac{r!}{x!(r-x)!} \cdot \frac{x!}{m!(x-m)!}c^{r-x}b^{x-m}a^m.$$

If we denote $(r-x)$ by k and $(x-m)$ by l, $k+l+m=r$ and the above becomes

$$\frac{r!}{k!\,l!\,m!}c^k b^l a^m \qquad . \qquad . \qquad . \qquad . \qquad . \qquad \text{(iii)}$$

Similarly we may write $(d+c+b+a)^r$ in the form $(d+\overline{c+b+a})^r$ of which, by appropriate change of symbols, the general term is

$$\frac{r!}{j!(r-j)!}d^j(c+b+a)^{r-j} = \frac{r!}{j!\,k!\,l!\,m!}d^j c^k b^l a^m \qquad . \qquad . \qquad . \qquad . \qquad \text{(iv)}$$

The same pattern is generally applicable to the expansion of an expression involving any number of basic terms, if $(j+k+l+m \ldots) = r$. For subsequent treatment of the theory of choice it is useful to recognise how many terms the expansion of an expression such as $(d+c+b+a)^r$ contains. This depends both on the number of *basic terms* involved and on the value of r. We shall here denote the number of terms in the expansion to the power r by the general symbol $^m S_r$, which is $^2 S_r$ for a binomial, $^3 S_r$ for a trinomial, etc. Thus the number of terms $^3 S_2$ in the expansion of $(a+b+c)^2$ is six, as is evident by counting the items on the right below :

$$(a+b+c)^2 = a^2 + b^2 + c^2 + 2ab + 2ac + 2bc.$$

Similarly, $^4 S_2 = 10$ from the identity

$$(a+b+c+d)^2 = a^2 + b^2 + c^2 + d^2 + 2ab + 2ac + 2ad + 2bc + 2bd + 2cd.$$

We may tabulate such results as follows :

Power r	$^1 S_r$	$^2 S_r$	$^3 S_r$	$^4 S_r$	$^5 S_r \ldots$
0	1	1	1	1	1 . . .
1	1	2	3	4	5 . . .
2	1	3	6	10	15 . . .
3	1	4	10	20	35 . . .
4	1	5	15	35	70 . . .
. . .	. . .	. . .	. . .	. . .	

The numbers in the rows and columns are familiar figurate series. On reading successive figures in the column headed ^2S_r, we see that this is also the series $^1F_{r+1}$, e.g. $^2S_4 = 5 = {}^1F_5$. By inspection of other columns, we also see that

$$^3S_4 = 15 = {}^2F_5; \quad {}^4S_3 = 20 = {}^3F_4; \quad {}^5S_4 = 70 = {}^4F_5, \text{ etc.,}$$

and, in general,

$$^mS_r = {}^{m-1}F_{r+1} = {}^rF_m.$$

Hence, from (ii) in 1.02

$$^mS_r = \frac{(r+1+\overline{m-1}-1)^{(m-1)}}{(m-1)!}$$

$$= \frac{(r+m-1)^{(m-1)}}{(m-1)!} \qquad \cdot \qquad \cdot \qquad \cdot \qquad \cdot \qquad \cdot \qquad (v)$$

In accordance with (iii) in 1.02 we may write this in the alternative form :

$$^mS_r = \frac{(r+m-1)!}{(m-1)!\,(r+1-1)!} = \frac{(r+m-1)!}{r!(m-1)!} \qquad \cdot \qquad \cdot \qquad \cdot \qquad (vi)$$

A formal derivation of (v) and (vi) follows from the rule defined by (i) and (iv) in 1.02. If we expand $(c+b+a)^r$ in two stages as above, we first obtain $(r+1)$ terms of the type

$$r_{(x)}\, c^{r-x}(b+a)^x.$$

In the expansion of $(b+a)^x$ there are likewise $(x+1)$, i.e. $^1F_{x+1}$, terms. The total number of terms of the complete expansion is the sum of all such expressions over the range $x=0$ to $x=r$ inclusive, i.e.

$$^3S_r = \sum_{x=0}^{x=r} {}^1F_{x+1} = \sum_{y=1}^{y=r+1} {}^1F_y = {}^2F_{r+1} = {}^{3-1}F_{r+1}.$$

If we now repeat the same process of successive expansion on $(d+c+b+a)^r$ we first get $(r+1)$ terms of the form $r_{(x)}\, d^{r-x}\,(c+b+a)^r$. The expansion of $(c+b+a)^r$ yields as above $^2F_{r+1}$ terms, and the total number of terms in the complete expansion is therefore given by

$$^4S_r = \sum_{x=0}^{x=r} {}^2F_x = \sum_{y=1}^{y=r+1} {}^2F_y = {}^3F_{r+1} = {}^{4-1}F_{r+1}.$$

We can continue this process indefinitely, so that in general

$$^mS_r = {}^{m-1}F_{r+1}.$$

Example.—How many terms has the expansion $(c+b+a)^4$? Here $m=3$ and $r=4$ in (vi) above, so that $(r+m-1) = 6$ and $(m-1) = 2$,

$$\therefore {}^mS_r = {}^3S_4 = \frac{6!}{4!\,2!} = 15.$$

The powers involved are :

a^4, b^4, c^4 $\cdots$	3
$a^3b, a^3c, b^3a, b^3c, c^3a, c^3b$ $\cdots$	6
a^2b^2, a^2c^2, b^2c^2 $\cdots$	3
a^2bc, ab^2c, abc^2 $\cdots$	3
Total $\quad\cdots$	15

In connection with the theory of choice, it is also instructive to know the sum of the coefficients. For the binomial, we have as in (iii) of 1.04,

$$\sum \frac{r!}{x!(r-x)!} = (1+1)^r = 2^r.$$

For the multinomial expansion, the corresponding expression is

$$\sum \frac{r!}{u!\, v!\, w! \, \ldots} = (1+1+1+\ldots)^r.$$

If $(1+1+1\ldots) = m$, i.e. if there are m basic terms,

$$(1+1+1\ldots)^r = m^r.$$

$$\therefore \sum \frac{r!}{u!\, v!\, w! \, \ldots} = m^r \qquad . \qquad . \qquad . \qquad . \qquad . \qquad \text{(vii)}$$

EXERCISE 1.07

1. Expand :
$$(p+q+r)^4; \quad (2u+3v+4w)^3.$$

2. Give the terms of the expansion $(\tfrac{1}{2} + \tfrac{1}{4} + \tfrac{1}{4})^3$.

3. Find the numerical values of successive terms of
$$(\tfrac{1}{2} + \tfrac{1}{3} + \tfrac{1}{6})^4.$$

4. Compare the results of expanding term by term
    (a) $(\tfrac{1}{2} + \tfrac{1}{2})^4$ and $(\tfrac{1}{2} + \tfrac{1}{4} + \tfrac{1}{4})^4$.
    (b) $(\tfrac{3}{4} + \tfrac{1}{4})^3$ and $(\tfrac{3}{4} + \tfrac{1}{8} + \tfrac{1}{8})^3$.

5. Find what term of the expansion of the following has the highest numerical value :
$$(\tfrac{2}{3} + \tfrac{1}{6} + \tfrac{1}{6})^4; \quad (\tfrac{1}{2} + \tfrac{3}{10} + \tfrac{1}{5})^5.$$

1.08 The Multinomial Expansion in Factorial Powers

If r, a, b, c, etc., are all positive integers it is possible to expand $(a+b+c+d\ldots)^{(r)}$ as a series analogous to the expansion of $(a+b+c+d\ldots)^r$. The binomial expansion of $(a+b)^{(r)}$ conforms to the same pattern as that of $(a+b)^r$ except in so far as factorial powers replace ordinary exponents. Hence the method of expansion by successive binomialisation employed in 1.07 is applicable with a comparable result, e.g. :

$$(a+b+c+d\ldots)^{(r)} = (a + \overline{b+c+d\ldots})^{(r)}$$
$$= \sum_{x=0}^{x=r} \frac{r!}{x!(r-x)!} a^{(r-x)}(b+c+d\ldots)^{(x)}.$$

If $r = (m+l+k+j\ldots)$, the general term of the complete expansion is

$$\frac{r!}{m!\, l!\, k!\, j! \, \ldots} a^{(m)} b^{(l)} c^{(k)} d^{(j)} \ldots \qquad . \qquad . \qquad . \qquad . \qquad \text{(i)}$$

**** Formally, the number of terms in the expansion $(a + b + c \ldots)^{(r)}$ is the same as the number of terms in the expansion $(a + b + c \ldots)^r$; but any term of the form exhibited in (i) will be equal to zero if $m > a$ or $l > b$ or $k > c$, etc. (see p. 38). The greatest possible values of j, k, l, m are respectively d, c, b, a. If $r > (a + b + c + d \ldots)$ all terms of the expansion vanish. If $r = (a + b + c + d \ldots)$, it follows that $(m + l + k + j \ldots) = (a + b + c + d \ldots)$. As stated any term will vanish if $j > d$ and it must also vanish if $j < d$ because this is then possible only if $k > c$ or $l > b$ or $m > a$. Hence j and d must be equal, if the term as a whole does not vanish; and by the same token $k = c$, $l = b$, $m = a$, so that there is only one residual term when $r = (a + b + c + d \ldots)$, viz.:

$$\frac{r!}{a!\,b!\,c!\,d!}\, a^{(a)} b^{(b)} c^{(c)} d^{(d)} = r! \qquad\qquad\qquad \text{(ii)}$$

Example.—$(2 + 3 + 4)^{(9)} = 9^{(9)} = 9! = \dfrac{9!}{2!\,3!\,4!}\, 2^{(2)} 3^{(3)} 4^{(4)}.$

In 1.07 we have used the symbol ^mS_r for the number of terms in the complete expansion $(a + b + c \ldots)^r$ in which the number of basic terms $(a, b, c,$ etc.$)$ is m. For the binomial expansion with only two basic terms we write this ^2S_r and employ $^2S_{(r)}$ in 1.06 for the number of *residual* terms in the expansion $(a + b)^{(r)}$. For the number of residual terms in the expansion $(a + b + c \ldots)^{(r)}$ with m basic terms we accordingly use the symbol $^mS_{(r)}$. To label a residual term of the form exhibited in (i) it suffices to refer to the factor $a^{(m)} b^{(l)} c^{(k)} d^{(j)} \ldots$ which we may here call the *efficient* in contradistinction to the coefficient

$$\frac{r!}{m!\,l!\,k!\,j!}.$$

A consideration of the following cases will indicate how it is possible to evaluate $^mS_{(r)}$.

Case 1. When $r = 0$ or 1 and when r does not exceed any one of the basic terms, no terms of the expansion itself vanish, so that $^mS_r = {}^mS_{(r)}$. In accordance with (vi) in 1.07, we then put

$$^mS_{(r)} = \frac{(r + m - 1)!}{r!(m - 1)!} \quad \text{or} \quad \frac{(r + m - 1)^{(m-1)}}{(m - 1)!} \qquad\qquad \text{(iii)}$$

Example.—The number of residual terms in $(4 + 7 + 9)^{(4)}$ is

$$\frac{(4 + 3 - 1)^{(2)}}{2!} = \frac{6.5}{2} = 15.$$

Case 2. When r exceeds only one basic term, here denoted a, we proceed as follows. We first note that the total number of terms including those that vanish is obtainable by summation of the numbers in the expansion of the second factor of all efficients of the form $a^{(x)}(b + c + d \ldots)^{(r-x)}$ obtained by expanding $(a + \overline{b + c + d} \ldots)^{(r)}$. The factor $(b + c + d \ldots)^{(r-x)}$ contains $(m - 1)$ basic terms. We may therefore denote the operation by

$$^mS_{(r)} = \sum_{x=0}^{x=r} {}^{m-1}S_{r-x} = \sum_{y=0}^{y=r} {}^{m-1}S_y.$$

Now all efficients of the form $a^{(x)} (b + c + d \ldots)^{(r-x)}$ will vanish in the range $x > a$. Thus the residual terms will be those in which $(r - x) = y$ lie in the range $y = r$ to $y = (r - a)$ corresponding to $x = 0$ to $x = a$.

**** Omit on first reading.

$$\therefore \ ^mS_{(r)} = \sum_{y=(r-a)}^{y=r} {}^{m-1}S_y = \sum_{y=(r-a)}^{y=r} {}^{m-2}F_{y+1}$$

$$= \sum_{p=(r-a+1)}^{p=(r+1)} {}^{m-2}F_p$$

$$= \sum_{p=1}^{p=(r+1)} {}^{m-2}F_p - \sum_{p=1}^{p=(r-a)} {}^{m-2}F_p$$

$$= {}^{m-1}F_{r+1} - {}^{m-1}F_{r-a} \ . \qquad . \qquad . \qquad . \qquad \text{(iv)}$$

$$= {}^mS_r - \frac{(r+m-a-2)^{(m-1)}}{(m-1)!} \qquad . \qquad . \qquad . \qquad \text{(v)}$$

Example.—The number of residual terms in the expansion of $(2+5+7+9)^{(4)}$ is given by

$$\frac{(4+4-1)^{(4-1)}}{(4-1)!} - \frac{(4+4-2-2)^{(4-1)}}{(4-1)!} = \frac{7.6.5}{3.2.1} - \frac{4.3.2}{3.2.1} = 31.$$

In this case the efficients of the four vanishing terms are : $2^{(3)}.5$, $2^{(3)}.7$, $2^{(3)}.9$, and $2^{(4)}$.

Case 3. If r exceeds each of two out of m basic terms, here denoted a and b, the value of $^mS_{(r)}$ is as before obtainable by summation of the residual terms of the expansion of each expression of the form $(b+c+d \ . \ . \ .)^{(v)}$ in the range $y = r$ to $y = (r-a)$ inclusive. To evaluate the sum we have to distinguish between two possibilities :

(i) If $r > (a+b)$, so that $(r-a) > b$ and y therefore exceeds b throughout the range r to $(r-a)$, we can make use of the foregoing formula for the residual terms of an expansion involving only one basic term (in this expression b) less than y itself. Since the expression itself contains $(m-1)$ basic terms we write in accordance with (iv) :

$$^{m-1}S_{(y)} = {}^{m-2}F_{y+1} - {}^{m-2}F_{y-b}.$$

The required sum is

$$^mS_{(r)} = \sum_{y=(r-a)}^{y=r} {}^{m-1}S_{(y)}$$

$$= \sum_{y=(r-a)}^{y=r} {}^{m-2}F_{y+1} - \sum_{y=(r-a)}^{y=r} {}^{m-2}F_{y-b}$$

$$= \sum_{p=r-a+1}^{p=r+1} {}^{m-2}F_p - \sum_{p=r-a-b}^{p=r-b} {}^{m-2}F_p$$

$$= {}^{m-1}F_{r+1} - {}^{m-1}F_{r-a} - {}^{m-1}F_{r-b} + {}^{m-1}F_{r-a-b-1}$$

$$= {}^mS_r - \frac{(r+m-a-2)^{(m-1)}}{(m-1)!} - \frac{(r+m-b-2)^{(m-1)}}{(m-1)!}$$

$$+ \frac{(r+m-a-b-3)^{(m-1)}}{(m-1)!} \qquad . \qquad \text{(vi)}$$

Example.—The number of residual terms of $(2+3+6)^{(6)}$ is given by

$$\frac{(6+3-1)^{(2)}}{2!} - \frac{(6+3-2-2)^{(2)}}{2!} - \frac{(6+3-3-2)^{(2)}}{2!} + \frac{(6+3-2-3-3)^{(2)}}{2!}$$

$$= 28 - 10 - 6 + 0 = 12.$$

(ii) If $r < (a+b)$, so that $b > (r-a)$ not all expressions of the form $(b+c+d \ . \ . \ .)^{(v)}$ will contain a term less than y itself. We can therefore divide the range into two parts : from $y = b+1$ to $y = r$ inclusive with $^{m-1}S_{(y)}$ residual terms in the expansion of each expression throughout the range ; from $y = (b - a)$ to $y = b$ with $^{m-1}S_y$ residual terms in the expansion of expressions therein. Consequently, we now put

$$
\begin{aligned}
{}^mS_{(r)} &= \sum_{y=(b+1)}^{y=r} {}^{m-1}S_{(y)} + \sum_{y=(r-a)}^{y=b} {}^{m-1}S_y \\
&= \sum_{y=b+1}^{y=r} {}^{m-2}F_{y+1} - \sum_{y=b+1}^{y=r} {}^{m-2}F_{y-b} + \sum_{y=r-a}^{y=b} {}^{m-2}F_{y+1} \\
&= \sum_{y=r-a}^{y=r} {}^{m-2}F_{y+1} - \sum_{y=b+1}^{y=r} {}^{m-2}F_{y-b} \\
&= \sum_{p=-a+1}^{p=r+1} {}^{m-2}F_p - \sum_{p=1}^{p=r-b} {}^{m-2}F_p \\
&= {}^{m-1}F_{r+1} - {}^{m-1}F_{r-a} - {}^{m-1}F_{r-b} + {}^{m-1}F_0 \\
&= {}^mS_r - \frac{(r+m-a-2)^{(m-1)}}{(m-1)!} - \frac{(r+m-b-2)^{(m-1)}}{(m-1)!} \qquad . \qquad . \quad \text{(vii)}
\end{aligned}
$$

We thus drop out the last term of (vi) when $r \leqslant (a + b)$.

Example.—The number of residual terms of the expansion $(4 + 5 + 6 + 7)^{(6)}$ is

$$
\frac{(6+4-1)^{(3)}}{3!} - \frac{(6+4-4-2)^{(3)}}{3!} - \frac{(6+4-5-2)^{(3)}}{3!} = 84 - 4 - 1 = 79.
$$

The efficients of the five vanishing terms are $4^{(5)}5$, $4^{(5)}6$, $4^{(5)}7$, $4^{(6)}$, $5^{(6)}$.

. . . .

The method of *Case 3* is adaptable to any number of basic terms less than r by successive summation with due regard to the considerations advanced under (i) and (ii) above.★★★★

EXERCISE 1.08

1. Write down successive terms of
$$(u + v + w)^{(4)} ; \quad (u + 2v + 3w)^{(3)}.$$

2. What is the coefficient of the term involving $p^{(3)} q^{(2)} r^{(2)}$ in the expansion of $(p + q + r)^{(7)}$.

3. Find the number of residual and vanishing terms in
$$(2 + 3 + 4)^{(5)} ; \quad (2 + 5 + 6)^{(4)} ; \quad (3 + 5 + 6)^{(4)}.$$

4. How many vanishing terms are there in the expansion of
$$(7 + 3 + 2 + 1)^{(5)} ; \quad (2 + 3 + 4 + 5)^{(12)}$$
$$(5 + 4 + 6 + 7)^{(5)} ; \quad (1 + 2 + 3 + 4 + 5)^{(4)}.$$

5. What is the net numerical value of the coefficient of $p^3 q^2 r^2$ in the expansion of $(p + q + r)^{(8)}$ in *ordinary* powers of p, q, r ?

1.09 EXPONENTIAL AND LOGARITHMIC APPROXIMATIONS

Certain approximations are of frequent utility for the derivation of statistical formulæ, and it is important to memorise them.

(a) When p is very small :

$$(1 + p)^x \simeq e^{px} \quad and \quad (1 - p)^x \simeq e^{-px} \qquad \text{. . . . (i)}$$

(b) When x is very small :

$$\left. \begin{array}{c} \log_e (1 + x) \simeq (x - \tfrac{1}{2}x^2) \\ and \\ \log_e (1 - x) \simeq - (x + \tfrac{1}{2}x^2) \end{array} \right\} \qquad \text{. . . . (ii)}$$

(c) When x is very large :

$$x! \simeq \sqrt{2\pi x} . x^x . e^{-x} \qquad \text{. . . . (iii)}$$

. . . . . .

(a) The proof of (i) depends on the limits for $n = \infty$:

$$\left(1 + \frac{1}{n}\right)^n \simeq e \quad and \quad \left(1 - \frac{1}{n}\right)^n \simeq e^{-1},$$

$$\therefore \left(1 \pm \frac{1}{n}\right)^{nx} \simeq e^{\pm x}.$$

Put $p^{-1} = n$, so that $np = 1$ and

$$(1 + p)^{nx} \simeq e^x, \quad (1 + p)^{npx} \simeq e^{px},$$
$$\therefore (1 + p)^x \simeq e^{px}.$$

This is equivalent to writing

$$\log_e (1 + p)^x \simeq px,$$
$$\therefore x \log_e (1 + p) \simeq px,$$
$$\therefore \log_e (1 + p) \simeq p.$$

Reference to tables of $\log_e n$ shows that :

$$\log_e 1\cdot05 = 0\cdot04879,$$
$$\log_e 1\cdot005 = 0\cdot00499,$$
$$\log_e 1\cdot001 = 0\cdot00100.$$

(b) The proof of (ii) depends on the expansion of $\log (1 \pm x)$ as a power series by recourse to *Maclaurin's* theorem. The procedure of the latter involves the initial assumption that we can expand the continuous function $y = f(x)$ as an infinite series of the form

$$y = A_0 + A_1 x + A_2 x^2 + A_3 x^3 + A_4 x^4 \ldots$$

The assumption is justifiable if the series derived by evaluating the arbitrary constants $A_0, A_1, \ldots A_n$ is uniformly convergent. When $x = 0$, $f(x) = A_0$, which we write as

$$A_0 = f(0).$$

For the first differential of $f(x)$ we have

$$\frac{d}{dx} f(x) = A_1 + 2A_2 x + 3A_3 x^2 + 4A_4 x^3;$$

when $x = 0$ we write this as

$$\frac{d}{dx} f(0) = A_1,$$

or more briefly $f^1(0) = A_1$ and in general for the value of the nth derivative we write $f^n(x)$, whose value we denote by $f^n(0)$ when $x = 0$. By successive differentiation we have

$$f^2(x) = 2A_2 + 3.2.A_3x + 4.3.A_4x^2 \ldots$$
$$f^3(x) = 3.2.1.A_3 + 4.3.2.A_4x \ldots$$
$$f^4(x) = 4.3.2.1.A_4 + \ldots.$$

Hence we may put

$$A_0 = f(0); \quad A_1 = f^1(0); \quad A_2 = \frac{f^2(0)}{2!}; \quad A_3 = \frac{f^3(0)}{3!}; \quad A_4 = \frac{f^4(0)}{4!} \ldots \text{ etc.}$$

By substitution of these values in the series for $y = f(x)$, we get

$$y = f(0) + x.f^1(0) + \frac{x^2}{2!}f^2(0) + \frac{x^3}{3!}f^3(0) + \frac{x^4}{4!}f^4(0) \ldots \text{ etc.} \tag{iv}$$

When $y = \log_e (1 + x)$,

$$\frac{dy}{dx} = f^1(1 + x) = \frac{1}{1 + x} = + (1 + x)^{-1},$$
$$f^2(1 + x) = -(1 + x)^{-2},$$
$$f^3(1 + x) = +2(1 + x)^{-3},$$
$$f^4(1 + x) = -3.2.(1 + x)^{-4},$$
$$f^5(1 + x) = +4.3.2.(1 + x)^{-5}.$$

When $x = 0$, $y = \log_e 1 = 0$, so that $f(0) = 0$. By substitution in the above we have

$$f^1(0) = 1, \quad f^2(0) = -1, \quad f^3(0) = 2!, \quad f^4(0) = -(3!), \quad f^5(0) = 4!, \text{ etc.,}$$

so that

$$\log_e (1 + x) = x - \frac{x^2}{2} + \frac{x^3}{3} - \frac{x^4}{4} + \frac{x^4}{5} \qquad . \qquad . \qquad . \qquad . \tag{v}$$

If x is less than unity, this must be convergent. For instance, if $x = 0.1$, or less than 0.1, its sum must be less than the repeating decimal $0.\dot{1}$. By recourse to tables of $\log_e n$ it is easy to test the error involved in rejecting all terms after the second, i.e. by using the relation (ii) above :

$$\log_e (1 + x) \simeq x - \frac{x^2}{2}.$$

The following show the level at which high precision is attainable :

x	$(x - \frac{1}{2}x^2)$	$log_e (1 + x)$	*Percentage error*
0.5000	0.3750	0.4055	$\simeq 5$
0.0500	0.0475	0.04879	$\simeq 2.5$
0.0100	0.00995	0.00995	< 0.1

. . . .

(c) *Stirling's Theorem*, defined by (iii) above, may be written in the more convenient form

$$x! \simeq x^{x + \frac{1}{2}}e^{-x}(2\pi)^{\frac{1}{2}},$$
$$\therefore \log_{10}(x!) \simeq (x + \frac{1}{2})\log_{10} x - x\log_{10} 2.7183 + \frac{1}{2}\log_{10} 6.2832$$
$$\simeq (x + \frac{1}{2})\log_{10} x - (0.4343)x + 0.3991 \qquad . \qquad . \qquad . \qquad . \tag{vi}$$

A rigorous proof of this approximation is tortuous, and it is scarcely profitable to undertake at this stage. We shall defer it to Chapter 6, since the student can easily verify that (a) it gives a very good approximation even when x is as small as 10 ; (b) the proportionate error involved decreases as x increases.

By applying (vi) above we find, for example :

$$\log 8! \simeq 8 \cdot 5 \ (0 \cdot 9031) - 8(0 \cdot 4343) + 0 \cdot 3991 = 4 \cdot 6010,$$
$$\therefore \ 8! \simeq 39 \cdot 9 \times 10^3.$$
$$\log 10! \simeq 10 \cdot 5 - 4 \cdot 343 + 0 \cdot 3991 = 6 \cdot 5561,$$
$$\therefore \ 10! \simeq 3599 \times 10^3.$$
$$\log 12! \simeq 12 \cdot 5 \ (1 \cdot 0792) - 12 \ (0 \cdot 4343) + 0 \cdot 3991 = 8 \cdot 6773,$$
$$\therefore \ 12! \simeq 475, 677 \times 10^3.$$

We may tabulate these results as follows :

	Actual value	Stirling approximation	Percentage error
8!	40,320	39,900	1·00
10!	3,628,800	3,599,000	0·83
12!	479,001,600	475,677,000	0·69

The *Stirling* formula provides a means of evaluating the coefficients of binomial or hyper-geometric series involving high powers of r in the expansion of $(p + q)^r$ or $(p + q)^{(r)}$. The following result is important :

$$\frac{r!}{x!(r-x)!} \simeq \frac{r^{r+\frac{1}{2}} \cdot e^{-r}(2\pi)^{\frac{1}{2}}}{x^{x+\frac{1}{2}} \cdot e^{-x} \cdot (2\pi)e^{-r+x} \cdot (r-x)^{r-x+\frac{1}{2}}}$$
$$= \frac{r^{r+\frac{1}{2}}}{x^{x+\frac{1}{2}}(2\pi)^{\frac{1}{2}}(r-x)^{r-x+\frac{1}{2}}} \quad . \quad . \quad . \quad . \quad \text{(vii)}$$

In particular, when $(p + q) = 1$ and $x = rp$ so that $(r - x) = rq$

$$\frac{r!}{x!(r-x)!} p^x q^{r-x} = \frac{r!}{(rp)! \ (rq)!} p^{rp} q^{rq}$$
$$\simeq \frac{r^{r+\frac{1}{2}} p^{rp} q^{rq}}{(rp)^{rp+\frac{1}{2}}(2\pi)^{\frac{1}{2}}(rq)^{rq+\frac{1}{2}}} = \frac{r^{r+\frac{1}{2}} p^{rp} q^{rq}}{r^{rp+rq+1}(2\pi)^{\frac{1}{2}}p^{rp+\frac{1}{2}}q^{rq+\frac{1}{2}}}$$
$$= \frac{r^{r+\frac{1}{2}}}{r^{r+1} p^{\frac{1}{2}} q^{\frac{1}{2}}(2\pi)^{\frac{1}{2}}} = \frac{1}{(2\pi r p q)^{\frac{1}{2}}},$$
$$\therefore \ \frac{r!}{(rp)! \ (rq)!} p^{rp} \cdot q^{rq} \simeq \frac{1}{\sqrt{2\pi r p q}} \quad . \quad . \quad . \quad \text{(viii)}$$

The last result is of fundamental importance in statistical theory ; and we shall have occasion to use it more than once.

  . . . . . .

Note on Approximate Solution of Differential Equations

The foregoing approximations crop up often in connection with the derivation of approximate solutions of differential equations in a form suitable for reference to tables of standard functions ; and it is here appropriate to mention a device which simplifies a solution involving a logarithmic term with that end in view. Whenever the solution of a differential equation involves a logarithmic term it is convenient to express the integration constant itself as a logarithm. For illustrative purposes, consider the equation :

$$\frac{dy}{dx} = \frac{Ay}{(B-x)} \quad . \quad . \quad . \quad . \quad . \quad . \quad \text{(ix)}$$

$$\therefore \frac{1}{y} dy = \frac{A}{(B-x)} dx$$

$$\therefore \int \frac{dy}{y} = \int \frac{A \cdot dx}{(B-x)}$$

$$\therefore \log_e y = -A \log_e (B-x) + c . \qquad . \qquad . \qquad . \qquad . \qquad \text{(x)}$$

We now put

$$c = \log K,$$

so that

$$\log_e y = \log_e K(B-x)^{-A}$$

$$\therefore y = K(B-x)^{-A}$$

$$= KB^{-A}\left(1 - \frac{x}{B}\right)^{-A}.$$

When $x = 0$, $y = KB^{-A}$, which is the ordinate through the origin, i.e.

$$y = y_0\left(1 - \frac{x}{B}\right)^{-A} \qquad . \qquad . \qquad . \qquad . \qquad . \qquad \text{(xi)}$$

This is an exact solution of (ix). If A is very large and B is large in comparison with admissible values of x, we may employ the approximation of (i) above, viz. :

$$y \simeq y_0 e^{\frac{Ax}{B}} . \qquad . \qquad . \qquad . \qquad . \qquad . \qquad \text{(xii)}$$

It may be convenient to make use of the alternative approximation (ii). We then write (x) as

$$\log_e y = -A \log_e B\left(1 - \frac{x}{B}\right) + \log_e K$$

$$= -A \log_e \left(1 - \frac{x}{B}\right) - A \log_e B + \log_e K$$

$$= -A \log_e \left(1 - \frac{x}{B}\right) + \log_e B^{-A} + \log_e K.$$

$$\therefore \log_e y - \log_e KB^{-A} = -A \log_e \left(1 - \frac{x}{B}\right)$$

$$\therefore \log_e\left(\frac{y}{KB^{-A}}\right) \simeq -A\left(\frac{-x}{B} - \frac{x^2}{2B^2}\right) = A\left(\frac{2Bx + x^2}{2B^2}\right)$$

$$\therefore \frac{y}{KB^{-A}} \simeq exp. \left\{A\left(\frac{2Bx + x^2}{2B^2}\right)\right\}$$

$$\therefore y \simeq KB^{-A} exp. \left\{A\left(\frac{2Bx + x^2}{2B^2}\right)\right\} *$$

When $x = 0$, $y = KB^{-A}e^0 = KB^{-A}$, so that

$$y \simeq y_0 exp. \left\{A\left(\frac{2Bx + x^2}{2B^2}\right)\right\} \qquad . \qquad . \qquad . \qquad . \qquad \text{(xiii)}$$

EXERCISE 1.09

(Exx. 1-4.) Make graphs of the following functions for the range $x = 0$ to $1\cdot0$, choosing a scale most suitable to display the differences between the curves (using tables of natural logarithms) :

* When t in e^t is unwieldy it is convenient to write the latter in the form exp. (t).

4

1. (i) x.   (ii) $x - \frac{1}{2}x^2$.   (iii) $\log_e(1+x)$.   (iv) $x - \frac{x^2}{2} + \frac{x^3}{3}$.

2. (i) $-x$.   (ii) $-\left(x + \frac{x^2}{2}\right)$.   (iii) $\log_e(1-x)$.

3. (i) $(1 \cdot 01)^x$.   (ii) $e^{0 \cdot 01x}$.   (iii) $1 + \frac{x}{100}$.

4. (i) $(1 \cdot 1)^x$.   (ii) $e^{\frac{x}{10}}$.

5. Compare the rapidity of approximation for $\log_e\left(\frac{14}{11}\right)$ by using
   (i) $\log_e(1+x)$ for $x = \frac{3}{11} = 0 \cdot 273$.
   (ii) $\log_e\left(\frac{1+x}{1-x}\right)$ for $x = 0 \cdot 12$.

6. Compare, as in 5, $\log_e\left(\frac{13}{11}\right)$, taking
   (i) $x = \frac{1}{12} = 0 \cdot 083$;   (ii) $x = 0 \cdot 04$.

7. Compare the results of using the binomial theorem to expand $(1 \cdot 01)^{10}$ in 2 stages (as $(1 \cdot 01)^{10}$ to 4 terms, and then this result again to the 10th power and to 4 terms) ; and the exponential series for e, to 10 terms :

$$e^x = 1 + x + \frac{x^2}{2!} + \frac{x^3}{3!} + \ldots \quad \text{for} \quad x = 1.$$

8. For $a = 0 \cdot 001$ and $n = 100$ compare the results of the following approximations with that of direct calculation by logs and with tables of powers of e :
   (i) $(1+a)^n \simeq 1 + na$.   (ii) $(1+a)^n \simeq e^{an}$.

9. Compare the values of

   (i) $\sum\limits_{2}^{10} \frac{1}{x}$; $\int_{1 \cdot 5}^{10 \cdot 5} \frac{1}{x} dx$.   (ii) $\sum\limits_{3}^{9} \frac{1}{1+x^2}$; $\left[\tan^{-1} x\right]_{2 \cdot 5}^{9 \cdot 5}$.

10. Express the following as infinite series :

   (i) $e^{\frac{-x^2}{2V}}$;   (ii) $\log_e\left(1 - \frac{x}{B}\right)^{-c}$;   (iii) $\frac{1}{2}\log_e\left(\frac{1+x}{1-x}\right)$;   (iv) $\tan^{-1} x$.

1.10 FINITE INTEGRATION

We can regard integration as the solution of a differential equation inasmuch as the evaluation of the left hand expression below is equivalent to finding the unknown A of the expression to the right of it :

$$A = \int y \cdot dx ; \quad \frac{dA}{dx} = y \ . \qquad . \qquad . \qquad . \qquad . \qquad \text{(i)}$$

Accordingly, we may speak of integration as the operation inverse to differentiation. We shall now see that the operation of finding the sum of a series of discrete terms involves the analogous performance of solving a *difference* equation and is in fact the operation *inverse* to Δ, i.e. :

$$S_n = \sum_{x=0}^{x=n} u_x ; \quad \Delta S_n = u_n.$$

Thus the figurate number of dimension d and rank n in Fig. 1 is the sum of all figurate numbers of dimension $(d-1)$ from zero rank to rank n while the figurate number of dimension $(d-1)$ and rank n (Fig. 13) is the result of performing the operation Δ on the figurate number of rank $(n+1)$ and dimension d, i.e.:

$$ {}^dF_n = \sum_{r=0}^{r=n} {}^{d-1}F_r, \quad and \quad {}^{d-1}F_{n+1} = \Delta {}^dF_n. $$

If we label the top row of the vanishing triangle (p. 33) as the series $v_0, v_1, v_2, \ldots v_n$ and the second row by $u_0, u_1, u_2, \ldots u_n$, the two series are connected by the relation $u_n = \Delta v_n$. So stated the relation exhibits how we generate the series whose general term is u_n from the series whose general term is v_n; but we are entitled to reverse the process. Clearly, there must also exist an operation which permits us to generate v_n from u_n. We shall denote it by Δ^{-1}, so that $v_n = \Delta^{-1}u_n$. Thus Δ^{-1} signifies *what we have to do to u_n in order to get v_n*. The operator Δ acting an v_n means : find the difference between v_n and its successor. The operator Δ^{-1} acting on u_n means : find the term of a series such that the difference between it and its successor is u_n itself.

The clue we get from Figs. 13-14 suggests that we should examine the meaning of the sum :

$$ \sum_{n=a}^{n=b} u_n = u_a + u_{a+1} + u_{a+2} + \ldots + u_{b-1} + u_b. $$

By definition, $u_n = \Delta v_n$ and $\Delta^{-1}u_n = v_n$, so that

$$ \sum_{n=a}^{n=b} u_n = \Delta v_a + \Delta v_{a+1} + \Delta v_{a+2} + \ldots + \Delta v_{b-1} + \Delta v_b. $$
$$ = (v_{a+1} - v_a) + (v_{a+2} - v_{a+1}) + (v_{a+3} - v_{a+2}) + \ldots + (v_b - v_{b-1}) + (v_{b+1} - v_b) $$
$$ = -v_a + v_{b+1} $$
$$ = \Delta^{-1}u_{b+1} - \Delta^{-1}u_a \quad . \qquad \qquad \qquad \text{(ii)} $$

Example 1.   Find the value of $\sum_{3}^{7} \dfrac{n(n+1)}{2}$.

In this case $u_n = \dfrac{n(n+1)}{2}$ and $\sum_{n=3}^{n=7} u_n = v_8 - v_3$. We have to find v_n, a function of n, which satisfies the relation $u_n = \Delta v_n$. We know that

$$ \Delta(n+1)^{(3)} = \overline{(n+1+1)}^{(3)} - (n+1)^{(3)} $$
$$ = (n+2)(n+1)n - (n+1)n(n-1) = 3n(n+1) $$
$$ \therefore \Delta \frac{(n+1)^{(3)}}{3.2} = \frac{n(n+1)}{2}, $$
$$ \therefore v_n = \frac{(n+1)^{(3)}}{3.2}, $$
$$ \therefore v_8 = \frac{9.8.7}{3.2} = 84 ; \quad v_3 = \frac{4.3.2}{3.2} = 4, $$
$$ \therefore v_8 - v_3 = 80 = \sum_{3}^{7} u_n. $$

Check. $\sum_{3}^{7} u_n = 6 + 10 + 15 + 21 + 28 = 80.$

Example 2.   Find $\sum_{0}^{4} 3^n$.

Here $3^n = u_n = \Delta v_n$. We have to solve the equation $\Delta^{-1}3^n = v_n$, whence we obtain the required sum as $v_{4+1} - v_0$. We know that

$$\Delta 3^n = 3^{n+1} - 3^n = 3^n(3-1) = 2(3^n),$$

$$\therefore \frac{\Delta 3^n}{2} = 3^n = u_n,$$

$$\therefore v_n = \frac{3^n}{2},$$

$$\therefore \sum_0^n 3^n = \frac{3^{n+1}}{2} - \frac{3^0}{2} = \frac{3^{n+1}-1}{2},$$

$$\therefore \sum_0^4 3^n = \frac{3^5-1}{2} = \frac{243-1}{2} = 121.$$

Check. $1 + 3 + 9 + 27 + 81 = 121.$

. . . .

****NOTE ON INVERSE OPERATIONS

The student who is familiar with the elements of the infinitesimal calculus will also be familiar with the notion that integration is the operation *inverse* to differentiation. In the same way the summation of terms of a series which is *discrete* in the sense already defined (p. 31) is the operation inverse to the difference operation denoted by Δ. In higher mathematics certain conventions with reference to operators are in general use ; but earlier symbolic devices for the more elementary operations of algebra and trigonometry are neither consistent *inter se* nor with later usage. A brief digression on the symbolism of inverse operation may therefore be helpful to the reader. To illustrate the modern usage, we may consider a simple operation of the calculus of finite differences denoted by the symbol E. The operator E steps up the rank of a term, e.g. $E(\frac{2}{3}F_n) \equiv \frac{2}{3}F_{n+1}$, or, in general,

$$E(u_n) \equiv u_{n+1} \equiv u_n + \Delta u_n.$$
$$\therefore E(u_{n+1}) \equiv u_{n+2} \equiv E(E . u_n).$$

It is the accepted convention to indicate by the index 2 the instruction that we are to perform the operation E twice, i.e. $u_{n+2} \equiv E^2(u_n)$; and, in general,

$$E^m(u_n) \equiv u_{n+m}, \quad \text{so that} \quad E^m(u_0) \equiv u_m.$$

This symbolism gives a new insight into Gregory's formula. We may write

$$E . u_n \equiv u_n + \Delta u_n = (1 + \Delta)u_n.$$

In conformity with this symbolism the operator $(1 + \Delta) \equiv E$, and

$$E^n u_0 \equiv (1 + \Delta)^n u_0.$$

The student will note that this is equivalent to Gregory's series, if we assume that the operator $(1 + \Delta)$ obeys the ordinary laws of arithmetic, as we do implicitly assume when we compute the amount accumulated by compound interest at 5 per cent. on principal P in n years in accordance with the formula $(1.05)^n . P$. On the assumption stated, we may expand the above as follows :

$$E^n(u_0) = u_n = (1 + n\Delta + n_{(2)}\Delta^2 + n_{(3)}\Delta^3 \ldots \text{etc.}) u_0$$
$$= u_0 + n\Delta u_0 + n_{(2)}\Delta^2 u_0 + n_{(3)}\Delta^3 u_0 \ldots \text{etc.}$$

**** Omit on first reading.

$$u_n = {}^5F_n$$

$$\Delta u_n = {}^4F_{n+1}$$

$$\Delta^2 u_n = {}^3F_{n+2}$$

$$\Delta^3 u_n = {}^2F_{n+3}$$

$$\Delta^4 u_n = {}^1F_{n+4}$$

$$\Delta^5 u_n = {}^0F_{n+5}$$

$$\Delta^6 u_n = 0$$

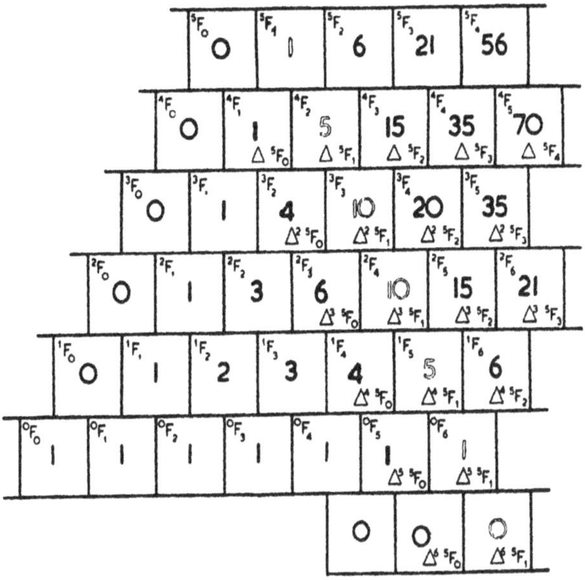

${}^5F_0$	${}^5F_1$	${}^5F_2$	${}^5F_3$	${}^5F_4$
0	1	6	21	56

${}^4F_0$	${}^4F_1$	${}^4F_2$	${}^4F_3$	${}^4F_4$	${}^4F_5$
0	1	5	15	35	70
	$\Delta\,{}^5F_0$	$\Delta\,{}^5F_1$	$\Delta\,{}^5F_2$	$\Delta\,{}^5F_3$	$\Delta\,{}^5F_4$

${}^3F_0$	${}^3F_1$	${}^3F_2$	${}^3F_3$	${}^3F_4$	${}^3F_5$
0	1	4	10	20	35
		$\Delta^2\,{}^5F_0$	$\Delta^2\,{}^5F_1$	$\Delta^2\,{}^5F_2$	$\Delta^2\,{}^5F_3$

${}^2F_0$	${}^2F_1$	${}^2F_2$	${}^2F_3$	${}^2F_4$	${}^2F_5$	${}^2F_6$
0	1	3	6	10	15	21
			$\Delta^3\,{}^5F_0$	$\Delta^3\,{}^5F_1$	$\Delta^3\,{}^5F_2$	$\Delta^3\,{}^5F_3$

${}^1F_0$	${}^1F_1$	${}^1F_2$	${}^1F_3$	${}^1F_4$	${}^1F_5$	${}^1F_6$
0	1	2	3	4	5	6
				$\Delta^4\,{}^5F_0$	$\Delta^4\,{}^5F_1$	$\Delta^4\,{}^5F_2$

${}^0F_0$	${}^0F_1$	${}^0F_2$	${}^0F_3$	${}^0F_4$	${}^0F_5$	${}^0F_6$
1	1	1	1	1	1	1
					$\Delta^5\,{}^5F_0$	$\Delta^5\,{}^5F_1$

					0	0
					$\Delta^6\,{}^5F_0$	$\Delta^6\,{}^5F_1$

$$\Delta^r\,{}^5F_n = {}^{5-r}F_{n+r}$$

Fig. 13. Successive application of the Delta operation to the series of Fig. 1.

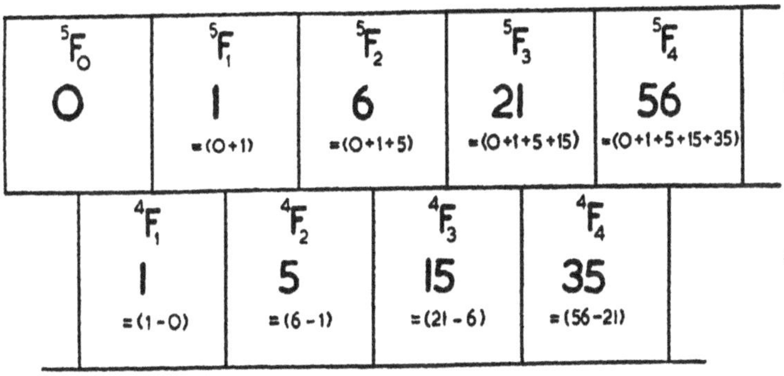

${}^5F_0$	${}^5F_1$	${}^5F_2$	${}^5F_3$	${}^5F_4$
0	1	6	21	56
	$=(0+1)$	$=(0+1+5)$	$=(0+1+5+15)$	$=(0+1+5+15+35)$

$${}^5F_n = \sum_{r=1}^{r=n} {}^4F_r$$

${}^4F_1$	${}^4F_2$	${}^4F_3$	${}^4F_4$
1	5	15	35
$=(1-0)$	$=(6-1)$	$=(21-6)$	$=(56-21)$

$${}^4F_{n+1} = \Delta\cdot{}^5F_n$$

$$\Delta^{-1}\cdot{}^4F_{n+1} = \Delta^{-1}\cdot\Delta\cdot{}^5F_n = {}^5F_n = \sum_{r=1}^{r=n} {}^4F_r$$

$$\therefore \sum_{r=1}^{r=n} {}^4F_r = \Delta^{-1}\cdot{}^4F_{n+1}$$

Fig. 14. Summation of Figurate Number Series as an Inverse Operation.

Accordingly, E^0 means that we have not yet performed the operation E on what follows. This usage is on all fours with the conventions of indices in elementary algebra, since n^1, n^2, n^3, etc. mean respectively : " perform the operation of multiplying n^0 ($= 1$) by n once, perform the operation on the result a second time, perform the operation on the second result a third time, *and so on* ". We may write n^0 in the form n^{1-1}, and more generally we may regard the operation denoted by the index $- x$ as the operation which *neutralises* the result of the x successive operations denoted by the index x.

The index in $sin^{-1} x$ signifies an inversion of the operation denoted by sin in this sense. *Sin x* signifies the length of the semichord subtended by the angle x in a circle of unit radius. *Sin^{-1} x* (*inverse* sine) means the angle which subtends a semichord of length x in a circle of unit radius, e.g. if $x = \frac{\pi}{2}$, $\sin x = 1$ and if $x = 1$, $\sin^{-1} x = \frac{\pi}{2}$. Unfortunately, the symbol $sin^n x$ for positive values of n is *not* consistent with the generalised index notation set forth above, since it does not signify successive performance of one and the same operation. The operator sin^n instructs us to perform two different operations one of semichord-specification and one of successive self-multiplication. To be consistent, we ought to write it in the form $(sin\ x)^n$.

The foregoing interpretation of the notation of the *inverse* operation implies that E^{-1} is the step-down operation, i.e. $E^{-1} u_n = u_{n-1}$, so that

$$E^{-2}(E^2 u_n) \equiv E^{-2}(u_{n+2}) \equiv u_{n+2-2} \equiv u_n.$$

If we were consistent with the symbolism of inverse operations, the symbol log^0 would signify that we have not yet performed the operation of replacing x by its logarithm ; and we should write $antilog_a x \equiv log_a^{-1} x$. If we write e^x in the form $exp.(x)$, $exp.^0(x)$ would signify that we have not yet performed the operation of raising e to the xth power. Since $x = log_e y$ when $exp.(x) = y$, we should write $exp.^{-1} exp.(x) = x = log_e y$, so that $exp.^{-1}(y) = log_e y$.

The identity of the operator E and the operator $(1 + \Delta)$, and hence the identity $E^x \equiv (1 + \Delta)^x$ carries with it the identity $(E - 1) \equiv \Delta$, and hence also $(E - 1)^x \equiv \Delta^x$. This permits us to obtain the expression

$$\Delta^x u_n = (E - 1)^x u_n$$
$$= E^x u_n - x \cdot E^{x-1} u_n + x_{(2)} E^{x-2} u_n \ldots \text{ etc.}$$
$$= u_{n+x} - x \cdot u_{n+x-1} + x_{(2)} u_{n+x-2} \ldots \text{ etc.}$$
$$\therefore \Delta^x u_0 = u_x - x \cdot u_{x-1} + x_{(2)} \cdot u_{x-2} - x_{(3)} \cdot u_{x-3} \ldots \text{ etc.}$$

The reader should be able to test this identity by recourse to any of the figurate series of 1.02.★★★★

1.11 Approximate Summation

In practice, we perform the operation of integration by relying on our knowledge of the form A must have in (i) to yield the function y as its differential coefficient ; and our ability to do so depends thus on the possibility of differentiating a wide range of functions. The method of summation set out above is analogous. It is, however, a method of much less extensive utility, because the operation Δ is of much more restricted range than differentiation in the infinitesimal domain. The main reason for this is that there is no operation in the finite domain analogous to

$$\frac{d}{dx} f(u) = \frac{du}{dx} \cdot \frac{d}{du} f(u). \qquad \qquad \text{(i)}$$

Exact summation of many discrete series is in fact impossible by recourse to the use of the inverse operation Δ^{-1}, and we have either to rely on *ad hoc* methods, such as that of figurate summation in 1.02, to achieve our end, or to content ourselves with a good approximation.   In practical statistics, a good approximation is all that we need ; and we shall mainly rely on an approximate method, when we have to sum discrete series of terms in subsequent chapters. The summation of a harmonic series will serve to illustrate the principle involved, and is apposite inasmuch as there is no exact summation formula for a H.P. on all fours with the familiar summation formulæ for an A.P. or a G.P.   The simplest H.P. is :

$$\frac{1}{0}, \frac{1}{1}, \frac{1}{2}, \frac{1}{3}, \frac{1}{4} \cdots$$

The general formula of this series is evidently

$$y_x = \frac{1}{x}. \qquad\qquad\qquad\qquad\qquad \text{(ii)}$$

The corresponding continuous function of y is the rectangular hyperbola $y = \frac{1}{x}$.   The curve of this function goes through every point corresponding to the terms of the harmonic series ; and we can make use of this fact to get an approximate summation formula for the H.P., i.e. to evaluate

$$\sum_{x=a}^{x=b} y_x = \sum_{a}^{b} \frac{1}{x} \qquad\qquad\qquad \text{(iii)}$$

The possibility of doing so depends on the geometry of the histogram, a visual device to exhibit the growth of a function which increases by discrete steps, as does an A.P. or a H.P. A histogram is an array of columns.   The height of a column represents the numerical value of y_x corresponding to a particular value of x.   Each column is of *unit width* on the same scale as y_x.   Its width thus represents a unit step Δx of x, marked off at the *mid-point of the base.* Thus $\Delta x = 1$ by definition, so that

$$y_x . \Delta x = y_x . \qquad\qquad\qquad\qquad \text{(iv)}$$

$$\sum_{x=a}^{x=b} y_x . \Delta x = \sum_{x=a}^{x=b} y_x \qquad\qquad\qquad \text{(v)}$$

The element $y_x . \Delta x$ is the area of the column corresponding to a particular value of x.   The total area of the histogram between, and including, the columns of height y_a and y_b is therefore numerically equivalent to the summation on the right of (v).

Fig. 15 shows that a continuous curve passing through the mid point of the upper extremity of each column of the histogram alternatively cuts off a triangular strip (*red*) from the top of a column and takes in a triangular strip (*black*) of nearly the same size from an adjacent column. Hence the area bounded between the ordinates at $x = a - \frac{1}{2}$ and $x = b + \frac{1}{2}$ is approximately equal to that of the histogram sector including at its extremities the columns of height y_a and y_b, i.e.

$$\int_{a-\frac{1}{2}}^{b+\frac{1}{2}} y . dx = \sum_{x=a}^{x=b} y_x \qquad\qquad\qquad \text{(vi)}$$

When x is very large, the correspondence between the *echelon* contour of the histogram and the smooth contour of the curve is very close ; and if b is also very large compared with a or *vice versa*, we may put.

$$\int_{a}^{b} y . dx = \sum_{a}^{b} y_x \qquad\qquad\qquad\qquad \text{(vii)}$$

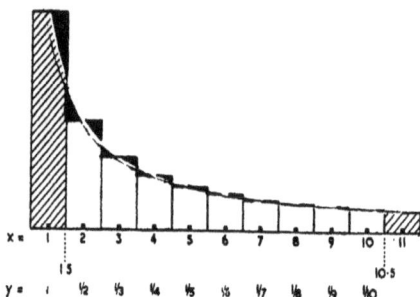

FIG. 15. The area of a histogram with column of *Unit* Base represents the exact value of the sum of the terms of a series within specified limits. Approximately, this is equivalent to the area enclosed by a smooth curve through the mid-points at the top of each column and ordinates respectively place one-half interval to the left and to the right of such limits. The accuracy of the approximation depends on the slope in the region involved.

The utility of (vi) and (vii) depends on the fact that we can often evaluate the expression on the left, when there is no available method for deducing a formula for the expression on the right. For instance,

$$\int_{a-\frac{1}{2}}^{b+\frac{1}{2}} \frac{1}{x} \cdot dx = \log_e (b + \tfrac{1}{2}) - \log_e (a - \tfrac{1}{2}).$$

Consider the nine terms of the Harmonic series cited above from $y = 0.5$ ($x = 2$) to $y = 0.1$ ($x = 10$) inclusive. By direct addition we find their sum as follows :

$$\tfrac{1}{2} + \tfrac{1}{3} + \tfrac{1}{4} + \tfrac{1}{5} + \tfrac{1}{6} + \tfrac{1}{7} + \tfrac{1}{8} + \tfrac{1}{9} + \tfrac{1}{10}$$

$$= 0.5000 + 0.3333 + 0.2500 + 0.2000 + 0.1667 + 0.1428 + 0.1250 + 0.1111 + 0.1000$$
$$= 1.9289.$$

Correct to four significant figures therefore :

$$\sum_{x=2}^{x=10} \frac{1}{x} = 1.929.$$

The corresponding area enclosed by the curve is bounded by ordinates at $x = 2 - \tfrac{1}{2} = 1.5$ and $x = 10 + \tfrac{1}{2} = 10.5$. Its area is, therefore,

$$\log_e (10.5) - \log_e (1.5) = 2.352 - 0.406$$
$$= 1.946.$$

The proportionate error is small, being $(1.946 - 1.929) \div 1.929$, which is less than 1 per cent. To show that it is still smaller if we extend the range from $x = 2$ to $x = 20$ is a useful exercise for the student. Since the function changes very rapidly in the region $x = 0$ to $x = 2$, we introduce a large source of error by including the first term, a circumstance which emphasises the need to use devices of this sort with due regard to the properties of the function in the region with which we are concerned. In particular, we cannot expect great precision if a region where the gradient is changing steeply is near either boundary of the distribution. If a function is symmetrical, or nearly so, we can get highest precision by placing the *origin where the axis of symmetry cuts the abscissa*.

EXERCISE 1.10

1. Find an expression for the sum of all terms from rank a to rank b of the following :

(a) $3 + 2x$;   (b) $3 + 2x^{(2)}$;   (c) $3 + 2x + 2x^2$.

2. Evaluate $\displaystyle\sum_1^5 (2x)^{(3)}$.

3. Sum to six terms :

$$3 + 12 + 48 + 192 + \ldots$$

4. Find the sum of n terms of

$$a + ab + ab^2 + ab^3 + \ldots$$

5. Evaluate :

$$\frac{1}{5!} + \frac{1}{6!} + \frac{1}{7!} + \frac{1}{8!} + \ldots \text{ to } \infty.$$

6. Sum :

$$1 + 7^{(1)} + 7^{(2)} + 7^{(3)} + \ldots \text{ to } \infty.$$

7.* Find an expression for the sum to n terms of the squares of the natural numbers by expressing n^2 as the difference of cubic terms.

8.* Find the sum to n terms of the cubes of the natural numbers by the method of the previous example.

9.* Seek a meaning for $x^{(-n)}$, and hence find a formula for $\Delta x^{(-n)}$.

10. By means of (9) evaluate the sum of the first six terms of

(a) $\dfrac{1}{1.2}, \ \dfrac{1}{2.3}, \ \dfrac{1}{3.4} \cdots$ (b) $\dfrac{1}{1.3}, \ \dfrac{1}{3.5}, \ \dfrac{1}{5.7} \cdots$ (c) $\dfrac{1}{2.4}, \ \dfrac{1}{4.6}, \ \dfrac{1}{6.8} \cdots$

CHAPTER 2

THE CALCULUS OF CHOICE

2.01 Definitions

Difficulty in dealing with problems of choice, i.e. *sampling*, arises less in connection with the solution of the mathematical problem than with the selection of the mathematical operation appropriate to its verbal formulation. It is therefore desirable at the outset to define in explicit terms conditions relevant to the enumeration and specification of samples. Also at the outset, it is well for the beginner to realise that difficulties which beset the verbal formulation will be less than otherwise forbidding, if study of the visual aids keeps in step with reading of the text.

Universe and Sample

Subsequent definitions presuppose a collection or collections of things (*items*) from which we extract the sample or samples under discussion. This collection is the *universe* of discourse. We shall adopt the letter n as a fixed convention to denote the number of items in a universe, with subscripts if necessary to distinguish one universe from another. A *sample* is a collection made up of items from the universe. We shall use the letter r for the total number of items in a sample, and accordingly speak of a sample so labelled as an r-fold sample. An r-fold sample is a sample with a niche for each of r items taken from the universe (or universes) of discourse. *Inter alia*, the extraction of such a sample may involve

(a) simultaneous or successive selection of r items from *one and the same* universe of n items ;

(b) simultaneous selection of one item from each of r universes respectively containing n_1, n_2, . . . n_r items ;

(c) successive selection of $r = mx$ items by successive withdrawal of m items from each of x universes.

Repetition

If we limit our definition of a sample to a simultaneous r-fold occurrence, a sample cannot contain more items than the universe or universes from which we extract the items themselves. For samples from a single universe r must then be numerically equal to or less than n. We are, however, entitled to use the term in a wider sense, as when we call the result of a 10-fold toss a sample of the behaviour of a coin. A sample in this extended sense admits the possibility of choosing the same item to fill more than one *niche*. When we no longer speak of sampling in the more usual and restricted sense of the term, such *repetitive* choice of items from a *single* universe presupposes reinstatement of any item chosen to fill a given niche before the possibility of withdrawal to take its place in a second niche : and the value of r has then no upper limit. According as we do or do not permit repetition of this sort, we have to distinguish successive acts of choice as *repetitive* and *restrictive* (i.e. non-repetitive). It goes without saying that

(a) *simultaneous* selection of r items from one and the same universe implies that identical constituents cannot be present in one and the same r-fold choice ;

(b) *successive* selection admits the possibility of replacing items chosen and therefore the possible repetition of the same item in an r-fold act of successive choice.

Since repetitive choice involving withdrawal from a single universe implies *replacement* of each item chosen before choice of another, repetitive choice is necessarily *successive ;* and since replacement implies the intervention of another occurrence between two single acts of choice *simultaneous* sampling from one universe is necessarily *restrictive* in the sense defined. If we choose to regard items of one universe as identical with items of another, the distinction between simultaneous and successive sampling is no longer fundamental from this viewpoint. Simultaneous selection from different universes containing identical items is equivalent to successive selection from one universe with replacement.

With due regard to the two types of sampling which we here distinguish as repetitive and restrictive, we should be alert to a distinction between two classes of models which provide a background for statistical analysis : (a) the *die* model ; (b) the *urn* and the *card pack* model. The tossing of dice or the tossing of a coin are examples of sampling in the sense that each score constitutes a sample of possible scores. Such sampling is necessarily the repetitive type, since the extraction of the sample, i.e. recording the score, leaves the universe unchanged. From this point of view, a penny is a 2-fold universe and a cubical die is a 6-fold universe. A triple toss of a die amounts to the same thing as a simultaneous single toss of 3 identical dice, and we can equally well regard the sample, i.e. the score, obtained as one way of : (a) selecting repetitively 3 items from a single finite universe of 6 different items ; (b) selecting a *single* item from each of 3 identical finite universes of 6 different items. If equipped with a trap-door lid to ensure one-way traffic (*outwards*), an urn containing balls, each with some distinctive mark, is a set-up which admits restrictive sampling alone. Simultaneous selection of cards from a pack is also restrictive, but successive selection need not be. According as we do or do not impose replacement of each card taken before withdrawal of another as a condition of choice, a card pack may thus be like or unlike a die model in the sense defined above. To use a card pack with relevance to a particular statistical situation, we have to make this condition or its contrary explicit.

A clear appreciation of the inherent difference between different types of models from this point of view is the more important, because statistical text-books commonly employ a *die* model (tossing of a coin) to illustrate sampling processes which involve withdrawal without replacement. That this is often permissible (p. 79), in so far as the universe and sample are respectively very large and relatively small, does not constitute a sufficient reason for adding to our difficulties by confusing the issues involved. The postulate of an *infinite* universe is often inapplicable to a statistical problem ; and we are on safe ground if, and only if, the model we visualise truly reproduces the relevant peculiarities of the system we are investigating.

Ordered Choice

In virtue of its *composition* one *r*-fold sample is the same as another if, and only if, each class of items represented in one consists of the same number of items in the other. If we regard the sample as a historical occurrence we can conceive of each niche as a *vacancy in a time sequence* and distinguish different ways of making up one and the same sample by the temporal rank assigned to a particular item. Thus one way of extracting a sample is the same as a second if, and only if, every item occupying a niche in one sample has the same class specification as the item occupying the corresponding niche in the other. If sampling is strictly simultaneous, this interpretation of the *number of ways of extracting* the sample does not at first sight tally with linear order in a time sequence ; but every linear arrangement of the cards placed face upwards in a row corresponds to a temporal sequence of exposing the faces one at a time. In this sense successive withdrawal without replacement and simultaneous withdrawal of the same number of cards admit of the same number of *different ways of extracting* a sample of given composition.

Enumeration of samples thus corresponds to what we also speak of as *combinations*, and different linear permutations or linear arrangements correspond to *different ways of extracting a sample*. By all possible ways of making up a particular sample, we therefore mean the number of linear permutations of its *r* constituent items. By a particular sample we mean a particular combination.

Classification

Implicit in the definition of a particular sample and hence of particular ways of extracting a sample is that each item of one and the same universe is in some way distinguishable from every other.* This is, of course, consistent with the recognition that they may be more or less alike in virtue of attributes they share ; and we can classify objects of a universe in more than one way, according to what attributes we employ. Thus we may classify the cards of a pack in two classes as picture cards and others, in three classes as picture cards, aces and others, in two classes as red cards and black cards, in four classes as cards with 2 or more pips, aces, black picture cards and red picture cards. The reader may find it helpful to set out other classifications of a full pack with the proportion of cards in each class. The only restriction of a classification relevant to our purpose is that it must be *exclusive* and *exhaustive* in the sense that

    (*a*) all items assigned to a class share an attribute which pertains to no item assigned to any other class ;

    (*b*) every item in the universe is assignable to one or other class.

In virtue of any such classification we can impose a particular *class structure* on a universe, and classify different samples or ways of extracting samples accordingly. It will be convenient to use the letters *a, b, c* . . . respectively for the number of items in the universe assigned to the classes *A, B, C* . . . , so that $(a + b + c \ldots) = n$. We shall likewise use *u, v, w* . . . respectively for the number of items of the classes *A, B, C* . . . in an *r*-fold sample, so that $(u + v + w \ldots) = r$. We shall also use *m* for the number of classes in the universe itself. If no two items share any *specified* attribute, being *unlike* in all respects relevant to our problem, each item belongs to a class by itself so that $a = 1 = b = c$, etc., and $n = (1 + 1 + 1 \ldots$ to *m* terms$) = m$.

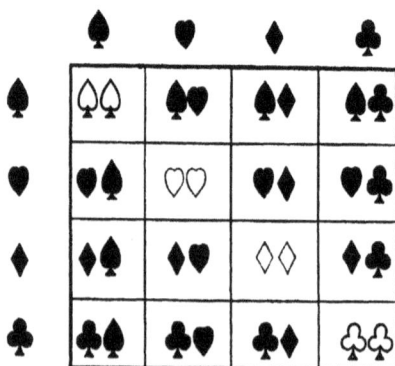

FIG. 16. Linear arrangements of 2 objects from a set of 4 :
with repetition (*all* pairs) . . . . $4 . 4 = 4^2$
without replacement (black pairs only) . . . . . $4 . 3 = 4^{(2)}$

* The reader will appreciate the importance of this qualification when we come to the derivation of (ii) and (iv) in 2.04.

2.02 ELEMENTARY THEOREMS OF ORDERED CHOICE

Elementary text-books cite three fundamental theorems of *linear* arrangement.

(i) *Objects unclassified and no object to be taken more than once.*

Consider the case of 5 letters a, b, c, d, e from which we may choose 3, as in the table below. There are 5 ways A, B, C, D, E of filling the first rank, 4 ways, I-IV, of filling the second, and 3 ways of filling the third corresponding to the number of items in each of the pigeon holes.

	A	B	C	D	E
I	a b c a b d a b e	b a c b a d b a e	c a b c a d c a e	d a b d a c d a e	e a b e a c e a d
II	a c b a c d a c e	b c a b c d b c e	c b a c b d c b e	d b a d b c d b e	e b a e b c e b d
III	a d b a d c a d e	b d a b d c b d e	c d a c d b c d e	d c a d c b d c e	e c a e c b e c d
IV	a e b a e c a e d	b e a b e c b e d	c e a c e b c e d	d e a d e b d e c	e d a e d b e d c

The total number of arrangements in the table is determinable by setting out : (a) all ways of filling the *first* place (5) ; (b) all remaining ways (4) of filling the *second* after fixing the first ; (c) all still remaining (3) ways of filling the last. This is evidently $5.4.3$; and in general the number of ways of arranging r of n unlike things in a row without repetition of any item is $n(n-1)(n-2) \ldots r$ factors. We write this as

$$^nP_r = n^{(r)} \qquad \qquad \text{(i)}$$

If we take all the items simultaneously, $r = n$, and

$$^nP_r = n^{(n)} = n! \qquad \qquad \text{(ia)}$$

We speak of this as the number of linear permutations of n things taken *all* at a time.

(ii) *Objects unclassified but a given object may be taken repeatedly.*

After filling the first place in n ways, we can still fill the second place in n ways. Hence we can fill the first two places in n^2 ways and in general we can fill r places in

$$n^r \text{ ways} \qquad \qquad \text{(ii)}$$

There is no special symbol for this. If $r = n$ it becomes n^n.

(iii) *The objects are classifiable into* m *groups with the same attributes, a of one group, b of a second, and so on. The problem is to determine in how many distinguishable ways we can arrange all of them taken together, if we take no account of differences between items of the same class.*

By definition $n = (a + b + c \ldots$ to m terms$)$. We denote the required number of such arrangements by ${}^{n}P_{a.b.c.} \ldots$ Now the a objects of one class are themselves all *distinguishable* in some way, and with due regard to such differences we can arrange them in $a!$ different ways corresponding to any distinguishable way of filling each residual niche allocated to an item of some other class. The total number of arrangements would then be : $a! \, {}^{n}P_{a.b.c.} \ldots$ We now do the same with b objects of a second class, so that the total number of distinguishable arrangements is : $a! \, b! \, {}^{n}P_{a.b.c.} \ldots$ and so on. Since all the objects of each group are in some way distinguishable, the total number of arrangements will be $n^{(n)}$, i.e. $n!$.

$$\therefore n! = a! \, b! \, c! \ldots {}^{n}P_{a.b.c.} \ldots$$

$$\therefore {}^{n}P_{a.b.c.} \ldots = \frac{n!}{a! \, b! \, c! \ldots m \text{ factors}} \quad . \quad . \quad . \quad \text{(iii)}$$

The expression *all of them taken together* implies the restriction that we can take one item once only. If we remove this restriction, there are distinguishable m different ways of filling each place and each way of filling each place is associable with each way of filling any other. So the number of n-fold arrangements we distinguish as different is

$$m \cdot m \cdot m \ldots n \text{ factors} = m^{n} \quad . \quad . \quad . \quad . \quad \text{(iv)}$$

Equipartition of Opportunity

Our preliminary definitions have specified the meaning we here attach to the number of *ways of extracting* a sample, *viz.*: the number of different linear arrangements of its constituent items. The device of Figs. 16-17 brings into focus a peculiarity of linear arrangement specially significant in connection with the role of such a specification in the realm of mathematical probability. These two figures refer to a universe of four items, *viz.*: a card pack consisting only of 1 *spade*, 1 *heart*, 1 *diamond* and 1 *club*; but the generality of the procedure is easy to recognise. If the condition of choice is replacement before subsequent withdrawal, each of the n^2 (here 4^2) ways of extracting a 2-fold sample in accordance with (ii) above corresponds to one of the pairs of an $n \cdot n$ lattice of the type shown in Fig. 16. To each row of the lattice corresponds one item of the universe ; and successive pigeon holes of the same row exhibit the result of drawing any one of n cards after first drawing the particular item indicated by the symbol in the left hand vertical margin. If choice is *not* repetitive, one such pair drops out of each row, leaving $(n - 1)$ pairs per row and a total of $n(n - 1) = n^{(2)}$ pairs in accordance with (i).

In visualising 2-fold samples of either type by this device, the underlying principle is that *each item first taken has an equal opportunity to associate with each remaining item.* We can extend the same method to the representation of linear arrangements of 3-fold samples or to larger samples by successive application as in Fig. 17. If choice is repetitive, there are n^2 ways of arranging the two items first taken to make up a 3-fold sample ; and there are n ways of taking the third. Accordingly, we lay out the n^2 ways of extracting a 2-fold sample in the left hand vertical margin and assign each such sample an equal opportunity to pair off with each of the n items in the universe reconstituted by replacement of cards extracted before withdrawal. The resulting lattice now has $n^2 \cdot n = n^3$ pigeon holes. If there is no replacement the lattice will have only $n^{(2)}$ instead of n rows ; and two triplets will drop out of every row, leaving $(n - 2)$ pairs per row or $n^{(2)} \cdot (n - 2) = n^{(3)}$ in all.

Every time we repeat the chessboard operation, we assign to each arrangement of $(r - 1)$ items already taken to make up an r-fold sample an equal opportunity to associate with each residual item of the universe. By identifying *ways of extracting* a sample with linear arrangements of its constituent items we therefore signify a method of enumeration based on *equipartition of opportunity for association*. The recognition (p. 94) that such equipartition of opportunity is implicit in the enumeration of linear arrangements, and hence of ways of extracting a sample as defined in 2.01, clarifies the notion of *randomisation* implicit in the connection between mathematical probability and probability as we use the word in common speech.

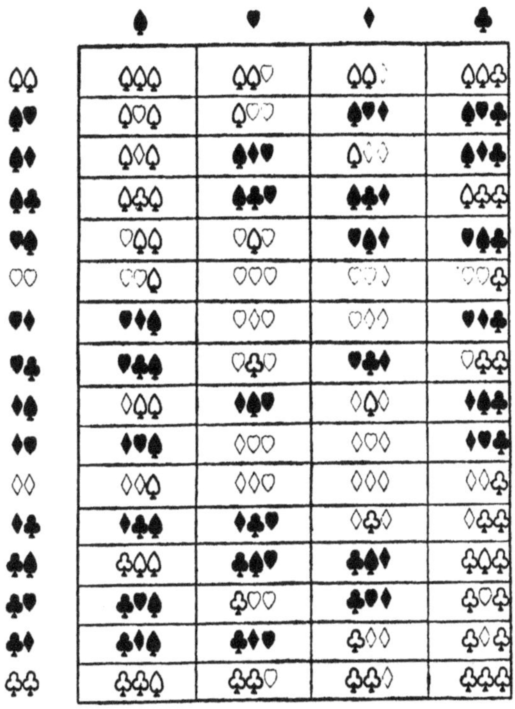

FIG. 17. Linear arrangements of 3 objects from a set of 4 :

with repetition . . . . $4^2 . 4 = 4^3$

without replacement (block pairs only) . . . . $4^{(2)} . 2 = 4^{(3)}$

EXERCISE 2.02

SET 1

Make chessboard diagrams to solve the following :

1. In how many recognisably different ways can two coins fall : (*a*) if both are pennies ; (*b*) if one is a penny and the other a half crown ?

2. In how many ways can three dice fall if thrown together ?

3. In how many ways is it possible to give 3 prizes to a class of 10 boys, without giving more than 2 to the same boy ?

4. In how many ways can 2 persons occupy 6 vacant seats of a bus ?

5. In how many ways can we select a consonant and a vowel out of an alphabet of 20 consonants and 6 vowels ?

6. In how many ways can 12 undergraduates and 12 undergraduettes form themselves into couples for a waltz ?

7. Having 5 pairs of gloves, in how many ways can a person select a right-hand and a left-hand glove which are not a pair ?

8. With 4 seals and 6 sorts of sealing wax, in how many ways is it possible to seal a letter ?

9. A cylindrical letter lock has 4 concentric rings of 6 letters. How many different unsuccessful attempts to open it is it possible to make ?

10. In how many ways is it possible to put 3 different letters in 4 different envelopes ?

EXERCISE 2.02

Set 2

1. In the Hindu-Arabic notation, how many numbers consist of 6 digits ?

2. Twenty-one schoolboys run a race for 4 prizes. In how many ways is it possible to allocate the prizes ?

3. How many different sums is it possible to donate from a purse which contains a pound note, a ten-shilling note, a half-crown, a florin, a shilling, a penny and a half-penny ?

4. In how many ways can 7 electors cast votes for 18 candidates ?

5. How many different arrangements of all the letters a, b, c, d, e, f begin with ab, if no letter occurs more than once ?

6. On a shelf there are 5 books in Latin, 4 in Hebrew and 8 in Greek. In how many ways can the books be arranged, keeping all the Latin together, all the Hebrew together and all the Greek together ?

7. In how many ways would it be possible to arrange the same books indiscriminately on the shelf ?

8. If 12 ladies and 12 men go to a ball, in how many ways can they take their places for a contre-danse ?

9. How many different signals of 4 flags is it possible to make with the flags of U.S.A., U.S.S.R., Britain, France, Holland, each of 3 Scandinavian nations : (i) on a single mast, and (ii) on a 3-masted ship ?

10. How many different sequences is it possible to ring upon 8 bells ? In how many of these will a particular bell ring last ?

EXERCISE 2.02

SET 3

1. In how many ways is it possible to arrange the letters of the words *parallelepiped, indivisibility* and *Mesopotamia* ?

2. In how many ways, irrespective of order, can we select 2 blocks of 4 letters from 3 *a*s, 3 *b*s, 2 *c*s and a *d* ?

3. Out of 15 consonants and 5 vowels, in how many ways is it possible to make an arrangement consisting of 3 different consonants and 3 different vowels ?

4. Out of the letters of the alphabet, in how many ways can we make an arrangement consisting of 4 different letters containing the 2 letters *a* and *b* ?

5. There are 10 *different* situations vacant. Four are for men and 3 for women, the remainder for either male or female candidates. In how many ways is it possible to fill the posts available ?

6. In how many ways can one make an arrangement of 4 letters from those of the words *choice* and *chance* ?

7. Eight men take their places in a boat with 8 oars. Two of them row only on stroke side, one of them only on the bow side ; the others on either side. In how many ways is it possible to dispose of the crew ?

8. In how many ways can 3 boys divide 12 oranges among themselves, if each takes 4 ?

9. In how many ways can a school of 90 boys divide themselves so that 24 play football, 22 play cricket, 30 practise music, 4 play squash and 10 take a country walk ?

10. A man has 10 shares in Guest Keen and Nettlefold, 12 in Imperial Chemical Industries, 7 in Unilever Ltd. and 5 in Austin Motors. In how many ways can he sell any or all of his shares ?

2.03 RELATION OF FIGURATE NUMBERS TO ELEMENTARY THEOREMS OF SELECTION IRRESPECTIVE OF ORDER

The accompanying chart (over page) shows the relation of the figurates of 1.01 and 1.02 to two problems of simple choice specified by the objects chosen without regard to the order.

(a) We may take *one, two, three* out of 5 letters of the alphabet successively without replacing letters taken or simultaneously, i.e. without using the same letter more than once. The number of different selections possible are

$$\text{for } one \text{ choice } (1 + 1 + 1 + 1 + 1) = \sum_1^5 {}^0F_r = {}^1F_5 = {}^1F_{5-1+1},$$

$$\text{for } two \text{ choices } (0 + 1 + 2 + 3 + 4) = \sum_1^4 {}^1F_r = {}^2F_4 = {}^2F_{5-2+1},$$

$$\text{for } three \text{ choices } (0 + 0 + 1 + 3 + 6) = \sum_1^3 {}^2F_r = {}^3F_3 = {}^3F_{5-3+1}.$$

5

In general, the number of selections of r out of n unlike objects is

$$^rF_{n-r+1} = \frac{\overline{(n-r+1+r-1)^{(r)}}}{r!} = \frac{n^{(r)}}{r!}.$$

As in (i) of 1.03 above

$$n_{(r)} = \frac{n^{(r)}}{r!} = \frac{n!}{r!\,(n-r)!}.$$

The Pascal number of row n column r in 1.03 therefore defines the number of *combinations* of n things taken r a time, i.e. how many r-fold samples of n things we can distinguish if we disregard the arrangement of the constituent items. We write this customarily as

$$^nC_r \quad or \quad \binom{n}{r},$$

$$^nC_r = \binom{n}{r} = \frac{n^{(r)}}{r!} = \frac{n!}{r!\,(n-r)!} \quad . \qquad . \qquad . \qquad . \qquad . \quad \text{(i)}$$

(b) We may take *one*, *two*, *three* out of 5 letters of the alphabet to make a set, repeating any letter as often as is consistent with the presented choice.

We now have the following results:

$$One\ choice\ (1+1+1+1+1) = \sum_1^5 {}^0F_r = {}^1F_5.$$

$$Two\ choices\ (1+2+3+4+5) = \sum_1^5 {}^1F_r = {}^2F_5.$$

$$Three\ choices\ (1+3+6+10+15) = \sum_1^5 {}^2F_r = {}^3F_5.$$

And in general the number of different selections of r letters out of n different ones is

$$^rF_n = \frac{(n+r-1)^{(r)}}{r!} = \frac{\overline{n+r-1!}}{r!\overline{n-1!}} \quad . \qquad . \qquad . \qquad . \quad \text{(ii)}$$

If the selection involves as many items as the number of items in the universe itself (i.e. $r = n$), we have the well-known expression

$$^nF_n = \frac{(2n-1)!}{n!\,(n-1)!} \quad . \qquad . \qquad . \qquad . \qquad . \quad \text{(iii)}$$

The relation (ii) above specifies either

(a) the choice of one item from each of r different collections of the same n different ones;
(b) the choice of r items out of n different ones on the understanding that we are free to replace any item chosen after recording the choice of it and hence to use it repeatedly.

Another elementary theorem of selection often cited is the *total number of combinations of* n *things taken* 1 *or* 2 *or* 3 ... *up to* n *at a time* without repetition. The solution is

$$^nC_1 + {}^nC_2 + {}^nC_3 \ldots {}^nC_n$$

The valuation of this expression depends on the fact that these form successive coefficients of a binomial expansion. In the expansion of $(1+1)^n$ every term is of the form $^nC_r.\,1^r.\,1^{n-r}$, and $1^r.\,1^{n-r} = 1$ for all values of n or r. In accordance with (ix) of 1.04, we thus obtain the identity

$$2^n = (1+1)^n = {}^nC_0 + {}^nC_1 + {}^nC_2 \ldots {}^nC_n.$$

SELECTION OF 1, 2, OR 3 LETTERS OUT OF 5 (a, b, c, d, e)

1. Without replacement

One choice	Two choices	Three choices
a b c d e	ab bc cd de ... ac bd ce ad be ae	abc bcd cde ... ... abd bce abe bde acd ace ade
$(1 + 1 + 1 + 1 + 1)$ $= {}^1F_5$	$(4 + 3 + 2 + 1)$ $= {}^2F_4$	$(6 + 3 + 1)$ $= {}^3F_3$

2. With unrestricted repetition

a b c d e	aa bb cc dd ee
	 ab bc cd de
	 ac bd ce
	 ad be
	 ae
		aaa bbb ccc ddd eee
		 aab bbc ccd dde
		 aac bbd cce
		 aad bbe dee
		 aae cdd
		     bcc cde
		 abb bcd
		 abc bce cee
		 abd
		 abe
		     bdd
		 acc bde
		 acd
		 ace bee
		 add
		 ade
		 aee

$(1 + 1 + 1 + 1 + 1)$ $= {}^1F_5$	$(5 + 4 + 3 + 2 + 1)$ $= {}^2F_5$	$(15 + 10 + 6 + 3 + 1)$ $= {}^3F_5$

Since ${}^nC_0 = 1$, we therefore have

$${}^nC_1 + {}^nC_2 \ldots {}^nC_n = 2^n - 1 \quad . \quad . \quad . \quad . \quad . \quad \text{(iv)}$$

The corresponding total for *repetitive* choice admits of solution by (i) in 1.02, *viz.* :

$$\sum_{r=1}^{r=n} {}^rF_n = \sum_{r=1}^{r=n} {}^{n-1}F_{r+1} = \sum_{x=2}^{x=n+1} {}^{n-1}F_x$$

Since ${}^nF_1 = 1$ for all values of n :

$$\sum_{r=1}^{r=n} {}^rF_n = \sum_{x=1}^{x=n+1} {}^{n-1}F_x - 1$$
$$= {}^nF_{n+1} - 1$$
$$= \frac{(2n)^{(n)}}{n!} - 1 \quad . \quad . \quad . \quad . \quad . \quad \text{(v)}$$

EXERCISE 2.03

1. Out of 20 male and 6 female candidates for a vacancy, what choice have we in selecting 3 men and 2 women ?

2. Out of a council of 42 Communists and 50 members of the Labour Party, in how many ways is it possible to choose a committee consisting of 4 Labour representatives and 4 Communists ?

3. The residue of a platoon consists of a captain, a lieutenant, a sergeant, and 80 other ranks. In how many ways can the captain choose a party of 10, including the lieutenant and excluding himself ?

4. There are 16 candidates for admission to a society with 2 vacancies. Each of 7 electors can either vote for 2 candidates, or plump for 1. In how many ways can they give their votes ?

5. A shop window exhibits 28 hats for sale. What choice has a purchaser who may buy any or all of them ?

6. In how many ways can 2 book stalls divide between them 200 copies of a time-table, 350 of a cookery book, 150 of the current issue of *Time*, and 100 copies of the *New Statesman and Nation* ?

7. How many hands of different make-up w.r.t. suit alone can a bridge player get from a full card pack ?

8. How many possible scores of a single throw of 2 dice exceed 5 ?

9. In how many ways, regardless of order, is it possible to choose from a full pack 3 different picture cards : (*a*) if one replaces each card chosen before taking another ; (*b*) if one takes up 3 cards simultaneously ?

10. In how many ways, regardless of order, is it possible to select from the digits 1 to 9 inclusive 2 like or unlike numbers whose sum is 6 ?

2.04 THE SEMANTICS OF CLASSIFIED CHOICE

We have seen (2.01) that we are entitled to impose on a universe of choice any class structure at will. Thus we may classify all the different cards in a full pack of 52 : (*a*) as members of different *suits ;* (*b*) as *picture cards* and *others ;* (*c*) as *kings, queens, knaves, aces* and *others*. As there stated, the only restriction relevant to enumeration in this context is that a manageable classification must be exhaustive and exclusive in the sense that

(i) items of a given class all possess some attribute A_x that members of other classes do not possess ;

(ii) every item in the universe is assignable to one or other of the *m* classes among which we distribute them.

In accordance with the particular class structure we impose on the items, we can also classify samples or arrangements. For instance, we may group samples of 5 from a full pack in 6 classes according as they contain 0, 1, . . . 5 spades, or we can divide them into 2 classes which respectively contain *no* spades and *at least one* spade.

We can speak of such a classification as *exclusive* with respect to the class structure assigned, only if the number of classes covers all possible numerical specifications of component items distinguished by the classes to which they belong. For instance, a *suit* classification of 2-fold samples from a full pack is exclusive, if it specifies 10 classes to distinguish samples of

2 spades	1 spade, 1 heart	1 heart, 1 club
2 hearts	1 spade, 1 diamond	1 diamond, 1 club
2 diamonds	1 spade, 1 club	– – – – – –
2 clubs	1 heart, 1 diamond	– – – – – –

Any one of the above classes of samples includes many different samples in the sense that individual items are distinguishable by criteria other than the possession of a relevant class attribute. For instance, the class of samples made up of 2 *spades*, include *inter alia :* (*a*) the ace and the king, (*b*) the seven and the ten, (*c*) the knave and the three.

Enumeration of classified samples may thus involve classification at two different levels : enumeration of *classes of samples* and enumeration of *samples which make up a class.* That is to say, we have to distinguish between two types of problem :

(*a*) enumeration of samples (combinations) or ordered ways of extracting samples (permutations) *consistent with a specification of the number of items of each class represented* therein, e.g. how many different samples of 10 cards or how many arrangements of 10 cards we can extract from a full pack of 52 subject to the condition that 3 of the cards chosen must be *spades*, 4 must be *hearts*, 2 must be *diamonds* and one a *club*.

(*b*) exhaustive enumeration of *classes of samples or of arrangements* consistent with a classification which assigns

(i) all samples to the same class if they contain the same number *u* of items of class *A*, the same number *v* items of class *B*, etc. ;

(ii) all arrangements to the same class if we assign to every rank in one arrangement an item which belongs to the same class as the item assigned to the *corresponding* rank in any other.

How many classes of 2-fold samples are extractable from a full pack, if we specify a class by the number of cards allocated to each suit represented in it is an example of a problem defined by (*b*). We have seen that the answer is 10. How many different 2-fold samples make up the class of samples distinguished by the fact that both the constituent cards are spades is an example of questions belonging to the alternative type (*a*). The number of 2-fold samples consistent with the limitation that both cards must be spades is the number of 2-fold samples of 13 items. If choice excludes repetition, this is $^{13}C_2 = 13^{(2)} \div 2!$. If the sample must consist of 1 spade with 1 of the 13 clubs, it may contain any one of $13 . 13 = 169$ samples. More generally, we may suppose that a universe consists of 3 classes of items, *A*, *B*, *C* respectively containing *a*, *b*, *c* items, and specify a class of *r*-fold samples each containing *u* items of class *A*, *v* of class *B* and *w* of class *C* in accordance with the following schema :

Total Number of Items in the Universe $n = (a + b + c)$

$A_1, A_2, A_3 \ldots A_a$	*a*	different items of class	*A*
$B_1, B_2, B_3 \ldots B_b$	*b*	*ditto*	*B*
$C_1, C_2, C_3 \ldots C_c$	*c*	*ditto*	*C*

Number of Items in the Specified Sample $(r = u + v + w)$,
when Choice is Restrictive

u of class A $^aC_u = \dfrac{a^{(u)}}{u!}$ *samples*

v of class B $^bC_v = \dfrac{b^{(v)}}{v!}$ *ditto*

w of class C $^cC_w = \dfrac{c^{(w)}}{w!}$ *ditto*

Any of the ^aC_u samples of class A can be associated with any of the ^bC_v samples of class B or the ^cC_w samples of class C. Hence the total number of different samples is

$$^aC_u \,^bC_v \,^cC_w = \frac{a^{(u)}b^{(v)}c^{(w)}}{u!\,v!\,w!} \qquad \cdots \qquad \text{(i)}$$

Now each sample consists of $r = (u + v + w)$ items which we can arrange in $r!$ ways. Hence the number of arrangements consistent with the specification is

$$r!\frac{a^{(u)}}{u!}\frac{b^{(v)}}{v!}\frac{c^{(w)}}{w!} = \frac{r!}{u!\,v!\,w!}a^{(u)}\,b^{(v)}\,c^{(w)} \qquad \cdots \qquad \text{(ii)}$$

This is the *general* term of the expansion $(a + b + c)^{(r)}$. If there were 4 classes the corresponding expression would be the general term of the expansion $(a + b + c + d)^{(r)}$, and so on.

In conformity with the schema of Figs. 18 and 19, we can arrive at this result without recourse to preliminary enumeration of the corresponding combinations ; and we can then visualise the derivation of (ii) by a different route. If items of the same class were indistinguishable, the number of recognisable permutations in an r-fold sample containing u items of class A, v of class B, etc., would be as given by (iii) in 2.02, *viz.* :

$$\frac{r!}{u!\,v!\,w!\,\ldots}$$

Corresponding to any class X of permutations deemed to be unrecognisable apart in this context there are in fact $a^{(u)}$ permutations attributable to picking u out of the a items of class A. If we recognise as distinct the items of class A, the single class X dissolves into $a^{(u)}$ distinguishable classes. In each of these classes v items of class B are disposable in $b^{(v)}$ ways. If we distinguish items of class B *inter se*, our single class X therefore resolves itself into $a^{(u)} . b^{(v)}$ classes. If we distinguish all items within any class, the single class X becomes $a^{(u)}b^{(v)}c^{(w)} \ldots$ classes. To get the total number of classes specified as above we have to multiply this product by the number of classes of which X is the type specimen. The result is, therefore, as given above :

$$\frac{r!}{u!\,v!\,w!\,\ldots}a^{(u)}\,b^{(v)}\,c^{(w)}\,\ldots$$

If repetitive choice is admissible (i) above no longer holds. The number of u-fold samples which we can then get by taking items of class A is ^uF_a in accordance with (ii) in 2.03 ; and the total number of samples consistent with the class specification is

$$^uF_a \cdot {}^vF_b \cdot {}^wF_c = \frac{(a + u - 1)^{(u)}}{u!} \cdot \frac{(b + v - 1)^{(v)}}{v!} \cdot \frac{(c + w - 1)^{(w)}}{w!} \qquad \cdots \qquad \text{(iii)}$$

Regardless of order, we may thus distinguish the following number of ways in which we may record the 6-fold toss of a cubical die if we specify that each of three faces turned uppermost carries less than three pips, each of two more than two and less than six, the remaining face uppermost being itself a six:

$$\frac{(2+3-1)^{(3)}}{3!} \cdot \frac{(3+2-1)^{(2)}}{2!} \cdot \frac{(1+1-1)^{(1)}}{1!} = 4 \cdot 6 \cdot 1 = 24.$$

The samples defined by (iii) are not necessarily composed of items of which no two are the same, as we presuppose if we assign $r!$ to the number of arrangements per sample. So we cannot derive the corresponding expression for (ii) by making use of (iii). We have to proceed by the alternative method, making use of (iii) in 2.02, that is to say, we suppose that all the u items of class A in the sample are initially indistinguishable, as are also all the v items of class B, or the w items of class C. In the customary jargon, this is equivalent to saying that the number of recognisable arrangements is that of r objects of which u are alike of one kind, v of a second and w of a third, i.e.

$$\frac{r!}{u!\,v!\,w!}$$

Let us now take stock of the fact that individual items of a class are in fact different. In accordance with (ii) of 2.02, we can order the a items of class A in each such arrangement in a^u ways, the b items of class B in b^v ways and the c items of class C in c^w ways. Hence the total number of arrangements is

$$\frac{r!}{u!\,v!\,w!} a^u b^v c^w \qquad \qquad \text{(iv)}$$

This is the general term of the expansion $(a+b+c)^r$. It needs no further discussion to see how we can generalise it for any number of classes.

Expressions (ii) and (iv), rarely cited in text-books, are entitled to rank as the two *fundamental theorems of mathematical probability*, respectively expressing in how many ways we can extract a sample specified by the numbers of items of each class represented therein in accordance with one or other of two postulates:

(a) *restrictive* choice, i.e. sampling from one and the same universe successively without replacement or simultaneously therefrom;

(b) *repetitive* choice, i.e. taking one item of an r-fold sample from each of r universes or successive sampling from one and the same universe *with* replacement.

**** We have now met with the coefficient of a particular term of a multinomial expansion in (ii) of 2.02 and with the general term of the multinomial expansion for factorial or ordinary exponents in (ii) and (iv) above. The enumeration of *classes of samples* or *classes of ways* of extracting a sample introduces us to another genus of problems w.r.t. which the multinomial expansion has a special significance. An examination of this problem leads to a more general treatment of which the elementary theorems of choice in 2.02 and 2.03 appear as special cases. To clarify the issue, it will be profitable to retrace our steps. Consider the choice of 10 items out of 100 of which

$$a = 20 \text{ belong to class } A \qquad c = 25 \text{ belong to class } C$$
$$b = 15 \quad ,, \quad ,, \quad B \qquad d = 30 \quad ,, \quad ,, \quad D$$
$$e = 10 \text{ belong to class } E$$

**** Omit on first reading.

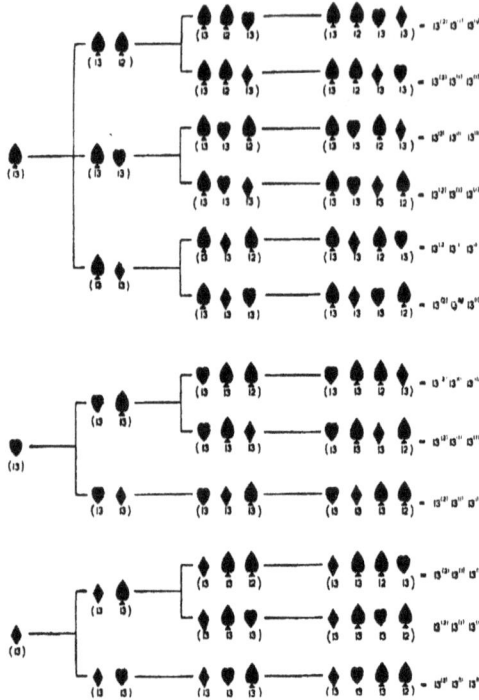

FIG. 18.   Different ways of taking from a full pack a sample of four cards of which two are spades, one is a heart and one a diamond. $\dfrac{4!}{2!\,1!\,1!}\,13^{(2)}\cdot 13^{(1)}\cdot 13^{(1)}$

One such class of samples is the class which consists of 3 items of class A, 2 of B, 1 of D and 4 of E. We may set out the specification of this class of samples as follows :

Class	A	B	C	D	E
No. of items chosen	3	2	0	1	4
No. of items in each class of the universe	a	b	c	d	e

The number of ways (arrangements) of extracting all samples of this class we have seen to be

(a) $\dfrac{10!}{3!\,2!\,0!\,1!\,4!}\,a^{(3)}\,b^{(2)}\,c^{(0)}\,d^{(1)}\,e^{(4)}$ non-repetitive choice.

(b) $\dfrac{10!}{3!\,2!\,0!\,1!\,4!}\,a^3\,b^2\,c^0\,d^1\,e^4$ repetitive choice.

The above expressions, respectively being terms of the expansions $(a+b+c+d+e)^{(10)}$ and $(a+b+c+d+e)^{10}$, each consist of two parts, a *coefficient* involving the factorials 10!, 3!, 2!, 4! and an *efficient* involving powers of $a, b, \ldots e$. The indices of the efficient add up to 10, as do the number of items chosen to make up a sample, and the index of each component a, b, etc., is numerically equivalent to the number of items chosen from the corresponding class. With due regard to a proviso stated below, there is therefore one term of the multinomial expansion, and one term only, corresponding to any specification of a class of samples in conformity with

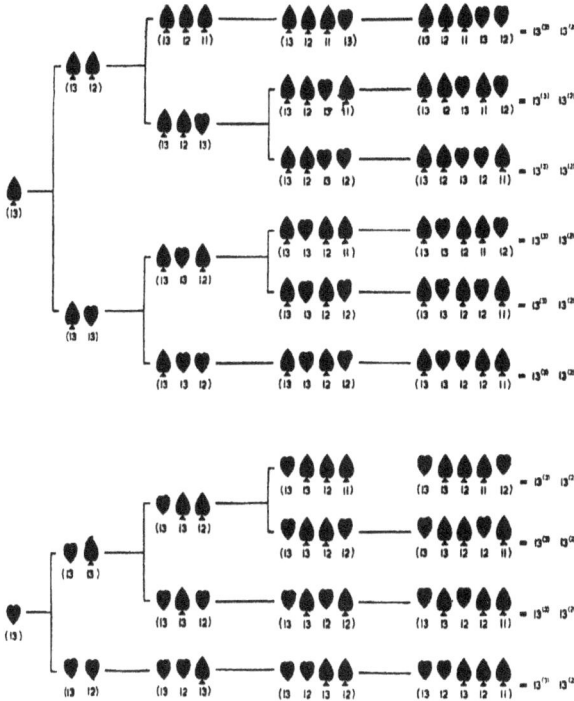

Fig. 19. Different ways of taking from a full pack a sample of five cards of which three are spades and two are hearts. $\dfrac{5!}{3!\,2!}\,13^{(3)}.\,13^{(2)}$.

the class structure of the universe; and there is likewise a unique specification which corresponds to each term. In other words, the enumeration of all classes of r-fold samples taken from a universe consisting of m classes is equivalent to the enumeration of the terms of the rth power of a multinomial of m basic terms.

If choice is *repetitive*, this means that the number of classes of r-fold samples which we can extract from a universe of n items assigned to m classes is the function defined by (v) in 1.07 as

$$^mS_r = \frac{(r+m-1)^{(m-1)}}{(m-1)!}$$
$$= \frac{(r+m-1)!}{(m-1)!\,r!} \qquad \qquad \text{(v)}$$

So far we have discussed classified choice without reference to unclassified choice; but it is possible to bring both within the same framework by recognising that our unclassified universe is merely a universe in which every item is in a class of its own. That is to say, we call a universe unclassified if the number of classes and the number of items is the same $(n = m)$. By this substitution we can therefore adapt (v) to cover the enumeration of samples in a universe of unlike things, *viz.* :

$$^nS_r = \frac{(n+r-1)!}{(n-1)!\,r!} \qquad \qquad \text{(vi)}$$

This agrees with the formula already derived in (ii) of 2.03. If $r = n$ this reduces to the particular form of (iii) in 2.03 :

$$^nS_n = \frac{(2n-1)!}{(n-1)!\,n!} \qquad \cdot \qquad \cdot \qquad \cdot \qquad \cdot \qquad \cdot \qquad \text{(vii)}$$

Simultaneous selection from a single universe or successive selection *without* replacement excludes choice of samples of which the number of items of a given class exceeds the number of corresponding items in the universe itself. *Restrictive* choice, so defined, therefore implies that u cannot exceed a, v cannot exceed b, w cannot exceed c and so on. In the expansion of the multinomial for a factorial exponent any term vanishes if u exceeds a, v exceeds b, etc., in the efficient $a^{(u)}\,b^{(v)}\,c^{(w)}, \ldots$ If the choice of the same item a second time is not permissible, the number of classes of samples therefore corresponds to the number of *residual* terms of the appropriate multinomial expansion in *factorial* powers. That is to say, the number of classes is given by

$$^mS_{(r)} \qquad \cdot \qquad \cdot \qquad \cdot \qquad \cdot \qquad \cdot \qquad \cdot \qquad \text{(viii)}$$

Provided that r does not exceed a, b, c, etc., the result given by (v) above holds good since $^mS_r = {}^mS_{(r)}$ for values of r consistent with this limitation. The numerical evaluation of $^mS_{(r)}$ depends on the nature of the case, as set forth in 1.09 above ; and we proceed in accordance with one or other method of 1.09 with due regard to its relevance.

If all the items are unlike ($m = n$), each class of items consists of one member. When this is so the appropriate expansion is $(1 + 1 + 1 \ldots n \text{ terms})^{(r)}$. By (vii) in 1.09 we then have

$$^nS_{(r)} = n_{(r)} = \frac{n!}{r!\,\overline{n-r}!}$$

$$= {}^nC_r \cdot \qquad \cdot \qquad \cdot \qquad \cdot \qquad \cdot \qquad \cdot \qquad \cdot \qquad \text{(ix)}$$

Thus the expression cited in 2.03 (i) is a particular case of a more general expression for the enumeration of classes of samples.

The expressions cited for restrictive and repetitive choice in (i) and (ii) of 2.02 respectively correspond to the sum of the terms of $(1 + 1 + 1 \ldots \text{ to } n \text{ terms})^{(r)}$ and $(1 + 1 + 1 \ldots \text{ to } n \text{ terms})^r$, i.e. $n^{(r)}$ and n^r. When $n = r$ these respectively become $n!$ and n^n.★★★★

EXERCISE 2.04

1. Out of a full pack how many different selections of 5 cards, irrespective of order, is it possible to make subject to the condition that 3 are diamonds : (*a*) if the player replaces each card before drawing another ; (*b*) if the player draws all 5 cards together ?

2. How many different 4-fold samples distinguished by order, if respectively extracted with and without replacement, is it possible to draw from a full pack if 2 are clubs, and 1 is a heart ?

3. How many different samples of 6 cards distinguished by order of choice are obtainable from a full pack with and without replacement if : (*a*) 4 are spades, 1 a heart and 1 a club ; (*b*) 3 are clubs, 2 are diamonds and 1 a spade ; (*c*) 3 are hearts and 3 are diamonds ?

4. Specify the number of selections, irrespective of order, consistent with the conditions stated in (3) above.

5. An urn contains 5 red balls, 3 black, 2 green, 4 blue and 1 yellow one. No two balls of a colour are of the same size. A player extracts 6 balls, find out : (a) how many selections, distinguishable by size alone, are possible if he draws the balls without replacing them, subject to the condition that 2 balls chosen are red, 3 blue and 1 green ; (b) how many samples, distinguished by size and order, are possible if he replaces each ball drawn before selecting the next, subject to the condition that 4 are red, 1 yellow and 1 blue.

6. In a triple toss of a die, how many ways are there of getting a score of 2 *aces* and a *six* ?

7. In a 6-fold toss of a die, how many ways are there of obtaining 2 *sixes*, 3 *fours* and 1 *two* ?

8. In how many ways is it possible to place 3 black and 2 white draughts on a board ?

9. In how many ways is it possible to have on a chessboard at the end of a game both kings, a white queen, 2 black knights, 3 black pawns and 1 white one ?

10. In how many ways can one choose with *replacement* 5 out of 10 tickets with consecutive numbers from 1 to 10, so that no ticket carries a number higher than 3 ?

2.05 PROPORTIONATE CHOICE

Mathematicians and statisticians do not all agree to use the same definition of mathematical probability. Broadly speaking, two classes of such definitions are current. One, which is purely formal and is open to the criticism that it has no obvious nor indeed *necessary* connection with the sort of judgment implied in stating that an event is more or less probable, rests on the ratio of the number of ways of taking a sample of specified composition to the number of ways of extracting all samples of the same size from the same universe or universes. The other involves an empirical notion concerning the frequency of an occurrence and suffers as such from the fact that the empirical content has no very obvious nor indeed necessary relation to the calculus invoked to deal with it. In so far as objection to the formal type of definition rests on the associations of the word itself, we can cut the Gordian knot at this stage by renaming the ratio specified by a purely formal definition, leaving for subsequent discussion its relevance to factual judgments as an issue for decision on the merits of the individual case. Accordingly, we shall here define a ratio to which we shall apply the emotively neutral but sufficiently suggestive epithet *electivity*, or *choosability* if the reader prefers a less pretentious expression. Our definition will be as follows : *the electivity of extracting an* r-*fold sample of a specified class from one or more universes is the ratio of the number of different ways of extracting such a sample to the number of different ways of extracting all possible* r-*fold samples from the same universe or universes.*

To leave no ambiguity about this definition it will be useful to restate considerations already advanced in 2.01, *viz.* :

(a) *Different ways* of extracting a sample of specified composition signifies in this context the total number of *linear arrangements* of the constituent items of each sample consistent with their specification ;

(b) for the purpose of specifying the number of samples in the relevant class we regard every item of one and the same universe as distinguishable from any other ;

(c) whether the extraction of the sample is consistent with *repetitive* or *restrictive* choice remains open for decision on the merits of the choice prescribed ;

(*d*) the words *sample, choice* and *item* in this context are open to the widest possible interpretation in so far as any possible occurrence constitutes an act of choice in this context, any numerically specified event is a sample or a class of samples of the behaviour of the universe and any numerically specifiable attribute is a relevant item.

In accordance with (*d*) we can speak of three tosses of a coin as a 3-fold sample, heads and tails being the items of a 2-fold universe. As stated in 2.01, the act of tossing a coin three times comes under the heading of repetitive choice, since the same item may turn up again and again. We can identify such an experiment or *trial* with *either* one act of choice from each of three identical 2-fold universes or the extraction of one 3-fold sample from a single 2-fold universe. If the latter, our sample is the record of a process of scoring equivalent to replacing each item before choosing another. In any case, we can extract only four different samples distinguished by their constituent items :

> (i) 3 heads 0 tails
> (ii) 2 heads 1 tail
> (iii) 1 head 2 tails
> (iv) 0 heads 3 tails

Each of these samples (or *events*) has its characteristic electivity. All possible linear sequences consistent with repetitive choice of three items from a 2-fold universe are $2 . 2 . 2 = 8$ in accordance with (ii) of 2.02. If the specified sample is 2 *heads and* 1 *tail*, the number of different possible arrangements is in accordance with (iv) of 2.04, i.e.

$$\frac{3!}{2! \, 1!} \, 1^2 . 1 = 3.$$

Hence the electivity of the sample is the ratio $\frac{3}{8}$.

Though electivity defined as above corresponds to mathematical probability as defined by many writers, it has no *necessary* connection with our expectation of the *proportionate frequency* of such a result in a large number of 3-fold tosses. It relies merely on the fact that a penny has two different faces : one head, one tail. This does not entail the assertion that a penny will come down heads about as often as tails in a large number of single tosses. Nor does the truth of the last assertion necessarily entail the consequence that the proportion of tosses of two heads and one tail in a large number of triple tosses would be about $\frac{3}{8}$. With suitable regulation of the electric current, an iron penny with faces of opposite polarity tossed in an alternating magnetic field could be induced to fall heads and tails alternately, and hence approximately as often one as the other. In a triple sequence, it would come down two heads and one tail or one head and two tails. In the long run the proportionate frequency of either occurrence would be about $\frac{1}{2}$.

At this stage we shall not examine the circumstances in which the assignment of a numerical value for the electivity of an event has more or less relevance to judgments about the frequency with which it occurs. The student of probability will approach such decisions with less effort if we do not blur the issues involved by preoccupation with algebraic manipulations essential to legitimate applications of the calculus of choice. We can best do this by first acquiring sufficient familiarity and facility with the more elementary mathematical techniques.

En passant we may pause to comment on a conundrum commonly cited and often with a view to reinforcing the reputation of the proponent for mathematical sagacity. The form it takes is a statement and question such as the following : I toss a penny ten times and find that it comes down heads each time. What is the probability that it will come down heads next time ? Retailers of this hardy perennial customarily assume that the correct answer is $\frac{1}{2}$, because the mathematical probability (i.e. electivity) of the single trial is $\frac{1}{2}$; but the relevance of this ratio to the behaviour of a penny is less than that of our information about its history, and the only knowledge we have about the history of the penny in question leads us to doubt whether it is relevant to its behaviour. Hence the only reasonable answer to the question itself is suspense of judgment.

The distinction emphasised in the foregoing remarks may serve to justify the extended use of the term *sample* in preference to *event* or *trial*. The word trial suggests an experiment to test a rule, and event suggests an occurrence in conformity with some natural law peculiar to a particular class of phenomena. In as much as the mathematical treatment of probability has its basis in the calculus of choice it is applicable to such aspects of occurrences as do not conform to natural laws peculiar to phenomena of a particular class. Indeed, we apply it with greatest assurance when we use it as a yardstick of neutrality to answer questions involving the putative existence of a law applicable to particular events. We say that a result might be due to pure chance, meaning merely that it is not certainly due to a particular condition associated with its occurrence; and pure chance in this context merely means laws of choice unrelated to the behaviour of any particular class of phenomena.

For purposes of a *unit sample* (single event or 1-fold trial) it is often convenient to divide items into two classes in as much as we score the choice of one as a *success*, the other as a *failure*. If the n items of a universe contain s of the first class and f of the second, there are s out of n ways of drawing a unit sample labelled a success and f out of n ways of drawing a unit sample labelled a failure. Accordingly, the electivity (p) of a success at a single trial is

$$\frac{s}{n} = \frac{s}{s+f}.$$

The electivity (q) of a failure is

$$\frac{f}{n} = \frac{f}{s+f} = \frac{(s+f)-s}{s+f} = 1 - \frac{s}{s+f}.$$

$$\therefore p = 1 - q \quad . \quad . \quad . \quad . \quad . \quad . \quad . \quad (i)$$

This relation is of fundamental importance.

In common parlance, sampling from a single universe usually implies *restrictive* choice as defined above; but sampling which permits *repetitive* choice is amenable to more elementary methods and the results are adaptable to the consideration of simultaneous sampling or successive sampling *without* replacement. We shall therefore start with *repetitive* sampling, and confine our discussion to sampling from a *single* universe.

With this restriction all possible ways of choosing an r-fold sample from an n-fold universe are given by (ii) in 2.02, i.e. n^r; and the number of possible ways of getting an r-fold sample consisting of x successes and $(r-x)$ failures is given by (iv) in 2.04, *viz.*:

$$\frac{r!}{x!\,\overline{r-x}!} \; s^x \cdot f^{r-x}.$$

Hence the electivity $E_{x \cdot r-x}$ of the prescribed r-fold sample containing x successes is

$$\frac{r!}{x!\,\overline{r-x}!} \cdot \frac{s^x \cdot f^{r-x}}{n^r} = \frac{r!}{x!\,\overline{r-x}!} \frac{s^x}{n^x} \cdot \frac{f^{r-x}}{n^{r-x}},$$

$$\therefore E_{x \cdot r-x} = \frac{r!}{x!\,(r-x)!} p^x \, q^{r-x} \quad . \quad . \quad . \quad . \quad (ii)$$

Thus the electivity of x successes in an r-fold sample is the general term of which the exponent of p is x in the expansion of the binomial $(p + q)^r$. Accordingly we may write for the case of a 4-fold sample

No. of Successes

No. of Successes	
4	$E_{4 \cdot 0} = p^4$
3	$E_{3 \cdot 1} = 4\,p^3 q$
2	$E_{2 \cdot 2} = 6\,p^2 q^2$
1	$E_{1 \cdot 3} = 4\,pq^3$
0	$E_{0 \cdot 4} = q^4$

As an example let us evaluate the electivities of extracting (a) 3 picture cards, (b) *no* picture cards among 5 cards successively taken from a full pack on the assumption that we replace each card taken before drawing another. In this context, choice of a picture card constitutes a success. There are 52 cards of which 12 are picture cards, hence

$$p = \frac{12}{52} = \frac{3}{13},$$

$$\therefore q = 1 - p = \frac{10}{13}.$$

The appropriate binomial is therefore $\left(\dfrac{3}{13} + \dfrac{10}{13}\right)^5$; and its general term is

$$\frac{5!}{x!\,(5 - x)!}\left(\frac{3}{13}\right)^x \left(\frac{10}{13}\right)^{5-x},$$

$$\therefore (a)\ E_{3 \cdot 2} = \frac{5!}{3!\,2!}\left(\frac{3}{13}\right)^3 \left(\frac{10}{13}\right)^2 = \frac{27000}{371293}.$$

and

$$(b)\ E_{0 \cdot 5} = \left(\frac{10}{13}\right)^5 = \frac{100000}{371293}.$$

As an alternative example we may consider the electivity of getting 5 heads and 1 tail in 6 tosses of a coin. In the nature of the case, the process of sampling is repetitive, and $p = \frac{1}{2} = q$. The appropriate binomial is $(\frac{1}{2} + \frac{1}{2})^6$. If we call heads successes, the required ratio $(E_{5 \cdot 1})$ is

$$\frac{6!}{5!\,1!}\left(\frac{1}{2}\right)^5 \left(\frac{1}{2}\right)^1 = \frac{6}{64} = \frac{3}{32}.$$

The electivity of a run of 10 heads in succession is $E_{10 \cdot 0}$, i.e.

$$\left(\frac{1}{2}\right)^{10} = \frac{1}{1024}.$$

Let us now impose the restriction that repetitive choice of the same item is *not* permissible. To emphasise the distinction we shall use $\bar{E}_{(x, r-x)}$ instead of $E_{x, r-x}$ for the electivity of x successes. By (i) in 2.02, all ways of choosing an r-fold sample from an n-fold universe are $n^{(r)}$. By (ii) in 2.04 the number of ways of extracting an r-fold sample containing x successes and $(r - x)$ failures is

$$\frac{r!}{x!\,(r - x)!}\, s^{(x)} \cdot f^{(r-x)},$$

$$\therefore \bar{E}_{(x, r-x)} = \frac{r!}{x!\,(r - x)!}\, \frac{s^{(x)} f^{(r-x)}}{n^{(r)}} \qquad \cdot \qquad \cdot \qquad \cdot \qquad \cdot \qquad \text{(iii)}$$

If choice is restrictive, the electivity of x successes in an r-fold sample from an n-fold universe is therefore the *ratio* of the general term containing the factor $s^{(x)}$ in the expansion of the *factorial binomial* $(s + f)^{(r)}$ *to the continued product* $n^{(r)}$.

The relation between (ii) and (iii) will be more clear when we have studied their derivation with the help of visual models, as below. Meanwhile, let us notice an important difference. If choice is repetitive, the electivity of a sample does *not* depend on the number of items in the universe. If choice is restrictive, the size of the universe appears explicitly in the appropriate expression. For exemplary purposes we may now repeat the last card pack example on the assumption that we either draw 5 cards at once or singly without replacing them. In accordance with (iii) above

$$E_{(3 \cdot 2)} = \frac{5!}{3! \, 2!} \, \frac{12^{(3)} \, 40^{(2)}}{52^{(5)}} = \frac{55}{833}.$$

$$E_{(0 \cdot 5)} = \frac{40^{(5)}}{52^{(5)}} = \frac{2109}{8330}.$$

What we usually mean by sampling in statistical theory is *de facto* restrictive. Accordingly, (iii) above has a wider application than (ii), though most elementary text-books of statistics employ (ii) extensively, even when the nature of the process necessarily excludes replacement. As stated more fully in 2.01, a justification for this is that it is often legitimate to consider the universe of statistics as indefinitely large and r as small by comparison. When this is so, the error involved in putting $s^{(r)} = s^r$, $f^{(r)} = f^r$ and $n^{(r)} = n^r$ is small. By these substitutions (iii) becomes (ii), and (ii) is therefore a good approximation. None the less, it is important to remember that it is at best an approximation and not always a good approximation, if the universe is explicitly finite, as is a human population, and the sample itself a sizeable fraction of the whole. It is therefore instructive to study the 4-fold sample distributions of Fig. 28 and to compare the values of E_3 and E_0 with those of $E_{(3)}$ and $E_{(0)}$ for the card pack samples above. In our notation

$$E_{3 \cdot 2} = 0 \cdot 0727 \, \ldots \qquad E_{(3 \cdot 2)} = 0 \cdot 0660 \, \ldots$$
$$E_{0 \cdot 5} = 0 \cdot 269 \, \ldots \qquad E_{(0 \cdot 5)} = 0 \cdot 253 \, \ldots$$

It is often possible to reduce the mathematical formulation of a statistical problem to a two-fold choice in accordance with the classification of items as *successes* or *failures*. So far we have considered the evaluation of the electivity of a specified sample or class of samples from this point of view ; but there is nothing in its definition to justify this limitation. The two distributions respectively defined by the expansions $(p + q)^r$ and $(s + f)^{(r)} \div n^{(r)}$ are each particular cases of a more general formulation. It will suffice to illustrate these more general *multinomial* distributions by reference to a 3-fold classification of the items of the universe. Our model universe will be a full card pack. The problem is to assign the electivity of a 6-fold sample containing one ace, two picture cards and three other cards which are *not* A, K, Q, J. Thus the classes relevant to our choice will be

(*a*) aces of any suit ;
(*b*) picture cards of any suit ;
(*c*) cards with 2-10 pips inclusive.

We shall denote the number of cards in each class by a, b, c as above, so that $a=4$, $b=12$, $c=36$ and $n = 52$. In accordance with the conventions adopted previously $u = 1$, $v = 2$, $w = 3$. We may denote the electivity of a repetitive choice of the specified class of samples by $E_{1 \cdot 2 \cdot 3}$ and that of a restrictive choice by $E_{(1 \cdot 2 \cdot 3)}$. Accordingly we have

(i) *with* replacement

$$E_{u \cdot v \cdot w} = \frac{r!}{u! \, v! \, w!} \frac{a^u \, b^v \, c^w}{n^r}$$

$$= \frac{r!}{u! \, v! \, w!} \left(\frac{a}{n}\right)^u \left(\frac{b}{n}\right)^v \left(\frac{c}{n}\right)^w \qquad \cdots \qquad \text{(iv)}$$

If we denote the fractions a/n, b/n, c/n respectively corresponding to the electivity of a single choice of items of each of the three classes by p, q, s,

$$E_{u \cdot v \cdot w} = \frac{r!}{u! \, v! \, w!} \, p^u \, q^v \, s^w.$$

This is the general term of the expansion $(p + q + s)^r$. For the above problem $p = \frac{4}{52} = \frac{1}{13}$, $q = \frac{3}{13}$ and $s = \frac{9}{13}$, so that

$$E_{1 \cdot 2 \cdot 3} = \frac{6!}{1! \, 2! \, 3!} \left(\frac{1}{13}\right) \left(\frac{3}{13}\right)^2 \left(\frac{9}{13}\right)^3 = \frac{393660}{4826709}$$

$$= 0 \cdot 082 \ (approx.)$$

(ii) *without* replacement

$$E_{(u \, , \, v \, , \, w)} = \frac{r!}{u! \, v! \, w!} \frac{a^{(u)} b^{(v)} c^{(w)}}{n^{(r)}} \qquad \cdots \qquad \cdots \qquad \text{(v)}$$

$$E_{(1 \, , \, 2 \, , \, 3)} = \frac{6!}{1! \, 2! \, 3!} \frac{4 \cdot 12^{(2)} \cdot 36^{(3)}}{52^{(6)}} = \frac{396}{4277}$$

$$= 0 \cdot 094 \ (approx.)$$

EXERCISE 2.05

1. In a normal family of 7 children (expected sex ratio $1:1$) what is the chance that there will be : (*a*) at least 1 boy ; (*b*) exactly 6 girls ?

2. In a normal family of 10 sibs what is the chance that : (*a*) all will be girls ; (*b*) exactly 8 will be girls ; (*c*) at least 1 will be a girl ?

3. What is the chance of scoring *at least one* double six in 4 tosses of a die ?

4. In a deal of 5 cards from a full pack what are the chances of getting : (*a*) at least 1 ace ; (*b*) 4 tens and a picture card ?

5. In a deal of 8 cards from a full pack what are the chances of getting : (*a*) no ace ; (*b*) no black card ; (*c*) no picture card ?

6. In a deal of 10 cards from a full pack what are the chances of getting : (*a*) at least 1 black card ; (*b*) at least 1 queen ; (*c*) 4 queens, 3 aces and any 2 other cards ?

7. What are the chances of getting 5 picture cards and 2 aces in a full hand at whist ?

8. What is the chance that a full hand at whist will contain neither a picture card nor an ace ?

9. If I empty out 5 balls from an urn containing 10 red, 5 black and 15 white ones, what is the chance that: (a) 3 will be red, 1 black and 1 white; (b) 3 will be white, 1 black and 1 red; (c) 3 will be black and 2 will be white?

10. One purse contains 5 half-crowns and 4 shillings; another contains 5 half-crowns and 3 shillings. One purse is taken at random and a coin drawn out. What is the chance that it will be a half-crown?

2.06 THE THREE FUNDAMENTAL RULES OF ELECTIVITY

The distributions defined by (ii)-(v) in 2.05 are deducible from rules which are themselves deducible from the chessboard set-up (Figs. 20-21) for specification of different ways of extracting a sample without recourse to (ii) and (iv) in 2.04. It will be convenient to state the first two with respect to samples of 2, and to extend them to larger samples when their meaning is clear.

1. The *Product* Rule. If the electivity of the choice of a single item of class A from one universe is p_a and the electivity of the choice of a single item of class B from a second universe is p_b, the electivity of getting an item of class A from the first and an item of class B from the second in a double draw is $p_a \cdot p_b$.

2. The *Addition* Rule. The electivity of a class of samples is the sum of the electivities of its constituent sub-classes, with the proviso that the specification of the latter is exclusive as defined in 2.04. Since the definition of electivity of a class implies its proportionate contribution to the whole, the addition rule signifies that the electivities of *all* the sets of an exclusive classification are together equal to unity.

3. The *Negation* Rule. The total absence of an item of a given class A and the presence of *at least one* item of the same class respectively define two sub-classes of samples of a 2-fold exhaustive and exclusive specification. Hence the sum of their electivities is unity. If a is the electivity of an r-fold sample containing *at least one* member of class A and b is the electivity of an r-fold sample containing no members of class A, $a = (1 - b)$.

It is important to notice that the *product* rule, as stated above, applies to simultaneous extraction of 1-fold samples from *different* universes. So stated the rule signifies the result of taking an item of class A from Universe I and an item of class B from Universe II; and it implies nothing about the proportion of items of class A in Universe II or of class B in Universe I. We can extend its application to sampling from one and the same universe, if we regard the single universe as constituted at our first choice as Universe I and the universe as constituted at our second choice as Universe II; but an exclusive and exhaustive specification of our choice then compels us to distinguish between: (a) choosing first an item of class A, then an item of class B; (b) choosing first an item of class B, then an item of class A. *With replacement*, the electivity of either event is $p_a p_b$. In accordance with the addition rule the electivity of a 2-fold sample consisting of one item of class A and one item of class B is then $p_a p_b + p_b p_a = 2p_a p_b$. On the other hand, the electivity of getting a 2-fold sample consisting of two items belonging to the same class is that of a single act of choice, since we can choose two members of class A in succession *only* by first extracting an item of class A and then again extracting an item of class A. If we put back the item first chosen, the proportion (p_a) of items of class A in the residual universe is the same as before, and the electivity of the event is p_a^2. The same is true if our sample is merely the record of an occurrence which may recur indefinitely often, as when we score the result of tossing a die.

6

If the product rule is true, the electivity of getting 2 heads with a double toss of a coin, or a single toss of 2 coins, is therefore $(\frac{1}{2})^2$, because the electivity of a head in a single toss is $\frac{1}{2}$, and the electivity of getting 2 heads is that of getting a head at the first toss and a head at the second. This is consistent with (ii) in 2.05. The appropriate binomial is $(\frac{1}{2} + \frac{1}{2})^2$, whence

$$E_{2.0} = (\tfrac{1}{2})^2 \quad E_{1.1} = 2(\tfrac{1}{2})^2 \quad E_{0.2} = (\tfrac{1}{2})^2$$

The value assigned to the electivity of getting a different result from successive tosses illustrates both rules. The class of samples composed of a head and a tail may be either of two sub-classes : (a) first a head, then a tail ; (b) first a tail, then a head. If rule (i) is true, the electivity of either is $(\frac{1}{2})^2$ and if rule (ii) is correct, the electivity $(E_{1.1})$ of the class which specifies *either* one *or* the other exclusively is the sum of their separate electivities, i.e. $(\frac{1}{2})^2 + (\frac{1}{2})^2 = 2(\frac{1}{2})^2$.

A 3-fold successive draw *with replacement* from a full pack of cards provides an illustration of the extension of the rule. The electivity of a heart at a single draw is $\frac{1}{4}$. In accordance with the extended product rule, the electivity of a run of 3 hearts is therefore that of getting a heart from each of 3 packs (A, B, C) at a single draw, i.e. $\frac{1}{4} \times \frac{1}{4} \times \frac{1}{4}$. The appropriate binomial is $(\frac{1}{4} + \frac{3}{4})^3$, whence $E_{3.0} = (\frac{1}{4})^3$. The choice of 2 hearts and a card of another suit in a 3-fold draw of this sort defines a class of samples which we can divide into 3 sub-classes, according as we draw one or other card of a given suit from one or other pack A, B, C. If the 3-fold sample contains 2 hearts, the extraction of a heart from 1 pack entails the extraction of a card of another suit from either of the remaining 2 packs. Hence we can make up our sample in only 3 ways, as below.

A	B	C	
Heart	Heart	Other	$\frac{1}{4} \cdot \frac{1}{4} \cdot \frac{3}{4} = (\frac{1}{4})^2 \cdot (\frac{3}{4})$
Heart	Other	Heart	$\frac{1}{4} \cdot \frac{3}{4} \cdot \frac{1}{4} = (\frac{1}{4})^2 \cdot (\frac{3}{4})$
Other	Heart	Heart	$\frac{3}{4} \cdot \frac{1}{4} \cdot \frac{1}{4} = (\frac{1}{4})^2 \cdot (\frac{3}{4})$

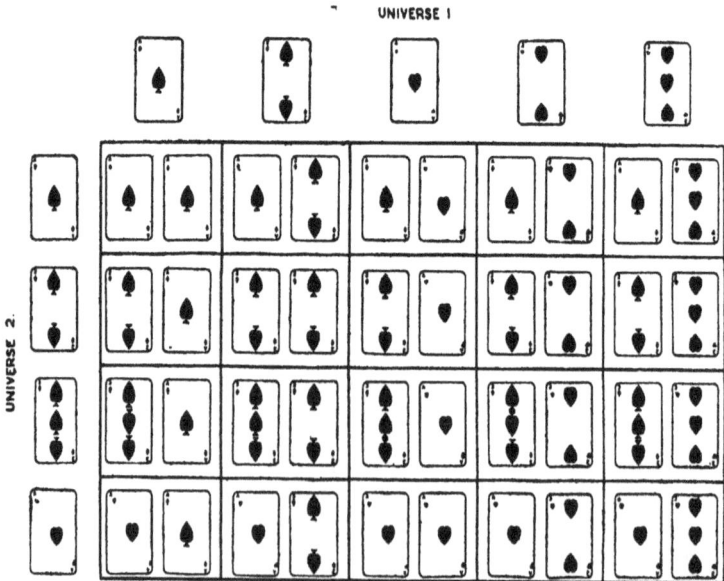

UNIVERSE 1

UNIVERSE 2.

FIG. 20. All possible ways of simultaneously choosing a card from one pack consisting of 2 spades and 3 hearts and a card from another pack consisting of 3 spades and 1 heart.

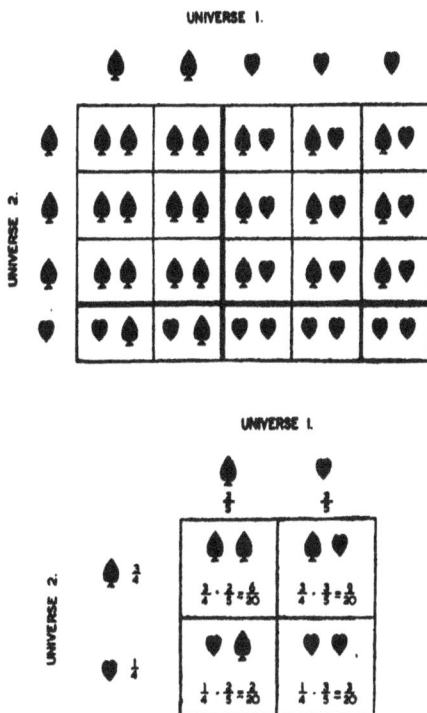

FIG. 21. The balance sheet of Fig. 20, when our concern is with *suit* alone.

From the binomial formula, the electivity of the specified class as a whole is $E_{2,1} = 3(\frac{1}{4})^2 \cdot (\frac{3}{4})$, and this is equal to the sum of the electivities assigned to the sub-classes as in the table above.

More generally, we may regard an r-fold sample as a super-sample made up of r samples of 1 item only, or alternatively as a super-sample made up of 2 samples, one containing 1 item only, the other containing $(r-1)$ items. In this way we see that the *product* and *addition* rules extend to samples of any kind. If p, q, r, s refer respectively to the electivities of classes A, B, C, D at a single draw, the electivity of a sample made up of x items of class A, y of class B, z of class C and 0 of class D *in that order* is therefore

$$p^x q^y r^z s^0 = p^x q^y r^z.$$

From the standpoint of our third rule we may regard the double toss of a die or the simultaneous toss of 2 dice as a sample divisible into 2 exclusive classes, according as

(i) *at least one* face turned up is a *six*;
(ii) *neither* face turned up is a *six*.

By the product rule the electivity of (ii) is

$$\left(\frac{5}{6}\right)^2 = \frac{25}{36}.$$

The negation rule states that the electivity of the alternate sample (i) is

$$1 - \frac{25}{36} = \frac{11}{36}.$$

If choice is repetitive, the more general form of the negation rule states that the electivity of getting *at least one* success in an r-fold sample is

$$1 - q^r \quad . \qquad . \qquad . \qquad . \qquad . \qquad . \qquad . \quad \text{(i)}$$

If choice is restrictive, the appropriate expression is

$$1 - \frac{f^{(r)}}{n^{(r)}} \quad . \qquad . \qquad . \qquad . \qquad . \qquad . \quad \text{(ii)}$$

These results have wide application in statistics.

In 2.02 we have seen that the chessboard device exhibits all linear arrangements of 2-fold samples from one universe or from two different universes. In conformity with our definition of p. 75, the electivity of a class of samples is therefore the ratio of the number of pigeon holes they occupy to the total number of pigeon holes in the chessboard lattice. The chessboard lay-out of Figs. 20-23 shows how the three rules stated are deducible without reference to the binomial formula for $E_{a.b}$; and that the latter is in fact deducible from them.

Fig. 20 sets out all possible ways of extracting a 2-fold sample by taking 1 card from each of 2 packs constituted as follows :

Left-hand pack (4 cards)	*Right-hand pack* (5 cards)
Ace of Spades	Ace of Spades
2 *ditto*	2 *ditto*
3 *ditto*	Ace of Hearts
Ace of Hearts	2 *ditto*
	3 *ditto*

By definition the electivities for a single draw from the two packs are

	Spades	*Hearts*	*Total*
From the left-hand pack	$\frac{3}{4}$	$\frac{1}{4}$	$(\frac{3}{4} + \frac{1}{4}) = 1$
From the right-hand pack	$\frac{2}{5}$	$\frac{3}{5}$	$(\frac{2}{5} + \frac{3}{5}) = 1$

There are 20 different linear arrangements of a 2-fold sample defined as above. For purposes of classification by suit, we can best visualise the result, if we represent each spade by the ace of spades and each heart by the ace of hearts, as in Fig. 21. By inspection of this figure we see that 6 out of the 20 arrangements consist of 2 spades, 3 of 2 hearts, 9 of a spade from the left pack with a heart from the right one and 2 of a heart from the left with a spade from the right one. We can thus draw up a table of electivities for alternative exclusive classifications demonstrating both the product and the addition rule as below :

$$\text{Both spades} \quad \tfrac{6}{20} = \tfrac{3}{4} \times \tfrac{2}{5} \ (\textit{Product rule}).$$
$$\text{Both hearts} \quad \tfrac{3}{20} = \tfrac{1}{4} \times \tfrac{3}{5} \ (\quad \text{,,} \quad \text{,,} \ ).$$
$$\text{Spade-heart} \quad \tfrac{9}{20} = \tfrac{3}{4} \times \tfrac{3}{5} \ (\quad \text{,,} \quad \text{,,} \ ).$$
$$\text{Heart-spade} \quad \tfrac{2}{20} = \tfrac{1}{4} \times \tfrac{2}{5} \ (\quad \text{,,} \quad \text{,,} \ ).$$
$$\text{Total} \quad \tfrac{6}{20} + \tfrac{3}{20} + \tfrac{9}{20} + \tfrac{2}{20} = 1.$$

Two cards of the same suit $\quad \frac{9}{20} = \frac{6}{20} + \frac{3}{20}$ (*Addition* rule).
Two unlike cards $\qquad\qquad \frac{11}{20} = \frac{9}{20} + \frac{2}{20}$ (,, ,,).

$$\text{Total} \quad \frac{9}{20} + \frac{11}{20} = 1.$$

No spades (both hearts) $\quad \frac{3}{20}.$
At least one spade $\qquad \frac{17}{20} = 1 - \frac{3}{20}$ (*Negation* rule).
No hearts (both spades) $\quad \frac{6}{20}.$
At least one heart $\qquad \frac{14}{20} = 1 - \frac{6}{20}$ (,, ,,).
Two like cards $\qquad\qquad \frac{9}{20}.$
Two unlike cards $\qquad\quad \frac{11}{20} = 1 - \frac{9}{20}$ (,, ,,).

The condensed chessboard diagram of Fig. 21 summarises these results in a way which is adaptable to successive application as in Fig. 24. Extracting an r-fold sample by taking 1 card from each of r packs is equivalent to taking r cards from 1 pack, if one replaces each card before picking out another. From Fig. 24 we see that the binomial series for the electivities of 0, 1, 2 . . . r hearts in an r-fold sample from a full pack on the assumption that there is replacement of each before drawing another is a necessary consequence of applying the two rules stated.

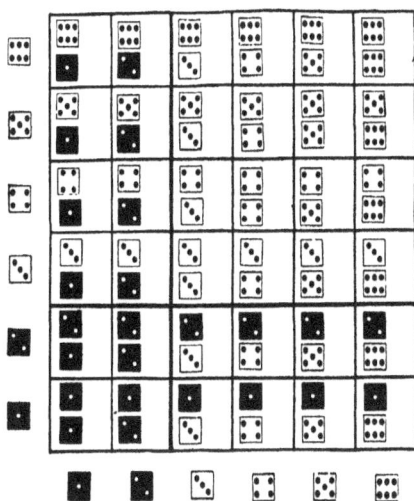

FIG. 22. All possible results of tossing 2 dice simultaneously or tossing 1 die twice.

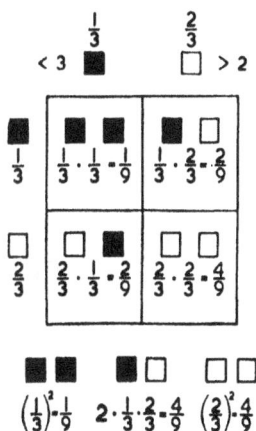

FIG. 23. The balance sheet of Fig. 22 when our only concern is whether the number of pips on one or other member of the pair exceed 2.

Figs. 20-21 refer to sampling simultaneously from two universes, but the device is equally applicable to repetitive sampling from a single universe, as illustrated by the lay-out in Figs. 22-23 for the results of a double toss of a die or single toss of two dice simultaneously. Thus we can classify the 36 possible results as

(a) both top faces with less than 3 pips, i.e. with either 1 or 2 pips—4 out of 36 ways.
(b) only one top face with less than 3 pips—16 out of 36 ways ;
(c) neither top face with less than 3 pips—16 out of 36 ways.

From the figure we thus see that the electivity of (a) is $4 \div 36 = \frac{1}{9}$. This itself illustrates both *product* and *addition* rules, since the result is exclusively classifiable as : (i) two scores of 1 ; (ii) a score of 1 followed by 2 ; (iii) a score of 2 followed by 1 ; (iv) two scores of 2. The electivity assignable to any face at a single toss is $\frac{1}{6}$. By the product rule that assignable to any of these four results is $\left(\frac{1}{6}\right)^2 = \frac{1}{36}$. By the addition rule that assignable to the class as a whole is

$$\tfrac{1}{36} + \tfrac{1}{36} + \tfrac{1}{36} + \tfrac{1}{36} = \tfrac{1}{9}.$$

At a single toss, the electivity assignable to getting a score of less than three, i.e. of getting *either* one *or* two, is $\frac{1}{6} + \frac{1}{6} = \frac{1}{3}$. By the product rule the electivity of getting two scores of less than three in succession is therefore $\frac{1}{3} \cdot \frac{1}{3} = \frac{1}{9}$, in agreement with the above.

Out of the 36 possible arrangements $16 + 16 = 32$ have at least one face turned up with more than 2 pips. For $36 - 32 = 4$ both faces have less than 3 pips. This is an exclusive 2-fold classification. In accordance with the negation rule, the electivity of a double toss involving *at least one* score greater than 2 is

$$1 - \left(\tfrac{1}{3}\right)^2 = \tfrac{8}{9} = \tfrac{32}{36}.$$

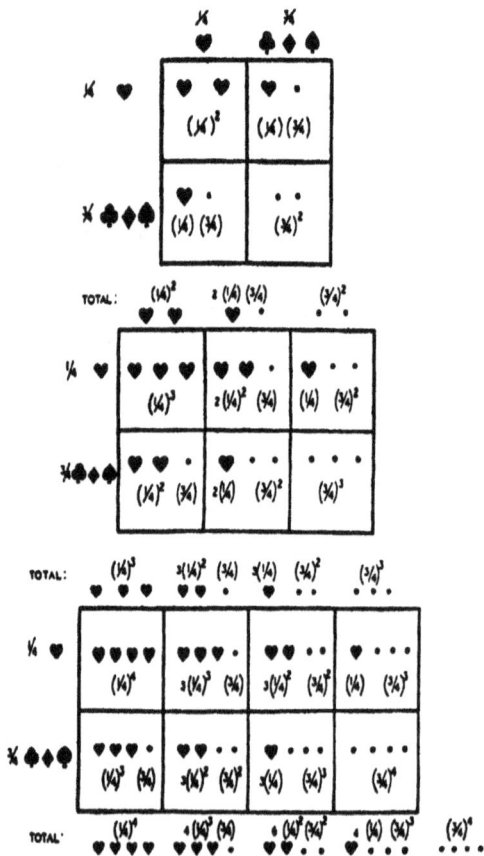

FIG. 24. The binomial law of the distribution of heart scores for choice of 2, 3 and 4 cards from a full pack as a chessboard application of the addition and product rules embodied in Figs. 21 and 23.

2.07 THE STAIRCASE MODEL

The chessboard diagram is applicable to simultaneous sampling from any 2 universes. Its relevance does not presuppose that the universes are identical like the universes of a 1-fold toss of 2 dice. Hence the product and addition rules (Figs. 21 and 23) apply to simultaneous choice from each of 2 card packs whose composition is *not* the same. This is important, because it indicates a way of visualising restrictive choice by suitable application of the same rules.

When we sample from only one universe, the chessboard diagram refers only to repetitive choice, i.e. to *die* models or to successive withdrawal from a card pack subject to replacement of each card taken before extracting another. To visualise the relevance of the *product* rule to the urn or card pack, if selection from a single universe is simultaneous, or if replacement is otherwise excluded, a 3-dimensional device is more manageable. That of Fig. 25 refers to a 2-fold choice without replacement from an 8-card pack made up of 1 *club*, 3 *hearts*, and 4 *diamonds*. Accordingly the electivities of the choice of a club, heart or diamond at a single draw are respectively

$$\tfrac{1}{8}; \ \tfrac{3}{8}; \ \tfrac{1}{2}.$$

Let us suppose that one of the two cards chosen at a double draw is a club. There remain in the pack only seven cards, 3 hearts and 4 diamonds. We may tabulate all three possibilities with respect to the residual composition of the pack as follows :

First Choice. (A)	Residual Pack.			Label.
	Clubs.	Hearts.	Diamonds.	
Club	0	3	4	B
Heart	1	2	4	C
Diamond	1	3	3	D

Corresponding to each first choice, there is a residual electivity w.r.t., the alternative draw, as below :

First Choice.		Residual Electivity of Second Choice.		
Suit.	Electivity.	Club.	Heart.	Diamond.
Club	$\tfrac{1}{8}$	0	$\tfrac{3}{7}$	$\tfrac{4}{7}$
Heart	$\tfrac{3}{8}$	$\tfrac{1}{7}$	$\tfrac{2}{7}$	$\tfrac{4}{7}$
Diamond	$\tfrac{1}{2}$	$\tfrac{1}{7}$	$\tfrac{3}{7}$	$\tfrac{3}{7}$

Since each first choice determines the range of alternative choice, we are at liberty to regard the double draw as an event which takes place *in two stages*. The first is the choice of a card of any one of three suits from a pack A. The second is the choice of a card from one of three packs, B, C, D prescribed by the nature of the first choice. The choice of a club from pack A constitutes a ticket which entitles one to draw a card from pack B, that of a heart from pack A a ticket permitting one to draw a card from pack C, and that of a diamond from pack A a ticket permitting one to draw a card from pack D. To regard the double draw in this way is equivalent to regarding any particular choice as the outcome of sampling from two universes. For instance, the choice of a heart followed by a club is equivalent to

(i) choosing a heart from pack A.
(ii) choosing a club from pack C.

So conceived the two acts of choice are *independent*. So the electivities of all possible pairs (*second* choice in italics) follows from the product rule for independent choice :

Club—*club*	$\frac{1}{8} \times 0 = 0.$
Heart—*club*	$\frac{3}{8} \times \frac{1}{7} = \frac{3}{56}.$
Diamond—*club*	$\frac{1}{2} \times \frac{1}{7} = \frac{1}{14}.$
Club—*heart*	$\frac{1}{8} \times \frac{3}{7} = \frac{3}{56}.$
Heart—*heart*	$\frac{3}{8} \times \frac{2}{7} = \frac{3}{28}.$
Diamond—*heart*	$\frac{1}{2} \times \frac{3}{7} = \frac{3}{14}.$
Club—*diamond*	$\frac{1}{8} \times \frac{4}{7} = \frac{1}{14}.$
Heart—*diamond*	$\frac{3}{8} \times \frac{4}{7} = \frac{3}{14}.$
Diamond—*diamond*	$\frac{1}{2} \times \frac{3}{7} = \frac{3}{14}.$

TOTAL $\left(\frac{3}{56} + \frac{1}{14} + \frac{3}{56} + \frac{3}{28} + \frac{3}{14} + \frac{1}{14} + \frac{3}{14} + \frac{3}{14}\right) = 1.$

In accordance with the addition rule the electivities for the double draw irrespective of order are

2 Clubs	0	Club and heart	$\frac{3}{56} + \frac{3}{56} = \frac{3}{28}.$
2 Hearts	$\frac{3}{28}$	Club and diamond	$\frac{1}{14} + \frac{1}{14} = \frac{1}{7}.$
2 Diamonds	$\frac{3}{14}$	Heart and diamond	$\frac{3}{14} + \frac{3}{14} = \frac{3}{7}.$

If choice were repetitive the corresponding ratios would be

2 Clubs	$\frac{1}{64}$	Club and heart	$\frac{3}{32}.$
2 Hearts	$\frac{9}{64}$	Club and diamond	$\frac{1}{8}.$
2 Diamonds	$\frac{1}{4}$	Heart and diamond	$\frac{3}{8}.$

Fig. 25 shows the electivity of the second choice in accordance with the application of the above rules when there is *no* repetition, *viz.* :

Club $0 + \frac{3}{56} + \frac{1}{14} = \frac{1}{8}.$
Heart $\frac{3}{56} + \frac{3}{28} + \frac{3}{14} = \frac{3}{8}.$

In spite of the fact that the choice of a club at the first draw excludes the possibility of choosing a club at the second, the electivity of a second choice is thus the same as that of a first choice. We can generalise this conclusion for a 2-fold classification as follows. Suppose the urn model has 2 sorts of balls the choice of which respectively score as *successes* and *failures*, the numbers being s and f, so that $(s + f) = n$ and $(s + f - 1) = (n - 1)$. In accordance with the preceding argument, we can use the product rule to derive the following electivities for a double draw of which the second choice is a success :

Second choice a success following a success $\dfrac{s(s - 1)}{n(n - 1)}.$

Second choice a success following a failure $\dfrac{s \cdot f}{n(n - 1)}.$

The two possibilities constitute an exclusive classification of a second success. So we may apply the addition rule :

$$\frac{s(s-1)}{n(n-1)} + \frac{sf}{n(n-1)} = \frac{s(s+f-1)}{n(n-1)} = \frac{s}{n}.$$

For the *restrictive* case the distribution of a 2-fold choice from a 3-class universe shown in Fig. 25 accords with the multinomial expansion

$$(1 + 3 + 4)^{(2)} \div 8^{(2)}.$$

For an r-fold choice from a 2-class universe, the corresponding binomial in factorial powers is $(s + f)^{(r)} \div n^{(r)}$ as given by (ii) in 2.06, and is deducible by successive application of the staircase model as shown in Fig. 26 (a, b, c). It is often more convenient to use the alternative form :

$$(s + f)^{(r)} \div n^{(r)} = (pn + qn)^{(r)} \div n^{(r)} \qquad . \qquad . \qquad . \qquad . \quad \text{(i)}$$

NOTE.—For an r-fold sample successive terms of this expansion, like those of $(p + q)^r$, when choice is repetitive, give the proportion of samples containing r, $(r-1)$, $(r-2)$ . . . 3, 2, 1, 0 *successes*. It is almost always more convenient to memorise the formula and its implication in reverse order, *viz.* : successive terms of $(q + p)^r$ for the repetitive case and $(qn + pn)^{(r)} \div n^{(r)}$ for the restrictive case respectively give the proportion of samples containing 0, 1, 2 . . . $(r-1)$, r successes.

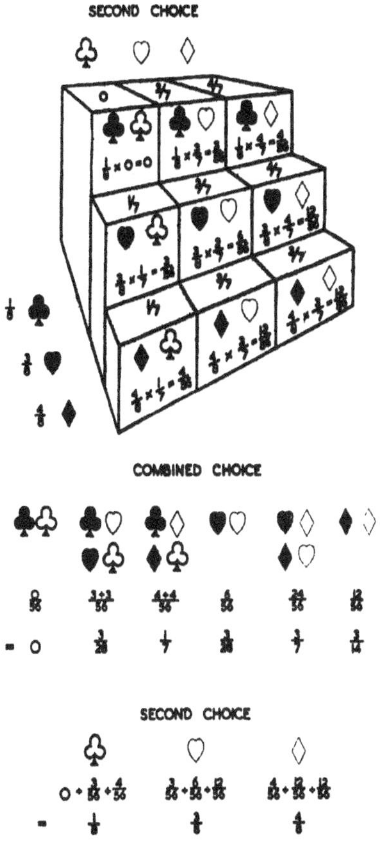

FIG. 25. All possible results of *simultaneous* choice of 2 cards from a 4-card pack made up of 1 club, 3 hearts and 4 diamonds by application of the product and addition rules embodied in Figs. 21, 23 and 24.

THE HYPERGEOMETRIC
(NON-REPLACEMENT) DISTRIBUTION

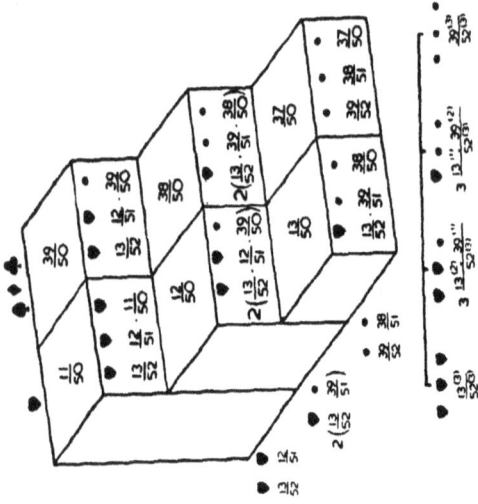

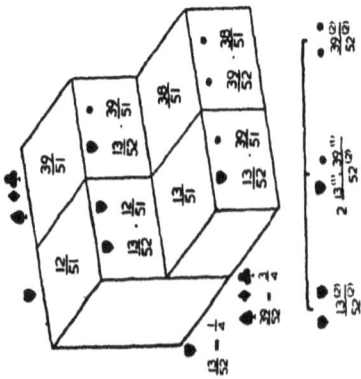

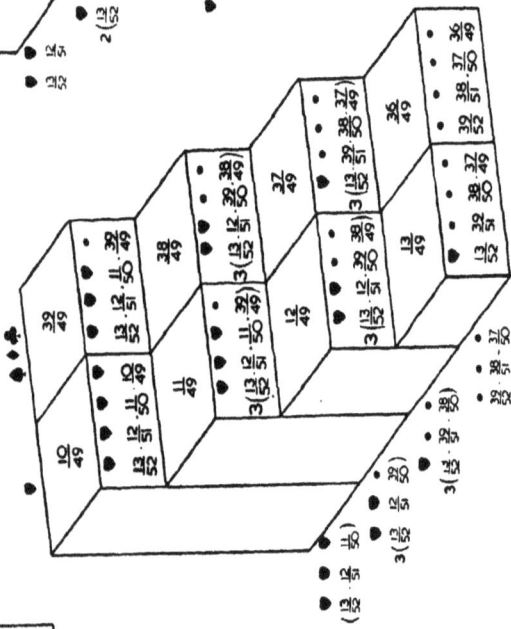

FIG. 26.   The Binomial Series with Factorial Indices (*Vandermonde's* Theorem) as the law of the distribution of heart scores for simultaneous choice of 2, 3, 4 cards from a full pack.

EXERCISES 2.06-2.07

1. Jones had 3 shares in a lottery with 3 prizes and 6 blanks. Smith has 1 share in a second with 1 prize and 2 blanks. Show that Jones has a better chance of winning a prize than Smith. In what ratio ?

2. If I draw 4 cards from a pack, what is the chance that I draw 1 of each suit ?

3. A bag contains 5 white and 4 black balls. If someone empties the bag by taking out the balls one by one, what is the chance that the first will be white, the second black, and so on alternately ?

4. A player takes 4 cards together from a full pack. What is the chance that 1 will be an ace and the remainder respectively a two, a three, and a four ?

5. A bag contains 5 red balls, 7 white balls, 4 green balls, and 3 black balls. If emptied one by one, what is the chance that all the red balls should be drawn first, then all the white ones, then all the green ones, and then all the black ones ?

6. If 9 ships out of 10 on an average return safe to port, what is the chance that at least 3 out of 5 ships expected will make good ?

7. Snooks and Jones play chess. On an average Snooks wins 2 games out of 3. Find the chance of Snooks winning 4 games and Smith winning 2 out of the first 6.

8. There are 10 tickets in a lottery, 5 of them numbered 1, 2, 3, 4, 5, and the other 5 blank. What is the chance of drawing a total score of 10 in 3 trials : (a) it the ticket drawn out is replaced at each trial ; (b) if the ticket is not replaced ?

9. Jones and Smith play chess. In the long run Jones wins 5 games out of 9. Find his chance of winning a majority of (a) 9 games ; (b) 3 games ; (c) 4 games.

10. In a throw of 2 dice, what is the chance that the score will be greater than 8 ?

11. What is the chance of throwing : (a) at least 1 ace in a double toss of a die ; (b) a double which contains neither an ace nor a six ?

12. In one throw with a pair of dice, what is the chance that : (a) there will be neither an ace nor a double ; (b) the score will be exactly 11 ?

13. What is the chance that 4 cards taken simultaneously from a well-shuffled full pack will contain : (a) 3 aces ; (b) 3 picture cards ; (c) at least 1 queen ?

14. What is the chance of picking the following pieces from a box of chessmen without replacement : 1 white king, 2 queens, 2 black bishops, 1 white rook, a black and 2 white knights, 3 black and 4 white pawns ?

15. What is the chance of getting all 4 aces and all 4 kings in a hand of bridge ?

2.08 ELECTIVITY, OPPORTUNITY AND CHANCE

**** Before we attempt to relate what we have hitherto called electivity to what others call probability, readers already familiar with current text-books may find it helpful to recall an alternative formulation. The *addition* rule given in 2.06 is sometimes stated in the following

**** Omit on first reading.

way, the precise wording here given being due to Uspensky * (p. 27) :

> *The probability for one of the mutually exclusive events* A_1, A_2, A_3 . . . *to materialise is the sum of the probabilities of these events.*

Uspensky here uses the symbol A_x merely as a label for an item (the xth) of a class A. It has no numerical meaning as such. For the numerical value of the probability of its occurrence, or as we here say, its *electivity*, in a *unit* sample, he uses the bracketed symbol (A_x). Any event, i.e. *item* of the class A labelled with a particular subscript is by definition an exclusive sub-class of A ; and the symbol (A) stands for the electivity of the class A as a whole. In this notation the foregoing rule is, therefore,

$$(A_1 + A_2 + A_3 \ldots A_x) = (A_1) + (A_2) + (A_3) \ldots (A_x).$$

The same author's formulation (p. 31) of the product rule for *compound* probability is sufficiently general to include the chessboard representation as a particular case of the staircase model :

> *The probability of simultaneous occurrence of A and B is given by the product of the unconditional probability of the event A by the conditional probability of B supposing that A actually occurred.*

This statement merits amplification, since it does not explicitly specify whether the rule applies to either or both possible arrangements of the two items. We may adapt and expand it for our purpose, and in our own terminology, in the two following propositions :

(a) the electivity of a 2-fold sample containing items of one and the same class is simply the product formed by multiplying the *initial* electivity of extracting one item of this class from the total universe by the *contingent* electivity of extracting another item from the residual universe ;

(b) the electivity of a 2-fold sample containing items of different classes A and B is the sum of the products respectively formed by multiplying

(i) the *initial* electivity of extracting a single item of class A from the total universe by the *contingent* electivity of extracting an item B from the residual universe ;

(ii) the *initial* electivity of extracting a single item of class B from the total universe by the *contingent* electivity of extracting an item A from the residual universe.

For the contingent electivity of B from a universe out of which we have already extracted an item of class A, Uspensky uses the symbol (B,A). With this notation we may write the two forms of the product rule thus :

$$(a) \qquad (AA) = (A)(A,A).$$
$$(b) \quad (AB + BA) = (A)(B,A) + (B)(A,B).$$

Needless to say, we may extend these rules to any size of sample, e.g.

$$(AAA) = (A)(A,A)(A,AA).$$

In the same notation, the addition and product rules are both implicit in the binomial representation of a 2-fold sample from a universe of 2 exclusive classes :

$$(A + B)(A + B) = (A)(A,A) + (A)(B,A) + (B)(A,B) + (B)(B,B) = 1.$$

If the two acts of choice are *independent*, i.e. if repetition is permissible, the residual and initial universes are identical ; and the contingent electivity of the second choice of an item of any class is identical with its initial electivity. Hence

$$(A,A) = (A); \ \ (B,A) = (B); \ \ (A,B) = (A); \ \ (B,B) = (B).$$
$$\therefore \ (A + B)(A + B) = (A)^2 + 2(A)(B) + (B)^2 = 1.$$

---

* *Introduction to Mathematical Probability* (1937), a most useful book.

In these expressions (A) and (B) respectively mean the proportionate contribution of items of class A and class B to *all* items of the universe. To say that the classification is *exclusive* thus signifies that

$$(A + B) = 1 = (A) + (B).$$

The symbols (A) and (B) in the binomial expression $(A + B)(A + B)$ above signify the electivities of 1-fold samples relevant to a binary classification which is exclusive in this sense ; and we may write the result

$$(A + B)(A + B) = (AA) + (AB + BA) + (BB) = 1.$$

In this form AA with electivity (AA), AB or BA with electivity $(AB + BA) = 2(AB)$, and BB with electivity (BB) together constitute 3 exclusive classes of samples. For 3-fold samples of items reducible to 2 classes, the corresponding classes of samples are : (i) AAA; (ii) AAB or ABA or BAA; (iii) ABB or BAB or BBA; (iv) BBB. The corresponding binomial for association of items consistent with *repetitive* choice is

$$(A + B)(A + B)(A + B)$$
$$= 1 = (AAA) + (AAB + ABA + BAA) + (ABB + BAB + BBA) + (BBB).$$

We have hitherto interpreted the binomial (or more generally, the multinomial) distributions in either form (usual or hypergeometric) given above by (ii) and (iii) in 2.05, as a compact statement about how much the different ways of extracting all r-fold samples included in each of a set of exclusive classes of samples respectively contribute to all possible r-fold samples consistent with the nature of the problem. About each linear arrangement of r items we can extract from the n-fold universe or universes specified thereby, we may rightly or wrongly assert that it belongs to a particular class of samples. If choice is *not* restrictive, the *proportion* of linear arrangements of 4 items consonant with a specification of $1A + 3B$ is then given by

$$(ABBB + BABB + BBAB + BBBA).$$

If $(A) = p$ and $(B) = q$,

$$(ABBB + BABB + BBAB + BBBA) = 4pq^3. \qquad \text{★★★★}$$

Of all 4-fold samples distinguished as different in virtue of both the different items contained in them and the different linear arrangements of the items without restriction w.r.t. repetition, the expression $4pq^3$ in the symbolism of 2.06 represents the proportion concerning which one can correctly assert : *this sample belongs to a set of which 3 items are of class B and 1 of class A*. The proportionate frequency with which we should assign it to its *correct niche in a classificatory framework*, if we always assumed that a sample belongs to a particular member X of a set of exclusive classes, is therefore the same as the electivity of samples which actually belong to the particular class X. This is the line of thought implicit in the definition of mathematical probability by those who reject any form of words involving explicit reference to the frequency of events in favour of a statement about the frequency of making correct judgments. Given certain data about a sample and a universe, e.g. that a sample consists of 4 items chosen from a universe of 52 items allocated to two classes, A and B respectively composed of 13 and 39 items, one can propose a particular proposition about any sample extracted from such a universe. For instance, one may propose the particular proposition that the sample consists of $1A$ and $3Bs$ as above. Of all possible ways of extracting an r-fold sample distinguished by its individual constituent items (i.e. $n^{(r)}$ if there is restrictive choice and n^r otherwise), some, let us say s, will conform to the specification asserted by the proposition, others (f) will not. For instance, there are in all 52^4 ways of laying out four cards of a full pack, if we record the result with no restriction w.r.t. repetition and $4 . 13 . 39^3$ is the number of which we can correctly assert that 1 is a heart and the other 3 belong to another suit. Of all different samples we can extract from such a universe, the ratio $s \div (s + f)$ therefore represents the proportion of samples correctly specified by the proposition : this sample belongs to class A. Accordingly, some writers identify this ratio with the probability that the proposition is correct.

So stated, the definition of probability does not go beyond what we have called the electivity of a sample ; but we have given no guarantee that a correct statement of this sort is a correct statement about how an act of choice *materialises*, i.e. how often we should be correct in asserting that a 4-fold sample selected with replacement from a full pack would contain only 1 heart. To have operational value as a basis of judgment, the ratio we call probability must refer back to how often we do, in fact, choose such a class of samples ; and this is not the same as specifying the numerical frequency of its occurrence among all different 4-fold samples in a *static* universe. When we speak of the relative frequency of making a correct judgment about 2 picture cards chosen simultaneously from a full pack we should therefore be clear about whether we are asserting one of two propositions between which there is no necessary connection :

(a) of $52^{(2)}$ ($= 2652$) different ways of arranging 2 cards from a full pack without repetition exactly $12^{(2)}$ ($= 132$) conform to the assertion that both bear a picture ;

(b) results of experiments on drawing 2 cards from a full pack justify the conclusion that pairs of which both are picture cards and pairs of which at least one number is not a picture card turn up in the ratio $132 : (2652 - 132) = 132 : 2520$.

If we define appropriately what we mean by different ways of extracting a sample, as above, the first statement is a logical tautology. The second cannot always be right, if only because the number of experiments performed limits the possibility of getting the prescribed ratio. The possibility of defining circumstances in which a connection between the two statements does in fact exist is the only justification for invoking mathematical probability in discussion about observations on natural occurrences. To employ mathematical technique as a tool of investigation in the domain of practical judgments with which we associate the term probability in common parlance, we have therefore to establish such a link.   Hitherto we have interpreted the definition of different ways of extracting a sample in terms of *linear arrangement*. At first sight this definition may seem arbitrary ; but the use of the chessboard or staircase device to set forth the possible arrangements resulting from extracting samples from different universes or from the same universe has already (p. 62) focused our attention on a property of linear order which is highly suggestive when we seek such a link.   We obtain a correct representation of every possible linear arrangement of 2-fold samples from a single universe by pairing off each item of the parent universe *once and once only* with each item of the residual universe ; and the same principle is implicit in successive application of the chessboard or staircase procedure to evaluate the proportion of linear arrangements of a given specification in samples of more than two items.   By defining the number of ways in which we can extract a sample as the number of different linear arrangements, we thus presuppose *equipartition of opportunity for association* between individual items.

Though the principle of equipartition of opportunity for association stated in this form has no necessary connection with the probability of making a correct judgment about the materialisation of occurrences, its recognition points the way out of a quagmire of ostensibly empirical, but logically circular, definitions involving statements about events which are *equally likely*.* Whether events do or do not occur in conformity with results derived from considerations suggested by the algebra of choice evidently depends on *the extent to which circumstances are more or less propitious to equipartition of opportunity for association of their constituent elements*. This suggestion

---

* *See* Aitken (*Statistical Mathematics*, pp. 6-11) for comments upon the unsatisfactory status of current definitions of randomness.

is not sufficient to lead us to the goal we have in view. For we have still to define what circumstances are more or less propitious or how we can make them so, when the extraction of a sample does not involve any prior knowledge of what items will in fact turn up. None the less, we have now in focus what all statisticians tacitly regard as a paramount issue ; and we have not involved ourselves in the vicious cycle of a definition of probability explicitly or implicitly containing the notion of equal likelihood. If we speak of the process of ensuring equipartition of opportunity for association as *randomisation*, our problem now takes the form : in what circumstances or with respect to what systems is randomisation practicable ? *En passant*, we should recognise that randomness implicit in the term so defined is a property of a system. Only on this understanding is it then appropriate to describe a *single* act of choice as random.

To find an answer to the question last stated we may profitably leave aside general considerations, and direct our attention to a particular case. The intuition which is common sense, *alias* everyday experience, leads us to hope that if we shuffle a pack of cards sufficiently thoroughly and sufficiently often, we shall in fact attain such equipartition of opportunity for association. We can justify this hope with some plausibility in so far as we have reason to suppose that the denomination of a card in no way affects whether it will retain its position relative to another when we mix the pack as thoroughly as we can ; but in any case we can put it to the test. In so far as the results of a very large number of such experiments show that pairs of two picture cards and pairs of which at least one member has no picture turn up in a simultaneous *double* draw from a full pack in the ratio 132 : 2520 or 1 : 19 (*approx.*), we are entitled to say that the calculus of electivity prescribes the *limiting* (i.e. long run) value of the relative frequency of such an occurrence. The following table shows the result (Pearson,* 1934) of counting the number of trumps in 3400 hands (of 13 cards) of whist and corresponding electivities calculated in accordance with the distribution defined by the factorial binomial $(13 + 39)^{(13)}$.

No. of Trumps in a Hand.	Observed No. of Hands.	Ditto expected to nearest Integer.
under 3	1021	1016
3–4	1788	1785
over 4	591	599

To say that experiment does confirm our supposition neither restricts the usefulness of the calculus of electivity to such trivial themes as evaluating the credentials of systems employed in games of chance, nor compels us to undertake extensive experiments on the distribution of other classes of events to which we can usefully apply it. Given any situation to which it is applicable, we can use such a situation as a model for the construction of a hypothesis to guide our judgments in the search for scientific laws. The justification for the use of such a model depends on what is implicit in the principle of equipartition of opportunity for association. A scientific law draws attention to some connection between the attributes of a system, e.g. the pressure of a gas and its volume at a given temperature, or the age of a human being and his or her liability to contract tuberculosis. We can often get the answer to a question about the association of two such attributes by putting it in the alternative form : what would happen *in the long run*, if there were no connection between them ? The card pack model provides us with an answer to questions of this sort.

* *Biometrika*, 16, p. 172.

FISHER'S TEA-CUP TEST

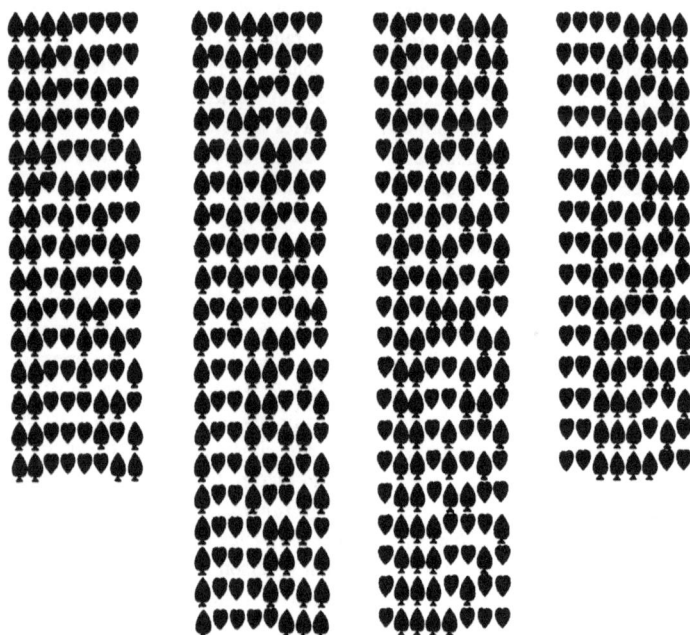

FIG. 27. This model set-up for Fisher's Experiment exhibits all the 70 possible ways of arranging 4 hearts and 4 spades together.

The statement that each item of the chessboard or staircase device has an equal opportunity to pair off with any other item of the residual universe, is equivalent to saying that the denomination of the card or die does not affect the frequency of its association with members having another class denomination. The only thing that does so, is the *number* of members of each class. The electivity of a sample—its mathematical probability—is therefore an index of *associative neutrality* or, as we might equally well say, an index of the *lawlessness* of the system. In so far as the observed behaviour of the card pack conforms to the law of equipartition stated above, the card pack is therefore

(a) a model for any system whose numerical properties depend solely on the numerical frequencies of the attributes concerned, e.g. as postulated by the Mendelian hypothesis ;

(b) a yardstick or " *null hypothesis* " for testing association between attributes of a system.

In the next chapter we shall take up the discussion of (a) in detail. To keep our feet on solid ground, we may, however, first consider an example of (b) taken from R. A. Fisher's book, *The Design of Experiments*. A lady claims—some do—that she can tell by sipping a cup of tea whether her hostess added the milk before or after pouring out the tea itself. To vindicate her credentials she agrees to submit to a test in which she has to identify 8 cups, 4 of which received milk first, the other 4 tea first, but otherwise like the rest. She can taste them in any order and receives no information other than the fact that 4 cups are of one sort and 4 of the other. If she identifies everyone correctly, has she justified her claims ? In common parlance, is the result of the test a fluke ?

The card model of her test is the assignment of 4 red and 4 black cards of a pack of 8 in a certain order w.r.t. the colour of the items alone. In all, the number of distinguishable arrangements of 8 cards of 2 colours (Fig. 27) is given by (iii) in 2.02, i.e.

$$\frac{8!}{4!\,4!} = 70.$$

The correct arrangement is thus one of 70, and its electivity is therefore $\frac{1}{70}$. In the long run an individual would therefore pick out the prescribed, i.e. *correct*, arrangement once in 70 trials from a repeatedly and well shuffled pack. Accordingly, we say that the odds are 69 : 1 against selecting the prescribed order, if there were *no association* between the subject's specification and the class to which an item belongs. Whether we accept this as grounds for entertaining much confidence in her claims is not itself a mathematical issue ; and it will be our theme in Chapter 5. Most of us would agree that the result at least offers sufficient grounds for further consideration of them.

Fisher's example thus illustrates how we can legitimately invoke the behaviour of a suitable model as a yardstick—or *null* hypothesis—for testing the connectedness of two sets of phenomena, in this case a person's judgment about a situation and the situation itself. It also draws

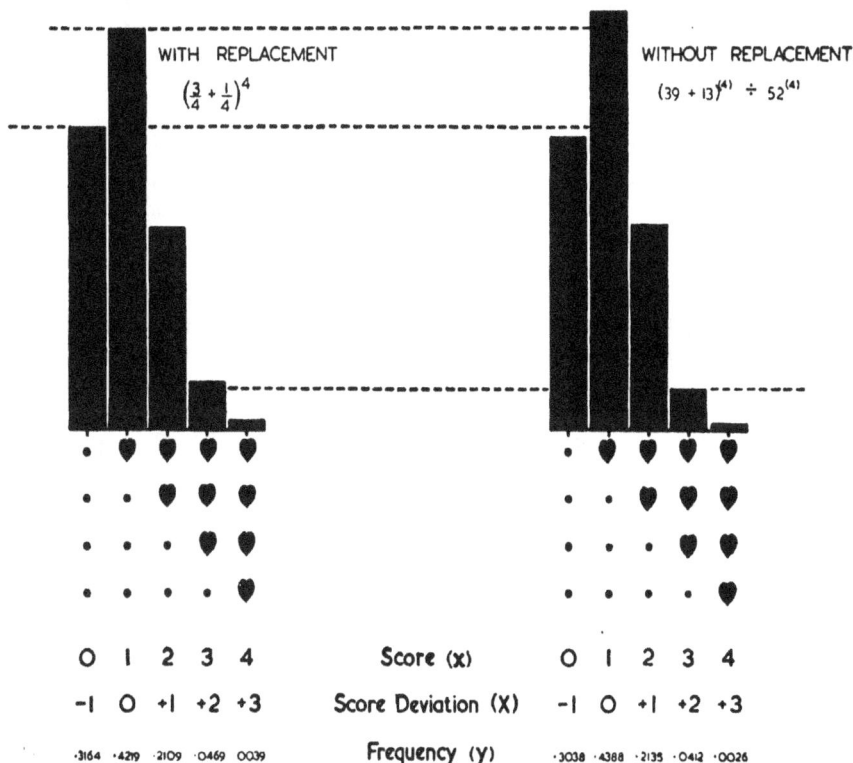

FIG. 28.   The Score Distribution for Hearts in 4-fold samples from a full pack depends on whether we do or do not replace each card taken before choosing another ; but the difference between the two distribution :
is not great if the sampling fraction is small.

7

attention to the necessity of *choosing the right model.* By informing the subject that there are 4 cups of each sort in the sample, we have in fact reduced the issue to one of *restrictive* choice ; but this would not be so if we withheld the knowledge that half the set were of one sort, half of the other. We could then regard the identification of each cup as a 1-fold sample judgment exclusively involving the alternatives of success and failure, as in 8 successive tosses of an un-biassed coin ; i.e. one whose long-run behaviour conforms to the distribution $(\frac{1}{2} + \frac{1}{2})^8$. The electivity of a run of 8 successes calculated on this basis is $(\frac{1}{2})^8 = \frac{1}{256}$. On the assumption that the subject's judgment is merely a *toss-up,* a correct identification of all the cups would therefore happen in the long run once in 256 tests of this sort. The odds would be 255 : 1 against the result, if the subject did not know how many of each class of cups she had to identify. In short, the conditions of an experiment or the nature of the attendant circumstances of an investigation alone decide what is the correct statistical model to use as yardstick of a *significant* result.

The choice of a correct model in the sense implied in this context endows the epithet *unbiassed* as applied to a coin or other die model with a particular significance. For our purpose a coin is unbiassed, if the long-run frequency of a particular class of samples, e.g. ten-fold tosses of which 3 yield heads and 7 yield tails, tallies with its electivity in a 2-class universe of 2 *items* only, subject to the replacement condition appropriate to die models in general. If the behaviour of a coin does not in fact conform to this requirement, we have to seek an interpretation of its behaviour in terms of another model whose behaviour we can guarantee. A coin which has a 40 per cent. heads score in an indefinitely large number of trials, behaves in accordance with the principle of associative neutrality or *independence* of successive sampling, if the distribution of *r*-fold samples of specified composition tallies with the long-run distribution of *r*-fold samples extracted by recording each item taken singly after a re-shuffle subsequent to replacement of its predecessor from a pack of 10 cards, of which 4 are red (*heads*) and 6 are black (*tails*) cards.

It is the standpoint of this book that Fisher's tea-cup problem gets into focus what is the essential link between choice and chance. In short, we ask a legitimate question about sampling only when we can set up a null hypothesis reducible to a model situation on all fours with a game of chance ; and our justification for invoking the algebra of choice in this context resides solely in the empirical circumstance of its observable relevance to such situations. In adopting this approach, we retrace our steps to Pascal and Fermat. That the connection between the calculus of choice and the calculus of chance has its roots in the milieu of the gaming table and the lottery is a truism to which the circumstances leading to a celebrated controversy between Pascal and Fermat bears testimony. The occasion which prompted it also furnishes a neat illustration of our negation rule for the *at least one* type of problem. The Chevalier de Meré had made a fortune by betting small favourable odds on getting at least one six in 4 tosses of a die, and lost it by betting small favourable odds on getting a double six in 24 double tosses. The negation rule gives the explanation :

Chance of getting at least one six in 4 tosses :

$$1 - (\tfrac{5}{6})^4 = 0 \cdot 517 \; (\textit{more} \text{ likely than not}).$$

Chance of getting a double six in 24 double tosses :

$$1 - (\tfrac{35}{36})^{24} = 0 \cdot 491 \; (\textit{less} \text{ likely than not}).$$

In so far as experiment justifies the view that the long-run frequency of an occurrence tallies with its electivity when we prescribe, or are entitled to assume, conditions favourable to equipartition of opportunity for association, the mathematical development of its laws can proceed regardless of our preference for an empirical or wholly formal definition of what we mean by probability ; but the distinction itself is relevant to the scope of two types of theoretical problems to which we commonly apply the term statistics. In contemporary usage we use the term statistics indifferently for :

(a) raw numerical data, as when we speak of trade statistics ;

(b) *summarising* indices which serve to indicate what such data tell us, if taken at face value, as when we specify population trends by a net reproduction rate or by a standardised death rate ;

(c) laws of the gross behaviour of populations of *particles*, such as molecules or genes, whose individual behaviour we cannot directly observe ;

(d) tests of the validity of our judgments about enumerable samples or recipes for classifying such samples with a view to fruitful judgments about the universe from which we extract them.

The theory of probability, whether defined empirically or formally, is necessarily relevant only to the last two (c) and (d). In this book our concern is mainly with (d) ; and only when (d) is our concern does the need to invoke a null hypothesis arise. The distinction between (c) and (d) is by no means trivial, if only because population samples, as conceived in the *Kinetic Theory of Gases* or the genetical *Theory of Inbreeding* refer to numbers of individual items (molecules or genes) too numerous to count. For this reason, we may regard the sample size itself as infinite. In any case, the focus of our interest in the behaviour of an assemblage of genes or molecules is how often an *observable* occurrence will happen ; and this is *de facto* an issue amenable to direct scrutiny. For reasons which we shall later examine, it is not easy to see how we can invoke direct observation to assess how often our judgments will be right or wrong if we reject a null hypothesis because it assigns a very low long-run frequency to an occurrence. In Chapter 5 we shall see that making any such judgment enlists either information of a sort which our foregoing symbolism cannot cover or additional assumptions that are not equally self-evident to all professional statisticians.

EXERCISE 2.08

EMPIRICAL DISTRIBUTIONS

1. The Swedish statistician Charlier records 1000 trials of a 10-card draw. In each trial he replaced each of the 10 cards taken before drawing another, the recorded score being the number of black cards drawn in a single trial. Compare the result shown below with the appropriate electivity distribution :

Score .	.	0	1	2	3	4	5	6	7	8	9	10
No. of trials .		3	10	43	116	221	247	202	115	34	9	0

(Aitken, *Statistical Mathematics*, p. 50.)

2. Uspensky records 7000 trials of a simultaneous 4-card draw from an otherwise full pack without picture cards, the score being the number of trials in which the 4 cards drawn from the 40-card pack were of different suits. Compare the net result with the electivity of the occurrence as computed from the appropriate multinomial expansion, his figures for 7 successive sets of 1000 trials being : 113, 113, 103, 105, 105, 118, 118.

(Uspensky, *Introduction to Mathematical Probability*, p. 110.)

3. The same author (*ibid.*) records 1000 results of a simultaneous 5-card draw classified by number :

5 different	503
2 alike, 3 different	436
2 different like pairs and one different	45
One triplet and 2 unlike	14
One pair and 1 triplet	2
4 alike	0

Compare this result with expectation on the assumption that the electivity distribution tallies with the long-run frequency distribution.

4. Compare, with the appropriate electivity distribution, the following data Edgeworth cites w.r.t. 4096 throws of 12 dice by Weldon, who scored a single throw of more than 3 pips as a success.

No. of Successes per Trial.	No. of Trials of 12 Tosses.	No. of Successes per Trial.	No. of Trials of 12 Tosses.
0	0	7	847
1	7	8	356
2	60	9	257
3	198	10	71
4	430	11	11
5	731	12	0
6	948		
		Total	4096

5. Uspensky cites results of a die game recorded by Bancroft H. Brown. The player has 2 dice. He wins unconditionally if he first scores 7 or 11 pips (*naturals*) at a simultaneous throw, and loses unconditionally if he throws *craps*, i.e. a total of 2, 3 or 12. Otherwise he goes on throwing until he either wins by repeating a previous score or loses by casting a 7. Compare the expectation of success with the recorded result, *viz.* : 4871 wins in 9900 games.

CHAPTER 3

HYPOTHESIS AND EXPECTATION

3·01 Electivity, Frequency and Expectation

WHAT we have provisionally called the electivity of a specified class of samples signifies its *proportionate* contribution to the number of·all possible classes of samples from the same universe in accordance with the principle of equipartition of opportunity for association of its constituent items. In so far as the process of selecting a sample fulfils this requirement, experience justifies the belief that the observable proportionate frequency of the act of choice should approach this figure in the long run. In the domain of large-scale experience we can therefore speak of the electivity of a sample of specified composition as its *theoretical frequency* or *expectation.**

When we are comparing samples w.r.t. a numerically specified attribute, e.g. how many hearts samples of ten cards from a full pack contain, we are commonly concerned with the following type of question : is the observed numerical discrepancy between its actual and theoretical frequency (so defined) *significant ;* or is it merely a fluke ? To say that it is a fluke or a matter of chance in this context means that an equally large discrepancy would not be an infrequent occurrence, if relevant features of the situation were strictly comparable to what happens in a game of chance. We are therefore less concerned with the theoretical frequency of a particular sample-score than with the theoretical frequency or expectation of a sequence of such scores within a certain range. There is no clear-cut usage which distinguishes the term expectation from theoretical frequency ; but it may be more expressive to use the former when we refer to the theoretical frequency of a set of numerical values within a specified range.

So far we have defined a class of r-fold samples by the number of constituent items of each class represented therein. If we know the size of the sample, the number of items in one class is fixed by the number of items in the others by subtraction from the total. If the classification is binary, one number therefore suffices to specify it. In the idiom of the game, we then speak of items of one class as *successes* in contradistinction to *failures ;* and the number of items labelled as successes (e.g. *red* cards taken from a full pack) is the sample score which tells us all we need to know about the composition of the sample. Thus the *score* 3 suffices to define a 4-fold sample of 3 red cards and 1 black one, if we distinguish only two classes, *red* (*successes*) and *black* (*failures*). In contradistinction to the *raw* score, i.e. the actual number of items of a particular class, we are equally entitled to define a sample by its proportionate contribution. Thus the *proportionate score* of red cards in a 4-fold sample containing 3 is $\frac{3}{4}$. In general the relation between the raw score (x) and the corresponding proportionate score (u) is given by $u = (x \div r)$.

Another way of scoring is of particular importance when we are concerned with the total electivity of one or other of a sequence of classes, i.e. the expectation that the score will fall within a certain range $\pm X$ (inclusive) about a fixed value F_x or that the proportionate score will fall within a corresponding range $\pm U$ about the corresponding fixed value F_u. In practice, we take the *most representative* score for F_x or F_u as the mid-point of the range. The most representative score of an r-fold sample in this context signifies the score of an r-fold sample whose proportionate composition is that of the parent universe. The most representative *proportionate score* (F_u) of a binary classification does not depend on the size of the sample, being equivalent to the proportion (p) of items scored as *successes* in the parent universe itself.

---

* The frequency of a class here signifies its *proportionate* contribution to the whole system. In this book we use frequency only in this sense, that is for fractions—never for whole numbers.

This is the electivity (p. 75) of a success in a sample of one item. If F_z is the most representative *raw score* in an r-fold sample, $F_z = rp$. With or without replacement, the most representative score is in fact the *mean value* * of such scores in all possible samples drawn from the same universe, as the following calculation shows w.r.t. the raw score.

(a) With replacement

The distribution of raw scores is

0	1	2	3 ...	r
q^r	$r \cdot q^{r-1} p$	$r_{(2)}\, q^{r-2} p^2$	$r_{(3)}\, q^{r-3} p^3$	p^r

The mean raw score value (M_z) is by definition the sum of the individual scores weighted by their proportionate frequencies :

$$M_z = \sum_{x=0}^{x=r} x \cdot r_{(z)}\, q^{r-z}\, p^z$$

$$= 0 \cdot q^r + 1 \cdot r \cdot q^{r-1} \cdot p + 2 \cdot \frac{r(r-1)}{2 \cdot 1} q^{r-2} \cdot p^2 + 3 \cdot \frac{r(r-1)(r-2)}{3 \cdot 2 \cdot 1} \cdot q^{r-3} p^3 \ldots$$

$$= rp \left[q^{r-1} + (r-1)q^{r-2} \cdot p + \frac{(r-1)(r-2)}{2 \cdot 1} q^{r-3} \cdot p^2 \ldots \right]$$

$$= rp\, (q+p)^{r-1}.$$

Since $(q+p) = 1$, we have

$$M_z = rp(q+p)^{r-1} = rp.$$

(b) Without replacement

As in 2.05 we put $p = s/n$ and $q = f/n$. The mean is then given by

$$\frac{0 \cdot f^{(r)}}{n^{(r)}} + 1 \cdot \frac{rf^{(r-1)}s}{n^{(r)}} + 2 \cdot \frac{r(r-1)}{2 \cdot 1} \frac{f^{(r-2)}s^{(2)}}{n^{(r)}} + 3 \cdot \frac{r(r-1)(r-2)}{3 \cdot 2 \cdot 1} \frac{f^{(r-3)}s^{(3)}}{n^{(r)}} \ldots$$

$$= \frac{rs}{n^{(r)}} \left[f^{(r-1)} + (r-1) \cdot f^{(r-2)} \cdot (s-1) + \frac{(r-1)(r-2)}{2 \cdot 1} f^{(r-3)} \cdot (s-1)^{(2)} \ldots \text{etc.} \right]$$

$$= \frac{rs}{n^{(r)}} \cdot (\overline{f + s - 1})^{(r-1)}$$

$$= \frac{rs \cdot (n-1)^{(r-1)}}{n^{(r)}} = \frac{rs}{n} = rp.$$

---

* We may write the arithmetic mean of the set of numbers 3, 4, 3, 5, 3, 2, 4 in any one of the following ways :

$$(a)\ \frac{3+4+3+5+3+2+4}{7}\,; \qquad (b)\ \frac{(1)(2) + (1)(5) + (2)(4) + (3)(3)}{7}\,;$$

$$(c)\ \tfrac{1}{7}(2) + \tfrac{1}{7}(5) + \tfrac{2}{7}(4) + \tfrac{3}{7}(3).$$

Symbolically, we shall represent (a) and (c) respectively by

$$(a)\ \frac{\Sigma x}{r}. \qquad (c)\ \Sigma y_z \cdot x.$$

When the mean of a score x or some function thereof such as x^2 is referable to a *theoretical* sampling distribution we shall always employ (c), for the sufficient reason that it represents for our purpose a large-scale result, realisable only as r in (a) becomes indefinitely large. We shall later (Chapter 7) have occasion to use (a) when referring to the properties of *observed* samples of score values.

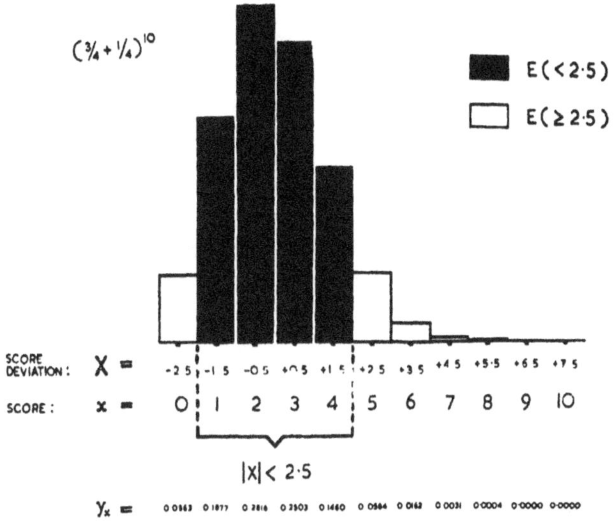

$(\tfrac{3}{4} + \tfrac{1}{4})^{10}$

■ E(<2·5)
□ E(≥2·5)

SCORE
DEVIATION: $X =$ -2·5 -1·5 -0·5 +0·5 +1·5 +2·5 +3·5 +4·5 +5·5 +6·5 +7·5

SCORE: $x =$ 0 1 2 3 4 5 6 7 8 9 10

$|X| < 2·5$

$y_x =$ 0·0563 0·1877 0·2816 0·2503 0·1460 0·0584 0·0162 0·0031 0·0004 0·0000 0·0000

FIG. 29. The expectation of getting at least 1 and not more than 4 hearts in a sample of 10 cards extracted from a full pack on the understanding that the player returns each card taken before extracting another.

Whenever we specify a range of raw scores on either side of a fixed value F_x by the differences $(x - F_x)$ our fixed value will therefore be the sample mean (M_x). We then refer to these differences as *score deviations* denoted by X, so that

$$X = x - M_x \quad or \quad x = X + M_x \qquad . \qquad . \qquad . \qquad . \qquad \text{(i)}$$

For 10-fold samples made up of one card drawn from each of ten full packs, we can thus label classes specified by the *heart* score with their appropriate frequencies (y_x) as in Table 1. The frequencies themselves are defined by the successive terms of the expansion $(\tfrac{3}{4} + \tfrac{1}{4})^{10}$, each term of which has as a factor a power of 3 divided by 4^{10} ($= 1048576$); and the mean heart score is $\tfrac{1}{4}(10) = 2·5$.

TABLE 1

Raw Score (x).	Score Deviation (X).	Proportionate Score (u_x).	Proportionate Score Deviation (U_x).	Frequency (y_x).
0	− 2·5	0·0	− 0·25	$1.3^{10}.4^{-10} = 0·0563$
1	− 1·5	0·1	− 0·15	$10.3^9.4^{-10} = 0·1877$
2	− 0·5	0·2	− 0·05	$45.3^8.4^{-10} = 0·2816$
3	+ 0·5	0·3	+ 0·05	$120.3^7.4^{-10} = 0·2503$
4	+ 1·5	0·4	+ 0·15	$210.3^6.4^{-10} = 0·1460$
5	+ 2·5	0·5	+ 0·25	$252.3^5.4^{-10} = 0·0584$
6	+ 3·5	0·6	+ 0·35	$210.3^4.4^{-10} = 0·0162$
7	+ 4·5	0·7	+ 0·45	$120.3^3.4^{-10} = 0·0031$
8	+ 5·5	0·8	+ 0·55	$45.3^2.4^{-10} = 0·0004$
9	+ 6·5	0·9	+ 0·65	$10.3^1.4^{-10} = 0·0000$
10	+ 7·5	1·0	+ 0·75	$1.4^{-10} = 0·0000$
				TOTAL 1·0000

We are now equipped with a compact way of defining the theoretical frequency of all samples within a certain score range or outside it. We shall use the symbol $E(>X)$ to signify the expectation that a deviation will be numerically *as large as X or larger*. It thus represents (Fig. 29) the total frequency of all samples with scores (x): (i) as great as $(M_x + X)$ or greater ; (ii) as little as $(M_x - X)$ or less. The symbol $E(<X)$ signifies the expectation that a deviation will be numerically less than X, hence that it will be in the range $\pm (X - 1)$ *inclusive ;* and this represents the total frequency of all samples with scores no greater than $(M_x + X - 1)$ and no less than $(M_x - X + 1)$. Thus $E(<2.5)$ for the above distribution is the theoretical frequency of a choice of one or other of the samples whose scores are $1, 2, 3, 4$ with corresponding deviations $- 1.5,\ -0.5, + 0.5$ and $+ 1.5$. Since we are here dealing with mutually exclusive acts of choice to which the *addition* rule applies

$$E(<2.5) = (10.3^9 + 45.3^8 + 120.3^7 + 210.3^6) \div 4^{10}$$
$$= \frac{907605}{1048576}.$$

We denote the expectation that a score deviation will lie outside this range by $E(>2.5)$. By the *negation* rule

$$E(>X) = 1 - E(<X).$$

For the distribution under discussion

$$E(>2.5) = 1 - \frac{907605}{1048576} = \frac{140971}{1048576}.$$

Thus the ratio of the frequency of a choice involving a heart score *deviation* in the range 1–4 inclusive to the frequency of a choice involving a heart score outside this range (i.e. a *raw* score of 0 or over 4) is 907605 : 140971, i.e. about 6·5 : 1.

In the idiom of the gaming table, we sometimes express this by saying that the odds in favour of getting 1, 2, 3, or 4 hearts in a 10-fold sample chosen as stated above are about $6\frac{1}{2}$ to 1. What we commonly call a *significant* discrepancy between the most representative figure prescribed by a hypothesis on the one hand (i.e. the one which corresponds to the mean result derived from an indefinitely large number of samples), and, on the other, a figure referable to a sample from which we propose to draw a conclusion concerning the truth of the hypothesis is a *difference so large as to involve heavy odds against its occurrence.* For reasons which will come up for subsequent discussion, it is customary to regard odds of 20 : 1 as large and odds of the order 400 : 1 as very large. This convention merely gives expression to an algebraic property of the way in which the odds against the occurrence of a proportionate deviation of a given size or greater increase as the deviation itself increases. Otherwise, it is arbitrary ; but there is nothing arbitrary about the exact specification of the odds involved. The very fact that betting odds which would dictate action on the basis of a conclusion drawn from a sample depend in the last resort on the optimism of the individuals concerned and the urgency of the situation involved makes it the more imperative to give precision to the frequency a particular hypothesis assigns to a class of occurrences.

To give this sort of numerical precision to the legitimate confidence with which we accept or reject a hypothesis, it is necessary to formulate (explicitly or implicitly) the die or urn model which reproduces its relevant features w.r.t. the following particulars :

(*a*) whether the items come from the same parent universe ;

(*b*) whether choice of individual items is restrictive, if the universe is single ;

(*c*) whether the sample is large or small as compared with the universe (or universes), and whether the latter is itself very large ;

(*d*) the nature of the classification involved, and hence of scoring the result.

3.02 Testing a Hypothesis

The simplest method of scoring a result arises when the classification of the universe is binary (e.g. hearts and *other* suits), and we define the classes of sample by the number of members of one or other classes present or by their corresponding proportions. This tallies with the situation which faces us when we want to answer a question such as the following : is the incidence of smallpox *significantly less* among individuals who have been vaccinated than among individuals who have not been vaccinated ? Recognition of a real difference in this sense is not the simplest type of problem involving scoring of the sort, because the construction of the appropriate null hypothesis involves the sort of guesswork which statisticians call *estimation*. To clarify what we mean by a significant result, it will be better to start with a hypothesis which leaves no room for guesswork. The Mendelian hypothesis will serve our purpose.

The mechanics of gases and the study of organic inheritance offer examples of hypotheses which are essentially statistical in the sense that they prescribe rules that claim validity only with respect to populations of molecules or large scale breeding operations and start from the assumption of associative neutrality in the terminology of the last chapter, i.e. in customary statistical jargon from *random association*. In the language of mathematical probability, the Mendelian hypothesis postulates that fertilisation is comparable to the extraction of a 2-fold sample (*zygote*) of balls (*gametes*) from different urns, the male and the female parent respectively. The class structure of the two universes defines the mode of inheritance. The simplest classification tallies with genetic differences arising from a single gene substitution. In that case, there can be at most only two different sorts of balls in either urn, and the maximum number of classes of 2-fold samples (*zygotes*) is 3, the two homozygotes and the single heterozygote. Since we are sampling from different universes, the condition of replacement is irrelevant, and an appropriate model is thus a two-faced die, e.g. a coin whose faces we may label A or a. A mating of two heterozygous parents is then comparable to spinning two coins each with a face A and a face a. The result of fertilisation is given by the binomial $(\frac{1}{2} + \frac{1}{2})^2$, i.e. $\frac{1}{4} AA : \frac{1}{2} Aa : \frac{1}{4} aa$. If two independent genes, i.e. genes on different chromosome pairs, are involved, the appropriate model is a tetrahedal (4-faced) die. If the parent is the double heterozygote, we label the faces of the die AB, Ab, aB, ab ; and the results of fertilisation are given by the expansion of $(\frac{1}{4} + \frac{1}{4} + \frac{1}{4} + \frac{1}{4})^2$ in accordance with the chessboard procedure.

A progeny of r individuals is itself a sample from a universe in which the classes of items and their *long-run* frequencies are defined by such considerations. That is to say, we assume the existence of an indefinitely large universe of zygotes—the result of all possible matings of a given type. Since the universe from which we sample is indefinitely large, the withdrawal of a small sample (p. 79) does not alter its composition appreciably. We can therefore regard choice as repetitive ; and testing a particular mode of inheritance then signifies comparison between the theoretical and observed frequencies of samples with different proportions of the prescribed zygotic classes. The assumed class structure of the parent universe of zygotes is our null hypothesis, and satisfactory agreement therewith constitutes our criterion of the validity of the mode of inheritance we infer from the results of a breeding experiment. If there are only two classes of zygotes, as in matings AA by Aa or aa by Aa, the sampling distribution corresponds to the terms of $(\frac{1}{2} + \frac{1}{2})^r$. For the mating of two heterozygotes the sampling distribution w.r.t. a specified homozygous class corresponds to the terms of $(\frac{3}{4} + \frac{1}{4})^r$.

Let us first consider the mating of a heterozygote by a homozygote, i.e. the familiar 1 : 1 ratio of a back-cross involving one gene difference. We hatch 12 eggs from a *blue* by *black* Andalusian cross, obtaining 7 blue chicks and 5 black ones. On the assumption that there is only one gene difference involved, the theoretical proportion of blue chicks is $\frac{1}{2}$. Thus the

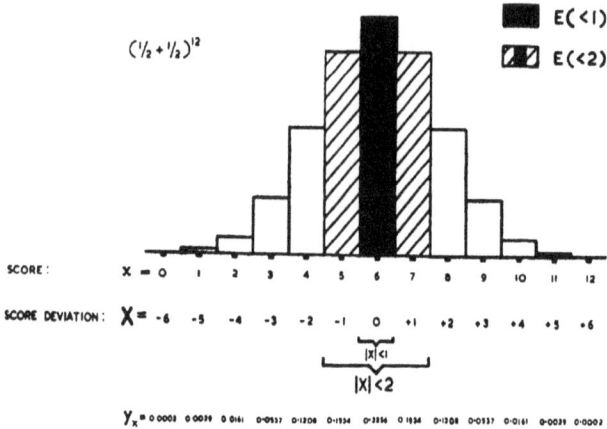

$(\tfrac{1}{2} + \tfrac{1}{2})^{12}$

Legend: ■ E(<1)   ▨ E(<2)

SCORE: X = 0   1   2   3   4   5   6   7   8   9   10   11   12

SCORE DEVIATION: $X =$ -6   -5   -4   -3   -2   -1   0   +1   +2   +3   +4   +5   +6

$|x| < 1$

$|x| < 2$

$Y_x =$ 0·0002   0·0029   0·0161   0·0537   0·1208   0·1934   0·2256   0·1934   0·1208   0·0537   0·0161   0·0029   0·0002

FIG. 30. The expectation of a raw score differing by less than 1 and of a raw score differing by less than 2 from the mean of all samples of 12 when the probability of success is $\frac{1}{2}$ and the replacement condition of Fig. 29 holds good. Success in this context may stand for choosing a red card from a full pack.

mean result of a large number of 12-fold samples of this specification should be 6 blue and 6 black. The question we have to settle is whether our observed result (7 : 5) is such as would occur rarely or otherwise, if the single gene interpretation is in fact correct. Since the universe of choice is indefinitely large, an appropriate model of the procedure is successive extraction *with replacement* of one ball from urn *A* containing only black balls, and one from urn *B* containing equal numbers of black and white ones. We label a sample black if both balls drawn are themselves black. We label it blue if one ball is black and the other white. For a sample of 12 zygotes (Fig. 30) the elective distribution is defined by $(\frac{1}{2} + \frac{1}{2})^{12}$, *viz.* :

Score (no. of blue chicks)	Theoretical Frequency
0	$(\frac{1}{2})^{12}$
1	$12(\frac{1}{2})^{12}$
2	$66(\frac{1}{2})^{12}$
3	$220(\frac{1}{2})^{12}$
- - - - - -	- - - - - -
10	$66(\frac{1}{2})^{12}$
11	$12(\frac{1}{2})^{12}$
12	$(\frac{1}{2})^{12}$

The numerical value of the deviation of the observed raw score (7) from the theoretical mean (6) is unity. An observed score value must either be equivalent to the theoretical mean or must deviate by at least 1 in either direction. Hence the expectation $E(< 1)$ that the score observed will neither exceed the theoretical mean nor fall short of it by *as much as* 1 is, of course, that of the theoretical mean itself (Fig. 30). This is

$$\frac{12 \cdot 11 \cdot 10 \cdot 9 \cdot 8 \cdot 7}{6 \cdot 5 \cdot 4 \cdot 3 \cdot 2 \cdot 1} \cdot \left(\frac{1}{2}\right)^{12} = \frac{924}{4096}$$

The expectation $E(> 1)$ that a deviation will be numerically as great as unity is the expectation that the observed deviation will *not* be zero, i.e. that the score will not be equivalent to the mean.

Thus the expectation of a deviation as large as ± 1 is

$$1 - \frac{924}{4096} = \frac{3172}{4096}.$$

In roughly three out of four samples a deviation should therefore be numerically as great as unity. In so far as this is not an uncommon occurrence, the observed result is consonant with the interpretation advanced.

Let us now suppose that only 4 of the 12 chicks hatched are blue. The deviation is numerically equal to 2. It therefore lies outside the range 6 ± 1. The expectation $(E < 2)$ of a score in the range 6 ± 1 *inclusive*, i.e. a score of 5, 6 or 7 is

$$12_{(5)} \cdot (\tfrac{1}{2})^{12} + 12_{(6)} \cdot (\tfrac{1}{2})^{12} + 12_{(7)} \cdot (\tfrac{1}{2})^{12}$$
$$= \frac{2(792) + 924}{4096} = \frac{2508}{4096}.$$

Thus the expectation $E(> 2)$ that a deviation will *not* lie in the range 6 ± 1 *inclusive* is

$$1 - \frac{2508}{4096} = \frac{1588}{4096}.$$

The odds that a deviation will be as great as 2 are therefore 1588 : 2508 or roughly 2 : 3. Again, we may say that the observed score does not deviate from the long run mean to an extent which entitles it to rank as an uncommon occurrence.

If our hypothesis is that the expectation for a particular genotype in a large sample is $\tfrac{1}{4}$ (simple recessive offspring of two heterozygous parents), the frequencies (Fig. 31) with which we may expect 0, 1, 2 . . . 11, 12 recessives in a sample of 12 accord with successive terms of the expansion $(\tfrac{3}{4} + \tfrac{1}{4})^{12}$. The theoretical mean score is $\tfrac{1}{4}(12) = 3$. The expectation $E(< 2)$ that a deviation will be in the range 3 ± 1 is the expectation (Fig. 31) of a score of 4, 3 or 2, i.e.

$$\frac{12!}{4!\,8!}\,(\tfrac{1}{4})^4 \cdot (\tfrac{3}{4})^8 + \frac{12!}{3!\,9!}\,(\tfrac{1}{4})^3 \cdot (\tfrac{3}{4})^9 + \frac{12!}{2!\,10!}\,(\tfrac{1}{4})^2 \cdot (\tfrac{3}{4})^{10}$$
$$= \frac{10,819,089}{16,777,216} = \frac{9}{14}\ (approximately).$$

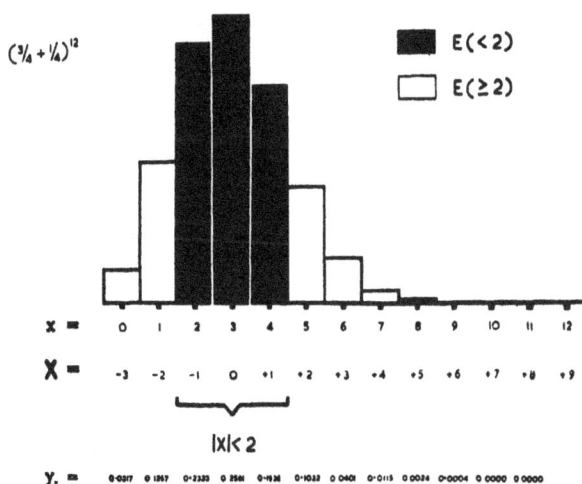

Fig. 31. The expectation of choosing at least 2 and no more than 4 hearts in a sample of 12 from a full pack subject to the replacement condition stated in the legend of Fig. 29.

So the theoretical frequency $E(\geqslant 2)$ of all deviations outside the range 3 ± 1 is approximately

$$1 - (\tfrac{9}{14}) = \tfrac{5}{14}.$$

Thus the odds are $9 : 5$ that we shall *not* get less than 2 or more than 4 recessives in a sample of 12. These are also the odds *against* getting a deviation *as great as* ± 2.

We may generalise these illustrative examples as follows. Our observed score in an r-fold sample is x. This differs by X from the most representative, i.e. *mean* value rp. If we want to know whether such a deviation is so *significant* as to throw doubt on our hypothesis, we ask ourselves whether so great a difference is a common or infrequent occurrence. Now the expectation of getting a difference numerically *as great as* X is the expectation that a deviation will *not* be numerically less than or equal to $(X - 1)$, hence that the score will lie in the range from $(M_x - \overline{X - 1})$ $(rp - X + 1)$ to $(M_x + X - 1) = (rp + X - 1)$ inclusive. The expectation that a deviation will lie within the range $rp \pm (X - 1)$ is given by

$$\sum_{x=(rp - X + 1)}^{x=(rp + X - 1)} r_{(x)} \cdot p^x \, q^{r-x} = E(< X)$$

The expectation that the score will deviate from the mean by an amount numerically as great as X, or even more, is given by $E(\geqslant X) = 1 - E(< X)$

If we know the size of the sample, this method tells us the exact odds that a deviation will not be as great or greater than $X = d$ (the observed discrepancy between observation and the limiting ratio). Calculation can be expedited by recourse to tables of the powers of integers and of binomial coefficients (for example, Warwick (1932), " Probability Tables for Mendelian Ratios," *Texas Agric. Exper. Station Bulletin* 463). When the sample is relatively large, for example, greater than 25, the direct method is very laborious, even if tables are available. The task can then be shortened by the approximate method explained in 3.04 below.

When testing a ratio in this way, we must remember that there is no hard and fast criterion of a *significant* result. The standard we adopt depends on what betting odds we are prepared to take on ; and our justification for doing so will be the theme of Chapter 5. Where we draw the line is partly a matter of taste and temperament, partly of good or bad judgment about the class of situations involved ; but such differences of personal judgment do not affect the value of an observation, if all the relevant data are on record, i.e. if the size of the sample as well as the observed ratio is accessible. Other workers, who might be inclined to take a more or less strenuous view of what odds justify acceptance or rejection of a particular hypothesis, can then draw their own conclusions.

When we dismiss a hypothesis for the reasons discussed above, we have to remember that any genetical hypothesis involves several suppositions. For instance, a $3 : 1$ ratio implies that both phenotypes are equally viable, and that the recognition of neither is appreciably affected by differences in the environment to which they are exposed. If our data force us to reject a $3 : 1$ ratio, they do not necessarily prove that we are *not* dealing with a single gene substitution. They may be consistent with this possibility, if we have not excluded the possibility that lack of homogeneity of the environment or the possibility that early mortality of one or other of the phenotypes contributes to the observed deviations. *Inter alia*, our confidence in rejecting the hypothesis that a single gene substitution is involved will be justified by the extent to which we can exclude such possibilities.

If a trait is determined by 2 *independent* recessive genes, the theoretical frequency of getting 0, 1, 2, . . . r double recessives from a cross between parents heterozygous for both genes is given by successive terms of $(\tfrac{15}{16} + \tfrac{1}{16})^r$.

If 2 dominant genes, which produce no effect except when both are present, determine a trait, the corresponding expression for the frequencies of 0, 1, 2, . . . dominant phenotypes is

$$(\tfrac{7}{16} + \tfrac{9}{16})^r.$$

When 2 or more gene differences are involved or if dominance is incomplete, we have to specify more than 2 phenotypes. If there are 3 phenotypes, we can calculate the frequencies of all possible numerical arrangements by the general term of the trinomial $(u + v + w)^r$, namely

$$\frac{r!}{j!\,k!\,l!}\; u^j\, v^k\, w^l.$$

Thus the theoretical frequency of getting an exact 1 : 2 : 1 ratio among 8 chicks hatched from eggs of a mating between 2 *blue* Andalusians, that is, the chance of getting 2 white, 4 blue, and 2 black would be

$$\frac{8!}{2!\,4!\,2!} \cdot (\tfrac{1}{4})^2 \cdot (\tfrac{1}{2})^4 \cdot (\tfrac{1}{4})^2 = \frac{420}{4096} = \frac{1}{10}\text{ (approx.).}$$

It is important to notice that this is not the expectation of getting 4 blue chicks in a sample of 8. This admits of all the following possibilities :

	Scores		Frequencies
Blue	Black	White	
4	4	0	70 ÷ 4096
4	3	1	280 ÷ 4096
4	2	2	420 ÷ 4096
4	1	3	280 ÷ 4096
4	0	4	70 ÷ 4096

Total 1120 ÷ 4096 = 35 ÷ 128.

Thus the theoretical frequency of a sample of 8 chicks of which 4 are blue is 35/128, which is, of course, identical with the corresponding term of the appropriate binomial $(\tfrac{1}{2} + \tfrac{1}{2})^8$, i.e.

$$8_{(4)} \cdot (\tfrac{1}{2})^8 = \tfrac{35}{128}.$$

If we are merely concerned with the proportion of chicks which are or are not blue, our classification is in fact *binary* and the binomial test given above suffices for the purpose ; but our hypothesis may demand that we give due weight to each of the residual classes. The binomial test then fails us ; and we have to look at the problem of significance in a different way. This will come up for treatment in Vol. II, where we deal with the *Chi Square* test.

EXERCISE 3.02

1. What are the proportionate possibilities of finding in a family of 7 : (a) 3 boys ; (b) 3 boys or 3 girls ; (c) at least 5 boys ; (d) at least 2 girls ; (e) not more than 4 girls ?

2. What is the theoretical frequency of exactly half the offspring of a cross R.H. being recessive, when the size of the sample is (a) 5, (b) 10, (c) 15, (d) 20 ?

3. From 12 eggs of a mating between a black Andalusian cock and a blue (heterozygous) Andalusian hen, 4 black and 8 blue chicks hatch out. What are the odds that a departure from the 1 : 1 ratio will be as great as this ?

4. From 12 eggs of a mating between offspring of a pure white (dominant) Leghorn cock and a black Minorca hen, 2 black and 10 white chicks hatch out. What are the odds against a departure from the 3 : 1 ratio no greater than this ?

5. Among 16 offspring of a mating between roan and red cattle 5 are roan and 11 are red. Compare the odds against a departure as great as this from (a) the 1 : 1 ratio, (b) the 3 : 1 ratio.

6. In the F_2 generation produced by mating black rabbits from parents respectively blue Beveren and chocolate Havana, the phenotypes are black (like the F_1), blue, chocolate, and lilac. On the assumption that the theoretical ratio is 9 : 3 : 3 : 1, what is the theoretical frequency of a litter of 8 containing at least 2 lilac types ?

7. Twelve chicks hatch from eggs produced by a mating of 2 blue Andalusians. What are the odds that : (a) the number of white chicks will be 3 ; (b) the number of blue chicks will be 6 ; (c) there will be 4 black, 4 white, 4 blue ?

8. If the theoretical expectation of white, roan and red calves of roan parentage is 1 : 2 : 1, what is the theoretical frequency of getting 1 white, 3 roan, and 2 red calves in 6 matings between a roan cow and a roan bull ?

9. From a backcross of F_1 walnut to single the long-run expectation is equal numbers of chicks with walnut, rose, pea and single combs. What is the theoretical frequency of getting chicks with combs as follows : 2 walnut, 5 pea, 4 rose, 1 single ?

10. What is the theoretical frequency of getting 3 pea and 2 walnut from 5 eggs of a mating between F_1 walnut and (a) pure pea, (b) heterozygous pea, (c) F_1 walnut, (d) single ?

11. In a litter of 7 produced by two F_1 black rabbits from a blue Beveren and chocolate Havana cross, what is the theoretical frequency of getting 3 black, 2 blue, 1 chocolate and 1 lilac ?

12. In a litter of 8 from the same two parents, what is the theoretical frequency of getting 2 lilacs ?

Answers

1. (a) $\frac{35}{128}$; (b) $\frac{70}{128}$; (c) $\frac{29}{128}$; (d) $\frac{120}{128}$; (e) $\frac{99}{128}$.

2. (a) 0 ; (b) $\frac{63}{256}$; (c) 0 ; (d) 46,189/262,144.

3. 372 : 627.

4. Approximately 53 : 115.

5. (a) Approximately 259 : 64 ; (b) approximately 96 : 334.

6. Approximately $\frac{1}{12}$.

7. (a) 1,082,565 : 3,111,739 or approximately 1 : 3 ; (b) 231 : 793 ; (c) 17,325 : 506,963.

8. $\frac{15}{128}$.

9. 10,395 : 2,097,152 or approximately 0.005.

10. (a) $\frac{1}{16}$; (b) 1,215 : 16,384 ; (c) 10,935 : 524,288 ; (d) $\frac{5}{512}$.

11. Approximately $\frac{1}{33}$.

12. $28 \times 15^6 \div 16^8$.

3.03 The Binomial Histogram

We have now seen how it is possible to test the significance of a discrepancy between the requirements of a hypothesis which postulates certain representative or theoretical frequencies and the results of observation of a sample, when the items involved are reducible to a *binary*

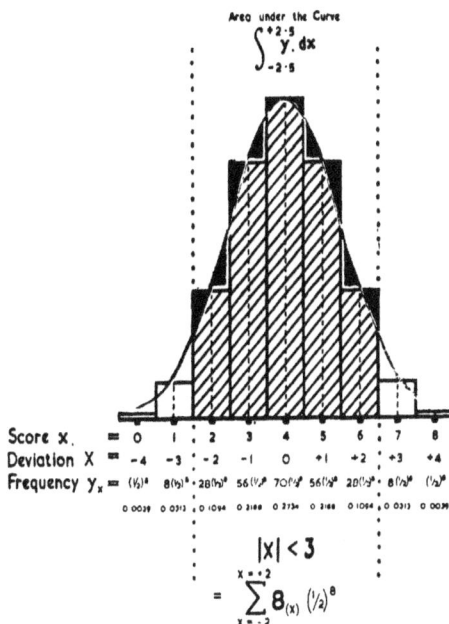

FIG. 32. Approximately we may sum the terms of a binomial within specified limits by regarding the total as an area, the numerical value of which is obtainable by integration with due regard to the half-interval difference exhibited in Fig. 15.

classification. The theoretical frequencies of all possible r-fold samples are then given by successive terms of the expansion of either $(q + p)^r$ or $(qn + pn)^{(r)} \div n^{(r)}$ in accordance with the properties of the model which reproduces the essential features of the hypothesis. The only difficulty involved is the laborious nature of the computation of individual terms and the sum of individual terms within a specified range, when r is large. Our next task will be to explore the possibility of reducing the labour without substantial loss of precision by recourse to the method of 1.10. Statistical theory is, in fact, largely occupied with defining *approximate* distributions which permit us to sidestep the labour of computing an exact result.

Fig. 32, which exhibits the binomial replacement distribution for 8-fold samples of cards classified by colour ($p = \frac{1}{2} = q$) recalls the principle of the method we have elsewhere (p. 55) applied to the summation of an H.P. We may write the equation of the curve alternatively : (a) in the form $y = f(x)$, as a function of the raw score ; (b) in the form $Y = F(X)$, as a function of the score deviation. As a function of the score deviation, the graph of Y is in this case symmetrical about the origin $X = 0$, when $x = M = rp$. Between two values $X = a$ and $X = b$, the area enclosed by the ordinates between the x-axis and the curve is :

$$A = \int_a^b Y . dX.$$

If b and a are situated at d units equidistant from $x = rp$ this is

$$\int_{rp-d}^{rp+d} y . dx \qquad \qquad \text{(i)}$$

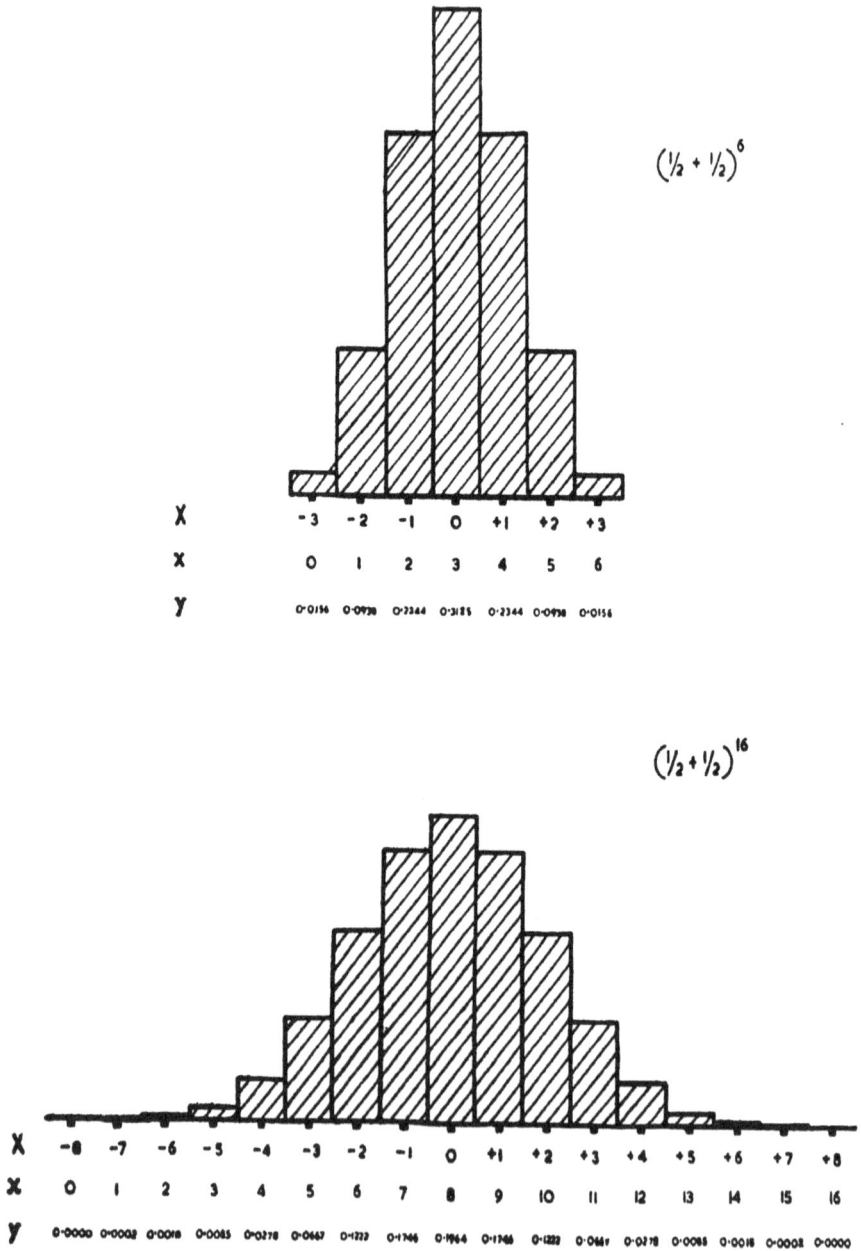

$$\left(\tfrac{1}{2} + \tfrac{1}{2}\right)^{6}$$

X		−3	−2	−1	0	+1	+2	+3
x		0	1	2	3	4	5	6
y		0·0156	0·0938	0·2344	0·3125	0·2344	0·0938	0·0156

$$\left(\tfrac{1}{2} + \tfrac{1}{2}\right)^{16}$$

X	−8	−7	−6	−5	−4	−3	−2	−1	0	+1	+2	+3	+4	+5	+6	+7	+8
x	0	1	2	3	4	5	6	7	8	9	10	11	12	13	14	15	16
y	0·0000	0·0002	0·0018	0·0085	0·0278	0·0667	0·1222	0·1746	0·1964	0·1746	0·1222	0·0667	0·0278	0·0085	0·0018	0·0002	0·0000

FIG. 33. Binomial histograms w.r.t. $(q + p)^r$ are always symmetrical when $p = \tfrac{1}{2} = q$. Otherwise they

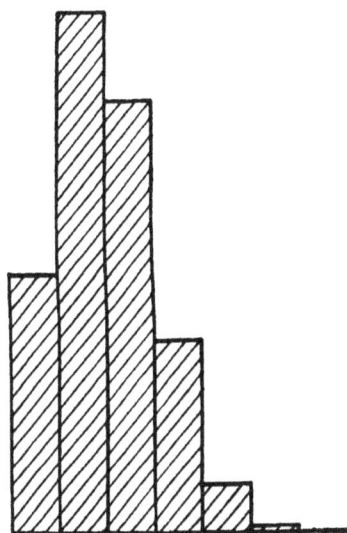

$$\left(\tfrac{3}{4} + \tfrac{1}{4}\right)^{6}$$

X		-1·5	-0·5	+0·5	+1·5	+2·5	+3·5	+4·5
x		0	1	2	3	4	5	6
y		0·1780	0·3560	0·2966	0·1318	0·0330	0·0044	0·0002

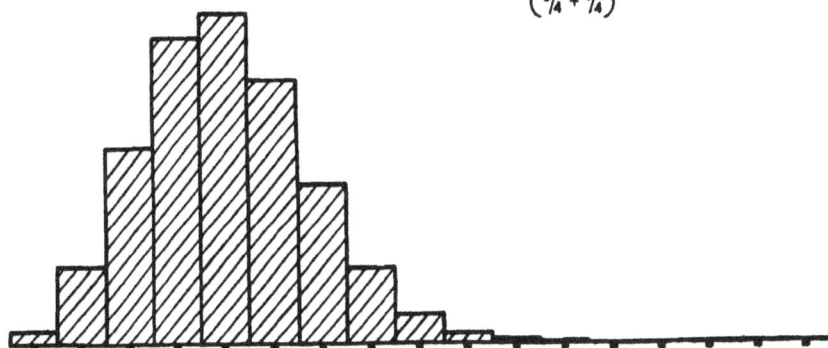

$$\left(\tfrac{3}{4} + \tfrac{1}{4}\right)^{16}$$

X	-4	-3	-2	-1	0	+1	+2	+3	+4	+5	+6	+7	+8	+9	+10	+11	+12
x	0	1	2	3	4	5	6	7	8	9	10	11	12	13	14	15	16
y	0·0100	0·0535	0·1336	0·2079	0·2252	0·1802	0·1101	0·0524	0·0197	0·0058	0·0014	0·0002	0·0000	0·0000	0·0000	0·0000	0·0000

are skew, but the skewness becomes less noticeable for corresponding values of p and q, as r increases.

8

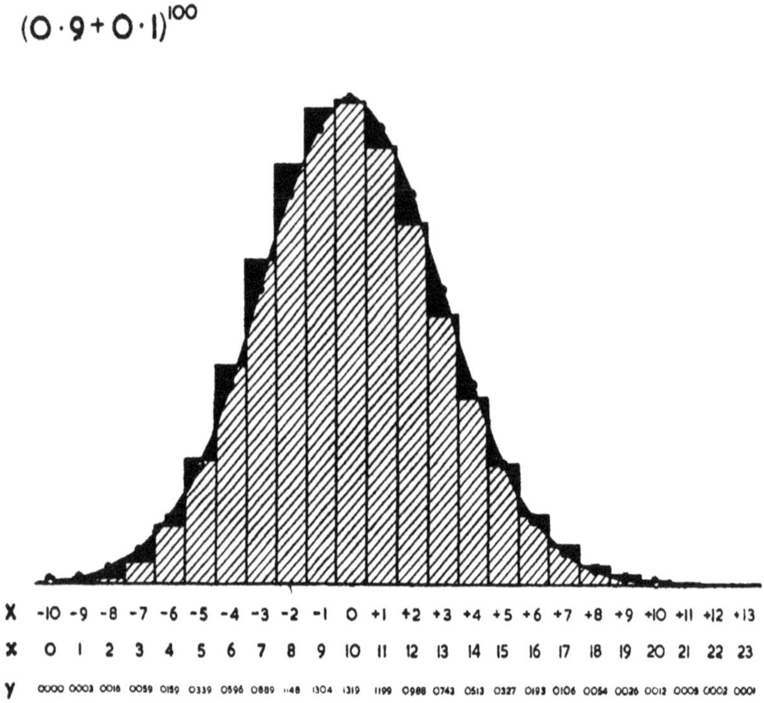

$$(0\cdot9 + 0\cdot1)^{100}$$

X	-10	-9	-8	-7	-6	-5	-4	-3	-2	-1	0	+1	+2	+3	+4	+5	+6	+7	+8	+9	+10	+11	+12	+13
x	0	1	2	3	4	5	6	7	8	9	10	11	12	13	14	15	16	17	18	19	20	21	22	23
y	0000	0003	0016	0059	0159	0339	0596	0889	·48	·304	·319	·199	0988	0742	0513	0327	0193	0106	0054	0026	0012	0009	0002	0001

FIG. 34. The symmetrical normal curve fitted to the skew binomial $(9/10 + 1/10)^{100}$ very closely reproduces the area enclosed between two given ordinates, because : (a) segments excluded on one side of the mean compensate for segments included on the other side of it ; (b) the ordinates of the histogram beyond the range $x = 0$ to $x = 2M$ are of trivial magnitude.

If X is the score deviation, $X = x - M$, $dX = dx$; and $X = d$ when $x = M + d$. If Y is a *symmetrical* function of X, the above is then equivalent to

$$2 . \int_0^d Y . dX \qquad \qquad \qquad \text{(ii)}$$

The exact expectation that a deviation will *not* be as great as d is given by

$$E(<d) \ \cdots \ \sum_{x = M-d+1}^{x = M+d-1} y_x \ . \qquad \qquad \text{(iii)}$$

$$\simeq \int_{x = M-d+\frac{1}{2}}^{x = M+d-\frac{1}{2}} y . dx$$

On the assumption that the distribution is symmetrical the above is equivalent to

$$2 \int_0^{d-\frac{1}{2}} Y . dX . \qquad \qquad \qquad \text{(iv)}$$

$$\therefore E(> d) \simeq 1 - 2 \int_0^{d-\frac{1}{2}} Y . dX . \qquad \qquad \text{(v)}$$

Examination of Figs. 33–34 brings into relief a highly important, and at first surprising, conclusion : as we increase r the distribution of corresponding score deviations becomes less skew for unequal values of p and q. Thus the contour of a distribution such as $(\frac{3}{4} + \frac{1}{4})^r$ differs little from the symmetrical distribution defined by $(\frac{1}{2} + \frac{1}{2})^r$, when r is over 30. By taking the mean raw score $M_x = rp$ as the origin of reference, we therefore ensure that the summation $E(< X)$ of frequencies within a given range tallies closely with the operation of integrating the corresponding curve between *appropriate* limits as explained in 1.10. Our next task will be to find a suitable formula for such a curve.

It will clarify certain necessary approximations, if we here notice one implication of the fact stated in the foregoing paragraph. If we set the origin of the distribution $(\frac{1}{2} + \frac{1}{2})^r$ at $x = rp$, the score deviation (X) range is $\pm rp$. More generally, for the binomial $(q + p)^r$, the range is from $X = - rp$ to $X = + rq$, e.g. from $- 10$ to $+ 90$ in Fig. 34 which refers to the binomial $(0.9 + 0.1)^{100}$. To say that the histogram becomes more symmetrical about the mean raw score value as r becomes very large, thus signifies that the bulk of the area of the histogram lies in the range $X = \pm rp$, when p is less than q. If so, the ratio $(X \div rp)$ *does not exceed unity for appreciably large values of the frequency corresponding to X.*

3.04 THE NORMAL CURVE

To evaluate the significance of a score deviation numerically equivalent to X, we have hitherto made use of the exact expressions :

$$E(\geqslant X) = 1 - E(< X) \qquad \qquad \text{(i)}$$

$$E(< X) = \sum_{x = (rp - X + 1)}^{x = (rp + X - 1)} y_x \qquad \qquad \text{(ii)}$$

$$y_x = \frac{r!}{x!\,(r - x)!}\, p^x \cdot q^{r-x} \qquad \qquad \text{(iii)}$$

The only difficulty we have encountered arises from the fact that the computation of y_x from (iii) is laborious, as is also the exact summation of (ii) if X is large. To sidestep this disability,

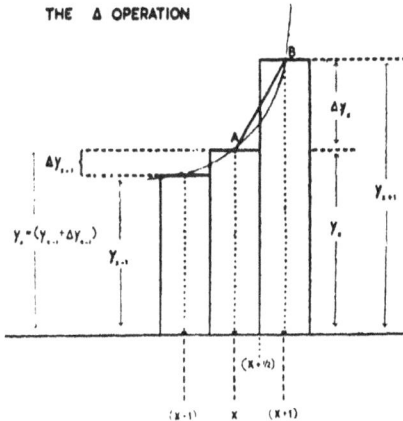

FIG. 35. The chord which connects the upper midpoints of two columns of the histogram corresponds closely to the slope of the smooth curve passing through them in the half interval.

we have now to find a function $y = f(x)$ whose curve passes actually or very nearly through the midpoints of the upper extremity of the histogram of y_x. From a theoretical point of view there are an infinite number of possibilities to choose from ; and the only criteria which need influence our choice of the form of the function $f(x)$ is whether it is : (a) easily calculable ; (b) sufficiently precise for our purpose.

We have already seen (p. 55) that the method of integration provides an approximate method for summation of *discrete* series when the form of the corresponding continuous function is recognisable, as is true of geometric series ; but it is not always easy to recognise a suitable form. In seeking for one we may take advantage of the geometrical properties of the histogram of the binomial series. We first note that $\Delta y_x = (y_{x+1} - y_x)$ is the increment of y_x per unit increase (Δx) of the *raw* score x in the interval between x and $x + 1$ (Fig. 35), i.e.

$$\Delta y_x = \frac{\Delta y_x}{\Delta x}.$$

By definition the differential coefficient dy/dx is the increment of y per unit increment of x in the immediate neighbourhood of the point whose co-ordinates are (x, y). At that point, it is the slope of the tangent drawn from the x-axis to the curve which represents y as a function of x. The difference Δy_x ($= \Delta y_x \div \Delta x$), which is the slope of the chord to the curve in the region x to $x + 1$, is approximately the slope of the tangent to the curve at the point whose co-ordinate is $(x + \frac{1}{2})$. When the number of terms of the series defined by y_x is very large, the mid-points cut by the corresponding curve lie very close and we may neglect the change of curvature in the interval x to $x + \frac{1}{2}$. Hence if Δy_x is a function $F(y_x, x)$ of y_x and/or x, we may put

$$\Delta y_x = F(y_x, x) \quad . \quad . \quad . \quad . \quad . \quad . \quad \text{(iv)}$$

$$\frac{dy}{dx} \simeq F(y, x) \quad . \quad . \quad . \quad . \quad . \quad . \quad \text{(v)}$$

We can often get a better approximation to the value of the differential coefficient of the curve which passes through the mid-points at the upper extremity of the columns of the histogram by defining the *central difference* $\Delta y_{x-\frac{1}{2}}$ (Fig. 36) as the difference between : (a) the mean value $(y_{x-\frac{1}{2}})$ of y_{x-1} and y_x ; (b) the mean value $(y_{x+\frac{1}{2}})$ of y_{x+1} and y_x :

$$\Delta y_{x-\frac{1}{2}} = y_{x+\frac{1}{2}} - y_{x-\frac{1}{2}}$$
$$= \tfrac{1}{2}(y_{x+1} + y_x) - \tfrac{1}{2}(y_x + y_{x-1})$$
$$= \tfrac{1}{2}(y_{x+1} - y_{x-1}) \quad . \quad . \quad . \quad . \quad . \quad \text{(vi)}$$

If $\Delta y_{x-\frac{1}{2}}$ is a function $\phi(y_x, x)$ of y_x and/or x, we then have

$$\frac{dy}{dx} \simeq \phi(y, x) \quad . \quad . \quad . \quad . \quad . \quad . \quad \text{(vii)}$$

If we have a general expression for y_x it is always possible to derive an expression for Δy_x and hence an approximate differential equation based on (v) or (vii). For example, the series of *triangular* numbers (0, 1, 3, 6, 10 . . .) is defined by the relation $y_{x+1} = (x + 1) + y_x$, so that $\Delta y_x = (x + 1)$. In accordance with (v) we put

$$\frac{dy}{dx} \simeq (x + 1)$$

$$\therefore \; y + k \simeq \int (x + 1)dx = \frac{x^2}{2} + x.$$

THE CENTRAL DIFFERENCE

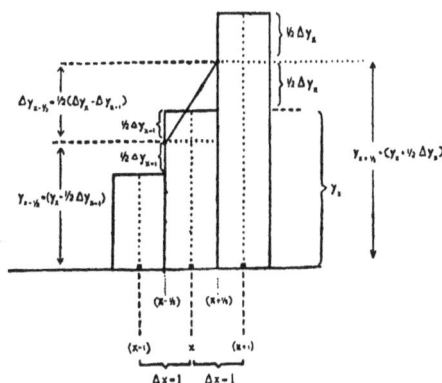

Fig. 36. As a rough and ready measure of the slope of the smooth curve through the upper mid-points of the histogram we may take the difference between the mean values of the ordinates on either side of the selected column.

When $x = 1$, $y_x = y_1 = 1$, so that

$$1 + k = \tfrac{1}{2} + 1.$$

$$\therefore k = \tfrac{1}{2}.$$

$$\therefore y = \frac{x^2}{2} + x - \tfrac{1}{2} = \frac{(x+1)^2}{2} - 1.$$

If $x = 101$, this gives $y = 5201$. The exact formula (p. 16) is $\dfrac{x(x+1)}{2}$, and when

$$x = 101, \quad y_x = 5151.$$

The proportionate error in this region is therefore about 1 per cent. In general, the proportionate error is of the order x^{-1} (i.e. 0.1 per cent. when $x = 1000$).

If we proceed in accordance with (vi) and (vii), we have $y_{x-1} = y_x - x$:

$$\Delta y_{x-\frac{1}{2}} = \tfrac{1}{2}(y_x + x + 1) - \tfrac{1}{2}(y_x - x)$$
$$= x + \tfrac{1}{2}.$$

Accordingly, we put

$$\frac{dy}{dx} \simeq x + \tfrac{1}{2}.$$

$$\therefore y + k = \frac{x^2 + x}{2}.$$

$$1 + k = \frac{1+1}{2}.$$

$$\therefore k = 0.$$

$$\therefore y = \frac{x(x+1)}{2}.$$

In this case, the central difference transformation yields the exact formula ; but it is important to realise that the above are isolated examples of many analogous methods of deriving a differential equation the solution of which is more or less adequate as a computing device. For instance, we might define $\Delta y_{x-\frac{1}{2}}$ as the difference between the *geometric* means of (a) y_{x+1} and y_x ; (b) y_x and y_{x-1}. Whether the form of the function $y = f(x)$ derived by any such method is satisfactory depends on the criteria stated above ; and which method of getting such a function if satisfactory in that sense is preferable depends on whether the parent differential equation admits of a simple solution. In what follows we shall therefore explore more than one method of deriving a descriptive curve for approximate summation of a segment of the binomial histogram. In doing so, we exonerate ourselves from any suspicion of arbitrariness w.r.t. the approximations we invoke ; and shall get a preview of a general method of curve-fitting for sampling distributions dealt with in Chapter 6.

To solve differential equations of the type which commonly arise in this context, it is useful to remember that we have frequently to integrate by the method of partial fractions. Hence it is wise to reduce all expressions involving improper fractions to their *proper* form at the outset and to avoid any simplification of such fractions in subsequent operations. To get Δy_x and $\Delta y_{x-\frac{1}{2}}$ in a suitable form, we therefore proceed to evaluate y_{x+1} and y_{x-1} as follows :

$$y_{x+1} = \frac{r!}{(x+1)!\,(r-x-1)!} \cdot p^{x+1} \cdot q^{r-x-1} = \frac{p(r-x)}{q(x+1)} \cdot \frac{r!}{x!\,(r-x)!} \cdot p^x \, q^{r-x}$$

$$= \frac{p(r-x)}{q(x+1)} \cdot y_x$$

$$= \left[\frac{pr}{q(x+1)} - \frac{px}{q(x+1)} \right] \cdot y_x$$

$$= \left[\frac{pr}{q(x+1)} + \frac{p}{q(x+1)} - \frac{p}{q} \right] \cdot y_x$$

$$\therefore \; y_{x+1} = \left[\frac{p(r+1)}{q(x+1)} - \frac{p}{q} \right] \cdot y_x \quad . \qquad . \qquad . \qquad . \qquad . \qquad . \qquad (viii)$$

$$y_{x-1} = \frac{r!}{(x-1)!\,(r-x+1)!} \cdot p^{x-1} \cdot q^{r-x+1} = \frac{q(x)}{p(r-x+1)} \cdot \frac{r!}{x!(r-x)!} \cdot p^x \, q^{r-x}$$

$$= \frac{qx}{p(r-x+1)} \cdot y_x$$

$$= \frac{q}{p} \left[\frac{r+1}{r-x+1} - 1 \right] \cdot y_x$$

$$= \left[\frac{q(r+1)}{p(r-x+1)} - \frac{q}{p} \right] \cdot y_x \quad . \qquad . \qquad . \qquad . \qquad . \qquad . \qquad (ix)$$

We have now the data to yield an expression for Δy_x and $\Delta y_{x-\frac{1}{2}}$ in accordance with (vi) above. From (viii), by definition,

$$\Delta y_x = y_{x+1} - y_x = \left[\frac{p(r+1)}{q(x+1)} - \frac{p}{q} \right] \cdot y_x - y_x$$

$$= y_x \cdot \left[\frac{p(r+1)}{q(x+1)} - \frac{p}{q} - 1 \right].$$

Since $(p+q) = 1$

$$\Delta y_x = y_x \cdot \left[\frac{p(r+1)}{q(x+1)} - \frac{1}{q} \right] . \qquad . \qquad . \qquad . \qquad (x)$$

By definition also

$$\Delta y_{z-1} = y_z - y_{z-1} = y_z - y_z \left[\frac{q(r+1)}{p(r-x+1)} - \frac{q}{p} \right]$$

$$= y_z \left[\frac{1}{p} - \frac{q(r+1)}{p(r-x+1)} \right] \qquad . \qquad . \qquad . \qquad \text{(xi)}$$

From (x) and (xi) in accordance with (vi)

$$\Delta y_{z-\frac{1}{2}} = \tfrac{1}{2}(\Delta y_z + \Delta y_{z-1})$$

$$= \tfrac{1}{2} y_z \left[\frac{p(r+1)}{q(x+1)} - \frac{q(r+1)}{p(r-x+1)} - \frac{1}{q} + \frac{1}{p} \right]$$

$$\therefore \Delta y_{z-\frac{1}{2}} = \tfrac{1}{2} y_z \left[\frac{p(r+1)}{q(x+1)} - \frac{q(r+1)}{p(r-x+1)} + \frac{q-p}{pq} \right] \qquad . \qquad . \qquad \text{(xii)}$$

The Difference Equation

In accordance with (iv) and (v) we put

$$\frac{dy}{dx} \simeq \left[\frac{p(r+1)}{q(x+1)} - \frac{1}{q} \right] \cdot y.$$

We now transfer the origin to $x = rp = M$, putting $X = x - rp$, so that $x = rp + X$; and $dy/dx = dY/dX$.

$$\therefore \frac{dY}{dX} \simeq \left[\frac{rp+p}{q(rp+1+X)} - \frac{1}{q} \right] \cdot Y.$$

When r is large we may always neglect p in comparison with rp, and if r is also large compared with $1/p$, we can neglect 1 by comparison with rp, putting

$$\frac{1}{Y} \cdot \frac{dY}{dX} \simeq \left[\frac{rp}{q(rp+X)} - \frac{1}{q} \right] = \frac{1}{q\left(1+\dfrac{X}{rp}\right)} - \frac{1}{q}$$

$$\therefore \log Y \simeq \frac{rp}{q} \cdot \log\left(1 + \frac{X}{rp}\right) - \frac{X}{q} + \log K$$

$$\therefore \log\left[Y \div K\left(1 + \frac{X}{rp}\right)^{\frac{rp}{q}} \right] \simeq -\frac{X}{q}$$

$$\therefore Y \simeq K\left(1 + \frac{X}{rp}\right)^{\frac{rp}{q}} \cdot exp\left(\frac{-X}{q}\right) \qquad . \qquad . \qquad . \qquad \text{(xiii)}$$

Over the greater part of the range of the curve of $Y = f(X)$, X will be small compared with rp, particularly if (as assumed) rp is large.* Hence we may try out the approximation (p. 47):

---

* Fig. 34 will help the student to visualise the validity of the assumption that $X \div rp$ is less than unity over the whole region which contributes appreciably to the total area of the histogram for large values of r. Since $q = 0.9$, so that $rp = 10$, X can have any positive value from 0 to 90; but values of Y for positive values of X greater than 10 are too small to represent on the scale of our chart.

$$\log\left(1 + \frac{X}{rp}\right) \simeq \frac{X}{rp} - \frac{X^2}{2r^2p^2}.$$

$$\therefore \log Y - \log K \simeq \frac{rp}{q}\left[\frac{X}{rp} - \frac{X}{2r^2p^2}\right] - \frac{X}{q}.$$

$$\therefore \log\left(\frac{Y}{K}\right) \simeq \frac{-X^2}{2rpq} = \frac{-X^2}{2V}.$$

$$\therefore Y \simeq K \cdot exp\left(\frac{-X^2}{2rpq}\right).$$

The product rpq, called the *variance* of the distribution, will recur frequently in what follows; and we shall represent it consistently by the symbol V. When $X = 0$, $Y = y_m$, the ordinate of the mean score. The foregoing thus reduces to the equation of the so-called *normal curve*

$$Y \simeq y_m \cdot exp\left(\frac{-X^2}{2V}\right) \qquad . \qquad . \qquad . \qquad . \qquad \text{(xiv)}$$

The Central Difference Equation

In accordance with (xii), we may also put

$$\frac{1}{y} \cdot \frac{dy}{dx} \simeq \frac{1}{2}\left[\frac{p(r+1)}{q(x+1)} - \frac{q(r+1)}{p(r-x+1)} + \frac{q-p}{pq}\right].$$

If as before $x = X + rp$, $(r - x + 1) = (r - rp - X + 1) = r(1 - p) - X + 1 = (rq + 1 - X)$, so that:

$$\frac{1}{Y} \cdot \frac{dY}{dX} \simeq \frac{1}{2}\left[\frac{p(r+1)}{q(rp+1+X)} + \frac{q(r+1)}{p(rq+1-X)} + \frac{q-p}{pq}\right].$$

$$\therefore \log\left(\frac{Y}{K}\right) \simeq \frac{p(r+1)}{2q} \cdot \log\left(1 + \frac{X}{rp+1}\right) + \frac{q(r+1)}{2p} \cdot \log\left(1 - \frac{X}{rq+1}\right) + \frac{(q-p)}{2pq} \cdot X \quad \text{(xv)}$$

$$\therefore \frac{Y}{K} \simeq \left[1 + \frac{X}{rp+1}\right]^{\frac{p(r+1)}{2q}} \cdot \left[1 - \frac{X}{rq+1}\right]^{\frac{q(r+1)}{2p}} \cdot e^{\frac{(q-p)X}{2pq}},$$

and as before $K = y_m$.

$$\therefore Y = y_m \cdot \left[\left(1 + \frac{X}{rp+1}\right)^{\frac{p(r+1)}{2q}} \cdot \left(1 - \frac{X}{rq+1}\right)^{\frac{q(r+1)}{2p}} e \cdot \frac{q-p}{2pq} \cdot X\right] \qquad . \qquad . \qquad \text{(xvi)}$$

When $p = \frac{1}{2} = q$, (xvi) reduces to

$$Y \simeq y_m\left[1 - \left(\frac{2X}{r+2}\right)^2\right]^{\frac{1}{2}(r+1)} \qquad . \qquad . \qquad . \qquad . \qquad \text{(xvii)}$$

If r is *large*, we may write this as

$$Y \simeq y_m \cdot \left[1 - \left(\frac{2X}{r}\right)^2\right]^{\frac{1}{2}r} \qquad . \qquad . \qquad . \qquad . \qquad \text{(xviii)}$$

If r is very large, we may use the approximation (p. 46)

$$(1 - a)^m \simeq e^{-am}.$$

$$\therefore Y \simeq y_m \cdot exp\left[\frac{-2X^2}{r}\right].$$

Since $V = rpq = \frac{1}{4}r$, when $p = \frac{1}{2} = q$, the above is equivalent to (xiv), i.e.

$$Y = y_m \cdot exp \left(\frac{-X^2}{2V} \right).$$

If r is large compared with $1/p$, we may write (xv) as

$$\log \left(\frac{Y}{K} \right) \simeq \frac{rp}{2q} \cdot \log \left(1 + \frac{X}{rp} \right) - \frac{rq}{2p} \cdot \log \left(1 - \frac{X}{rq} \right) + \frac{q-p}{2pq} X.$$

We can then employ the approximations :

$$\log \left(1 + \frac{X}{rp} \right) \simeq \frac{X}{rp} - \frac{X^2}{2r^2p^2},$$

$$\log \left(1 - \frac{X}{rq} \right) \simeq - \frac{X}{rq} - \frac{X^2}{2r^2q^2}.$$

$$\therefore \; \log \left(\frac{Y}{K} \right) \simeq \frac{rp}{2q} \left[\frac{X}{rp} - \frac{X^2}{2r^2p^2} \right] + \frac{rq}{2p} \left[\frac{-X}{rq} - \frac{X^2}{2r^2q^2} \right] + \frac{(q-p)}{2pq} \cdot X$$

$$\simeq \frac{-X^2}{2rpq} = \frac{-X^2}{2V}.$$

$$\therefore \; Y \simeq K \cdot exp \left(\frac{-X^2}{2V} \right).$$

and as before $K = y_m$, so that

$$Y \simeq y_m \cdot exp \left(\frac{-X^2}{2V} \right) \qquad \cdot \qquad \cdot \qquad \cdot \qquad \cdot \qquad \text{(xix)}$$

This again is the equation of the *normal curve*. The constant y_m in (xiv) and (xix) is the value of y_x corresponding to $X = 0$, $x = M_x = rp$ and $(r - x) = (r - rp) = r(1 - p) = rq$,

$$\therefore \; y_m = \frac{r!}{(rp)! \, (rq)!} \cdot p^{rp} \, q^{rq}.$$

Accordingly, by (viii) in 1.09,

$$y_m \simeq \frac{1}{(2\pi rpq)^{\frac{1}{2}}} = \frac{1}{(2\pi V)^{\frac{1}{2}}}.$$

We may thus write (xiv) or (xix) in the form

$$Y \simeq \frac{1}{(2\pi V)^{\frac{1}{2}}} \cdot exp \left(\frac{-X^2}{2V} \right) \qquad \cdot \qquad \cdot \qquad \cdot \qquad \cdot \qquad \text{(xx)}$$

The constant $V = rpq$ has a special significance w.r.t. the geometrical properties of the normal curve. If we differentiate once we get

$$\frac{dY}{dX} \simeq \frac{-1}{V \sqrt{2\pi V}} \cdot X \, exp \left(\frac{-X^2}{2V} \right).$$

When $\quad \dfrac{dY}{dX} = 0$, $X = 0$ or $exp \left(\dfrac{-X^2}{2V} \right) = 0$, so that $X = \pm \infty$.

That is to say, the curve has a turning point at $X = 0$ and extends asymptotically to the X-axis at infinity in both directions. On differentiating a second time, we get

$$\frac{d^2Y}{dX^2} \simeq \frac{-1}{V\sqrt{2\pi V}} \cdot \left[exp \left(- X^2 \div 2V \right) - \frac{X^2}{V} \cdot exp \left(- X^2 \div 2V \right) \right].$$

When we equate this to zero, we get

$$1 - \frac{X^2}{V} = 0,$$

$$\therefore X = \pm \sqrt{V}.$$

Thus the normal curve has points of inflection symmetrically distant $\pm \sqrt{V}$ units from the origin. It is customary to represent $\sqrt{V}$ by the Greek small *sigma* σ (i.e. $V = \sigma^2$). We call the quantity so defined the *standard deviation* of the distribution, being therefore given by

$$\sigma^2 = rpq \quad . \qquad . \qquad . \qquad . \qquad . \qquad . \qquad \text{(xxi)}$$

The usefulness of the normal curve as a computing device depends on how far it can give a good approximation to the sum of the terms of a binomial expansion in any specified range. By definition, the sum of all the individual terms of such an expansion is unity. It is therefore necessary to satisfy ourselves that the entire area of the normal curve is also unity, i.e.

$$2 \int_0^\infty Y \, . \, dX = 1.$$

To do so, we have now to investigate the value of the integral

$$\frac{2}{(2\pi V)^{\frac{1}{2}}} \int_0^\infty exp\left(\frac{-X^2}{2V}\right) . \, dX = \frac{2}{\pi^{\frac{1}{2}}} \int_0^\infty exp\left(\frac{-X^2}{2V}\right) . \, d\left(\frac{X}{\sqrt{2V}}\right)$$

$$= \frac{2}{\pi^{\frac{1}{2}}} \int_0^\infty e^{-a^2} . \, da = \frac{2}{\pi^{\frac{1}{2}}} . \, I.$$

$$\therefore \frac{2}{(2\pi V)^{\frac{1}{2}}} \int_0^\infty exp\left(\frac{-X^2}{2V}\right) . \, dX = \frac{2}{\pi^{\frac{1}{2}}} I \quad . \qquad . \qquad . \qquad . \qquad . \qquad \text{(xxii)}$$

Since I is a *definite* integral it is immaterial whether we write it in either form below

$$I = \int_0^\infty e^{-a^2} . \, da = \int_0^\infty e^{-a^2 x^2} . \, d(ax).$$

$$\therefore I = \int_0^\infty e^{-a^2 x^2} . \, a . \, dx.$$

$$\therefore I . e^{-a^2} . \, da = \int_0^\infty e^{-a^2(1+x^2)} . \, a . \, da . \, dx$$

$$= \frac{1}{2} \int_0^\infty e^{-a^2(1+x^2)} . \, d(a^2) . \, dx$$

$$= \frac{1}{2} \int_0^\infty e^{-(1+x^2)k} . \, dk . \, dx,$$

$$\therefore I \int_0^\infty e^{-a^2} \, da = \frac{1}{2} \int_0^\infty \left\{ \int_0^\infty e^{-(1+x^2)k} . \, dk \right\} . \, dx,$$

$$\therefore I^2 = \frac{1}{2} \int_0^\infty \left[\frac{-1}{1+x^2} . \, e^{-(1+x^2)k} \right]_0^\infty . \, dx$$

$$= \frac{1}{2} \int_0^\infty \frac{dx}{1+x^2} .$$

The integral above is a standard form, its value being :

$$\left[\tan^{-1} x \right]_0^\infty .$$

$$\therefore I^2 = \frac{\pi}{4}, \qquad\qquad \therefore I = \frac{\sqrt{\pi}}{2}, \qquad\qquad \therefore \frac{2}{\pi^{\frac{1}{2}}} I = 1.$$

Hence from (xxii)

$$\frac{2}{(2\pi V)^{\frac{1}{2}}} \int_0^\infty exp\left(\frac{-X^2}{2V} \right) . dX = 1 \qquad . \qquad . \qquad . \qquad . \qquad \text{(xxiii)}$$

The normal curve, or, strictly speaking, the family of normal curves individually specified by a particular numerical value of V or σ^2, is thus the limit to which the contour of the binomial histogram attains when r is indefinitely large. So far we have no guarantee of its numerical reliability for any assigned *finite* value of r ; and our best way of getting a clear idea of its uses or abuses as a computing tool is to examine what correspondence exists between Y-values calculated : (a) directly from the expansion of the binomial itself ; (b) from (xx) above. Fortunately, it is possible to curtail the work involved by recourse to statistical tables.

Equation (xx) expresses a relation which sidesteps the laborious calculations involving factorials of large numbers, being suitable for logarithmic computation as it stands. It has the additional advantage that it is possible to evaluate the integral between any assigned limits with a view to approximate specification of the area of a corresponding segment of the binomial histogram. To be able to read off directly the ordinate Y corresponding to any particular value of X in any one of a class of functions $Y = A.f(BX)$, it is not necessary to have a library of 2-way tables constructed for all possible values of A and B. For it calls for scarcely any additional effort to extract the required result from a single 2-way table of the function $y_c = f(c)$, identical in form with the foregoing when $A = 1 = B$. To use such a table in order to find the numerical value of Y corresponding to a particular numerical value of X, we first read off the value of Y corresponding to $c = BX$. The required result merely calls for a second act of multiplication, *viz.* $Y = Ay_c$.

Equation (xx) involves the constant V, of which the value ($= rpq$) depends both on the size (r) of the sample and the parameter p characteristic of the universe. As it stands, it is therefore unsuitable for 2-way tabulation ; but it is possible to shorten the labour of calculating the approximate value of Y corresponding to a particular value of $X = (x - rp)$ for any distribution by a simple scalar transformation, as indicated above. A single table of (xx) for the particular case which arises when $V = 1$ then suffices for the computation. Such a table cites corresponding values of X and Y for the normal curve of *unit variance*, i.e.

$$Y = \frac{1}{\sqrt{2\pi}} . e^{-\frac{1}{2}x^2}.$$

To remind ourselves that this is the equation of a particular class of normal distributions whose common characteristic is that $rpq = 1$, *i.e.* $r^{-1} = p(1 - p)$, we may write it in the form

$$y_c = \frac{1}{\sqrt{2\pi}} . e^{-\frac{1}{2}c^2} \qquad . \qquad . \qquad . \qquad . \qquad \text{(xxiv)}$$

We now substitute in (xx) above $X = c\sqrt{V} = c\sigma$, so that

$$c = \frac{X}{\sigma} \qquad . \qquad . \qquad . \qquad . \qquad \text{(xxv)}$$

Accordingly (xx) becomes

$$Y = \frac{1}{\sqrt{2\pi V}} . e^{-\frac{1}{2}r^2}.$$

This is equivalent to (xxiv) if we write

$$Y = \frac{y_o}{\sqrt{V}} = \frac{y_o}{\sigma}.$$

$$y_o = \sigma Y \qquad \qquad . \qquad . \qquad . \qquad . \qquad . \qquad \text{(xxvi)}$$

Below is a condensed table of the *Frequency Function* (xxiv) of the normal distribution

c $(= X \div \sigma)$.	y_o $(= \sigma Y)$.	c $(= X \div \sigma)$.	y_o $(= \sigma Y)$.
0·0	0·3989	2·0	0·0540
0·5	0·3521	2·5	0·0175
1·0	0·2420	3·0	0·0444
1·5	0·1295	3·5	0·0009

A single example suffices to illustrate the use of a frequency function table. Let us suppose that we wish to extract an approximate value for the ordinate y of the binomial $(\frac{1}{5} + \frac{4}{5})^{25}$ corresponding to a score $x = 17$. The mean value of the score is

$$rp = 25(0\cdot8) = 20$$
$$X = 17 - 20 = -3.$$

The standard deviation is given by

$$\sqrt{25(0\cdot8)(0\cdot2)} = \sqrt{4}.$$
$$\therefore \sigma = 2.$$

In accordance with (xxiv) we therefore put

$$c = (X \div \sigma) = -(3 \div 2) = -1\cdot5.$$

Against $c = 1\cdot5$, the table shows $Y = 0\cdot1295$. Hence from (xxvi) above

$$y_o = \sigma Y = 0\cdot1295$$
$$2Y = 0\cdot1295$$
$$Y = 0\cdot0648$$

The exact frequency by direct calculation is given by

$$\frac{25!}{17!\,8!} \cdot \left(\tfrac{4}{5}\right)^{17} \cdot \left(\tfrac{1}{5}\right)^{8} = 0\cdot0622.$$

The proportionate error in this case is $(0\cdot0648 - 0\cdot0622) \div 0\cdot0622 = (0\cdot0026 \div 0\cdot0622)$, i.e. about 4 per cent. In this case the error is largely attributable to the fact that the binomial distribution itself is *skew*. For a positive score deviation of 3, the exact frequency is

$$\frac{25!}{17!\,8!} \cdot \left(\tfrac{4}{5}\right)^{23} \cdot \left(\tfrac{1}{5}\right)^{2} = 0\cdot0706.$$

For a negative deviation of equivalent magnitude the exact frequency is, as above,

$$\frac{25!}{17!\,8!} \cdot \left(\tfrac{4}{5}\right)^{17} \cdot \left(\tfrac{1}{5}\right)^{8} = 0\cdot0622.$$

Thus the mean value of Y for a deviation of ± 3 is $\frac{1}{2}(0\cdot0622 + 0\cdot0706) = 0\cdot0664$. This mean value of Y differs from that obtained from the table of the *symmetrical* normal distribution for both $X = +3$ and $X = -3$ by less than 2 per cent. Whether the normal curve gives a more or less good fit to a binomial distribution $(q + p)^r$ depends partly on how large r is, and partly on whether p is large in comparison with q or *vice versa*.

The accompanying tables (2-4) exhibit

(a) the exact values of y for the distribution $(\frac{1}{2} + \frac{1}{2})^{16}$ side by side with corresponding figures extracted from the table of the normal curve and with approximate values based on the central difference formula (xvi);

(b) the exact values side by side with the corresponding normal values of $(\frac{1}{2} + \frac{1}{2})^{25}$ and $(\frac{1}{5} + \frac{4}{5})^{25}$;

(c) the exact values and normal table approximations for $(0\cdot9 + 0\cdot1)^{100}$.

TABLE 2

$\pm X.$	$16_{(x)}(\frac{1}{2})^{16}.$	C.D. Equation.	$(2\pi V)^{-\frac{1}{2}} exp\,(-X^2 \div V).$
0	0·1964	0·1964	0·1994
1	0·1746	0·1767	0·1760
2	0·1222	0·1277	0·1210
3	0·0667	0·0720	0·0647
4	0·0278	0·0302	0·0270
5	0·0085	0·0085	0·0088
6	0·0018	0·0013	0·0022
7	0·0002	0·0007	0·0004
8	0·0000	0·0000	0·0001

TABLE 3

	$(\frac{1}{2} + \frac{1}{2})^{25}.$				$(\frac{1}{5} + \frac{4}{5})^{25}.$		
	Frequency.				Frequency.		
$\pm X.$	$X \div \sigma.$	Exact Value $25_{(x)} . (\frac{1}{2})^{25}.$	Normal Curve.	$X.$	$(X \div \sigma).$	Mean Value $25_{(x)}(\frac{1}{5})^x(\frac{4}{5})^{25-x}.$	Normal Curve.
0·50	0·20	0·1555	0·1596	0·0	0·00	0·1954	0·1994
1·50	0·60	0·1333	0·1333	1·00	0·50	0·1745	0·1760
2·50	1·00	0·0977	0·0968	2·00	1·00	0·1229	0·1210
3·50	1·40	0·0611	0·0599	3·00	1·50	0·0664	0·0675
4·50	1·80	0·0323	0·0316	4·00	2·00	0·0265	0·0270
5·50	2·20	0·0144	0·0142	5·00	2·50	0·0078	0·0088
6·50	2·60	0·0053	0·0054				
7·50	3·00	0·0016	0·0017				
8·50	3·40	0·0004	0·0005				

TABLE 4

The Expansion of $(0.9 + 0.1)^{100}$

Terms whose frequencies exceed 0.0001

(a) Actual frequencies by evaluation of $100_{(x)}(0.1)^x \cdot (0.9)^{1-x}$

$x.$	$X.$	$Y_x.$	$x.$	$X.$	$Y_x.$
0	-10	0·0000	12	2	0·0988
1	-9	0·0003	13	3	0·0743
2	-8	0·0016	14	4	0·0513
3	-7	0·0059	15	5	0·0327
4	-6	0·0159	16	6	0·0193
5	-5	0·0339	17	7	0·0106
6	-4	0·0596	18	8	0·0054
7	-3	0·0889	19	9	0·0026
8	-2	0·1148	20	10	0·0012
9	-1	0·1304	21	11	0·0005
10	0	0·1319	22	12	0·0002
11	1	0·1199	23	13	0·0001

(b) Comparison with the Normal Distribution

$\pm X.$	Mean Binomial Value of $Y_x.$	Ordinate of the Normal Curve ($V = 3$).	$X \div \sqrt{V}.$
0	0·1319	0·1330	0·0
1	0·1251	0·1263	0·3
2	0·1068	0·1066	0·6
3	0·0816	0·0807	1·0
4	0·0554	0·0547	1·3
5	0·0333	0·0333	1·6
6	0·0176	0·0180	2·0
7	0·0083	0·0087	2·3
8	0·0035	0·0038	2·6
9	0·0015	0·0015	3·0
10	0·0006	0·0006	3·3

From a study of these and of other distributions of the same sort we see that two circumstances determine the goodness of fit : (a) the size (r) of the sample, (b) the skewness of the distribution, i.e. the ratio of p to q. For very unequal values of p and q, a good fit is obtainable only if r is relatively large. To state this imposes the obligation of defining a satisfactory numerical standard of *largeness* in this context ; and we shall examine the issue more fully in Chapter 5. Meanwhile, the student should be forewarned and forearmed against lightly assuming that the methods of 3.05 and of Chapter 4 are applicable to vital statistics of comparatively rare conditions, such as cancer death rates.

3.05 THE PROBABILITY INTEGRAL

Testing the significance of a hypothesis which involves a binary classification invokes a process of summation which we can represent by an area, hence approximately by the operation of integration in conformity with (viii) in 3.03, *viz.* :

$$E(\geqslant d) \simeq 1 - 2 \int_0^{d-\frac{1}{2}} Y \, . \, dX.$$

We may now write this as

$$E(\geqslant d) \simeq 1 - \frac{2}{\sqrt{2\pi V}} \int_0^{d-\frac{1}{2}} e^{-\frac{X^2}{2V}} \, . \, dX \qquad \qquad \text{(i)}$$

If $c^2 = (X^2 \div V)$, so that $c = X \div \sigma$,

$$\frac{dc}{dX} = \frac{1}{\sigma}.$$

$$\therefore \, dX = \sigma dc.$$

Also $c = (d - \frac{1}{2}) \div \sigma$ when $X = (d - \frac{1}{2})$, so that

$$E(\geqslant d) \simeq 1 - \frac{2\sigma}{\sqrt{2\pi V}} \int_0^{\frac{d-\frac{1}{2}}{\sigma}} e^{-\frac{1}{2}c^2} \, dc$$

$$\simeq 1 - \frac{2}{\sqrt{2\pi}} \int_0^{\frac{2d-1}{2\sigma}} e^{-\frac{1}{2}c^2} \, dc \qquad \qquad \text{(ii)}$$

The integrals of (i) and (ii) can be expanded as series suitable for numerical evaluation between specified limits. Whereas we can tabulate the definite integral in (i) only for particular values of V and hence of p and r in the binomial $(p + q)^r$, a single table suffices for the evaluation of the integral in (ii). Such tables, variously referred to as tables of the normal *distribution* function, the *probability integral* or *error function* supply requisite information for evaluating $E(\geqslant d)$ in various ways of which the most recent cites the distribution function defined by

$$F(h) = \frac{1}{2} + \frac{1}{(2\pi)^{\frac{1}{2}}} \int_0^h e^{-\frac{1}{2}c^2} \, dc \qquad \qquad \text{(iii)}$$

To use any table of this sort we have to make the substitution in accordance with (iv) of 3.03

$$h = \frac{d - \frac{1}{2}}{\sigma} \, . \qquad \qquad \text{(iv)}$$

The required value of $E(\geqslant d)$ is then given by the relation

$$E(\geqslant d) = 2 - 2F(h) \qquad \qquad \text{(v)}$$

Tables of the distribution function give the following values :

h	$F(h)$	$2 - 2F(h)$
0·675	0·7500	0·5000
1·000	0·8413	0·3172
1·500	0·9332	0·1332
2·000	0·9773	0·0454
2·500	0·9938	0·0124
3·000	0·9987	0·0026
3·500	0·9998	0·0005

To perform what we shall henceforth call a *c*-test of significance we make use of (iv) and (v) as follows. Let us suppose that a cross of hybrid yellow and green peas yields 140 green and 116 yellow seeded progeny, and that we wish to decide whether the result is consistent with the assumption that such a cross would yield equal numbers in the long run. The appropriate binomial is $(\frac{1}{2} + \frac{1}{2})^{256}$, whence

$$V = 256 \times \tfrac{1}{2} \times \tfrac{1}{2} = 64.$$
$$\therefore \sigma = 8.$$

The theoretical expectation of either class is 128 and the observed deviation (d) is ± 12, so that

$$h = \frac{12 - \frac{1}{2}}{8} = 1.44.$$

Tables of $F(h)$ give $F(1.44) \simeq 0.93$, whence $E(> 12) \simeq 2 - 2(0.93) = 0.14$. This means that the odds are 86 : 14 or about 6 to 1 against a discrepancy as large as 12.

The usual procedure is to neglect the refinement of (iv) in favour of the approximation

$$h \simeq \frac{d}{\sigma}.$$

In this case we should then put $h = 1.5$ and for $E(> 12)$ from the foregoing table we obtain the value 0.133. As an indication of the order of significance involved this is good enough ; but it is important to realise that the error involved by neglect of the half interval in setting the correct limits of integration may be large, if the sample is small. The following numerical illustrations are instructive. Column (a) shows the value of $E(> d)$ computed by means of (iv) from tables of $F(h)$, and column (c) shows the value computed by the more usual procedure which neglects the half interval refinement. The middle column (b) shows the exact value of $E(> d)$ based on summation of the terms of the appropriate binomial.*

TABLE 5

(a) Distribution of $(\frac{1}{2} + \frac{1}{2})^{16}$.

d.	$\dfrac{d}{\sigma}$.	$\dfrac{d - \frac{1}{2}}{\sigma}$.	(a).	(b).	(c).
1	0.5	0.25	0.8020	0.8036	0.6171
2	1.0	0.75	0.4538	0.4544	0.3173
3	1.5	1.25	0.2113	0.2100	0.1336
4	2.0	1.75	0.0801	0.0766	0.0455
5	2.5	2.25	0.0246	0.0210	0.0124
6	3.0	2.75	0.0060	0.0040	0.0027

(b) Distribution of $(\frac{4}{8} + \frac{4}{8})^{25}$.

d.	$\dfrac{d}{\sigma}$.	$\dfrac{d - \frac{1}{2}}{\sigma}$.	(a).	(b).	(c).
1	0.5	0.25	0.8020	0.8046	0.6171
2	1.0	0.75	0.4538	0.4556	0.3173
3	1.5	1.25	0.2113	0.2098	0.1336
4	2.0	1.75	0.0801	0.0770	0.0455
5	2.5	2.25	0.0246	0.0241	0.0124

* It is useful to have a name for the ratio $c = (X \div \sigma)$ Statistical writers refer to it variously as the *critical ratio* or as the *standard score*.

EXERCISE 3.05

1. Examine Mendel's original data with regard to offspring of hybrid parents, as given below, for significant departures from the 3 : 1 ratio :

Form of Seed	5474 *round*	1850 *wrinkled*
Colour of Seed	6022 *yellow*	2001 *green*
Colour of Unripe Pods	428 *green*	152 *yellow*
Length of Stem	787 *tall*	277 *dwarf.*

2. Examine the following results of workers who repeated the second set of Mendel's experiments cited above :

Investigator	Yellow Seeds	Green Seeds
Correns	1,394	453
Tschermak	3,580	1,190
Hurst	1,310	445
Bateson	11,902	3,903
Lock	1,438	514
Darbishire	109,090	36,186
TOTALS	128,714	42,691

3. Hybrids of pure *Blue* (coat colour) and *Silver-Fawn* mice are blue. A cross between the blue hybrids give the following results :

Blue 46   *Silver-Fawn* 17

Is this result consistent with the 3 : 1 ratio ?

(Bateson : *Mendel's Principles of Heredity.*)

4. Investigate the following result : *Black* hybrids of pure *Black* and *Chocolate* mice crossed *inter se* gave : *Black* 76 ; *Chocolate* 24.

5. Examine the segregation of each pair of allelic genes with regard to the 3 : 1 ratio, in the following crosses between mice :

(i) *Black* (BbDd) hybrids of pure *Blue* (BBdd) and pure *Chocolate* (coat colour) crossed *inter se* gave the following progeny :

Black (BD)	*Blue* (Bd)	*Chocolate* (bD)	*Silver-Fawn* (bd)
44	17	17	5

(ii) *Black hybrids* of pure *Black* (BBDD) and *Silver-Fawn* (bbdd) crossed *inter se* gave the following progeny :

Black	*Blue*	*Chocolate*	*Silver-Fawn*
67	21	20	5

(Bateson : *Mendel's Principles of Heredity.*)

9

6. (i) *Walnut Comb* hybrids of a cross between fowls with *Rose* (RRpp) and *Pea* (rrPP) combs were mated *inter se* to give the following results :

Walnut (RP)	*Pea* (rP)	*Rose* (Rp)	*Single* (rp)
279	132	99	45

Is this result consistent with the 3 : 1 ratio for the segregation of each of the allelomorphic pairs (P-p and R-r) ?

(ii) The *Walnut* hybrids obtained in the initial cross above were mated back to the double recessive (single comb) giving

Walnut	*Pea*	*Rose*	*Single*
664	705	664	716

Examine these results for significant departures from the 1 : 1 ratio with regard to each pair of alleles.

(Bateson's results quoted in Babcock and Clausen : *Genetics in Relation to Agriculture.*)

7. Test the following results of mating coloured hybrids between two *Cream-white* stocks (*Matthiola*) for a 9 : 7 ratio of coloured to *Cream-white* :

Purple 30 ; Red 8 ; Cream-white 24.

8. In another experiment the coloured progeny of a cross between two *Cream-white* Sweet Peas (*Lathyrus*) were crossed *inter se* to give

Purple 1634 ; Red 498 ; White 1593.

Are these results consonant with the expected 9 : 7 ratio ?

(Bateson : *Mendel's Principles of Heredity.*)

9. In the same experiment the pollen characteristics *Long* and *Round* were also investigated from the following figures :

Long 2844   Round 881

Are these results compatible with the 3 : 1 ratio ?

(*Ibid.*)

10. Sweet Peas with *light* leaf axils were crossed with plants having *dark* ones. The hybrids crossed *inter se* gave

Dark 654   Light 231

Examine these results for significant departures from the 3 : 1 ratio.

(*Ibid.*)

11. The *purple-starchy* hybrid from a cross between a *purple-sweet* (PPss) and a *white-starchy* (ppSS) maize was selfed and the following types of grains obtained :

Purple-starchy	*Purple-sweet*	*White-starchy*	*White-sweet*
1861	614	584	217

Does the result conform with a theoretical 3 : 1 ratio for either or both of the two pairs of factors ?
(East and Hayes, quoted in Babcock and Clausen : *Genetics in Relation to Agriculture.*)

12. In a number of experiments with *Drosophila*, the female offspring of crosses *Purple-Vestigial* male to *Wild Type* female were back-crossed to *Purple-Vestigial* males and the following results were obtained. Investigate separately the validity of the 1 : 1 hypothesis for non-crossover and for crossover phenotypes.

No.	Non-crossovers.		Crossovers.	
	Purple-Vestigial.	Wild Type.	Purple.	Vestigial.
1	178	202	16	16
2	152	227	13	14
3	91	100	18	13
4	69	104	12	8
5	165	150	17	19
6	191	216	18	17
7	140	149	20	15
8	116	122	9	4

(Bridges and Morgan : *The Second Chromosome of Drosophila.*)

13. In another experiment *Purple* males were crossed to *Wild Type* females, and the offspring mated *inter se*. From the results given below, test : (*a*) Mendel's 3 : 1 ratio for offspring of the hybrids ; (*b*) the equality of the sexes among each phenotype.

Wild Type.		Purple.	
Male.	Female.	Male.	Female.
81	18	32	35
54	47	40	33

(*Ibid.*)

14. The offspring of the cross *Purple-Vestigial* male and *Wild Type* female, when back-crossed to *Purple-Vestigial* females yielded the following progeny. There was no crossing over.

Purple-Vestigial.	Wild Type.
62	52
113	141
131	96
34	28
89	68
33	22
90	112

Does our record of any one experiment disclose significant deviation from equality of the two classes ?

(*Ibid.*)

16. The *Wild Type* female progeny of the cross between *Purple* male and *Vestigial* female were back-crossed to *Purple-Vestigial* males with offspring as exhibited below :

Non-crossovers.		Crossovers.	
Purple.	Vestigial.	Purple Vestigial.	Wild Type.
157	178	26	21
200	165	12	14
198	176	23	23
242	195	19	26
252	227	34	38
198	178	26	20

Detect significant deviations, if any, from the 1 : 1 ratio : (*a*) among crossover phenotypes ; (*b*) among non-crossovers.

(Ibid.)

3.06   THE MEANING OF VARIANCE

In deriving an expression for computing an approximate value of $E(\geqslant d)$, we have encountered a statistical constant (V) of great importance. So far we have defined the variance (V) of a distribution and its square root called the standard deviation (σ) by the relation

$$V = \sigma^2 = rpq \qquad \qquad \text{(i)}$$

This specifies V with reference to a distribution of which the frequencies are successive terms of the binomial $(p + q)^r$. An alternative definition of V as the weighted *mean square deviation* is consistent with (i) and is calculable for any distribution in conformity with our convention that y is the frequency of a particular score x, i.e.

$$V = \Sigma y(x - M_x)^2 = \Sigma y \cdot X^2 \qquad \qquad \text{(ii)}$$

To show that (i) and (ii) are consistent, it is first necessary to show the relation of V to the *mean square score* or *second zero moment* (V_0) and to the mean score (M_x), *viz.*

$$V = \Sigma y \cdot (x - M_x)^2.$$
$$= \Sigma y(x^2 - 2M_x x + M_x^2).$$
$$= \Sigma y \cdot x^2 - 2M_x \Sigma y \cdot x + M_x^2 \Sigma y.$$

Since $V_0 = \Sigma y \cdot x^2$ and $M_x = \Sigma y \cdot x$,

$$V = V_0 - 2M_x^2 + M_x^2 \Sigma y.$$

By definition $\Sigma y = 1$, so that

$$V = V_0 - M_x^2 \qquad \qquad \text{(iii)}$$

For a binomial distribution we have

$$M_x^2 = (rp)^2 = r^2 p^2,$$
$$\therefore V = \Sigma y \cdot x^2 - r^2 p^2 \qquad \qquad \text{(iv)}$$

Since $y = r_{(x)} p^x q^{r-x}$

$$\sum_{x=0}^{x=r} y \cdot x^2 = V_0 = \sum_{x=0}^{x=r} x^2 r_{(x)} p^x q^{r-x}.$$

Now $x^2 = x(x - 1) + x$,

$$\therefore V_0 = \Sigma \left[x(x-1) + x \right] r_{(x)} p^x q^{r-x}$$
$$= \Sigma x(x-1) \cdot r_{(x)} p^x q^{r-x} + \Sigma x \cdot r_{(x)} p^x q^{r-x}.$$

The term on the extreme right is the mean raw score, i.e. rp (see p. 102).

$$\therefore V_0 = \left\{ \sum_0^r x(x-1) \frac{r!}{x!(r-x)!} p^x q^{r-x} \right\} + rp$$

$$= \left\{ \sum_0^r \frac{r!}{(x-2)!\,(r-x)!} p^x q^{r-x} \right\} + rp$$

$$= \left[r(r-1)p^2 \sum_{x=0}^{x=r} \frac{(r-2)!}{(x-2)!\,(r-x)!} p^{x-2} q^{r-x} \right] + rp.$$

If we put $k = (x-2)$ and $l = (r-2)$ so that $k = (r-2) = l$ when $x = r$ and $k = -2$ when $x = 0$,

$$V_0 = \left[r(r-1)p^2 \sum_{k=-2}^{k=l} \frac{l!}{k!\,(l-k)!} p^k q^{l-k} \right] + rp.$$

For terms involving $k = -2$ or $k = -1$ we have a coefficient involving the reciprocal of the factorial of a negative integer, and (Ex. 5, p. 12)

$$\frac{1}{(-n)!} = 0.$$

Hence these two terms will vanish, and

$$V_0 = \left[r(r-1)p^2 \sum_{k=0}^{k=l} \frac{l!}{k!\,(l-k)!} p^k q^{l-k} \right] + rp.$$

The factor involving the summation sign is now $(p+q)^l = 1$.

$$\therefore V_0 = r(r-1)p^2 + rp.$$

Hence from (iv) above

$$V = rpq + r^2 p^2 - r^2 p^2 = rpq.$$

Thus (ii) and (i) are consistent; but (ii) defines a *scatter index* of any collection of items to which we can assign a score, and is not restricted to sampling distributions as such, whether of the binomial type or otherwise. The constant V is of importance in statistical theory partly on account of two properties formulated by theorems respectively associated with the names of Bernoulli and of Tchebychev.

We may here state *Bernoulli's theorem* in the following form :

If a score deviation X_a in sample A has the same theoretical frequency as a score deviation X_b in a smaller sample B, the *proportionate* score deviation $U_a = (X_a \div r_a)$ is numerically less than the proportionate score deviation $U_b = (X_b \div r_b)$. The expectation that a proportionate deviation will not exceed a certain value ϵ thus increases as we increase the size of the sample.

The validity of this assertion is inherent in the c-test. If raw score deviations as large as X_a from a sample of r_a items and X_b from a sample of r_b items have the same expectation

$$\frac{X_a - \frac{1}{2}}{\sigma_a} = \frac{X_b - \frac{1}{2}}{\sigma_b},$$

$$\therefore \frac{X_a - \frac{1}{2}}{(r_a pq)^{\frac{1}{2}}} = \frac{X_b - \frac{1}{2}}{(r_b pq)^{\frac{1}{2}}},$$

$$\therefore \frac{X_a - \frac{1}{2}}{\sqrt{r_a}} = \frac{X_b - \frac{1}{2}}{\sqrt{r_b}}.$$

If $r_a = k^2 \cdot r_b$, $k > 1$, since $r_a > r_b$; and

$$\frac{X_a - \frac{1}{2}}{k\sqrt{r_b}} = \frac{X_b - \frac{1}{2}}{\sqrt{r_b}},$$

$$\therefore X_a = kX_b + \tfrac{1}{2}(1 - k),$$

$$\therefore U_a = \frac{X_a}{r_a} = \frac{X_a}{k^2 r_b} = \frac{X_b}{kr_b} + \frac{(1-k)}{2k^2 r_b} = \frac{U_b}{k} + \frac{1-k}{2k^2 r_b}.$$

Now $(1 - k)$ is negative since $k > 1$,

$$\therefore U_a < \frac{U_b}{k} < U_b.$$

****So stated, Bernoulli's theorem refers only to sampling which conforms to the binomial pattern. *Tchebychev's theorem* (Fig. 37), which sets an upper limit to the expectation that a deviation will be as large as $h\sigma$ is noteworthy because its proof entails no assumption concerning the nature of the distribution, being implicit in (ii) above. We assume that the range of score deviations X extends from $-u$ to v, so that its range is divisible as follows :

(i) $X = -u$ to $-(h\sigma + 1)$; $|X| > h\sigma$.

(ii) $X = -h\sigma$ to $h\sigma$; $|X| \leqslant h\sigma$.

(iii) $X = (h\sigma + 1)$ to $+v$; $|X| > h\sigma$.

By definition (p. 108), the expectation that a deviation will be numerically as large as $h\sigma$ is given by

$$E(\geqslant h\sigma) = \sum_{-u}^{-(h\sigma+1)} y + \sum_{(h\sigma+1)}^{v} y \qquad \cdots \qquad \text{(v)}$$

By (ii) above

$$\sigma^2 = \sum_{-u}^{-(h\sigma+1)} y \cdot X^2 + \sum_{-h\sigma}^{h\sigma} y \cdot X^2 + \sum_{(h\sigma+1)}^{v} y \cdot X^2.$$

$$\therefore \sum_{-u}^{-(h\sigma+1)} y \cdot X^2 + \sum_{(h\sigma+1)}^{v} y \cdot X^2 \leqslant \sigma^2. \qquad \cdots \qquad \text{(vi)}$$

Since every value of X^2 in the range defined by the expression on the left of (vi) is greater than $h^2\sigma^2$

$$\sum_{-u}^{-(h\sigma+1)} y \cdot h^2\sigma^2 + \sum_{(h\sigma+1)}^{v} y \cdot h^2\sigma^2 < \sum_{-u}^{-(h\sigma+1)} y \cdot X^2 + \sum_{(h\sigma+1)}^{v} y \cdot X^2$$

$$h^2\sigma^2 \left\{ \sum_{-u}^{-(h\sigma+1)} y + \sum_{(h\sigma+1)}^{v} y \right\} < \sum_{-u}^{-(h\sigma+1)} y \cdot X^2 + \sum_{(h\sigma+1)}^{v} y \cdot X^2.$$

From (v) and (vi) above

$$h^2\sigma^2 \cdot E(\geqslant h\sigma) < \sum_{-u}^{-(h\sigma+1)} y \cdot X^2 + \sum_{(h\sigma+1)}^{v} y \cdot X^2.$$

$$\therefore h^2\sigma^2 \cdot E(\geqslant h\sigma) < \sigma^2.$$

$$\therefore E(\geqslant h\sigma) < \frac{1}{h^2} \qquad \cdots \qquad \text{(vii)}$$

The last relation constitutes the theorem. In particular, an upper limit to the expectation that X will be equal to or greater than 3σ is given by

$$E(\geqslant 3\sigma) < \tfrac{1}{9}.$$

For the normal distribution $E(\geqslant 3\sigma) \simeq \frac{1}{370}$; but sample distributions are not necessarily normal. Tchebychev's theorem tells us that there exists *no* distribution for which the net frequency of score deviations exceeding 3 times the standard deviation can be as great as $\frac{1}{9}$.****

****Omit on first reading.

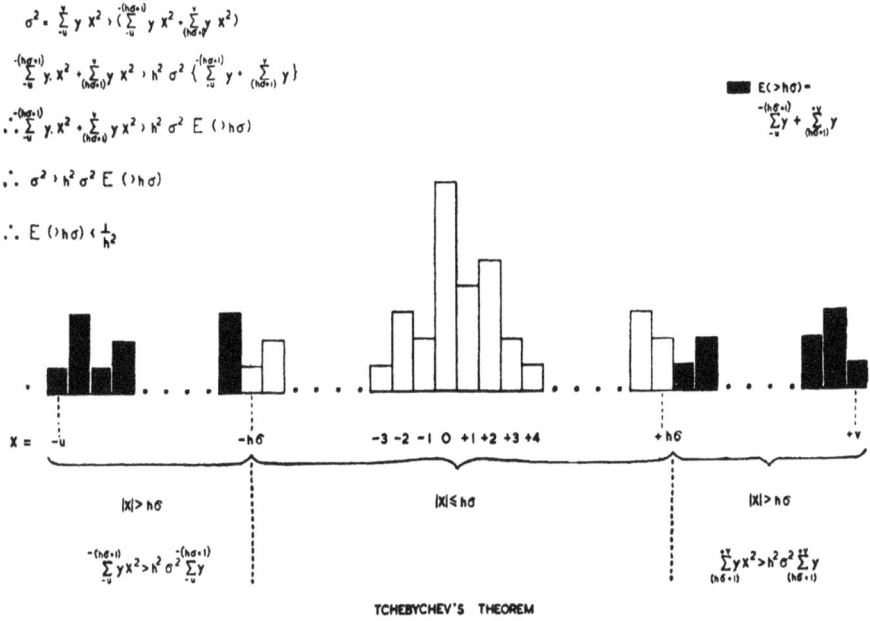

$\sigma^2 = \sum\limits_{-u}^{v} y\, x^2 , (\sum\limits_{-u}^{-(h\sigma+1)} y\, x^2 , \sum\limits_{(h\sigma+1)}^{v} y\, x^2)$

$\sum\limits_{-u}^{-(h\sigma+1)} y\, x^2 , \sum\limits_{(h\sigma+1)}^{v} y\, x^2 , h^2\sigma^2 \{\sum\limits_{-u}^{-(h\sigma+1)} y , \sum\limits_{(h\sigma+1)}^{v} y\}$

$\therefore \sum\limits_{-u}^{-(h\sigma+1)} y\, x^2 , \sum\limits_{(h\sigma+1)}^{v} y\, x^2 , h^2\sigma^2\, E\,(>h\sigma)$

$\therefore \sigma^2 , h^2\sigma^2\, E\,(>h\sigma)$

$\therefore E\,(>h\sigma) < \dfrac{1}{h^2}$

$E(>h\sigma) = \sum\limits_{-u}^{-(h\sigma+1)} y + \sum\limits_{(h\sigma+1)}^{v} y$

X = −u   −hσ   −3 −2 −1 0 +1 +2 +3 +4   +hσ   +v

|x| > hσ |x| ≤ hσ |x| > hσ

$\sum\limits_{-u}^{-(h\sigma+1)} y x^2 > h^2\sigma^2 \sum\limits_{-u}^{-(h\sigma+1)} y$ $\sum\limits_{(h\sigma+1)}^{v} y x^2 > h^2\sigma^2 \sum\limits_{(h\sigma+1)}^{v} y$

TCHEBYCHEV'S THEOREM

FIG. 37. Tchebychev's theorem shows that the region outside a score range expressible as a multiple (h) of the standard deviation can never exceed a fraction itself specifiable in terms of h.

EXERCISE 3.06

1. Find the standard deviation of the following collection of scores:

$\tfrac{1}{2}$; 0; 1; $\tfrac{1}{2}$; 3; 5; 3; 3; 4; 2; $\tfrac{1}{4}$.

2.* If $u = x \div r$ is the *proportionate* score of a distribution of *raw* scores (x) defined by $(p + q)^r$, show that the variance of the proportionate score is given by

$$V_u = (pq \div r).$$

3. Show that the variance of the following distributions calculated in accordance with the weighted mean square deviations formula accords with the corresponding numerical value of rpq:

$$(\tfrac{1}{2} + \tfrac{1}{2})^{16}; \quad (\tfrac{3}{4} + \tfrac{1}{4})^{6}; \quad (\tfrac{1}{3} + \tfrac{2}{3})^{5}; \quad (\tfrac{4}{5} + \tfrac{1}{5})^{25}.$$

4.* Show that the mean and the variance of the natural numbers from 1 to n by recourse to the formula in No. 4 of Ex. 1.04 for the sum of their squares are respectively

$$M = \frac{n+1}{2}; \quad V = \frac{n^2 - 1}{12}.$$

Check each formula by direct calculation w.r.t. the first 10 integers.

5.* Show that the variance of the non-replacement distribution $(s + n)^{(r)} \div n^{(r)}$ is given by

$$rpq\left(1 - \frac{r}{n}\right).$$

3.07 The Poisson Distribution

In deriving the equation of the normal curve, we have assumed that we can put $rp + 1 \simeq rp$. This means that rp is much greater than unity, and hence that r is large compared with the reciprocal of p. Otherwise we are not entitled to make this substitution, and the appropriate continuous distribution for large values of r is not symmetrical. In any case, we may put

$$q^{r-x} = (1-p)^{r-x} = (1-p)^r \cdot (1-p)^{-x}.$$

For large values of r in accordance with the approximation given on p. 46 we may write this as

$$q^{r-x} \simeq e^{-rp} \cdot e^{px} \qquad . \qquad . \qquad . \qquad . \qquad . \qquad . \qquad (i)$$

For the frequency y of a raw score x defined by the appropriate term of $(q + p)^r$, we therefore have

$$y \simeq \frac{r!}{(r-x)! \, x!} \, p^x \cdot e^{-rp} \cdot e^{px}$$

$$\simeq \frac{r!}{(r-x)! \, r^x} \frac{(rp)^x \cdot e^{-rp} \cdot e^{px}}{x!} \qquad . \qquad . \qquad . \qquad . \qquad (ii)$$

By Stirling's theorem (p. 48)

$$\frac{r!}{(r-x)!} \simeq \frac{r^{r+\frac{1}{2}} \cdot e^{-r} \cdot \sqrt{2\pi}}{(r-x)^{r-x+\frac{1}{2}} \cdot e^{x-r} \cdot \sqrt{2\pi}} = \frac{r^{r+\frac{1}{2}}}{(r-x)^{r-x+\frac{1}{2}} \cdot e^x} \cdots$$

$$\frac{r!}{(r-x)! \, r^x} \simeq \frac{r^{r+\frac{1}{2}} \cdot e^{-x}}{(r-x)^{r-x+\frac{1}{2}} \cdot r^x} \qquad . \qquad . \qquad . \qquad . \qquad . \qquad (iii)$$

We may now write

$$(r-x)^{r-x+\frac{1}{2}} = r^{r-x+\frac{1}{2}} \cdot \left(1 - \frac{x}{r}\right)^{r-x+\frac{1}{2}},$$

$$\therefore (r-x)^{r-x+\frac{1}{2}} \cdot r^x = r^{r+\frac{1}{2}} \cdot \left(1 - \frac{x}{r}\right)^{r-x+\frac{1}{2}}.$$

In the neighbourhood of the mean $(M = rp)$, x is small compared with r, since the present assumption is that r is not great compared with the reciprocal of p. If therefore r itself is large $(r - x + \frac{1}{2}) \simeq r$ in the neighbourhood of the mean. On that understanding

$$\left(1 - \frac{x}{r}\right)^{r-x+\frac{1}{2}} \simeq \left(1 - \frac{x}{r}\right)^r \simeq e^{-x},$$

$$\therefore (r-x)^{r-x+\frac{1}{2}} \cdot r^x \simeq r^{r+\frac{1}{2}} \, e^{-x} \qquad . \qquad . \qquad . \qquad . \qquad . \qquad (iv)$$

By substitution of (iv) in (iii), we have

$$\frac{r!}{(r-x)! \, r^x} \simeq 1.$$

Hence from (ii)

$$y = \frac{(rp)^x \cdot e^{-rp} \cdot e^{px}}{x!} = \frac{M^x \cdot e^{-M} \cdot e^{px}}{x!}.$$

For all values of x in the neighbourhood of the mean, px is necessarily very small, so that $e^{px} \simeq e^0$, i.e. $e^{px} \simeq 1$,

$$\therefore y \simeq \frac{M^x \cdot e^{-M}}{x!} \qquad . \qquad . \qquad . \qquad . \qquad . \qquad (v)$$

Subject to the same condition

$$rpq = rp(1 - p) \simeq rp.$$
$$\therefore V \simeq M \quad \text{and} \quad \sigma \simeq \sqrt{M} \quad . \quad . \quad . \quad . \quad . \quad \text{(vi)}$$

Hence we may write the *Poisson distribution* defined by (v) in the alternative form

$$y \simeq \frac{V^x . e^{-V}}{x!}.$$

This formula or (v) serves for approximate evaluation of the theoretical frequency 0, 1, 2, . . . successes when p is very small. We then have

$$y_0 = e^{-M} ; \quad y_1 = e^{-M} . M.$$
$$y_2 = e^{-M} . \frac{M^2}{2!} ; \quad y_3 = e^{-M} . \frac{M^3}{3!}.$$

Accordingly we have

$$\sum_{x=0}^{x=r} y_x = e^{-M}\left(1 + M + \frac{M^2}{2!} + \frac{M^3}{3!} \cdots\right).$$

When r is indefinitely large,

$$\sum_{x=0}^{x=r} y_x = e^{-M} . e^{M} = 1.$$

The Poisson formula is therefore consistent with the condition that the sum of the frequencies of all possible score values is unity. The following example from Weatherburn (p. 49) illustrates its use as an approximate description of an actual sampling distribution. One thousand consecutive issues of a periodical recorded deaths (x) of centenarians as follows :

Recorded deaths	.	.	.	0	1	2	3	4	5	6	7	8
No. of issues	.	.	.	229	325	257	119	50	17	2	1	0
Poisson frequencies ($M = 1.5$)	.			223.1	334.7	251.0	125.5	47.1	14.1	3.5	0.8	0.2

3.08 VARIANCE OF A NON-REPLACEMENT DISTRIBUTION

We have hitherto assumed that our universe is indefinitely large in comparison with our sample, by the same token that extraction of the sample does not appreciably change the composition of the universe, and hence that the condition of replacement is irrelevant to a specification of the sampling distribution. Fig. 28 in Chapter 2 gives us some solid ground for the assumption last stated, but we are not entitled to rely on it until we have investigated the properties of the non-replacement distribution defined by (iii) in 2.05, *viz.*

$$y_x = r_{(x)} . \frac{s^{(x)} f^{(r-x)}}{n^{(r)}} \quad . \quad . \quad . \quad . \quad . \quad \text{(i)}$$

Before proceeding to examine in what circumstances, if any, (i) is reducible to the normal equation, it is necessary to evaluate the two constants, M (the mean score) and V (the variance). We have already seen that the mean raw score of the binary hypergeometric distribution is the same as the mean (rp) of the binomial distribution for repetitive choice. To get the variance we proceed as in 3.06 above. In accordance with (iii) in 3.06 we have

$$V = V_0 - M^2 = V_0 - r^2p^2$$

and

$$V_0 = \sum_{x=0}^{x=r} y_r . x^2 = \sum_{x=0}^{x=r} x(x-1).y_x + \sum_{x=0}^{x=r} x.y_x$$

$$= \left[\sum_{x=0}^{x=r} x(x-1)y_x\right] + rp.$$

$$\therefore V_0 - rp = \sum_{x=0}^{x=r} x(x-1) . \frac{r!}{x!\,(r-x)!}\,\frac{s^{(x)}\,f^{(r-x)}}{n^{(r)}}$$

$$= \frac{r(r-1).s(s-1)}{n^{(r)}} \sum_{x=0}^{x=r} \frac{(r-2)!}{(x-2)!\,(r-x)!}\,(s-2)^{(x-2)}\,f^{(r-x)}.$$

$$= \frac{r(r-1).s(s-1)}{n^{(r)}} . (s-2+f)^{(r-2)}$$

$$= \frac{r(r-1).s(s-1)}{n^{(r)}} . (n-2)^{(r-2)}$$

$$= \frac{r(r-1).s(s-1)}{n(n-1)}.$$

Since $np = s$

$$V_0 - rp = rp . \frac{(r-1)(s-1)}{n-1}.$$

$$\therefore V_0 = rp\left\{\frac{rs - r - s + n}{n-1}\right\}.$$

$$\therefore V = V_0 - r^2p^2 = rp\left\{\frac{rs - r - s + n - rpn + rp}{n-1}\right\}.$$

Since $rs = rpn$ and $(n-s) = f = nq$

$$V = rp\left\{\frac{nq - r(1-p)}{n-1}\right\}$$

$$= \frac{rpq\,(n-r)}{n-1}$$

$$= rpq\left(\frac{n}{n-1}\right)\left(1 - \frac{r}{n}\right) \quad . \quad . \quad . \quad . \quad . \quad \text{(ii)}$$

When n is large, we may consider $n \simeq (n-1)$, so that

$$V = rpq\left(1 - \frac{r}{n}\right).$$

If we write F for the *sampling fraction* $(r \div n)$

$$V = rpq\,(1-F) \quad . \quad . \quad . \quad . \quad . \quad \text{(iii)}$$

Even when r is large, F is small if n is itself much larger than r and (iii) then reduces to (i) in 3.06, the variance formula for the replacement distribution.

EXERCISES 3.07–3.08

1.* Compare the exact values of the following frequency distributions with values respectively assigned by Poisson's expression and by the table of the ordinates of the normal curve with due regard to the half-interval correction.

(i) $(\frac{1}{10} + \frac{9}{10})^{10}$.

(ii) $(0\cdot05 + 0\cdot95)^8$.

(iii) $(0\cdot01 + 0\cdot99)^{20}$.

2. Compute the variance for the distribution of (a) *hearts*; (b) *red* cards; (c) *picture* cards in samples of 4, 5, 6 and 7 cards *simultaneously* extracted from a full pack.

3.09 THE NORMAL AS THE LIMIT OF THE HYPERGEOMETRIC DISTRIBUTION

If a pack of n cards consists of $s = (np)$ cards the choice of which constitutes a success and $f = (nq)$ cards the choice of which constitutes a failure, the raw score distribution for an r-fold sample is as given by (i) in 3.08 with mean rp and variance defined by (iii) in 3.08, *viz.*

$$V = rpq\,(1 - F).$$

By Ex. 3 of 1.06 (p. 39), we have

$$y_{x+1} = \frac{r - x}{x + 1} \frac{s^{(x+1)} \cdot f^{(r-x-1)}}{s^{(x)} \cdot f^{(r-x)}} \cdot y_x.$$

By definition

$$s^{(x+1)} = s(s - 1)(s - 2) \ldots (s - x + 1)(s - x) = (s - x) \cdot s^{(x)}$$

$$f^{(r-x)} = f(f - 1)(f - 2) \ldots (f - r + x + 2)(f - r + x + 1) = (f - r + x + 1) \cdot f^{(r-x-1)}.$$

$$\therefore y_{x+1} = \frac{(r - x)(s - x)}{(x + 1)(f - r + x + 1)} \cdot y_x.$$

If the sample is fairly large so that $(r - 1) \simeq r$

$$y_{x+1} \simeq \frac{(r - x)(np - x)}{(x + 1)(nq - r + x)} \cdot y_x.$$

$$\therefore \frac{\Delta y_x}{y_x} \simeq \frac{(r - x)(np - x)}{(x + 1)(nq - r + x)} - 1$$

$$= \frac{n(rp - q) + r - (n + 1)x}{(nq - r) + (nq - r + 1)x + x^2} \cdot \quad \cdot \quad \cdot \quad \cdot \quad \text{(i)}$$

When n is large and r is large, we may put $(nq - r + 1) \simeq (nq - r)$, $(n + 1) \simeq n$ and $(rp - q) \simeq rp$; and if n is also large in comparison with p^{-1}, $(nrp + r) \simeq nrp$

$$\therefore \frac{\Delta y_x}{y_x} \simeq \frac{n(rp - x)}{(nq - r) + (nq - r)x + x^2} \quad \cdot \quad \cdot \quad \cdot \quad \text{(ii)}$$

As usual, we may write $rp = M$, and it will be convenient to put

$$C = nq - r \quad \cdot \quad \cdot \quad \cdot \quad \cdot \quad \cdot \quad \text{(iii)}$$

$$\therefore \frac{\Delta y_x}{y_x} \simeq \frac{-n(x - M)}{C + Cx + x^2}.$$

On transferring the origin to the mean by the customary substitution which specifies the *score deviation* distribution, i.e. $X = (x - M)$, and $x = (X + M)$, so that

$$x^2 = M^2 + 2MX + X^2,$$

we have

$$\therefore \frac{\Delta Y_x}{Y_x} \simeq \frac{-nX}{(C + CM + M^2) + (C + 2M)X + X^2} \quad \cdot \quad \cdot \quad \text{(iv)}$$

For economy we now write

$$A = (C + CM + M^2) \quad and \quad B = (C + 2M) \qquad . \qquad . \qquad . \qquad \text{(v)}$$

Whence from (iii)

$$B - M = nq - r + M = nq - r + rp = (n - r)q \qquad . \qquad . \qquad . \qquad \text{(vi)}$$

In accordance with the assumption that n is large by comparison with r, the following relations between the constants simplify subsequent work

$$B^2 - 4A = C(C - 4) = (B - 2M)(B - 2M - 4) \simeq (B - 2M)^2 . \qquad . \qquad \text{(vii)}$$

$$B = (nq - r + 2rp) = nq - r(1 - 2p) = nq - r(q - p) \qquad . \qquad . \qquad \text{(viii)}$$

By a substitution already invoked we use $V = rpq$ for the variance of the *normal* distribution, and

$$MB = Mqn - Mqr + Mrp \simeq V(n - r) + M^2.$$

$$\therefore \quad \frac{MB - M^2}{n} \simeq V\left(1 - \frac{r}{n}\right).$$

If, as in 3.08, we use $F = r \div n$ to denote the sampling fraction, we have the following expression for the true variance of the hypergeometric distribution as defined by (iii) in 3.08, *viz.*

$$\frac{MB - M^2}{n} \simeq V(1 - F) \qquad . \qquad . \qquad . \qquad . \qquad . \qquad \text{(ix)}$$

With these values of the relevant constants we can treat (iv) as a differential equation, so that

$$\frac{dY}{Y} = \frac{-nX . dX}{A + BX + X^2} \qquad . \qquad . \qquad . \qquad . \qquad . \qquad \text{(x)}$$

If the factors * of $(A + BX + X^2)$ are $(X - a)$ and $(X + b)$

$$a \quad \cdots \frac{-B + \sqrt{B^2 - 4A}}{2} \quad and \quad -b \, \cdots \, \frac{-B - \sqrt{B^2 - 4A}}{2} .$$

Whence from (vii) and (ix) above, we have

$$a \simeq \frac{-B + (B - 2M)}{2} = -M \quad and \quad b = \frac{B + (B - 2M)}{2} = (B - M) \qquad . \qquad \text{(xi)}$$

$$ab = M^2 - MB = -nV(1 - F) \qquad . \qquad . \qquad . \qquad \text{(xii)}$$

We may now write (x) in the form

$$\frac{dY}{Y} \simeq \frac{-nX . dX}{(X - a)(X + b)} = \frac{n}{(b + a)} \left\{ \frac{a}{a - X} - \frac{b}{b + X} \right\} dX.$$

$$\therefore \quad \log Y \simeq \frac{na}{b + a} \log (a - X) - \frac{nb}{b + a} \log (b + X) + \log K.$$

$$\therefore \quad \log Y \simeq \frac{na}{b + a} \log \left(1 - \frac{X}{a}\right) - \frac{nb}{b + a} \log \left(1 + \frac{X}{b}\right) + \log k \qquad . \qquad . \qquad \text{(xiii)}$$

In accordance with the assumption that r is large and that n is also large compared with r, both a and b in (xi) and (xii) are in general large compared with X over the greater part of the

---

* From an algebraic point of view alone, the choice of opposite signs is immaterial, but as we shall see necessary to give the constants a and b a geometrical meaning in (xvi) below, descriptive of the range.

range. We can therefore use the logarithmic approximation

$$\log Y \simeq -\frac{na}{b+a} \cdot \left[\frac{-X}{a} - \frac{X^2}{2a^2}\right] - \frac{nb}{b+a}\left[\frac{X}{b} - \frac{X^2}{2b^2}\right] + \log k.$$

$$\therefore \log\left(\frac{Y}{k}\right) \simeq \frac{nX^2}{2ab} \qquad \cdot \qquad \cdot \qquad \cdot \qquad \cdot \qquad \text{(xiv)}$$

From (xii) above

$$ab = -nV(1-F).$$

Thus (xiv) reduces to

$$\log\left(\frac{Y}{k}\right) \simeq \left(\frac{-X^2}{2V(1-F)}\right).$$

$$\therefore Y \simeq k \cdot exp\left(\frac{-X^2}{2V(1-F)}\right)$$

When $X = 0$, so that $Y = Y_0$, $Y = k$ and

$$Y \simeq Y_0 \cdot exp \cdot \left(\frac{-X^2}{2V(1-F)}\right) \qquad \cdot \qquad \cdot \qquad \cdot \qquad \cdot \qquad \text{(xv)}$$

The last equation is identical with the normal, except in so far as $V(1 - F)$, which is the variance of the hypergeometric score distribution itself, replaces $V = rpq$, the variance of the normal distribution, and if, as we here assume, n is in fact large by comparison with r, $V(1-F) \simeq V$. Hence (xv) is equivalent to the normal equation, in accordance with our initial assumptions. In other words, the non-replacement raw score distribution for large values of r does not differ sensibly from the replacement raw score distribution if the sample extracted is a small fraction of the universe itself.

**** It is suggestive to notice the family likeness of (xiv) above to (xix) in 3.04 derived for the replacement distribution by recourse to the central difference equation when $p = \frac{1}{2} = q$. With the substitution of $u = -na \div (b + a)$ and $v = -nb \div (b + a)$, we may write (xiii) as

$$Y \simeq k\left(1 - \frac{X}{a}\right)^u\left(1 + \frac{X}{b}\right)^v \qquad \cdot \qquad \cdot \qquad \cdot \qquad \cdot \qquad \text{(xvi)}$$

When $a = b$, so that $u = v$, this reduces to the same form as (xviii) in 3.04, viz.

$$Y \simeq k\left(1 - \frac{X^2}{a^2}\right)^u \qquad \cdot \qquad \cdot \qquad \cdot \qquad \cdot \qquad \text{(xvii)}$$

When $Y = 0$ in (xvi), $X = a$ or $-b$. These co-ordinates therefore define the entire range of the distribution, so that

$$\int_{-b}^{a} Y \, dx = 1 \qquad \cdot \qquad \cdot \qquad \cdot \qquad \cdot \qquad \cdot \qquad \text{(xviii)}$$

We may profitably anticipate further insight into the properties of (xvi) by a substitution involving change of scale and origin, viz.: $X = (a + b)z - b$, so that

$$\left(1 + \frac{X}{b}\right) = \frac{(a+b)z}{b}; \quad \left(1 - \frac{X}{a}\right) = \left(\frac{a+b}{a}\right)(1-z); \quad dX = (a+b)dz.$$

$$\therefore \left(1 - \frac{X}{a}\right)^u\left(1 + \frac{X}{b}\right)^v \cdot dX = \frac{k(a+b)^{u+v+1}}{a^u \cdot b^v} \cdot z^v (1-z)^u \cdot dz.$$

$$\therefore Y \cdot dX = C \cdot z^v(1-z)^u \, dz.$$

**** Omit on first reading.

Since $z = (X + b) \div (a + b)$

$$\int_m^n Y \, . \, dX = C \, . \, \int_{\frac{(m+b)}{(a+b)}}^{\frac{(n+b)}{(a+b)}} z^v (1 - z)^u \, dz \qquad . \qquad . \qquad . \qquad \textbf{(xix)}$$

For the entire area bounded by the curve, $z = 1$ when $n = a$ and $z = 0$ when $m = -b$. Hence from (xviii)

$$C \int_0^1 z^v (1 - z)^u \, dz = 1$$

$$\therefore \int_0^1 z^v (1 - z)^u \, dz = \frac{1}{C} \qquad . \qquad . \qquad . \qquad . \qquad \textbf{(xx)}$$

The definite integral on the left belongs to the class known as *Beta functions*, dealt with in 6.07 below.★★★★

THE RECOGNITION OF A TAXONOMIC DIFFERENCE

4.01 STATEMENT OF THE PROBLEM

A VERY common type of question to which it has been customary to apply the mathematical theory of probability is one of which the following is an example : *is vaccination effective against smallpox ?* It would be unnecessary to invoke such considerations if it were true that : (*a*) no vaccinated person ever gets smallpox ; (*b*) a high proportion of persons not themselves vaccinated do get smallpox. As things are, neither statement would be true of the population of our own country. In so far as we are justified in concluding that vaccination is beneficial, we therefore have to base a verdict on the possibility of giving an affirmative answer to the question : *is the incidence of smallpox higher among persons who have not been than among persons who have been vaccinated?* Alternatively : is the proportion of infected persons lower in a population sample of vaccinated persons than in a population sample of persons who have not been vaccinated ?

So stated, the problem is essentially like that of deciding whether there is the same proportion of black balls in two urns each containing white balls and black ones, when our only source of information is such as we can derive by taking a sample of balls from each. In such a situation our null hypothesis is that the two urns are indeed identical. Our first task is therefore to explore the implications of testing this hypothesis.

Hitherto we have confined our attention to the sampling distribution of *raw scores*, i.e. the theoretical frequencies of getting 0, 1, 2 . . . *r* items of a specified class of items in an *r*-fold sample. When we are comparing samples of different sizes (r_a and r_b), our basis of comparison must be the *proportions* rather than the actual numbers of items of a given class. Corresponding to 0, 1, 2 . . . *r* items of a particular class, e.g. black balls from an urn containing black and white balls in the proportions p and $q = (1 - p)$, the proportions of items of the same class in the *r*-fold sample will be

$$0, \frac{1}{r}, \frac{2}{r}, \frac{3}{r} \ldots \frac{r}{r}(= 1).$$

In any *r*-fold sample containing x black balls, the proportion of black balls will be $x \div r$, and the theoretical frequencies of samples containing $x \div r$ black balls is therefore given by corresponding terms (p. 89) of the expansion $(qn + pn)^{(r)} \div n^{(r)}$. If the number ($n$) of balls in the urn is very large compared with r, we may use (p. 77) the more convenient expression $(q + p)^r$, so that the theoretical frequency of a sample in which the proportion of black balls is $x \div r$ is $r_{(x)} \cdot p^x \cdot q^{r-x}$. Evidently, the sampling distribution of $x \div r$ is the same as the sampling distribution of x itself.

To keep our feet on the ground, let us consider a particular experiment of this sort. From two urns we draw samples as below. For simplicity we shall assume that we *replace* each ball drawn before choosing another, since the class of problems we later discuss presupposes a universe indefinitely large by comparison with the sample chosen :

	Size of Sample	No. of Black Balls	Proportion *Ditto*
Urn A . . .	12	9	¾
Urn B . . .	8	4	½

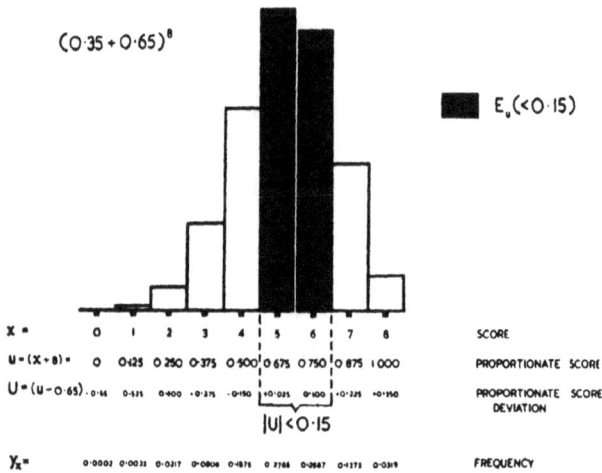

FIG. 38. Expectation that the proportionate score deviation for an 8-fold sample will be less than ± 0·15 when the probability of success is 13 in 20.

Our problem is to decide whether the proportion (p) of black balls in urn A is the same as the proportion of black balls in urn B. If we did not know that the two samples came from different urns, we might state the question in the alternative, and algebraically equivalent, form : could two such samples have come from the same urn ?

If we knew the composition of such an urn, i.e. the numerical value of p, we could state the theoretical frequency of getting either a sample of r_a balls among which $p_a = x_a \div r_a$ is the proportion of black ones or a sample of r_b balls among which the proportion of black ones is $p_b = x_b \div r_b$, in accordance with the binomial sampling distribution of $x \div r$. We could then answer either of the following questions : (a) how often will it happen that the proportion of black balls in a sample of r_a balls will differ from p by as much as $\pm (p - p_a)$; (b) how often will it happen that the proportion of black balls in a sample of r_b balls will differ from p by as much as $\pm (p - p_b)$. Should the theoretical frequency of either event be very small, we should have reason to doubt the truth of the hypothesis that one or other sample came from an urn in which the proportion of black balls is p.

Actually, we do not know the exact numerical value of p ; and we can construct a *null hypothesis* only if we make an *estimate* of it. Provisionally, we shall first make an estimate on the basis of all the information at our disposal. In conformity with the condition stated above we shall also assume that the total number of balls in the urn is very large or—what comes to the same thing from the algebraic viewpoint—that we replace each ball taken before drawing another. If both samples come from the same urn, we may treat each as part of a single sample of $(12 + 8) = 20$ balls out of which $(9 + 4) = 13$ are black. The initial assumption we thus explore is that the proportion (p) of black balls in the putative common urn identical with A and B is $(13 \div 20) = 0.65$, so that $q = (1 - p) = 0.35$. For samples of 8 and of 12 balls taken from such an urn successive terms of $(0.35 + 0.65)^8$ and $(0.35 + 0.65)^{12}$, as in Figs. 38 and 39, respectively, give the theoretical frequencies of samples containing 0, 1, 2, 3 . . . 8 and 0, 1, 2, 3 . . . 12 balls.

Since the proportion of black balls in sample B is 0·5, the observed deviation from the mean value $(0.5 - 0.65)$ is numerically equal to 0·15. Thus part of our problem is to decide how often it will happen that the actual proportions of black balls in an 8-fold sample taken from

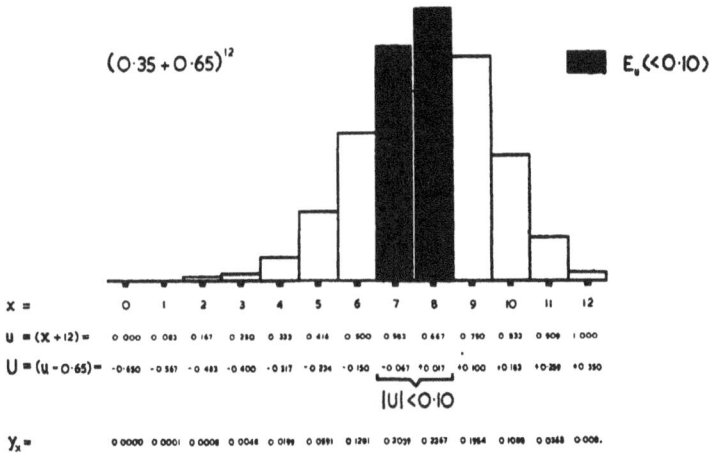

$(0.35 + 0.65)^{12}$

$E_x(<0.10)$

$X =$	0	1	2	3	4	5	6	7	8	9	10	11	12
$u = (X \div 12) =$	0·000	0·083	0·167	0·250	0·333	0·416	0·500	0·583	0·667	0·750	0·833	0·909	1·000
$U = (u - 0.65) =$	-0·650	-0·567	-0·483	-0·400	-0·317	-0·234	-0·150	-0·067	+0·017	+0·100	+0·183	+0·259	+0·350

$|U| < 0.10$

| $y_x =$ | 0·0000 | 0·0001 | 0·0008 | 0·0048 | 0·0199 | 0·0591 | 0·1291 | 0·2039 | 0·2367 | 0·1954 | 0·1088 | 0·0368 | 0·006. |

Fig. 39. Expectation that the proportionate score deviation for a 12-fold sample will be less than ± 0·10 when the probability of success is 13 in 20.

an urn containing an indefinitely large number of black and white balls in the ratio 13 : 7 will differ from $13 \div (13 + 7) = 0.65$ by as much as 0·15. To answer this we first ask how often the proportionate deviation will be less than ± 0·15, i.e. how often the proportion of black balls will be greater than 0·500 and less than 0·800. Fig. 38 shows that this is so only when the number of black balls in the sample is 5 or 6. The total frequency of such samples is

$$\frac{8!}{5!\,3!} (0.35)^3(0.65)^5 + \frac{8!}{6!\,2!} (0.35)^2(0.65)^6 = 0.2786 + 0.2587 = 0.54.$$

Thus the theoretical frequency of samples containing as few as 4 or as many as 7 black balls, i.e. the frequency of samples in which the proportion of black balls differs numerically from 0·65 by as much as 0·15 is

$$1 - 0.54 = 0.46.$$

If our null hypothesis is correct, the odds *against* a deviation of this magnitude are therefore 27 : 23. In other words, we should expect to score a deviation as great as the observed deviation nearly as often as to score a smaller one. So far as it involves the sample from urn B there is thus no sufficient reason to suspect that our null hypothesis is incorrect.

Let us now consider the sample from urn A. The observed proportion of black balls is 0·75 which differs numerically from 0·65 by 0·10. Again, the proportion of black balls (Fig. 39) in only two sorts of samples (score 7 or 8) lies inside this range. From Fig. 39 we see that the theoretical frequency of such samples is

$$\frac{12!}{7!\,5!} (0.35)^5(0.65)^7 + \frac{12!}{8!\,4!} (0.35)^4(0.65)^8 = 0.2039 + 0.2367 = 0.441.$$

Thus the expectation of a proportionate deviation numerically as great as 0.10 is

$$1 - 0.441 = 0.559.$$

Approximately, the odds in favour of a deviation as great as the observed one are therefore 5 : 4. In the long run we should therefore expect deviations as large as the observed deviation to be somewhat more frequent than smaller ones. So this result conveys no reason to suspect that our null hypothesis is false.

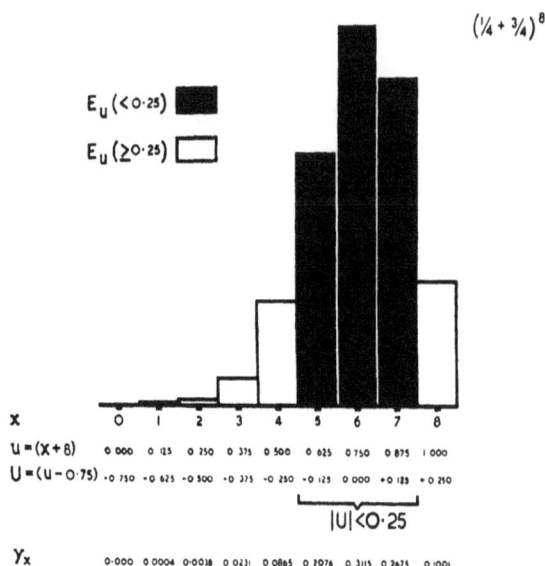

FIG. 40. Expectation that the proportionate score deviation for an 8-fold sample will be at least as great as 0·25 when the probability of success is 0·25. The appropriate model is the choice of 8 cards from a full pack subject to replacement, hearts being successes.

We shall later see that the form we have chosen for our null hypothesis is *not the best* we can do in such a situation ; but whatever else we do, we have to make an estimate of p. To do this we have pooled all available information which is relevant if our null hypothesis is true. Meanwhile, we are entitled to explore other possibilities. For instance, we might prefer to assume that the observed proportion of black balls in urn A is as observed (0·75) in the 12-fold sample taken therefrom. In that case we are concerned only with the question : how often might we get from an urn so constituted an 8-fold sample like the one from urn B ? Again, we might prefer to assume that the observed proportion (0·50) of black balls in the sample from urn B is the actual proportion of black balls therein. If so, our problem is to decide how often we might get from it a 12-fold sample, such as the sample from urn A.

Let us now examine each of these hypotheses in turn.

(*a*) If the observed proportion of black balls in the sample taken from urn A is the actual proportion therein, the frequency distribution of 8-fold samples. such as the sample from urn B, is given by the terms of $(\frac{1}{4} + \frac{3}{4})^8$. The proportion of black balls in a sample from B is 0·5 which differs numerically from 0·75 by 0·25. Our problem is now therefore to determine the observed frequency of 8-fold samples in which the proportion of black balls lies *inside* the range 0·75 ± 0·25 as is true if the score is 5–7 inclusive. Fig. 40 shows that

$$E(< 0·25) = 0·2076 + 0·3115 + 0·2675 = 0·7866$$
$$E(> 0·25) = 1 - 0·7866 = 0·2134.$$

Thus the odds are approximately 79 : 21 or 4 to 1 against getting a deviation as great as the observed, if we assume that the composition of the putative common universe tallies with that of of urn A. These are not highly unfavourable odds against the validity of our null hypothesis.

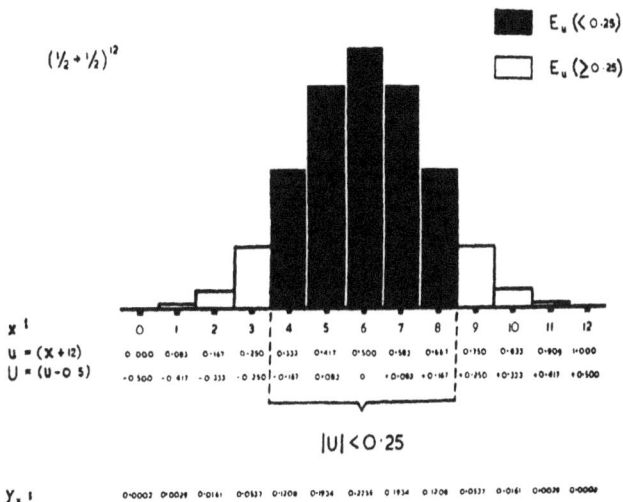

FIG. 41. Expectation that the proportionate score for a 12-fold sample will be at least as great as 0·25 when the probability of success is 0·5. The appropriate model is the choice of 12 cards from a full pack subject to replacement, a red card being a success.

(b) If the observed proportion of black balls in the sample from urn B is the actual proportion therein, the frequency distribution of 12-fold samples, such as the sample from urn A, is given by successive terms $(\frac{1}{2} + \frac{1}{2})^{12}$. The proportion of black balls in sample A is 0·75, and this differs numerically from the assumed proportion (0·50) in urn B by 0·25. In accordance with our present assumption we have therefore to ask how often the proportion of black balls in a 12-fold sample would lie inside the range 0·50 ± 0·25, as is true (Fig. 41), if the raw score is 4–8 inclusive. Fig. 41 shows that

$$E_u(< 0·25) = 0·1208 + 0·1934 + 0·2256 + 0·1934 + 0·1208 \simeq 0·85.$$
$$E_u(\geqslant 0·25) \simeq 0·15.$$

Thus the odds are less than 6 to 1 against a deviation of the observed magnitude. In other words, we might expect the occurrence of a deviation as great as the observed one in about one-seventh of a very large number of samples. This is not remarkably infrequent ; and the observed composition of the sample from urn A is in that sense consistent with the assumption that urn A has the same proportionate constitution as the sample from urn B.

Some assumption about how far either sample or both samples taken together yield a representative value of a sufficient parameter of the universe, in this case p, is necessarily implicit in any treatment of the recognition of a real difference. This raises the question : have we any reason to prefer one or other estimate of p so far discussed ? Bernoulli's theorem of 3.06 supplies the answer. The estimate p_{ab} $(= 0·65)$ of p which we get by pooling all our data refers to a larger sample than either of the estimates derived from observation of one sample alone ; and Bernoulli's theorem tells us that large score deviations occur with greater frequency in smaller samples than in larger ones from the same universe of choice. In the long run, the estimate p_{ab} which we get by pooling all our data will therefore be less liable to mislead us than an estimate based on one or other of the individual samples.

4.02 SAMPLING DISTRIBUTION OF A PROPORTION BY THE C-TEST

In 4.01 above we have made a first approach to the recognition of a real difference by adopting the null hypothesis that

(a) our samples are referable to a single parent universe ;

(b) the pooled proportion (p_{ab}) of items of a specified class in the combined sample is the best estimate at our disposal of their actual proportion p in the putative parent universe.

The problem then reduces to the evaluation of the expectation that the proportionate deviation in

(i) a sample A of a items among which the observed proportion is p_a will be as great as $p_{ab} - p_a$;

(ii) a sample B of b items among which the observed proportion is p_b will be as great as $p_{ab} - p_b$.

So stated the issue is precisely on all fours with that of 3.02 except in so far as : (a) the distribution involved is that of a *proportionate* score deviation in contradistinction to that of the score deviation itself ; (b) the assumed mean proportionate score (p_{ab}) is merely an *estimate* of p the corresponding parameter of the putative parent universe.   For the present, we shall assume that p_{ab} is a good estimate of p, in the sense that no large error arises from the substitution of p_{ab} and $q_{ab} = (1 - p_{ab})$ for p and $q = (1 - p)$ in the binomial $(q + p)^r$ definitive of the proportionate score distribution of r-fold samples taken from our universe.   In 3.05 we have seen that it is possible to sidestep the work of computing $E(\geqslant X)$ by recourse to the table of the probability integral when r is large.   We shall therefore examine the use of the c-test w.r.t. the significance of a proportionate score deviation $E(\geqslant U)$ as one way of finding an answer to the question raised in 4.01.   Later (p. 217) we shall see that it is not an *efficient* method of doing so, in the statistical sense of the term.

Meanwhile, it is necessary to remind ourselves that any justification for doing so depends on the fact that one and the same value of $y = r_{(x)} \cdot p^x \cdot q^{r-x}$ specifies for an r-fold sample the frequency of a particular raw score (x), the corresponding proportionate score $(u_x = x \div r)$, the corresponding score deviation $(X = x - rp)$ and the corresponding proportionate score deviation, which we may write as

$$U_x = \frac{X}{r} = u_x - p.$$

Accordingly we may express the critical ratio c (p. 128) as any of the following :

(i) $c^2 = \dfrac{(x - rp)^2}{rpq} = \dfrac{(x - rp)^2}{V}.$ (ii) $c^2 = \dfrac{X^2}{rpq} = \dfrac{X^2}{V}.$

(iii) $c^2 = \dfrac{(u_x - p)^2}{(pq \div r)}.$ (iv) $c^2 = \dfrac{U_x^2}{(pq \div r)}.$

Let us now examine the meaning of the ratio $(pq \div r)$ in (iii) and (iv) above.   The quantity $rpq = V$ in (i) and (ii) we have seen to be equivalent to the *mean square score deviation* given by (iv) in 3.06, i.e.

$$V = \left\{ \sum_{x=0}^{x=r} r_{(x)} \cdot p^x \cdot q^{r-x} \cdot x^2 \right\} - r^2 p^2.$$

In this equation rp is the mean score, the mean proportionate score being p (p. 102).   In accordance with (iii) in 3.06 the *mean square proportionate score deviation* (V_u) is therefore given by

$$V_u = \left\{ \sum_{x=0}^{x=r} r_{(x)} \cdot p^x \cdot q^{r-x} \cdot \left(\frac{x}{r}\right)^2 \right\} - p^2,$$

$$\therefore r^2 V_u = V = rpq,$$

$$\therefore V_u = (pq \div r) \qquad . \qquad . \qquad . \qquad . \qquad . \qquad . \qquad . \qquad (v)$$

Hence we may write (iii) and (iv) above in the form

$$c^2 = \frac{(u_x - p)^2}{V_u} = \frac{U_x^2}{V_u}.$$

In accordance with previous usage, we denote the square root of the variance of the proportionate score as

$$\sigma_u = \sqrt{V_u} = \sqrt{\frac{pq}{r}} \qquad . \qquad . \qquad . \qquad . \qquad . \qquad (vi)$$

$$c = \frac{u_x - p}{(pq \div r)^{\frac{1}{2}}} \qquad . \qquad . \qquad . \qquad . \qquad . \qquad (vii)$$

The observed proportionate score (u_x) of the sample A is p_a; and the assumed value of p with which we are here concerned is the pooled value p_{ab}. If we neglect the half interval *correction for continuity* defined by (iv) in 3·05 we take as the upper limit of integration from zero to h:

$$h = \frac{p_a - p_{ab}}{\sigma_u} = \frac{p_a - p_{ab}}{[p_{ab}(1 - p_{ab}) \div a]^{\frac{1}{2}}} \qquad . \qquad . \qquad . \qquad (viii)$$

Alternatively, to test the significance of the difference $p_b - p_{ab}$ we put

$$h = \frac{p_b - p_{ab}}{[p_{ab}(1 - p_{ab}) \div b]^{\frac{1}{2}}} \qquad . \qquad . \qquad . \qquad . \qquad (ix)$$

Whether it is necessary to make the correction for continuity depends on the size of the sample. When we are concerned with the expectation that the score or score deviation will respectively be as great as x or X, we put

$$h = \frac{X - \frac{1}{2}}{\sqrt{rpq}}.$$

When we are concerned with the expectation that the proportionate score deviation will be as great as U_x, we may write this as

$$h = \frac{r\left(\frac{X}{r} - \frac{1}{2r}\right)}{\sqrt{rpq}} = \frac{U_x - \frac{1}{2r}}{\sqrt{(pq \div r)}}$$

$$= \frac{U_x - \frac{1}{2r}}{\sigma_u} \qquad . \qquad . \qquad . \qquad . \qquad . \qquad . \qquad (x)$$

Hence (viii) becomes

$$h = \frac{(p_a - p_{ab}) - \frac{1}{2a}}{[p_{ab}(1 - p_{ab}) \div a]^{\frac{1}{2}}}.$$

Example. Greenwood and Yule (1915) cite the following results w.r.t. cholera inoculation

	Attacked	TOTAL
Inoculated	3	279
Not inoculated . .	66	539
All	69	818

To perform the *c*-test in accordance with the foregoing proportional procedure we may summarise the above thus :

	Per Cent. Attacked	Size of Sample
Sample A (*inoculated*) . . .	1·1	279
Sample B (*not inoculated*) .	12·2	539
Pooled sample	8·4	818

To say that inoculation is *not* effective is to say that samples distinguished merely by the fact that individuals have or have not submitted to inoculation are samples from the same universe or from identical universes in which the proportions of attacked and exempt are the same. As our best estimate of the proportion of attacked in the putative common universe, we take the pooled value, i.e.

$$p_{ab} = 0.084$$
$$(p_a - p_{ab}) = 0.011 - 0.084 = -0.073$$
$$(p_b - p_{ab}) = 0.122 - 0.084 = 0.038$$

To test the significance of the first deviation (-0.073) we have to find the standard deviation of the distribution of $p \simeq p_{ab}$ for a sample of 279 items. In accordance with (vi) above, since $pq \simeq (0.084)(1 - 0.084)$

$$\sigma_u^2 = \frac{(0.084)(0.916)}{279} \simeq 0.00028$$
$$\therefore \sigma_u = \sqrt{0.00028} \simeq 0.017.$$

Since 279 is a fairly large sample, we neglect the correction for continuity and put

$$h \simeq \frac{0.073}{0.017} \simeq 4.3.$$

The deviation is thus about 4·3 times the assumed standard deviation ; and it is not necessary to refer to the probability integral table for assurance that such an occurrence is very rare, if the assumption is correct. To test the significance of the alternative deviation we put

$$\sigma_u^2 = \frac{(0.084)(0.916)}{539} \simeq 0.00014,$$
$$\therefore \sigma_u \simeq 0.012.$$

Accordingly we put

$$h \simeq \frac{0.038}{0.012} > 3.0.$$

In this case the deviation is over three times its assumed standard deviation and is again highly significant ; but the second test is unnecessary, and an insignificant result would be immaterial in view of the result of the preceding test.

Normal Integral of the Proportionate Score Distribution

Since there is a one to one correspondence between the proportionate score and the raw score, we have assumed that it is equally legitimate to apply the c-test to the distribution of the former, if appropriate to that of the latter ; but the reader may wish for more formal assurance that this is so.   Accordingly we shall denote as the area defined by the boundaries $X \div \sqrt{V_x} = 0$ and $X \div \sqrt{V_x} = h$:

$$\left[A \right]_{X=0}^{X-h\sqrt{V_x}} = (2\pi V_x)^{-\frac{1}{2}} \int_0^{h\sqrt{V_x}} exp \left(-\frac{X^2}{2V_x} \right) dX.$$

If U is the proportionate score deviation corresponding to X, $dX = d(rU) = rdU$; and if $V_u = (pq \div r)$ is the variance of the proportionate score distribution, $V_u = V_x \div r^2$, so that $V_x = r^2 . V_u = r^2 \sigma_u^2$.   When $h = (X \div \sigma_x)$, i.e. $X = h\sqrt{V_x}$, we therefore have $rU = rh\sigma_u$ and $U = h\sigma_u$.   The above is then equivalent to

$$\left[A \right]_{U=0}^{U-h\sigma_u} = (2\pi V_u)^{-\frac{1}{2}} r^{-1} \int_0^{U-h\sigma_u} exp \left(-\frac{U^2}{2V_u} \right) d(rU)$$

$$= \frac{1}{\sigma_u \sqrt{2\pi}} \int_0^{U-h\sigma_u} exp \left(-\frac{U^2}{2V_u} \right) dU.$$

We now put $c = U \div \sigma_u$, so that c is the proportionate score in its *standard* form, i.e. expressed as so many times the s.d. of the proportionate score distribution.   When $U = h\sigma_u$, we then have $c = h$, and $dU = d(\sigma_u . c) = \sigma_u dc$, so that

$$\left[A \right]_{c=0}^{c-h} = \frac{1}{\sqrt{2\pi}} \int_{c-0}^{c-h} e^{-\frac{1}{2}c^2} dc.$$

It is perhaps worthwhile to comment on the outside limits of the integral for the distribution. The range of the binomial histogram for the raw score extends from $- rp$ to rq, hence from $- \infty$ to $+ \infty$ when r is indefinitely large.   Irrespective of the value we assign to r the limits of the proportionate score distribution extend from $- p$ to q, but as all frequencies referable to a score less than $- p$ or greater than q are zero, it does not affect the value of the definite integral for the whole area of the distribution if we extend the range from $- \infty$ to $+ \infty$.   A pitfall which calls for comment is the implications of the scalar change from X to $U = (X \div r)$, when we derive the frequency equation of the proportionate score in its normal form.   We then have to remember that the frequency Y_u of the proportionate score whose numerical value is U is the frequency Y_x of the raw score deviation whose numerical value is $X = rU$, so that

$$\therefore \ Y_u = Y_x = (2\pi rpq)^{-\frac{1}{2}} exp \left(- X^2 \div 2rpq \right)$$
$$= (2\pi rpq)^{-\frac{1}{2}} exp \left(- rU^2 \div pq \right).$$

For the variance of the proportionate score distribution, we have $V_u = (pq \div r)$, so that $rpq = r^2 V_u$. Thus the normal equation of the proportionate score *frequency* is *

$$Y_u = \frac{1}{\sqrt{2\pi r^2 V_u}} exp \left(\frac{- U^2}{2V_u} \right). \tag{xi}$$

*See remarks in 4.08 on pp. 183-6 below.

EXERCISE 4.02

1. Examine the following data of Greenwood and Yule w.r.t. typhoid inoculation :

	Attacked	*Not* Attacked
Inoculated	56	6,759
Not inoculated . . .	272	11,396
Total .	328	18,155

(Cited Fisher, *Statistical Methods*, p. 85.)

2. Test the following data cited by Kendall (Vol. I, p. 302) w.r.t. inoculation against cholera on a tea estate

	Attacked	*Not* Attacked
Inoculated	431	5
Not inoculated . . .	291	9
Total .	722	14

3. Test the following data cited by Kendall (Vol. I, p. 307) from an official report on the Spahlinger treatment of cattle exposed to infection :

	Severe Tuberculosis (including Fatal Cases)	Mildly affected or Immune
With vaccine treatment . .	6	13
Without *ditto* . . .	8	3
Total . .	14	16

4. Examine the following data also cited by Kendall (Vol. I, p. 304) :

	Teeth Normal	With Tooth Maloccluded
Breast-fed . . .	4	16
Bottle-fed . . .	1	21
Total . .	5	37

5. In *U.S. Public Health* (1936), Vol. 51, p. 443, Dr. Selwyn Collins cites the following data w.r.t. a smallpox follow-up of 8000 families in 18 States over one year :

	Population at Risk	Cases of Smallpox
No history of vaccination or prior attack . . .	16,603	16
Vaccinated over 7 years earlier	11,793	1
Ditto less than 7 years earlier . . .	8,769	0
Previous attack	1,157	0
Previously attacked or vaccinated at some time . .	21,719	1

Examine these figures w.r.t. relevance of vaccination or previous attack to subsequent risk, and discuss sources of erroneous interpretation to which treatment of such pooled data is open.

4.03 A RAW SCORE DIFFERENCE SAMPLING DISTRIBUTION FOR SMALL EQUAL SAMPLES

At a later stage (Chapter 5) we shall examine the logical implications of the fact that our null hypothesis in this context prescribes no exact numerical value of the parameter p which we assume to be very nearly the same as p_{ab} in making the tests of 4·01 and 4.02. We shall then see that it is possible to give precision to what uncertainty invests the c-test on this account. Apart from this, the provisional test dealt with in 4.02 introduces only one feature not dealt with in the previous chapter, namely, that the distribution with which we are concerned is that of the proportionate score $(x \div r)$ or its deviation from the mean (p) in contradistinction to that of the raw score (x) or its deviation from the mean score (rp). There is in fact a one-to-one correspondence between proportionate score and raw score, the frequency of a given raw score being therefore that of the corresponding proportionate score. Hence the only novelty of the method set forth in 4.01 is the verbal interpretation of two c-ratios which are in fact numerically identical. What we call in one case the ratio $(X \div \sigma_x)$ of the score deviation (X) to the standard deviation $(\sigma_x = \sqrt{rpq})$ of the raw score distribution, we may alternatively call the ratio of the proportionate score deviation $U_x = (X \div r)$ to the standard deviation $(\sigma_u = \sqrt{pq \div r})$ of the proportionate score distribution.

The form of words used in the last sentence merits comment, because a standard deviation or a variance in this context is a parameter of *a distribution*. Hence to speak of the *s.d.* of a score or the *s.d.* of a proportion is misleading, and particularly liable to cause confusion in connection with the use of variance or *s.d.* formulæ as scatter indices of the class frequencies of items to which we can assign a measurement or ordinal rank specification in a particular sample. This will become more apparent at a later stage when we extend to *representative scoring (vide 4.10 infra)* the mathematical theory of probability here applied to *taxonomic* scoring w.r.t. a binary classification. We shall then have to draw a sharp distinction between two classes of frequency distributions, *viz.* : (*a*) r-fold sampling distributions with which we are solely concerned in this and in previous chapters ; (*b*) probability (or *unit sample*) distributions which specify a particular universe of many classes. When we are concerned with a binary classification of attributes, such as *hearts* and *other* cards, the probability distribution of the universe is simply $q + p$, the terms of which refer respectively to the proportion of items or occurrences of terms labelled as successes and failures.

If we test the significance of a difference as prescribed in 4.02, our c-ratio involves an error of estimation both in the numerator and in the denominator. Though subsequent examination of the problem of estimation as such will indicate the possibility of assessing the frequency with which errors of judgment will arise from so doing, the recognition that the procedure of 4.02 does in fact do so serves to focus attention on the possibility of an alternative which entails *no* error in the specification of the numerator itself. To keep within the framework of sampling distributions elucidated in the previous chapter, we have committed ourselves to an unnecessarily circuitous formulation of our problem, and in consequence to a procedure which entails the performance of two *c-tests* to answer a single question. We have formulated our problem in a way which commits us to examine whether the structure of *each of two* samples is consistent with the null hypothesis ; but an appropriate null hypothesis is indeed amenable to more direct statement. Instead of asking how often proportionate scores of each of two samples will deviate to a greater or less extent from the parameter p of a putative common universe of choice, we shall now ask : how often does the difference between score values of samples from the same universe or from identical universes attain a certain numerical magnitude ? We can answer this question only if we know the *sampling distribution of a score difference*. Nothing we have

HEART SCORE DIFFERENCES – PAIRS OF
2 - FOLD SAMPLES
$(p = \frac{1}{4}, q = \frac{3}{4})$

HEART SCORE DIFFERENCES – PAIRS OF 2 - FOLD SAMPLES $(p = \frac{1}{4}, q = \frac{3}{4})$

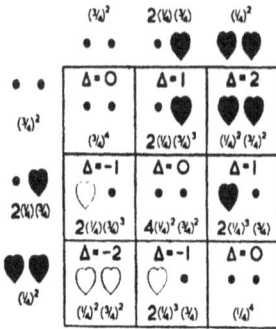

FIG. 42. Heart Score Differences—
pairs of 2-fold samples from a full
pack subject to replacement. Each
cell centre shows the value of the
score difference (Δ) above and the
frequency below.

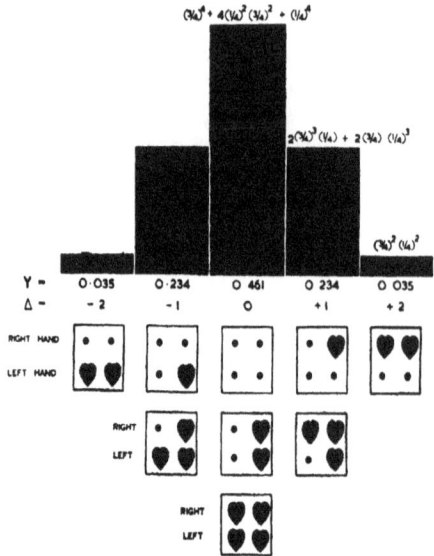

FIG. 43. The balance sheet of Fig. 42.

HEART SCORE DIFFERENCES – PAIRS OF 4 - FOLD SAMPLES $(p = \frac{1}{4}, q = \frac{3}{4})$

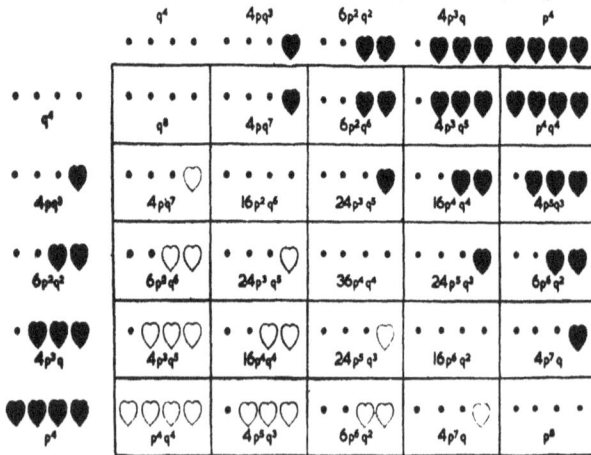

FIG. 44. Heart Score Differences—pairs of 4-fold samples ($p = \frac{1}{4}$; $q = \frac{3}{4}$), subject to replacement.

learnt so far tells us that it is exactly, or approximately, referable to a binomial, to a normal or
to an otherwise specifiable pattern. It is an issue *sui generis*, and one which therefore invites
investigation on its own merits.

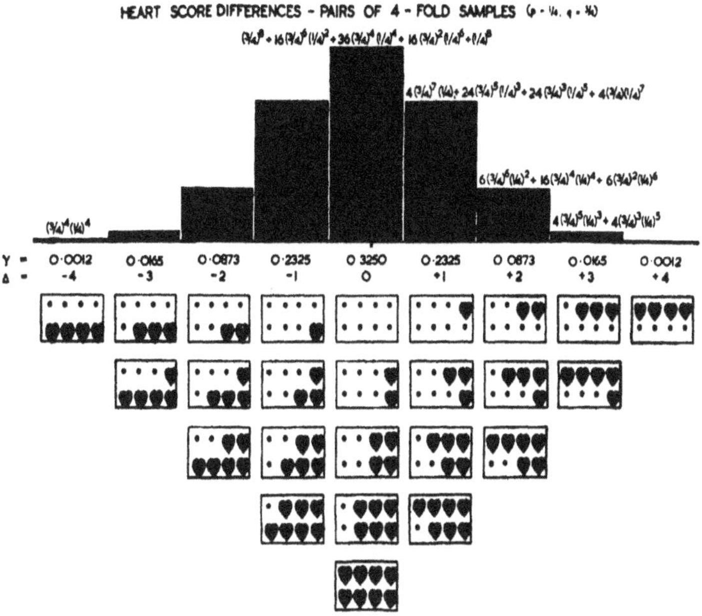

FIG. 45.   The balance sheet of Fig. 44.

Our model set-up for a preliminary investigation of this sort will be two full card packs A (*right* hand) and B (*left* hand). The difference we shall first explore is the difference between the right hand *raw* score and the left hand *raw* score, i.e. the result of subtracting the B score from the A score. To start with, we shall assume that each trial involves withdrawal of an equal number of cards from the two packs. This imposes a limitation which it will be necessary to remove at a later stage, but simplifies the task of visualisation as a prelude to a more general treatment of the problem. We shall assume replacement of each card drawn before withdrawal of another from the same pack in conformity with the assumption that we intend to use the results of our analysis in connection with samples which are relatively small compared with their parent universes. Accordingly, we can use the customary binomial expansion for the distribution of scores of samples taken from one and the same pack. For instance, 2-fold samples from either turn up with *heart* scores of 0, 1 and 2 defined by successive terms of $(\frac{3}{4} + \frac{1}{4})^2$, and 4-fold samples turn up from either pack with *heart* scores of 0, 1, 2, 3, 4 defined by successive terms of $(\frac{3}{4} + \frac{1}{4})^4$. Figs. 42 and 44 show the chessboard lay-out for pairs of 2-fold and for pairs of 4-fold samples, citing the frequency of independent association of a sample from a pack A with a particular heart score and a sample from pack B with a particular heart score in conforming with the product rule of 2.06. In conformity with the addition rule we can collect all such results, and express them as in Figs. 43 and 45 which respectively show the sampling distributions of the heart score differences for pairs of 2-fold and 4-fold samples.

Two features of the distributions of score differences for pairs of 2-fold and 4-fold samples exhibited in Figs. 43 and 45 are generally characteristic of score difference distributions of pairs of samples of the same size, as is evident from the structure of the chessboard diagram :

TABLE 1

Frequency Distributions of Heart Score Differences for pairs of 2-fold and 4-fold Samples

Score Difference X.	y_x Pairs of 2-fold Samples.	$(\frac{1}{2} + \frac{1}{2})^4$.	y_x Pairs of 4-fold Samples.	$(\frac{1}{2} + \frac{1}{2})^8$.
− 4	—	...	0·0012	0·0039
− 3	....	—	0·0165	0·0313
.. 2	0·0352	0·0625	0·0873	0·1094
− 1	0·2343	0·2500	0·2324	0·2188
0	0·4609	0·3750	0·3254	0·2734
1	0·2343	0·2500	0·2324	0·2188
2	0·0352	0·0625	0·0873	0·1094
3	.	. ..	0·0165	0·0313
4		...	0·0012	0·0039

(a) Since they are always symmetrical about a mean difference of *zero*, i.e. positive and negative deviations which are numerically equivalent occur with equal frequency, the mean score difference is also zero.

(b) The number of terms in the distribution is $2r + 1$. Now the binomial histogram is exactly symmetrical only if $p = \frac{1}{2} = q$; and the only binomial distribution which could exactly correspond to one of the type under discussion is therefore the expansion of $(\frac{1}{2} + \frac{1}{2})^{2r}$. The accompanying table (Table 1) shows that the frequency distributions of Figs. 43 and 45 are not identical with successive terms of the expansion $(\frac{1}{2} + \frac{1}{2})^{2r}$, being in fact more *steep* in the middle of the range. That it is in fact less flat means that the use of the binomial for an approximate evaluation of significance levels would give a too high expectation for large deviations. If we did so we should not therefore err by overstating the odds against a difference being as large as observed. Thus reference to the half of the table which exhibits frequencies for pairs of 4-fold samples shows that

$$E(\geqslant 3) = 2(0·0012 + 0·0165) = 0·0354.$$

The true odds against getting a heart score difference numerically as great as 3 in a draw of 4 cards from each of 2 full packs are therefore roughly $965 : 35$ or over $27 : 1$. If we acted on the assumption that the distribution approximately tallies with the terms of $(\frac{1}{2} + \frac{1}{2})^8$, we should infer an expectation of $2(0·0039 + 0·0313) = 0·0704$, or odds of roughly $93 : 7$ or less than $14 : 1$ against the occurrence.

The study of these examples does not at first sight encourage the hope that the distribution of score differences is reducible to a *normal* type ; and hence that the assessment of a significance difference is amenable to a *c*-test. On the other hand, we have seen that the normal distribution approximates most closely to the exact distribution of a small sample, when $p = \frac{1}{2} = q$. This prompts us to explore the *colour* score difference, in contradistinction to the *heart*-score difference distribution, i.e. the frequency with which the difference between the number of *red* (heart *or* diamond) cards chosen assumes particular values. From the chessboard diagrams of Figs. 42 and 44 we can derive the *colour* score (red-black) difference frequencies by making the substitution $\frac{1}{2}$ for both $\frac{1}{4}$ and $\frac{3}{4}$, *viz.*

Score Difference (d)	For Samples of 2	For Samples of 4
− 4	—	$1(\tfrac{1}{2})^8$
− 3	—	$8(\tfrac{1}{2})^8$
− 2	$1(\tfrac{1}{2})^4$	$28(\tfrac{1}{2})^8$
1	$4(\tfrac{1}{2})^4$	$56(\tfrac{1}{2})^8$
0	$6(\tfrac{1}{2})^4$	$70(\tfrac{1}{2})^8$
1	$4(\tfrac{1}{2})^4$	$56(\tfrac{1}{2})^8$
2	$1(\tfrac{1}{2})^4$	$28(\tfrac{1}{2})^8$
3	—	$8(\tfrac{1}{2})^8$
4	—	$1(\tfrac{1}{2})^8$

When $p = \tfrac{1}{2} = q$, the frequencies of score difference of samples of equal size thus correspond exactly to successive terms of the expansion $(\tfrac{1}{2} + \tfrac{1}{2})^{2r}$. This gives us some encouragement for entertaining the hope that a normal distribution would correctly describe that of a score difference involving *unequal* values of p and q, if the samples were large. If so, the binomial which we should expect to give the best fit for the distribution of the heart score difference would be a binomial $(\tfrac{1}{2} + \tfrac{1}{2})^r$ having the same *variance* w.r.t. the raw score distribution. We have therefore to ask : what is the variance (V_d) of the distributions exhibited in Figs. 43 and 45. It will suffice to consider the latter, *viz.* that of pair differences w.r.t. 4-fold samples. If we denote the frequency of a difference d by y_d, (iii) in 3·06 becomes

$$V_d = \sum_{d=-4}^{d=+4} y_d . d^2 - M_d{}^2.$$

Since the mean difference (M_d) is zero

$$V_d = \sum_{d=-4}^{d=+4} y_d . d^2.$$

The summation is easiest to perform if we reverse the order of terms in one margin of the chess-board set-up of Fig. 44 as in Table 1A.

TABLE 1A

Score		0	1	2	3	4
	Frequency	$(0·75)^4$	$4(0·75)^3(0·25)$	$6(0·75)^2(0·25)^2$	$4(0·75)(0·25)^3$	$(0·25)^4$
4	$(0·25)^4$	$D = -4$ $(0·75)^4(0·25)^4$	$D = -3$ $4(0·75)^3(0·25)^5$	$D = -2$ $6(0·75)^2(0·25)^6$	$D = -1$ $4(0·75)(0·25)^7$	$D = 0$ $(0·25)^8$
3	$4(0·25)^3(0·75)$	$D = -3$ $4(0·75)^5(0·25)^3$	$D = -2$ $16(0·75)^4(0·25)^4$	$D = -1$ $24(0·75)^3(0·25)^5$	$D = 0$ $16(0·75)^2(0·25)^6$	$D = +1$ $4(0·75)(0·25)^7$
2	$6(0·25)^2(0·75)^2$	$D = -2$ $6(0·75)^6(0·25)^2$	$D = -1$ $24(0·75)^5(0·25)^3$	$D = 0$ $36(0·75)^4(0·25)^4$	$D = +1$ $24(0·75)^3(0·25)^5$	$D = +2$ $6(0·75)^2(0·25)^6$
1	$4(0·25)(0·75^3)$	$D = -1$ $4(0·75)^7(0·25)$	$D = 0$ $16(0·75)^6(0·25)^2$	$D = +1$ $24(0·75)^5(0·25)^3$	$D = +2$ $16(0·75)^4(0·25)^4$	$D = +3$ $4(0·75)^3(0·25)^5$
0	$(0·75)^4$	$D = 0$ $(0·75)^8$	$D = +1$ $4(0·75)^7(0·25)$	$D = +2$ $6(0·75)^6(0·25)^2$	$D = +3$ $4(0·75)^5(0·25)^3$	$D = +4$ $(0·75)^4(0·25)^4$

A pattern now emerges if we apply the device already used to derive Pascal's triangle from the Figurate number table of p. 26. That is to say, we slide the second column downwards one row, the third column downwards two rows, and so on, the result being as shown in Table 1B.

TABLE 1B

Difference	Frequencies
-4	$(0 \cdot 75)^4(0 \cdot 25)^4$
-3	$4(0 \cdot 75)^5(0 \cdot 25)^3 + 4(0 \cdot 75)^3(0 \cdot 25)^5$
-2	$6(0 \cdot 75)^6(0 \cdot 25)^2 + 16(0 \cdot 75)^4(0 \cdot 25)^4 + 6(0 \cdot 75)^2(0 \cdot 25)^6$
-1	$4(0 \cdot 75)^7(0 \cdot 25) + 24(0 \cdot 75)^5(0 \cdot 25)^3 + 24(0 \cdot 75)^3(0 \cdot 25)^5 + 4(0 \cdot 75)(0 \cdot 25)^7$
0	$(0 \cdot 75)^8 + 16(0 \cdot 75)^6(0 \cdot 25)^2 + 36(0 \cdot 75)^4(0 \cdot 25)^4 + 16(0 \cdot 75)^2(0 \cdot 25)^6 + (0 \cdot 25)^8$
1	$4(0 \cdot 75)^7(0 \cdot 25) + 24(0 \cdot 75)^5(0 \cdot 25)^3 + 24(0 \cdot 75)^3(0 \cdot 25)^5 + 4(0 \cdot 75)(0 \cdot 25)^7$
2	$6(0 \cdot 75)^6(0 \cdot 25)^2 + 16(0 \cdot 75)^4(0 \cdot 25)^4 + 6(0 \cdot 75)^2(0 \cdot 25)^6$
3	$4(0 \cdot 75)^5(0 \cdot 25)^3 + 4(0 \cdot 75)^3(0 \cdot 25)^5$
4	$(0 \cdot 75)^4(0 \cdot 25)^4$

From inspection of the first column of frequency terms, we see that the sum is in fact

$$(0 \cdot 75)^4 \cdot (0 \cdot 25 + 0 \cdot 75)^4 = (0 \cdot 75)^4.$$

Similarly, the sum of the terms in the second column is

$$4(0 \cdot 75)^3 \cdot (0 \cdot 25)(0 \cdot 25 + 0 \cdot 75)^4 = 4(0 \cdot 75)^3(0 \cdot 25).$$

In this way we arrive at the following table of column totals:

Column	Total
1	$(0 \cdot 75)^4 \times (0 \cdot 25 + 0 \cdot 75)^4$
2	$4(0 \cdot 75)^3(0 \cdot 25) \times (0 \cdot 25 + 0 \cdot 75)^4$
3	$6(0.75)^2(0.25)^2 \times (0.25 + 0.75)^4$
4	$4(0 \cdot 75)(0.25)^3 \times (0 \cdot 25 + 0.75)^4$
5	$(0.25)^4 \times (0.25 + 0.75)^4$

To elucidate the pattern of weighted square differences, we may now use $p = 0 \cdot 25$ and $q = 0 \cdot 75$ throughout. Since $(d - 4)^2 = (4 - d)^2$, we then see that the sum of the weighted squares in col. 1 is

$$q^4 \sum_{d=0}^{d=4} (d - 4)^2 \frac{4!}{d!\,(4 - d)!}\, p^{4-d} q^d.$$

Similarly, the sum of the weighted squares in col. (ii) is

$$4pq^3 \sum_{0}^{4} (d - 3)^2 \frac{4!}{d!\,(4 - d)!}\, p^{4-d} \cdot q^d.$$

The total of the weighted scores is therefore the sum of

$$q^4 \cdot \sum_{0}^{4} (d - 4)^2 \cdot 4_{(d)} \cdot p^{4-d} \cdot q^d;$$

$$4pq^3 \sum_{0}^{4} (d - 3)^2 \cdot 4_{(d)} \cdot p^{4-d} \cdot q^d;$$

$$6p^2q^2 \sum_{0}^{4} (d - 2)^2 \cdot 4_{(d)} \cdot p^{4-d} \cdot q^d;$$

$$4p^3q \sum_{0}^{4} (d - 1)^2 \cdot 4_{(d)} \cdot p^{4-d} \cdot q^d;$$

$$p^4 \sum_{0}^{4} (d - 0)^2 \cdot 4_{(d)} \cdot p^{4-4} \cdot q^4.$$

We may write this total as

$$\sum_{x=0}^{x=4} 4_{(x)} \cdot q^x p^{4-x} \cdot \sum_{d=0}^{d=4} (d-x)^2 \cdot 4_{(d)} \cdot p^{4-d} \cdot q^d \qquad . \qquad . \qquad . \qquad \text{(i)}$$

Since $(x-d)^2 = (d-x)^2 = (d^2 - 2xd + x^2)$, the general term of the *second* summation is equivalent to

$$\sum_{d=0}^{d=4} (d^2 - 2xd + x^2) \cdot 4_{(d)} \; p^{4-d} \; q^d \qquad . \qquad . \qquad . \qquad \text{(ii)}$$

Since x is constant in each such expression, we may bring it outside the summation sign, so that (ii) becomes

$$\sum_{d=0}^{d=4} d^2 \cdot 4_{(d)} \cdot p^{4-d} \cdot q^d - 2x \sum_{d=0}^{d=4} d \cdot 4_{(d)} \cdot p^{4-d} \cdot q^d + x^2 \sum_{d=0}^{d=4} 4_{(d)} \cdot p^{4-d} \cdot q^d.$$

From (iv) in 3.06

$$\sum_{d=0}^{d=4} d^2 \cdot 4_{(d)} \cdot p^{4-d} \cdot q^d = 4pq + (4q)^2 = 4pq + 16q^2 \qquad . \qquad . \qquad \text{(iii)}$$

Also, as we have seen (p. 102) in 3.01

$$\sum_{d=0}^{d=4} d \cdot 4_{(d)} \, p^{4-d} \cdot q^d = 4q \qquad . \qquad . \qquad . \qquad . \qquad \text{(iv)}$$

By definition :

$$\sum_{d=0}^{d=4} 4_{(d)} \cdot p^{4-d} \cdot q^d = (p+q)^4 = 1 \; . \qquad . \qquad . \qquad . \qquad \text{(v)}$$

Hence (ii) becomes

$$4pq + 16q^2 - 8qx + x^2.$$

Thus (i) reduces to

$$\sum_{x=0}^{x=4} 4_{(x)} \cdot p^{4-x} \cdot q^x \cdot (4pq + 16q^2 - 8qx + x^2)$$

$$= (4pq + 16q^2) \sum_{x=0}^{x=4} 4_{(x)} \cdot p^{4-x} \cdot q^x - 8q \sum_{x=0}^{x=4} x \cdot 4_{(x)} \cdot p^{4-x} \cdot q^x + \sum_{x=0}^{x=4} x^2 \cdot 4_{(x)} \cdot p^{4-x} \cdot q^x.$$

By means of the relations in (iii)-(v) above, this becomes

$$4pq + 16q^2 - 32q^2 + 4pq + 16q^2 = 8pq.$$

The student should without difficulty be able to generalise this result, i.e. to show that for pairs of r-fold samples from a universe specified by p successes and q failures, the variance of the difference distribution is $2rpq$. For the particular case under consideration $r = 4$ and $p = \frac{1}{4}$, so that

$$V_d = 8(0{\cdot}25)(0{\cdot}75) = 1{\cdot}5.$$

For the binomial $(\frac{1}{2} + \frac{1}{2})^r$ of which the variance of the raw score distribution is 1.5, the numerical value of r is given by

$$rpq = (0.5)^2\, r = 1.5,$$
$$\therefore\ r = 6.$$

Accordingly, we shall now examine how far the heart score *difference* distribution w.r.t. 4-fold samples tallies with the terms of $(\frac{1}{2} + \frac{1}{2})^6$ for corresponding *deviations* as exhibited in Table 1c.

TABLE 1c

Score Difference or Deviation.	Individual Frequencies.		Cumulative Frequencies.	
	Pairs of 4-fold Samples $(p = \frac{1}{4})$.	Single 6-fold Samples $(p = \frac{1}{2})$.	Pairs of 4-fold Samples $(p = \frac{1}{4})$.	Single 6-fold Samples $(p = \frac{1}{2})$.
− 4	0·001	0·000	0·001	0·000
− 3	0·016	0·016	0·018	0·016
− 2	0·087	0·094	0·105	0·109
− 1	0·233	0·234	0·338	0·343
0	0·325	0·313	0·662	0·656
1	0·233	0·234	0·895	0·891
2	0·087	0·094	0·982	0·985
3	0·016	0·016	0·999	1·000
4	0·001	0·000	1·000	1·000

The correspondence exhibited by these figures is rewarding, especially if we compare the columns setting out the summated frequencies which specify the expectation that a difference or deviation will be as great as a particular value shown in the column at the extreme left.

EXERCISE 4.03

1. By the sliding chessboard method of 4.05 show that the variance of the raw score difference distribution for pairs of r-fold samples taken from the same universe with replacement is $2rpq$.

2. For samples of equal size, find an expression for the weighted mean cube of the raw score difference (*third moment*).

3. Evaluate the weighted mean of the 4th power of the raw score difference w.r.t. equal samples from the same universe.

4. Use the method of 4.03 to show that the mean difference of two sample raw scores taken from the same universe without replacement is the difference between the sample means.

5. By the same method, show that for the distribution of *score deviation* differences w.r.t. samples respectively composed of a and b items the variance is $(a + b)pq$.

4.04 MEAN SCORE DIFFERENCE W.R.T. SMALL SAMPLES OF DIFFERENT SIZE

For simplicity, we have hitherto discussed samples of equal size; but it would be inconvenient to restrict comparison of samples by insisting on this limitation. Within the framework of this assumption, our method of scoring the result of an individual trial is immaterial, and the raw score is the most direct method of recording the result. For reasons which will now appear, other methods of scoring are more convenient when samples of the same pair are *not* of the same size.

Any significance test of the sort we have examined so far presupposes the possibility of defining the expectation that an observed quantity will lie within or outside a certain range of values on either side of its mean. To do this we must know where the mean lies; and we know that the mean raw score difference is zero when the samples are of equal size. Table 2, which sets out the frequency distribution of raw score differences for pairs of 3-fold and 5-fold samples, illustrates the truism that the range of negative score differences is greater than the range of positive score differences if B is the larger sample. Thus the mean raw score difference is not necessarily zero, and in general will not be zero, if the samples are *unequal*.

TABLE 2

Raw Score Differences for Pairs of 3-fold and 5-fold Samples with Corresponding Frequencies

		SAMPLE A (3 items)			
	Score (x)	0	1	2	3
	Frequency (y)	q^3	$3pq^2$	$3p^2q$	p^3
		0	1	2	3
0	q^5	q^8	$3pq^7$	$3p^2q^6$	p^3q^5
1	$5pq^4$	-1 / $5pq^7$	0 / $15p^2q^6$	1 / $15p^3q^5$	2 / $5p^4q^4$
2 (SAMPLE B, 5 items)	$10p^2q^3$	-2 / $10p^2q^6$	-1 / $30p^3q^5$	0 / $30p^4q^4$	1 / $10p^5q^3$
3	$10p^3q^2$	-3 / $10p^3q^5$	-2 / $30p^4q^4$	-1 / $30p^5q^3$	0 / $10p^6q^2$
4	$5p^4q$	-4 / $5p^4q^4$	-3 / $15p^5q^3$	-2 / $15p^6q^2$	-1 / $5p^7q$
5	p^5	-5 / p^5q^3	-4 / $3p^6q^2$	-3 / $3p^7q$	-2 / p^8

(First data row B=0: differences 0, 1, 2, 3.)

In what follows, we shall therefore explore the difference distribution of the score deviation (X), the proportionate score ($u_x = x \div r$) and the proportionate score deviation ($U_x = X \div r$) in samples of *unequal* size; and it will be helpful if we first recall what we do when we score the result of taking a single sample from either pack by one or other method, as it affects the way in which we score a pair of differences for unequal samples taken from two identical packs. The schema below (Table 2A) refers illustratively to pairs of which one member (A) contains 3 cards, the other (B) 5 cards.

TABLE 2A

SAMPLE A (3 items)				SAMPLE B (5 items)			
Raw Score (x_a).	Score Deviation (X_a).	Proportionate Score (U_a).	Proportionate Score Deviation (U_a).	Raw Score (x_b).	Score Deviation (X_b).	Proportionate Score (u_b).	Proportionate Score Deviation (U_b).
0	$0-3p$	0	$0-p$	0	$0-5p$	0	$0-p$
1	$1-3p$	$\frac{1}{3}$	$\frac{1}{3}-p$	1	$1-5p$	$\frac{1}{5}$	$\frac{1}{5}-p$
2	$2-3p$	$\frac{2}{3}$	$\frac{2}{3}-p$	2	$2-5p$	$\frac{2}{5}$	$\frac{2}{5}-p$
3	$3-3p$	1	$1-p$	3	$3-5p$	$\frac{3}{5}$	$\frac{3}{5}-p$
—	—	—	—	4	$4-5p$	$\frac{4}{5}$	$\frac{4}{5}-p$
—	—	—	—	5	$5-5p$	1	$1-p$

For purposes of general discussion, it is now customary to speak of the independent variable of a frequency distribution as a *variate*. Hitherto, we have consistently reserved the symbol y for frequencies, and have variously labelled the single variate of a frequency distribution as x, X, u_x and U_x. We now have to deal with two variates ; and it will be convenient to use A and B for variates respectively referable to samples of a and b items. Hence the range of raw A scores is from o to a, and that of B from o to b. If we now retain y for the frequency of the sampling distribution of a difference, we shall need respectively symbols v and w for the frequencies of the two variates A and B to the frequency of whose difference $(D = A - B)$ in this context y itself refers. We shall denote the mean of the difference distribution by M_d. In accordance with definitions elsewhere (pp. 102 and 132) we define the means (M_a and M_b) of A and B and the variances (V_a and V_b) of their sampling distributions by

$$M_a = \sum_{k=o}^{k=a} v_k A_k \; ; \quad M_b = \sum_{l=o}^{l=b} w_l B_l \qquad \qquad \text{(i)}$$

$$V_a = \left\{ \sum_{k=o}^{k=n} v_k A_k^2 \right\} - M_a^2 \; ; \quad V_b = \left\{ \sum_{l=o}^{l=b} w_l B_l^2 \right\} - M_b^2 \qquad \text{(ii)}$$

By definition

$$\sum_{k=o}^{k=a} v_k = 1 \; ; \quad \sum_{l=o}^{l=b} w_l = 1 \qquad \qquad \text{(iii)}$$

The distribution of the difference extends over the whole chessboard of Table 3 in two dimensions ; and the determination of the weighted means thus involves a *double* summation which we may perform *either* by first adding up all the elements in a row and then adding up all the row totals *or* by first adding up all the items in a column and then adding up all the column totals. We write this in the form

$$M_d = \sum_{k=o}^{k=a} \sum_{l=o}^{l=b} y_{kl} \cdot D_{kl} = \sum_{k=o}^{k=a} \sum_{l=o}^{l=b} v_k \cdot w_l \, (A_k - B_l) \qquad \text{(iv)}$$

TABLE 3

Difference Distribution w.r.t. Independent Variates

Variate		A_o	. . . .	A_k	. . . .	A_a	Total
	Frequency	v_o	. . . .	v_k	. . . .	v_a	frequency
B_o	w_o	D_{oo} $(A_o - B_o)$ $y_{oo} = v_o \cdot w_o$	. .	D_{ko} $(A_k - B_o)$ $y_{ko} = v_k \cdot w_o$	. .	$D_{ao} = (A_a - B_o)$ $y_{ao} = v_a \cdot w_o$	w_o
		. . . .	. . .	. . . .	. . .	. . . .	
B_l	w_l	$(D_{ol} = (A_o - B_l)$ $y_{ol} = v_o \cdot w_l$	. .	$(D_{kl}$ $(A_k$ $B_l)$ $y_{kl} = v_k \cdot w_l$	. .	D_{al} $(A_a - B_l)$ $y_{al} = v_a \cdot w_l$	w_l
		. . . .	. . .	. . . .	. . .	. . . .	
B_b	w_b	$D_{ob} = (A_o$ $B_b)$ $y_{ob} = v_o \cdot w_b$	. .	D_{kb} $(A_k - B_b)$ $y_{kb} = v_k \cdot w_b$	. .	D_{ab} $(A_a - B_b)$ $y_{ab} = v_a \cdot w_b$	w_b
Total Frequency		v_o		v_k		v_a	

A similar operation comes into a large number of statistical problems and it is important to visualize each step. If we first sum the items of a row, we may set out the preliminary addition for the lth row as follows :

$$\left\{ y_{ol} \cdot D_{ol} + y_{1l} \cdot D_{1l} + y_{2l} \cdot D_{2l} \ldots + y_{kl} \cdot D_{kl} \ldots + y_{al} D_{al} \right\} =$$

$$\left\{ v_o w_l (A_o - B_l) + v_1 w_l (A_1 - B_l) + v_2 w_l (A_2 - B_l) \ldots + v_k w_l (A_k - B_l) \ldots + v_a w_l (A_a - B_l) \right\}$$

$$= \sum_{k=0}^{k=a} v_k \cdot w_l (A_k - B_l).$$

We may thus write the lth row weighted total difference as

$$\sum_{k=0}^{k=a} v_k \cdot w_l A_k - \sum_{k=0}^{k=a} v_k \cdot w_l B_l.$$

Within the lth row B_l and w_l remain constant, so that this expression is equivalent to

$$w_l \cdot \sum_{k=0}^{k=a} v_k A_k - w_l B_l \sum_{k=0}^{k=a} v_k.$$

By (i) and (iii) above this is : $w_l \cdot M_a - w_l B_l$.

We now have to add all these row totals from the oth to the bth row inclusive to get the final result, i.e.

$$M_d = \sum_{l=0}^{l=b} (w_l M_a - w_l B_l) = \sum_{l=0}^{l=b} w_l M_a - \sum_{l=0}^{l=b} w_l B_l.$$

Since M_a is a constant term

$$M_d = M_a \sum_{l=0}^{l=b} w_l - \sum_{l=0}^{l=b} w_l B_l = M_a - M_b \quad . \quad . \quad . \quad . \quad \text{(v)}$$

FIG. 46. SUM OR DIFFERENCE OF TWO INDEPENDENT VARIATES

$$\sum_{k=0}^{k=a} \sum_{l=0}^{l=b} v_k w_l (A_k \pm B_l) = M_a \pm M_b$$

FREQUENCY	FREQUENCY	v_0	v_1	. . .	v_k	. . .	v_a	TOTAL
FREQUENCY	VARIATE	A_0	A_1	. . .	A_k	. . .	A_a	
w_0	B_0	$w_0 v_0(A_0 \pm B_0)$	$w_0 v_1(A_1 \pm B_0)$	. . .	$w_0 v_k(A_k \pm B_0)$	. . .	$w_0 v_a(A_a \pm B_0)$	$w_0 \sum_{k=0}^{k=a} v_k A_k \pm w_0 B_0 \sum_{k=0}^{k=a} v_k$ $= w_0 M_a \pm w_0 B_0$
w_1	B_1	$w_1 v_0(A_0 \pm B_1)$	$w_1 v_1(A_1 \pm B_1)$	. . .	$w_1 v_k(A_k \pm B_1)$	. . .	$(w_1 v_a(A_a \pm B_1)$	$w_1 \sum_{k=0}^{k=a} v_k A_k \pm w_1 B_1 \sum_{k=0}^{k=a} v_k$ $= w_1 M_a \pm w_1 B_1$
—	—	—	—		—		—	—
—	—	—	—		—		—	—
—	—	—	—		—		—	—
—	—	—	—		—		—	—
—	—	—	—		—		—	—
—	—	—	—		—		—	—
w_l	B_l	$w_l v_0(A_0 \pm B_l)$	$w_l v_1(A_1 \pm B_l)$	. . .	$w_l v_k(A_k \pm B_l)$	. . .	$w_l v_a(A_a \pm B_l)$	$w_l \sum_{k=0}^{k=a} v_k A_k \pm w_l B_l \sum_{k=0}^{k=a} v_k$ $= w_l M_a \pm w_l B_l$
—	—	—	—		—		—	—
—	—	—	—		—		—	—
—	—	—	—		—		—	—
—	—	—	—		—		—	—
—	—	—	—		—		—	—
—	—	—	—		—		—	—
w_b	B_b	$w_b v_0(A_0 \pm B_b)$	$w_b v_1(A_1 \pm B_b)$	. . .	$w_b v_k(A_k \pm B_b)$	. . .	$w_b v_a(A_a \pm B_b)$	$w_b \sum_{k=0}^{k=a} v_k A_k \pm w_b B_b \sum_{k=0}^{k=a} v_k$ $= w_b M_a \pm w_b B_b$
		$v_0 A_0 \sum_{l=0}^{l=b} w_l$ $\pm v_0 \sum_{l=0}^{l=b} w_l B_l$	$v_1 A_1 \sum_{l=0}^{l=b} w_l$ $\pm v_1 \sum_{l=0}^{l=b} w_l B_l$	. . .	$v_k A_k \sum_{l=0}^{l=b} w_l$ $\pm v_k \sum_{l=0}^{l=b} w_l B_l$	. . .	$v_a A_a \sum_{l=0}^{l=b} w_l$ $\pm v_a \sum_{l=0}^{l=b} w_l B_l$	$M_a \sum_{l=0}^{l=b} w_l \pm \sum_{l=0}^{l=b} w_l B_l = M_a \pm M_b$
	TOTAL	$= v_0 A_0 \pm v_0 M_b$	$= v_1 A_1 \pm v_1 M_b$	. . .	$= v_k A_k \pm v_k M_b$	. . .	$= v_a A_a \pm v_a M_b$	$= \sum_{k=0}^{k=a} v_k A_k \pm M_b \sum_{k=0}^{k=a} v_k = M_a \pm M_b$

FIG. 47. THE SQUARE OF THE SUM OR DIFFERENCE OF TWO INDEPENDENT VARIATES

$$\sum_{k=0}^{k=a}\sum_{l=0}^{l=b} v_k w_l (A_k \pm B_l)^2$$

FREQUENCY		v_0	...	v_k	..	v_a	
FREQUENCY	VARIATE	A_0	...	A_k	.	A_a	TOTAL
w_0	B_0	$v_0 w_0 (A_0 \pm B_0)^2$ $= v_0 w_0 A_0^2 \pm 2v_0 w_0 A_0 B_0$ $+ v_0 w_0 B_0^2$	...	$v_k w_0 (A_k \pm B_0)^2$ $= v_k w_0 A_k^2 \pm 2v_k w_0 A_k B_0$ $+ v_k w_0 B_0^2$	....	$v_a w_0 (A_a \pm B_0)^2$ $= v_a w_0 A_a^2 \pm 2v_a w_0 A_a B_0$ $+ v_a w_0 B_0^2$	$w_0 \sum_{k=0}^{k=a} v_k A_k^2 \pm 2w_0 B_0 \sum_{k=0}^{k=a} v_k A_k$ $+ w_0 B_0^2 \sum_{k=0}^{k=a} v_k$ $= w_0 V_0(A) \pm 2w_0 B_0 M_a + w_0 B_0^2$
—	—	—	—	—	—	—	—
—	—	—	—	—	—	—	..
—	—	—	—	—	—	—	—
—	—	—	—	—	—	—	—
—	—	—	...	—	—	—	—
—	—	—	—	—	—	—	—
—	—	—	—	—	—	—	—
w_l	B_l	$v_0 w_l (A_0 \pm B_l)^2$ $= v_0 w_l A_0^2 \pm 2v_0 w_l A_0 B_l$ $+ v_0 w_l B_l^2$	....	$v_k w_l (A_k \pm B_l)^2$ $= v_k w_l A_k^2 \pm 2v_k w_l A_k B_l$ $+ v_k w_l B_l^2$	....	$v_a w_l (A_a \pm B_l)^2$ $= v_a w_l A_a^2 \pm 2v_a w_l A_a B_l$ $+ v_a w_l B_l^2$	$w_l \sum_{k=0}^{k=a} v_k A_k^2 \pm 2w_l B_l \sum_{k=0}^{k=a} v_k A_k$ $+ w_l B_l^2 \sum_{k=0}^{k=a} v_k$ $= w_l V_0(A) \pm 2w_l B_l M_a + w_l B_l^2$
—	—	—	—	—	—	—	—
—	—	—	—	—	—	—	—
—	—	—	—	—	—	—	—
—	—	—	—	—	—	—	—
—	—	—	—	—	—	—	—
—	—	—	—	—	—	—	—
—	—	—	—	—	—	—	—
—	—	—	—	—	—	—	—
w_b	B_b	$v_0 w_b (A_0 \pm B_b)^2$ $= v_0 w_b A_0^2 \pm 2v_0 w_b A_0 B_b$ $+ v_0 w_b B_b^2$	...	$v_k w_b (A_k \pm B_b)^2$ $= v_k w_b A_k^2 \pm 2v_k w_b A_k B_b$ $+ v_k w_b B_b^2$	....	$v_a w_b (A_a \pm B_b)^2$ $= v_a w_b A_a^2 \pm 2v_a w_b A_a B_b$ $+ v_a w_b B_b^2$	$w_b \sum_{k=0}^{k=a} v_k A_k^2 \pm 2w_b B_b \sum_{k=0}^{k=a} v_k A_k$ $+ w_b B_b^2 \sum_{k=0}^{k=a} v_k$ $w_b V_0(A) + 2w_b B_b M_a + w_b B_b^2$
							$V_0(A) \sum_{l=0}^{l=b} w_l \pm 2M_a \sum_{l=0}^{l=b} w_l B_l$ $+ \sum_{l=0}^{l=b} w_l B_l^2$ $= V_0(A) \pm 2M_a M_b + V_0(B)$
	TOTAL	$v_0 A_0^2 \sum_{l=0}^{l=b} w_l \pm 2v_0 A_0 \sum_{l=0}^{l=b} w_l B_l$ $+ v_0 \sum_{l=0}^{l=b} w_l B_l^2$ $= v_0 A_0^2 \pm 2v_0 A_0 M_b + v_0 V_0(B)$	.	$v_k A_k^2 \sum_{l=0}^{l=b} w_l \pm 2v_k A_k \sum_{l=0}^{l=b} w_l B_l$ $+ v_k \sum_{l=0}^{l=b} w_l B_l^2$ $= v_k A_k^2 \pm 2v_k A_k M_b + v_k V_0(B)$	...	$v_a A_a^2 \sum_{l=0}^{l=b} w_l \pm 2v_a A_a \sum_{l=0}^{l=b} w_l B_l$ $+ v_a \sum_{l=0}^{l=b} w_l B_l^2$ $= v_a A_a^2 \pm 2v_a A_a M_b + v_a V_0(B)$	$\sum_{k=0}^{k=a} v_k A_k^2 \pm 2M_b \sum_{k=0}^{k=a} v_k A_k$ $+ V_0(B) \sum_{k=0}^{k=a} v_k$ $= V_0(A) \pm 2M_a M_b + V_0(B)$

So far we have used M_d for the mean difference of two scores, A and B, without reference to the nature of the score. We shall now indicate the particular type of score difference by parenthesis, e.g. for the mean *raw* score differences

$$M_d(x) = M_a(x) - M_b(x).$$

If the two samples come from the same universe $M_a(x)$, the mean of a-fold samples, is ap; and $M_b(x)$, the mean of b-fold samples, is bp; so that

$$M_d(x) = (a - b)p \qquad . \qquad . \qquad . \qquad . \qquad \text{(vi)}$$

which is zero only when the samples are equal $(a = b)$. Similarly, we have

$$M_a(X) = 0 = M_b(X),$$
$$\therefore M_d(X) = 0 \qquad . \qquad . \qquad . \qquad . \qquad \text{(vii)}$$

The mean proportionate score of samples of any size from the same universe is p, so that

$$M_d(u_x) = 0 = M_d(U_x) \qquad . \qquad . \qquad . \qquad . \qquad \text{(viii)}$$

Thus the mean difference w.r.t. either the score deviation or the proportionate score in samples of unequal size is in either case *zero*. One of the parameters of a null hypothesis relevant to the recognition of a real difference is therefore *unique*, if we frame it in terms of the distribution between raw score deviations or proportionate scores.

4.05 SCORE DIFFERENCE SAMPLING DISTRIBUTION FOR SMALL UNEQUAL SAMPLES

For the reason stated in the concluding sentence of 4.04, we shall confine our examination of score difference distributions for pairs of samples of which the members do *not* contain the same number of items to differences between *score deviations* and between *proportionate scores*. By (vii) and (viii) of 4.04 the mean difference is then zero.

TABLE 4

Score Deviation Differences for Pairs of 3-fold and 5-fold Samples with corresponding Frequencies $(p = \tfrac{1}{2} = q)$

		SAMPLE A (3 items)			
Deviation (X)		-1.5	-0.5	0.5	1.5
Frequency (Y)		q^3	$3pq^2$	$3p^2q$	p^3
-2.5	q^5	1 q^8	2 $3pq^7$	3 $3p^2q^6$	4 p^3q^5
-1.5	$5pq^4$	0 $5pq^7$	1 $15p^2q^6$	2 $15p^3q^5$	3 $5p^4q^4$
SAMPLE B (5 items) -0.5	$10p^2q^3$	-1 $10p^2q^6$	0 $30p^3q^5$	1 $30p^4q^4$	2 $10p^5q^3$
0.5	$10p^3q^2$	-2 $10p^3q^5$	-1 $30p^4q^4$	0 $30p^5q^3$	1 $10p^6q^2$
1.5	$5p^4q$	-3 $5p^4q^4$	-2 $15p^5q^3$	-1 $15p^6q^2$	0 $5p^7q$
2.5	p^5	-4 p^5q^3	-3 $3p^6q^2$	-2 $3p^7q$	-1 p^8

HEART SCORE DEVIATION DIFFERENCES – PAIRS OF 4-FOLD AND 8-FOLD SAMPLES

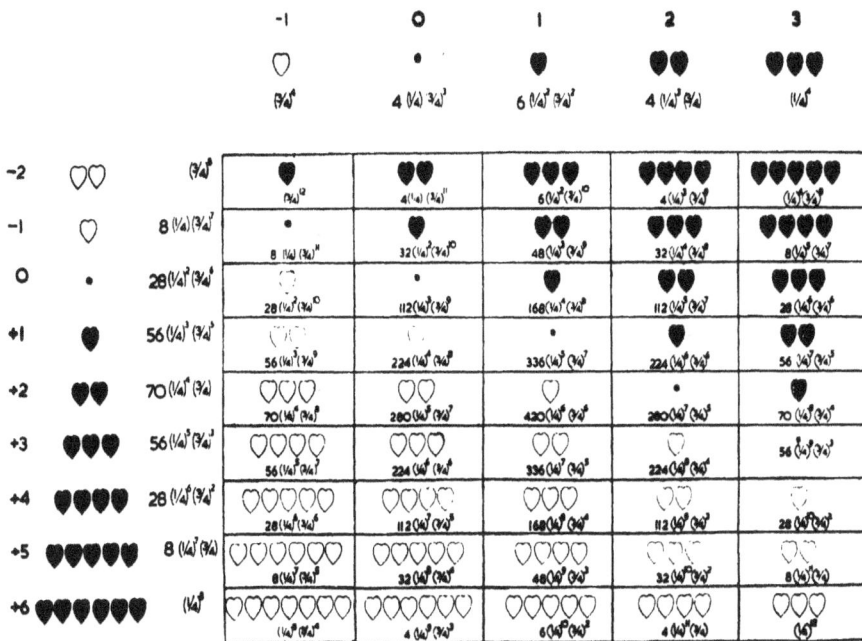

$(p = \tfrac{1}{4},\ q = \tfrac{3}{4})$

		−1	0	1	2	3
		$(\tfrac{3}{4})^4$	$4(\tfrac{1}{4})(\tfrac{3}{4})^3$	$6(\tfrac{1}{4})^2(\tfrac{3}{4})^2$	$4(\tfrac{1}{4})^3(\tfrac{3}{4})$	$(\tfrac{1}{4})^4$
−2	$(\tfrac{3}{4})^8$	$(\tfrac{3}{4})^{12}$	$4(\tfrac{1}{4})(\tfrac{3}{4})^{11}$	$6(\tfrac{1}{4})^2(\tfrac{3}{4})^{10}$	$4(\tfrac{1}{4})^3(\tfrac{3}{4})^9$	$(\tfrac{1}{4})^4(\tfrac{3}{4})^8$
−1	$8(\tfrac{1}{4})(\tfrac{3}{4})^7$	$8(\tfrac{1}{4})(\tfrac{3}{4})^{11}$	$32(\tfrac{1}{4})^2(\tfrac{3}{4})^{10}$	$48(\tfrac{1}{4})^3(\tfrac{3}{4})^9$	$32(\tfrac{1}{4})^4(\tfrac{3}{4})^8$	$8(\tfrac{1}{4})^5(\tfrac{3}{4})^7$
0	$28(\tfrac{1}{4})^2(\tfrac{3}{4})^6$	$28(\tfrac{1}{4})^2(\tfrac{3}{4})^{10}$	$112(\tfrac{1}{4})^3(\tfrac{3}{4})^9$	$168(\tfrac{1}{4})^4(\tfrac{3}{4})^8$	$112(\tfrac{1}{4})^5(\tfrac{3}{4})^7$	$28(\tfrac{1}{4})^6(\tfrac{3}{4})^6$
+1	$56(\tfrac{1}{4})^3(\tfrac{3}{4})^5$	$56(\tfrac{1}{4})^3(\tfrac{3}{4})^9$	$224(\tfrac{1}{4})^4(\tfrac{3}{4})^8$	$336(\tfrac{1}{4})^5(\tfrac{3}{4})^7$	$224(\tfrac{1}{4})^6(\tfrac{3}{4})^6$	$56(\tfrac{1}{4})^7(\tfrac{3}{4})^5$
+2	$70(\tfrac{1}{4})^4(\tfrac{3}{4})^4$	$70(\tfrac{1}{4})^4(\tfrac{3}{4})^8$	$280(\tfrac{1}{4})^5(\tfrac{3}{4})^7$	$420(\tfrac{1}{4})^6(\tfrac{3}{4})^6$	$280(\tfrac{1}{4})^7(\tfrac{3}{4})^5$	$70(\tfrac{1}{4})^8(\tfrac{3}{4})^4$
+3	$56(\tfrac{1}{4})^5(\tfrac{3}{4})^3$	$56(\tfrac{1}{4})^5(\tfrac{3}{4})^7$	$224(\tfrac{1}{4})^6(\tfrac{3}{4})^6$	$336(\tfrac{1}{4})^7(\tfrac{3}{4})^5$	$224(\tfrac{1}{4})^8(\tfrac{3}{4})^4$	$56(\tfrac{1}{4})^9(\tfrac{3}{4})^3$
+4	$28(\tfrac{1}{4})^6(\tfrac{3}{4})^2$	$28(\tfrac{1}{4})^6(\tfrac{3}{4})^6$	$112(\tfrac{1}{4})^7(\tfrac{3}{4})^5$	$168(\tfrac{1}{4})^8(\tfrac{3}{4})^4$	$112(\tfrac{1}{4})^9(\tfrac{3}{4})^3$	$28(\tfrac{1}{4})^{10}(\tfrac{3}{4})^2$
+5	$8(\tfrac{1}{4})^7(\tfrac{3}{4})$	$8(\tfrac{1}{4})^7(\tfrac{3}{4})^5$	$32(\tfrac{1}{4})^8(\tfrac{3}{4})^4$	$48(\tfrac{1}{4})^9(\tfrac{3}{4})^3$	$32(\tfrac{1}{4})^{10}(\tfrac{3}{4})^2$	$8(\tfrac{1}{4})^{11}(\tfrac{3}{4})$
+6	$(\tfrac{1}{4})^8$	$(\tfrac{1}{4})^8(\tfrac{3}{4})^4$	$4(\tfrac{1}{4})^9(\tfrac{3}{4})^3$	$6(\tfrac{1}{4})^{10}(\tfrac{3}{4})^2$	$4(\tfrac{1}{4})^{11}(\tfrac{3}{4})$	$(\tfrac{1}{4})^{12}$

FIG. 48. Chessboard diagram for Heart-score Deviation Differences w.r.t. samples of 4 and 8 taken from a full pack, subject to replacement.

Table 4 shows the chessboard set-up for the score-deviation difference distribution w.r.t. pairs of 3-fold and 5-fold samples. When $p = \tfrac{1}{2} = q$, the frequencies of the differences tally exactly with successive terms of the binomial $(\tfrac{1}{2} + \tfrac{1}{2})^{3+5}$, as we see by collecting coefficients in cells with the same difference value :

Score Deviation Difference $D(X)$	Frequency (Y_d)
− 4	$= 1(\tfrac{1}{2})^8$
− 3	$(5 + 3)(\tfrac{1}{2})^8 = 8(\tfrac{1}{2})^8$
− 2	$(10 + 15 + 3)(\tfrac{1}{2})^8 = 28(\tfrac{1}{2})^8$
− 1	$(10 + 30 + 15 + 1)(\tfrac{1}{2})^8 = 56(\tfrac{1}{2})^8$
0	$(5 + 30 + 30 + 5)(\tfrac{1}{2})^8 = 70(\tfrac{1}{2})^8$
1	$(10 + 30 + 15 + 1)(\tfrac{1}{2})^8 = 56(\tfrac{1}{2})^8$
2	$(10 + 15 + 3)(\tfrac{1}{2})^8 = 28(\tfrac{1}{2})^8$
3	$(5 + 3)(\tfrac{1}{2})^8 = 8(\tfrac{1}{2})^8$
4	$= 1(\tfrac{1}{2})^8$

For the set-up of Fig. 48 which refers to *heart*-score deviation differences w.r.t. 4-fold and 8-fold samples, $p\ (= \tfrac{1}{4})$ and $q\ (= \tfrac{3}{4})$ are unequal ; and the distribution is *not* a true binomial ; but a table of coefficients (as below) shows that the corresponding difference distribution for the colour (red-black) score deviation difference $D(X)$ would tally exactly with the terms of

$$(\tfrac{1}{2} + \tfrac{1}{2})^{4+8} = (\tfrac{1}{2} + \tfrac{1}{2})^{12} :$$

Difference D (X)		Sum of Coefficients	
(a) $p = \frac{1}{4}$	(b) $p = \frac{1}{4}$		
5	6	1	$= 1$
4	5	$4 + 8$	$= 12$
3	4	$6 + 32 + 28$	$= 66$
2	3	$4 + 48 + 112 + 56$	$= 220$
1	2	$1 + 32 + 168 + 224 + 70$	$= 495$
0	1	$8 + 112 + 336 + 280 + 56$	$= 792$
-1	0	$28 + 224 + 420 + 224 + 28$	$= 924$
-2	-1	$56 + 280 + 336 + 112 + 8$	$= 792$
-3	-2	$70 + 224 + 168 + 32 + 1$	$= 495$
-4	-3	$56 + 112 + 48 + 4$	$= 220$
-5	-4	$28 + 32 + 6$	$= 66$
-6	-5	$8 + 4$	$= 12$
-7	-6	1	$= 1$

The student should be able to show that the results last stated are quite general, i.e. the score deviation difference distribution w.r.t. pairs of a-fold and b-fold samples is given by successive terms of the expansion of $(\frac{1}{2} + \frac{1}{2})^{a+b}$, when $p = \frac{1}{2} = q$. When p and q are not equal, the distribution does *not* exactly correspond with the terms of the expansion of $(q + p)^{a+b}$.

TABLE 5

Proportionate Score Differences for Pairs of 4-fold and 8-fold Samples with corresponding Frequencies

SAMPLE A (4 *items*)

u_a		0	0·250	0·500	0·750	1·000
	y	q^4	$4pq^3$	$6p^2q^2$	$4p^3q$	p^4
0	q^8	0 q^{12}	0·250 $4pq^{11}$	0·500 $6p^2p^{10}$	0·750 $4p^3q^9$	1·000 p^4q^8
0·125	$8pq^7$	$-0\cdot125$ $8pq^{11}$	0·125 $32p^2q^{10}$	0·375 $48p^3q^9$	0·625 $32p^4q^8$	0·875 $8p^5q^7$
0·250	$28p^2q^6$	$-0\cdot250$ $28p^2q^{10}$	0 $112p^3q^9$	0·250 $168p^4q^8$	0·500 $112p^5q^7$	0·750 $28p^6q^6$
0·375	$56p^3q^5$	$-0\cdot375$ $56p^3q^9$	$-0\cdot125$ $224p^4q^8$	0·125 $336p^5q^7$	0·375 $224p^6q^6$	0·625 56^7q^5
0·500	$70p^4q^4$	$-0\cdot500$ $70p^4q^8$	$-0\cdot250$ $280p^5q^7$	0 $420p^6q^6$	0·250 $280p^7q^5$	0·500 $70p^8q^4$
0·625	$56p^5q^3$	$-0\cdot625$ $56p^5q^7$	$-0\cdot375$ $224p^6q^6$	$-0\cdot125$ $316p^7q^5$	0·125 $224p^8q^4$	0·375 $56p^9q^3$
0·750	$28p^6q^2$	$-0\cdot750$ $28p^6q^6$	$-0\cdot500$ $112p^7q^5$	$-0\cdot250$ $168p^8q^4$	0 $112p^9q^3$	0·250 $28p^{10}q^2$
0·875	$8p^7q$	$-0\cdot875$ $8p^7q^5$	$-0\cdot625$ $32p^8q^4$	$-0\cdot375$ $48p^9q^3$	$-0\cdot125$ $32p^{10}q^2$	0·125 $8p^{11}q$
1·00	p^8	$-1\cdot000$ p^8q^4	$-0\cdot750$ $4p^9q^3$	$-0\cdot500$ $6p^{10}q^2$	$-0\cdot250$ $4p^{11}q$	0 p^{12}

SAMPLE B (8 *items*)

TABLE 6

Proportionate Score Differences for Pairs of 3-fold and 5-fold Samples with corresponding Frequencies

		SAMPLE A (3 items)			
Proportionate Score		0	$\frac{5}{15}$	$\frac{10}{15}$	1
	Frequency	q^3	$3pq^2$	$3p^2q$	p^3
0 — q^5		0 q^8	$\frac{5}{15}$ $3pq^7$	$\frac{10}{15}$ $3p^2q^6$	1 p^3q^5
$\frac{3}{15}$ — $5pq^4$		$-\frac{3}{15}$ $5pq^7$	$\frac{2}{15}$ $15p^2q^6$	$\frac{7}{15}$ $15p^3q^5$	$\frac{12}{15}$ $15p^4q^4$
$\frac{6}{15}$ — $10p^2q^3$		$-\frac{6}{15}$ $10p^2q^6$	$-\frac{1}{15}$ $30p^3q^5$	$\frac{4}{15}$ $30p^4q^4$	$\frac{9}{15}$ $10p^5q^3$
$\frac{9}{15}$ — $10p^3q^2$		$-\frac{9}{15}$ $10p^3q^5$	$-\frac{4}{15}$ $30p^4q^4$	$\frac{1}{15}$ $30p^5q^3$	$\frac{6}{15}$ $10p^6q^2$
$\frac{12}{15}$ — $5p^4q$		$-\frac{12}{15}$ $5p^4q^4$	$-\frac{7}{15}$ $15p^5q^3$	$-\frac{2}{15}$ $15p^6q^2$	$\frac{3}{15}$ $5p^7q$
1 — p^5		-1 p^5q^3	$-\frac{10}{15}$ $3p^6q^2$	$-\frac{5}{15}$ $3p^7q$	0 p^8

SAMPLE B (5 items) — row labels at left.

Tables 5-6 respectively show the distributions for the proportionate score difference w.r.t. samples of 4 and 8 and samples of 3 and 5. Distributions of proportionate score differences and of differences between proportionate score deviations are necessarily equivalent in all circumstances, as is evident from the equation

$$U_a - U_b = (u_a - p) - (u_b - p) = u_a - u_b.$$

There is an interesting difference between the distribution of raw score deviation differences and that of proportionate score deviation differences, or, what comes to the same thing, differences between proportionate scores. When one sample A consists of a items and another (*larger*) sample B consists of b items, the character of the distribution of proportionate score differences depends on whether

(i) b is an exact multiple of a, as in Table 5;

(ii) b shares a common factor with a;

(iii) b is prime to a, as is 5 to 3 in Table 6.

We need consider only the two extreme cases. The student can examine (ii) as an exercise. The distribution exhibited in Table 5 and Fig. 49 refers to samples of 4 and 8. It has 17 terms like the expansion of $(\frac{1}{2} + \frac{1}{2})^{16}$; but they do not correspond to those of the latter when $p = \frac{1}{2} = q$, nor can they exactly correspond with the terms of the expansion of $(\frac{1}{2} + \frac{1}{2})^{12}$. The proportionate colour score difference distribution ($p = \frac{1}{2} = q$) of Fig. 49 is obtainable by summing the coefficients of Table 5. The total is 4096. To get the actual frequencies of each proportionate difference, $D(u_x)$, it is therefore necessary to divide by that number the figures in the middle column below:

$D(u_x)$ (*Negative* Range).	Frequency ($\times 4096$)	$D(u_x)$ (*Positive* Range.)
0·000	646	0·000
− 0·125	600	0·125
− 0·250	480	0·250
− 0·375	328	0·375
− 0·500	188	0·500
− 0·625	88	0·625
− 0·750	32	0·750
− 0·875	8	0·875
− 1·000	1	1·000

Comparison of the above table with a similar table of frequencies of the proportionate score differences w.r.t. pairs of 3-fold and 5-fold samples reveals a striking difference. In the foregoing table w.r.t. samples of 4 and 8, admissible differences increase or decrease by equal steps on either side of zero and the frequencies of successive values of $D(u_x)$ fall off steadily on either side of a peak value. Neither statement is true when a is prime to b. From the entries of Table 6, we obtain the following result, when $p = \frac{1}{2} = q$:

$D(u_x)$ Negative Range (in Fifteenths)	Frequency	$D(u_x)$ Positive Range (in Fifteenths)
− 0	0·0078	0
− 1	0·1172	1
− 2	0·0586	2
− 3	0·0195	3
− 4	0·1172	4
− 5	0·0117	5
− 6	0·0391	6
− 7	0·0586	7
− 8	*0·0000*	8
− 9	0·0391	9
− 10	0·0117	10
− 11	*0·0000*	11
− 12	0·0195	12
− 13	*0·0000*	13
− 14	*0·0000*	14
− 15	0·0039	15

This distribution is : (*a*) distinctly *bimodal* (Fig. 50), having a trough at the mean value (*zero*) ; (*b*) *oscillatory* inasmuch as certain frequency values within the range *of equally spaced differences* are zero. The student will find it valuable to make a chessboard similar to that of Table 6 for pairs of 5-fold and 7-fold samples (Fig. 50). He (or she) will then discover that such discontinuities do not disappear, nor does the bimodality of the distribution, the frequency for the mean difference (*zero*) being always $(q^{a+b} + p^{a+b})$ if a is prime to b.

This would discourage our hopes of finding a *c*-test for proportionate score differences, if we lost sight of the fact that the usefulness of the normal distribution depends on defining the *approximate area of a specified range of the frequency histogram* rather than defining the approximate value of any particular ordinate. No unimodal curve resembling the normal even superficially could fulfil the second function for the distribution of proportionate score differences w.r.t. samples which are *co-prime ;* but such distributions may closely approximate to the binomial type, if we *group individual values of the difference in equally spaced intervals*, as in Fig. 51.

★★★★ An interesting feature of the co-prime distributions illustrated by the entries of Table 6 is that one and the same value of the difference (*d*) other than zero occurs in only a single cell.

★★★★ Omit on first reading.

PROPORTIONATE COLOUR SCORE DIFFERENCES FOR PAIRS OF 4-FOLD AND 8-FOLD SAMPLES
$$(p = \tfrac{1}{2} = q)$$

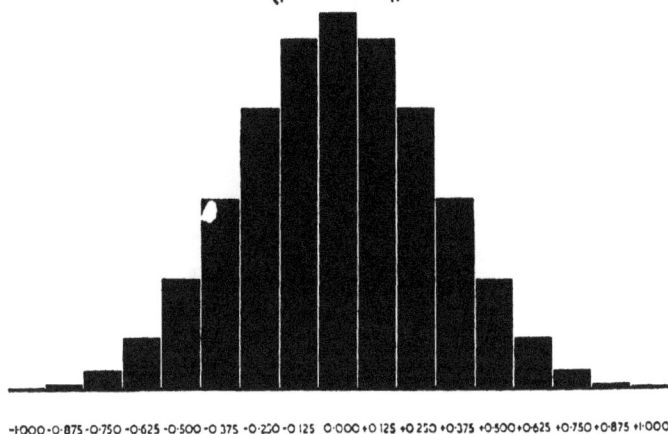

-1·000 -0·875 -0·750 -0·625 -0·500 -0·375 -0·250 -0·125 0·000 +0·125 +0·250 +0·375 +0·500 +0·625 +0·750 +0·875 +1·000

FIG. 49. The score is the difference between the proportion of *red* cards in 4-fold and 8-fold samples respectively taken from full packs with replacement.

If we denote the raw score of the *k*th column by *k* and the raw score of the *l*th row by *l*, we have by definition

$$\frac{k}{a} - \frac{l}{b} = d = \frac{bk - al}{ab}.$$

If $d = (x \div ab)$, *x* is necessarily a whole number, since *a* and *b* the total number of items and *k* and *l* the *raw* scores are all integers. We thus derive the *Diophantine* equation (i.e. equation admitting solution in *whole* numbers alone) containing three constants *a*, *b*, *x* and two variables *k*, *l* for a particular value of the difference

$$bk - al = x.$$

The Diophantine solution of this equation defines whether the sample values *a* and *b* are consistent with one or more solutions of *x*. The three relevant properties of an equation of this form are

(i) it admits of *no* solution, unless *x* is either zero or an exact multiple of *H*, the H.C.F. of *a* and *b* ;

(ii) for the particular case $x = 0$ it admits of two solutions, *viz.* : $k = 0, l = 0$ and $k = a, l = b$;

(iii) possible solutions for the two variables when there is no restriction on the magnitudes of *k* or *l* are

$$(k, l); \quad \left(k + \frac{a}{H}, \ l + \frac{b}{H}\right); \quad \left(k + \frac{2a}{H}, \ l + \frac{2b}{H}\right); \quad \left(k + \frac{3a}{H}, \ l + \frac{3b}{H}\right), \text{ etc.}$$

PROPORTIONATE COLOUR SCORE DIFFERENCE DISTRIBUTION (p = ½ = q) SAMPLE NUMBERS CO-PRIME

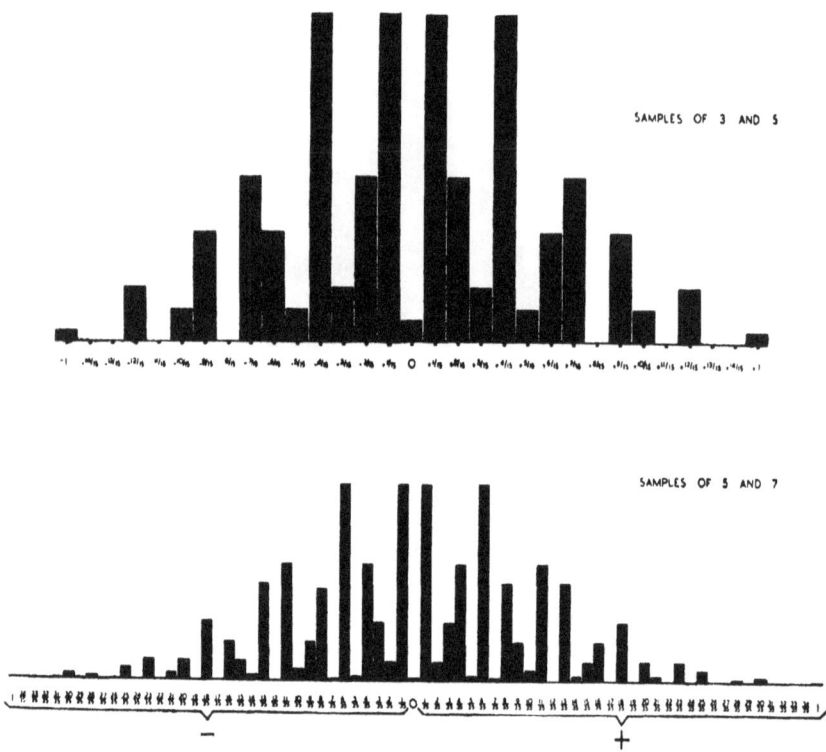

FIG. 50.   Proportion Colour Score Difference Distribution ($p = \frac{1}{2} = q$) Sample Numbers Co-Prime. *Above* samples of 3 and 5, *below* samples of 5 and 7.

In Table 10, which illustrates the intermediate case referred to above (a and b being different multiples of a common factor), $a = 4$ and $b = 6$, so that $H = 2$. For the particular value $d = \frac{8}{24}$, $x = 8$ and
$$6k - 4l = 8.$$

Since $H = 2$ is an exact divisor of $8 = x$, this admits of solution, of which by trial and error one is $k = 2$, $l = 1$. Hence other solutions with no restriction on k and l are

$$\left(2 + \frac{4}{2} = 4, \ 1 + \frac{6}{2} = 4\right); \ \left(2 + \frac{2.4}{2} = 6, \ 1 + \frac{2.6}{2} = 7\right);$$

$$\left(2 + \frac{3.4}{2} = 8, \ 1 + \frac{3.6}{2} = 10\right), \text{ etc.}$$

In fact, only the first is admissible, since k cannot exceed 4, nor can l exceed 6. Hence the only cells in which $d = \frac{8}{24}$ will be found as an entry are $(2, 1)$ and $(4, 4)$.

When a is prime to b, $H = 1$ and only one solution consistent with the restriction $k < a$ is admissible except when $x = 0$, since the lowest pair of values other than k and l defined as above is $(k = a, l = b)$. Since H is always an exact divisor of x when $H = 1$, the only restriction on

GROUPED PROPORTIONATE COLOUR SCORE DIFFERENCE DISTRIBUTIONS – CO-PRIME SAMPLES

$(p = \frac{1}{2} = q)$

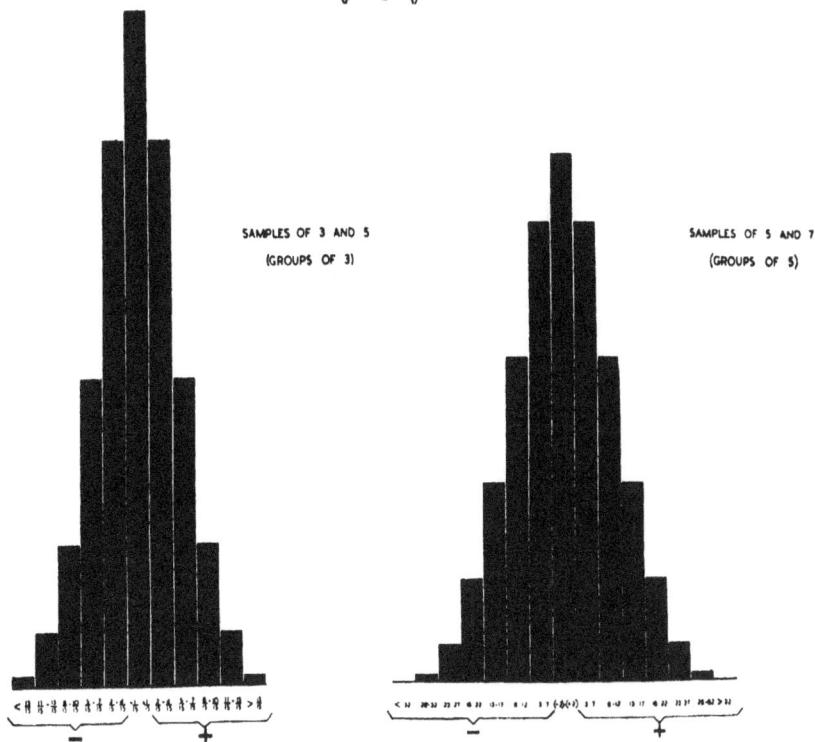

SAMPLES OF 3 AND 5
(GROUPS OF 3)

SAMPLES OF 5 AND 7
(GROUPS OF 5)

FIG. 51.   Grouped Proportionate Colour Score Difference Distributions—Co-Prime Samples

$(p = \frac{1}{2} - q)$.

The Score Differences of Fig. 50 are here grouped in equal intervals, and the Histogram Contours begin to suggest the *normal* form.

the existence of one solution of the Diophantine equation concerns the limiting value of k.   In Table 5, $ab = 15$, and we note that $x = 7$ occurs in cell $(2, 1)$ in accordance with $5k - 3l = 7$. The equation $5k - 3l = x = 8$ we may write as

$$l = \frac{5k - 8}{3}.$$

Thus l is fractional for all values of k from 0 to 3 inclusive, and there is no cell with an entry $\frac{8}{15}$. On the other hand, $5k - 9 = 31l$ admits of solution for a positive value $\frac{9}{15}$ in the range $l < 5$, $k < 3$, *viz.* $k = 3$, $l = 2$ in accordance with the entry for cell $(3, 2)$.★★★★

We may sum up the result of our enquiry at this stage as follows :

.   (i) For paired samples of equal size $(a = r = b)$ distributions of the difference of raw scores, score deviations or proportionate scores are alike unimodal and symmetrical ; and correspond to the terms of the binomial $(\frac{1}{2} + \frac{1}{2})^{2a}$, when $p = \frac{1}{2} = q$;

(ii) when a and b, the number of items of which the two samples are respectively composed are unequal, the raw score difference distribution is not symmetrical; but the score deviation difference distribution is necessarily symmetrical when $p = \frac{1}{2} = q$ and is given by $(\frac{1}{2} + \frac{1}{2})^{a+b}$;

(iii) proportionate score difference distributions of pairs of samples of unequal size are also symmetrical when $p = \frac{1}{2} = q$; but never conform exactly to a truly binomial pattern;

(iv) when b is prime to a, the proportionate score difference distribution is always multi-modal and oscillatory; but the combined frequencies of equal numbers of adjacent terms may roughly correspond to successive terms of a binomial expansion.

With these indications, we may proceed with some encouragement to examine the applicability of the binomial distribution to score deviation differences or proportionate score differences w.r.t. large samples. It will simplify our task if we first explore the meaning of the index $(a + b)$ in the expansion $(p + q)^{a+b}$;

4.06 VARIANCE OF A DIFFERENCE

When $p = \frac{1}{2} = q$, successive terms of $(\frac{1}{2} + \frac{1}{2})^r$ give the frequencies (y) of raw scores $x = (X + \frac{1}{2}r) = 0, 1, 2 \ldots r$ in r-fold samples. If r is fairly large, we have seen (pp. 111-114) that the contour of the corresponding histogram tallies closely with the symmetrical curve of the equation

$$Y = (2\pi V)^{-\frac{1}{2}} . exp\left(- X^2 \div 2V\right) \quad . \quad . \quad . \quad . \quad \text{(i)}$$

In this equation $V = rpq$. When p and q are not equal, the distribution is skew; but the skewness becomes less noticeable as r becomes larger. For very large values of r the normal equation gives a good fit to the distribution of X, if we then give V its appropriate value. For large samples of a and of b items respectively taken from the same universe, we may thus put

$$Y_a = (2\pi V_a)^{-\frac{1}{2}} . exp\left(- X_a^2 \div 2V_a\right); \quad V_a = apq.$$
$$Y_b = (2\pi V_b)^{-\frac{1}{2}} . exp\left(- X_b^2 \div 2V_b\right); \quad V_b = bpq.$$

We have seen that the distribution of the *score deviation difference* $D = (X_a - X_b)$ corresponds to successive terms of the binomial $(\frac{1}{2} + \frac{1}{2})^{a+b}$; when $p = \frac{1}{2} = q$. For fairly large values of $(a + b)$, the curve which fits this distribution is given by appropriate specification of the symbols in (i) above, viz.:

$$Y_d = (2\pi V_d)^{-\frac{1}{2}} . exp\left(- D^2 \div 2V_d\right). \quad . \quad . \quad . \quad \text{(ii)}$$

In this equation

$$V_d = (a + b)pq = apq + bpq.$$

When $p = \frac{1}{2} = q$, we therefore have

$$V_d = V_a + V_b \quad . \quad . \quad . \quad . \quad . \quad \text{(iii)}$$

The reasoning which has led us to (ii) and (iii) depends on the assumption that $p = \frac{1}{2} = q$; but the derivation of (i) encourages the suspicion that the same relations may hold good when p and q are not equal. If so, we can sidestep the laborious process of evaluating the significance of a score deviation by the methods of 4.03 and 4.05 by recourse to a c-test for which the appropriate ratio is

$$D_{ab} \div \sqrt{(V_a + V_b)} \quad . \quad . \quad . \quad . \quad . \quad \text{(iv)}$$

It will clear the decks for further action with this end in view, if we first seek a general proof of (iii). It requires no elaborate analysis to show that (iii) is implicit in the chessboard principle of equipartition of opportunity, and is therefore applicable to the sampling distribution of the difference w.r.t. any two variates which are *independent*. To evaluate (iii) we shall employ the symbolism of 4.04, and define V_d in accordance with (iii) in 3.06 as

$$V_d = \left\{ \sum_{k=0}^{k=a} \sum_{l=0}^{l=b} w_l v_k (A_k - B_l)^2 \right\} - M_d^2 \qquad \qquad \text{(v)}$$

More briefly we may write this in conformity with (iii) of 3.06 as

$$V_d = V_0 - M_d^2 \qquad \qquad \text{(vi)}$$

It follows (Fig. 47, p. 165) that

$$V_0 = \sum_{k=0}^{k=a} \sum_{l=0}^{l=b} w_l v_k (A_k^2 - 2A_k B_l + B_l^2)$$

$$= \sum_{k=0}^{k=a} \sum_{l=0}^{l=b} w_l v_k A_k^2 - 2 \sum_{k=0}^{k=a} \sum_{l=0}^{l=b} w_l v_k A_k B_l + \sum_{k=0}^{k=a} \sum_{l=0}^{l=b} w_l v_k B_l^2.$$

We proceed with the double summation of each term in two stages as in 4.04, putting first

$$V_0 = \sum_{l=0}^{l=b} w_l \sum_{k=0}^{k=a} v_k A_k^2 - 2 \sum_{l=0}^{l=b} w_l B_l \sum_{k=0}^{k=a} v_k A_k + \sum_{l=0}^{l=b} w_l B_l^2 \sum_{k=0}^{k=a} v_k.$$

By means of (i) and (ii) in 4.04, this becomes

$$V_0 = (V_a + M_a^2) \sum_{l=0}^{l=b} w_l - 2M_b M_a + (V_b + M_b^2) \sum_{k=0}^{k=a} v_k.$$

By (iii) in 4.04 we therefore derive

$$V_0 = V_a + M_a^2 - 2M_b M_a + V_b + M_b^2$$
$$= V_a + V_b + (M_a - M_b)^2.$$

Hence from (vi):
$$V_d = V_a + V_b + (M_a - M_b)^2 - M_d^2.$$

Hence from (v) in 4.04
$$V_d = V_a + V_b. \qquad \qquad \text{(ix)}$$

By the same method (p. 165) we can show that the variance V_s of the sampling distribution of the sum $(A + B)$ is also given by
$$V_s = V_a + V_b \qquad \qquad \text{(x)}$$

By successive application of the chessboard device we can likewise show that the variance of the sampling distribution of the sum of three variates (e.g. score sums for samples from three card packs) would be $(V_a + V_b + V_c)$, and in general w.r.t. variates $A, B, C \ldots Z$

$$V_s = V_a + V_b + V_c \ldots V_z \qquad \qquad \text{(xi)}$$

To use (xi) correctly we have to pay due attention to the nature of the sampling distribution to which it refers. If we are concerned with the raw score difference or the score-deviation difference (*vide* (viii) in 4.04)
$$V_d = a p_a q_a + b p_b q_b \qquad \qquad \text{(xii)}$$

If we are concerned with the proportionate score distribution:

$$V_d = \frac{p_a \cdot q_a}{a} + \frac{p_b \cdot q_b}{b} \qquad \qquad \text{(xiii)}$$

When $p_a = p_b$ (xii) becomes
$$V_d = (a + b)pq \qquad \qquad \text{(xiv)}$$

and (xiii) becomes

$$V_d = pq\left(\frac{1}{a} + \frac{1}{b}\right) = \frac{(a + b)pq}{ab} \qquad \qquad \text{(xv)}$$

The last two equations respectively define the variance of the distribution of a raw score and of a proportionate score of $(a + b)$ fold samples in accordance with the binomial law, and we have seen (4·02) that the integrand of the probability integral which describes the limiting contour of such a distribution of either sort has the same form with due regard to the definition of V_d. If therefore a normal curve is capable of describing a difference distribution, we anticipate the integrand will assume the form :

$$(2\pi V_d)^{-\frac{1}{2}} \cdot exp - (D^2 \div 2V_d) \quad . \quad . \quad . \quad . \quad . \quad \text{(xvi)}$$

The result of our preliminary investigation of the *colour* score deviation difference distribution, we have found to accord with the expansion of $(\frac{1}{2} + \frac{1}{2})^{a+b}$. Let us consider the heart score deviation difference distribution $(p = \frac{1}{4})$ for 4-card and 8-card samples in the light of the new clue we have gleaned. In accordance with (xiv) the variance of the score deviation difference distribution is

$$V_d = (8 + 4)(0·25)(0·75) = (1·5)^2$$
$$\therefore \ \sigma_d = 1·5.$$

The heart score deviation differences themselves increase by unit steps, and the cumulative frequencies of Table 7 exhibit the total area of a histogram including at the left-hand extreme the column referable to the negative maximum heart score deviation difference $(d = -7)$ of Fig. 48 and the column referable to the specified value of d at the other extreme. A glance at Fig. 32 suffices to remind us that the boundary ordinates of the corresponding area of the continuous normal distribution are $-\infty$ and $(d + \frac{1}{2})$. To make a just comparison between cumulative values of the distribution of Fig. 48 and the areas cited in the table of the normal integral, we therefore have to make a half interval correction for c as defined on p. 127. Accordingly, we compare with the actual cumulative frequency for the specified values of d for the exact distribution exhibited in Fig. 48, the area cited in the table of the normal integral opposite : $(d + \frac{1}{2}) \div \sigma_d$. Since $\sigma_d = 1.5$, it follows that $c = 1$ when $d = 1$; and $c = 3$, when $d = \pm 4$, etc. The student will find it helpful to check the figures in Table 7 against those of Table 2 in the Appendix to Part I of Kendall's invaluable treatise.

TABLE 7

Heart Score Deviation Difference w.r.t. Samples of 4 and 8.
$(p = \frac{1}{4}, q = \frac{3}{4})$

Difference (d).	Frequency.	Cumulative Frequency.	Normal Integral up to $(d + \frac{1}{2}) \div \sigma_d = 1·5$.
− 7	0·0000	0·0000	0·0000
− 6	0·0001	0·0001	0·0001
− 5	0·0014	0·0015	0·0013
− 4	0·0090	0·0105	0·0099
− 3	0·0379	0·0484	0·0480
− 2	0·1073	0·1557	0·1587
− 1	0·2055	0·3612	0·3696
0	0·2638	0·6250	0·6204
1	0·2201	0·8451	0·8413
2	0·1140	0·9591	0·9520
3	0·0349	0·9940	0·9901
4	0·0057	0·9997	0·9987
5	0·0004	1·0000	0·9999

With due regard to the small size of the two samples to which Fig. 48 refers, the correspondence exhibited in Table 7 is striking. Even more striking is the correspondence between appropriate normal and exact values of the frequencies of proportionate *colour* score differences

TABLE 8A

Proportionate Score Differences w.r.t. Samples of 4 and 8
$(p = \frac{1}{2} = q)$

Score Difference.	Cumulative Frequency of Score Difference.	Normal Integral $\sigma = 0.3062$.
$- 1.000$	0.0002	0.0013
$- 0.875$	0.0022	0.0046
$- 0.750$	0.0100	0.0138
$- 0.625$	0.0315	0.0357
$- 0.500$	0.0774	0.0768
$- 0.375$	0.1575	0.1582
$- 0.250$	0.2747	0.2739
$- 0.125$	0.4212	0.4206
0.000	0.5789	0.5794
$+ 0.125$	0.7254	0.7261
$+ 0.250$	0.8246	0.8418
$+ 0.375$	0.9227	0.9232
$+ 0.500$	0.9686	0.9643
$+ 0.625$	0.9901	0.9862
$+ 0.750$	0.9979	0.9954
$+ 0.875$	0.9999	0.9987
$+ 1.000$	1.0000	0.9997

w.r.t. pairs of 4-fold and 8-fold samples of Fig. 49 as shown in Table 8A. Here it is necessary to apply (xv) above, i.e.

$$\sigma_d^2 = \frac{(4 + 8)(0.5)^2}{32} = \frac{3}{32}.$$

$$\therefore \sigma_d = 0.3062.$$

In this case the differences increase or decrease by $\Delta = \pm 0.125$. If $d = + x(0.125)$ is the specified column bounding the positive limit of the histogram, the boundary ordinate of the corresponding continuous distribution will be $(0.125)x + \frac{1}{2}\Delta d = (0.125)x + 0.0625$, and the appropriate c-value is

$$\frac{(0.125)x + 0.0625}{0.3062}.$$

Table 8B refers to the condensed histogram in the right half of Fig. 51, which shows the results of grouping frequencies for discrete values of the proportionate score distribution in the lower half of Fig. 50. Here again $p = \frac{1}{2} = q$. The samples respectively consist of 5 and 7 items, so that we may substitute in (xv)

$$\sigma_d^2 = \frac{(5 + 7)(0.5)^2}{35} = \frac{3}{35}.$$

$$\therefore \sigma_d = 0.2928.$$

Columns of the condensed histogram of Fig. 51 refer to groups of 5 score values, so that the central values increase or decrease by an interval of one-seventh (0.1429) and the appropriate half interval correction is 0·0714.   For the c value corresponding to the xth interval to the right of the mean, we therefore use

$$\frac{(0·1429)x + 0·0714}{0·2928}.$$

Scrutiny of Tables 7, 8A and 8B encourages us to start on the last lap of our course with confident anticipation of the end in view, i.e. a general proof that the distribution of score deviation differences and proportionate score differences w.r.t. samples of the size customarily employed in statistical enquiries are normally distributed with variance defined by (xii) above.

TABLE 8B

Grouped Proportionate Score Differences w.r.t. Samples of 5 and 7
$(p - \frac{1}{2} - q)$

Central Value of Grouped Scores.	Cumulative Frequency of Grouped Scores.	Normal Integral $\sigma = 0·2928.$
− 0·8571	0·0031	0·0003
− 0·7145	0·0168	0·0140
− 0·5716	0·0534	0·0445
− 0·4287	0·1242	0·1147
− 0·2858	0·2404	0·2411
− 0·1429	0·4057	0·4191
0·0000	0·5942	0·5812
+ 0·1429	0·7595	0·7593
+ 0·2858	0·8757	0·8856
+ 0·4287	0·9465	0·9558
+ 0·5716	0·9831	0·9863
+ 0·7145	0·9968	0·9967
+ 0·8571	0·9997	0·9996

EXERCISE 4.06

1. Give the numerical values of the standard deviation of the replacement distribution of both the score deviation difference and the proportionate score difference for pairs of 3-fold and 7-fold samples from a full pack w.r.t. :

(*a*) the colour score ;

(*b*) the diamond score ;

(*c*) the picture-card score ;

(*d*) the honours-card score.

2. Repeat Ex. 1 w.r.t. samples respectively composed of 4 and 6 cards.

3. Evaluate the proportionate colour score difference distribution for samples of 7 and 11, assuming replacement.

4. Embody the result of 3 in a unimodal histogram like those of Fig. 51 and test its conformity to the normal distribution with due regard to the appropriate half-interval correction.

5. Test the distributions of Ex. 2 by the method of Tables 7 and 8.

4.07 The Exact Difference Distribution

Mathematical analysis evokes quite unnecessary alarm and despondency, if the student approaches it with the wrong idea that a *proof* is a statement about the way in which the mathematician arrives at a generalisation. In fact, a proof is nothing of the sort. It sets out the connection between a result already surmised to be true for reasons commonly concealed, and often forgotten, with a view to exhibiting its connection with other known results and hence with a view to recognition of its *legitimate scope and limitations*. As the number of mathematical generalisations increases, the status which a particular result occupies in the entire corpus of current knowledge perennially invites reconsideration, and the pathway of discovery becomes less and less retraceable in a jungle of more and more *rigorous* demonstration. If one grasps this conception of the true nature of a proof, the devious steps which lead to the end result cease to evoke the discouraging sentiment that one is an unwilling witness of a conjuring trick which one can never hope to perform.

It is our standpoint that statistical theory, as concerned with sampling, derives its practical rationale from the calculus of choice and chance, as first expounded by Pascal ; but the application of the calculus of choice and chance to large sampling confronts us with formidable problems of computation. Consequently, it is the constant preoccupation of theoretical statistics to substitute for exact statements which would involve laborious evaluation approximate formulæ to provide sufficiently safe guidance for practical judgments. The rationale of such approximations is referable to purely mathematical considerations having no necessary connection with the laws of choice as such ; and it is therefore easy to lose sight of the nature of the problem which invokes such operations in a welter of symbols with no direct relation to it. We can keep our feet on solid ground only if we constantly restate the problems in *exact* terms as a prelude to the derivation of an approximate solution ; and we can keep ourselves alert to the limitations of such solutions only if we do so.

In this context we shall therefore approach the theme of the concluding sentence of 4.06 by a preliminary generalisation which we shall have to discard for practical use, at a later stage. We shall in fact seek an exact statement of the difference distribution in terms of the calculus of choice only to find that it would be useless as an instrument of computation ; but doing so will give us a clearer insight into the meaning of a formula which is a suitable calculating device. To this end, it will be necessary to refine the symbols at our disposal. One and the same difference D may occur in several cells of a chessboard set-up as those of Figs. 42, 43 and 48. So far, we have added frequencies referring to such a difference term by term, without seeking for a general expression (y_d) for a particular value of the difference D common to several cells. We shall now denote the frequency term of a cell in the kth column and the lth row by $y(k, l)$. Our problem is to express y_d as the sum of all such terms as refer to one and the same difference D. In Table 9 the score difference of column 3, row 1 is $+ 2$ and its frequency is $12p^4q^3$, but a difference of 2 also occurs in cells $(2, 0)$ and $(4, 2)$, the total frequency y_2 for all score differences of $+ 2$ being given by

$$y_2 = y(2, 0) + y(3, 1) + y(4, 2)$$
$$= 6p^2q^5 + 12p^4q^3 + 3p^6q.$$

For the respective values in the cell (k, l) of the two variates A (referable to a items per sample), and B (referable to b items per sample), we shall here simply write k and l, their difference D being $(k - l)$. In accordance with the product rule, the chessboard exhibits the result

$$y(k, l) = \frac{a!}{k! (a - k)!} p^k q^{a-k} \cdot \frac{b!}{l! (b - l)!} p^l q^{b-l} \quad . \qquad . \qquad . \quad \text{(i)}$$

We can simplify the task of handling an expression which involves the two variables k and l by using the relation implicit in the definition of the difference, *viz.* :

$$l = k - D.$$

Hence we can reduce all terms which add up to y_d for a fixed value D to expressions involving only the variate A :

$$y(k, l) = \frac{a!}{k! \, (a - k)!} \, p^k \, q^{a-k} \cdot \frac{b!}{(k - D)! \, (b - k + D)!} \, p^{k-D} \, q^{b-k+D} \qquad . \qquad . \quad \text{(ii)}$$

To get an exact expression for y_d the frequency of the particular value D in (ii) we now have to decide which expressions of this type collectively add up to y_d. A glance at the chessboard (Fig. 44) gives us the clue. Cells in which $D = 0$ and cells with the same *positive* value of D go downwards by unit steps diagonally from the column specified by $k = D$ to the column specified by $k = a$; and cells with the same *negative* value go downwards from $k = 0$ to $k = b + D$. For the raw score difference distribution of Table 9, we therefore have two rules of summation :

(a) for *positive* values of D and when $D = 0$:

$$y_d = \sum_{k=D}^{k=a} y(k, l).$$

(b) for *negative* values of D :

$$y_d = \sum_{k=0}^{k=(b+D)} y(k, l).$$

Thus the summation for positive values of D terminates at the maximum value of k and that of negative values starts at the minimum value of k. At first sight, there seems to be no general rule. To bring the two classes of differences into line, the summation must start and end at the same k value ; and this is possible only if we extend the range of summation of $y(k, l)$ terms from D to 0 for positive values of D and from $(b + D)$ to a for negative values of D. We ask therefore what would this entail ?

First suppose D is positive. Then $(k - D)$ must be negative for all values of k less than D. Now any term specified by (ii) contains as a factor the reciprocal of $(k - D)!$; and we know that the *reciprocal of the factorial of a negative integer is zero* (Ex. 5, p. 12). This means that we could include in our total all values of $y(k, l)$ from $y(0, l)$ to $y(D, l)$ without making any difference to the result. So long as D is positive, we can thus write

$$\sum_{k=D}^{k=a} y(k, l) = \sum_{k=0}^{k=a} y(k, l) . \qquad . \qquad . \qquad . \qquad . \quad \text{(iii)}$$

The maximum value of k in any cell for which D is negative is $(b + D)$. If k is greater than $b + D$, the expression $(b - k + D)$ will be negative and the reciprocal of $(b - k + D)!$ will therefore be zero. This means that any term of the form defined by (ii) will vanish, if $k > b + D$. So there is no reason why we should not extend the summation from $k = (b + D)$ to $k = a$ in conformity with the equation

$$\sum_{k=0}^{k=(b+D)} y(k, l) = \sum_{k=0}^{k=a} y(k, l) . \qquad . \qquad . \qquad . \quad \text{(iv)}$$

TABLE 9

Frequency of a Raw Score Difference of $+2$ between pairs of 4-fold and 3-fold Samples

SAMPLE A (4 items)

		Score (k)	0	1	2	3	4
		(l) Frequency	q^4	$4pq^3$	$6p^2q^2$	$4p^3q$	p^4
	0	q^3	0	1	2 $\quad 6p^2q^5$	3	4
	1	$3pq^2$	.1	0	1	2 $\quad 12p^4q^3$	3
SAMPLE B (3 items)	2	$3p^2q$	-2	-1	0	1	2 $\quad 3p^6q$
	3	p^3	-3	-2	-1	0	1

For the sum of all values of $y(k, l)$ which add up to y_d referable to the particular value D of the raw score difference, we can therefore write

$$y_d = \sum_{k=0}^{k=a} \frac{a!}{k!\,(a-k)!}\, p^k\, q^{a-k} \cdot \frac{b!}{(k-D)!\,(b-k+D)!}\, p^{k-D}\, q^{b-k+D} \qquad . \qquad . \quad \text{(v)}$$

The foregoing remarks apply to the *raw* score difference distribution, and hence only to *whole number* values of D. Score *deviations* corresponding to a particular raw score may be fractional, and the corresponding proportionate score will always be a proper fraction except when it is zero or numerically equal to unity. Consequently, score deviation differences and proportionate score differences introduce a new issue. It will suffice for our purpose if we consider the proportionate difference (d) which must be fractional unless $d = 0$ or ± 1. By definition

$$d = \frac{k}{a} - \frac{l}{b} = \frac{bk - al}{ab} \qquad . \qquad . \qquad . \qquad . \quad \text{(vi)}$$

$$l = \frac{bk - abd}{a} \qquad . \qquad . \qquad . \qquad . \quad \text{(vii)}$$

$$y(k, l) = \frac{a!}{k!\,(a-k)!} \cdot p^k \cdot q^{a-k} \frac{b!}{\left(\dfrac{bk - abd}{a}\right)!\left(\dfrac{ab - bk + abd}{a}\right)!} p^{\frac{bk - abd}{a}}\, q^{\frac{ab - bk + abd}{a}} \quad \text{(viii)}$$

This raises an interesting issue, on which elementary books are not commonly explicit. The fact that the algebraic definition of a factorial number, i.e. the continued product $n(n-1)$ $(n-2)\ldots$, is consistent with the value of the *Gamma function* (6.05) for negative and positive integers is also consistent with the view that the factorial is a special case of the Gamma function ; but this leaves open the question : what meaning should we attach to the factorial of a fraction in the domain of its original specification ? It is not inconsistent with the infinite number of terms in the binomial expansion, alike for a positive or negative fractional and for an integral negative whole number index to regard its reciprocal as zero, like that of the factorial

TABLE 10

Proportionate Score Difference Distribution for pairs of 4-fold and 6-fold Samples

SAMPLE A (4 *items*)

		Proportionate Score Frequency	0 q^4	$\frac{6}{24}$ $4pq^3$	$\frac{12}{24}$ $6p^2q^2$	$\frac{18}{24}$ $4p^3q$	$\frac{24}{24}$ p^4
	0	q^6	0 q^{10}	$\frac{6}{24}$ $4pq^9$	$\frac{12}{24}$ $6p^2q^8$	$\frac{18}{24}$ $4p^3q^7$	$\frac{24}{24}$ p^4q^6
	$\frac{4}{24}$	$6pq^5$	$-\frac{4}{24}$ $6pq^9$	$\frac{2}{24}$ $24p^2q^8$	$\frac{8}{24}$ $36p^3q^7$	$\frac{14}{24}$ $24p^4q^6$	$\frac{20}{24}$ $6p^5q^5$
	$\frac{8}{24}$	$15p^2q^4$	$-\frac{8}{24}$ $15p^2q^8$	$-\frac{2}{24}$ $60p^3q^7$	$\frac{4}{24}$ $90p^4q^6$	$\frac{10}{24}$ $60p^5q^5$	$\frac{16}{24}$ $15p^6q^4$
SAMPLE B (6 *items*)	$\frac{12}{24}$	$20p^3q^3$	$-\frac{12}{24}$ $20p^3q^7$	$-\frac{6}{24}$ $80p^4q^6$	0 $120p^5q^5$	$\frac{6}{24}$ $80p^6q^4$	$\frac{12}{24}$ $20p^7q^3$
	$\frac{16}{24}$	$15p^4q^2$	$-\frac{16}{24}$ $15p^4q^6$	$-\frac{10}{24}$ $60p^5q^5$	$-\frac{4}{24}$ $90p^6q^4$	$\frac{2}{24}$ $60p^7q^3$	$\frac{8}{24}$ $15p^8q^2$
	$\frac{20}{24}$	$6p^5q$	$-\frac{20}{24}$ $6p^5q^5$	$-\frac{14}{24}$ $24p^6q^4$	$-\frac{8}{24}$ $36p^7q^3$	$-\frac{2}{24}$ $24p^8q^2$	$\frac{4}{24}$ $6p^9q$
	$\frac{24}{24}$	p^6	$-\frac{24}{24}$ p^6q^4	$-\frac{18}{24}$ $4p^7q^3$	$-\frac{12}{24}$ $6p^8q^2$	$-\frac{6}{24}$ $4p^9q$	0 p^{10}

negative integer. If we were to adopt this unorthodox definition of the factorial of a fraction in its lowest terms, all expressions such as (viii) will vanish unless each of the following factors is either a positive integer or zero $(0! = 1)$:

$$\frac{bk - abd}{a} ; \quad \frac{ab - bk + abd}{a}.$$

It is therefore instructive to visualise, with the help of Table 10, the conditions which define the disappearance of terms involving a value of k which is inconsistent with the particular value d of the cell (k, l). This table refers to a distribution w.r.t. which $a = 4$ and $b = 6$. The difference $+ \frac{8}{24}$ occurs in two cells (2, 1) and (4, 4); and for this entry $abd = 8$, so that

$$\frac{bk - abd}{a} = \frac{3k - 4}{2} = l$$

$$\frac{ab - bk + abd}{a} = \frac{16 - 3k}{2} = b - l.$$

We shall now write out values of l and $b - l$ for terms involving all possible values of k consistent with $d = \frac{+8}{24}$:

k	l	$b - l$
0	-2	8
1	$-\frac{1}{2}$	$6\frac{1}{2}$
2	1	5
3	$2\frac{1}{2}$	$3\frac{1}{2}$
4	4	2

It is at once apparent that only two pairs of values of k and l referable to the cells $(2, 1)$ and $(4, 4)$ respectively satisfy the condition that neither b nor $(b - l)$ is negative or fractional. If we therefore make the unorthodox assumption that reciprocals of $l!$ and of $(b - l)!$ are zero if either l or $(b - l)$ is fractional, as is true if either is negative, we get the right result for the total frequency of a difference $d = \dfrac{+8}{24}$ by the appropriate substitution in the expression comparable to (v) above, viz. :

$$y_d = \sum_{k=0}^{k=a} \frac{a!}{k!\,(a-k)!}\, p^k \cdot q^{a-k} \cdot \frac{b!}{\left(\dfrac{bk - abd}{a}\right)!\left(\dfrac{ab - bk + abd}{a}\right)!}\, p^{\frac{bk - abd}{a}} \cdot q^{\frac{ab - bk + abd}{a}} \quad \text{(ix)}$$

To carry out the last summation in accordance with the procedure which is valid for (v) we have in effect to employ an *ad hoc* definition of the factorial of a fractional number inconsistent with the accepted interpretation of factorial as values of the Gamma function. In any case, both (ix) and (v) are unsuitable for computation w.r.t. large samples. Happily, dilemmas of this sort do not trouble the pure mathematician content to operate within the domain of the continuum ; but they are no less challenging, because we can sidestep them by so doing. To say this signifies that we should at all times submit results established in the rarefied atmosphere of the infinite to the arbitrament of arithmetical investigation, when the method of proof offers no certain indication of the order of approximation involved in the outcome *vis-a-vis* the finite samples of practical statistics. It is on this understanding that we now proceed to exhibit the difference distribution of two normally distributed proportionate scores of samples from the same universe. To do so, we must take cognisance of our findings at the end of 4.02.

4.08 THE DIFFERENCE DISTRIBUTION IN A HYPOTHETICAL CONTINUUM

The issue raised by the derivation of (xi) in 4.02 is one which we shall need to consider in Vol. II, where the fiction of *probability density* will call for treatment at length. Here it is sufficient to draw attention to a distinction which does not emerge in the derivation of the normal equation as the limit of the binomial *frequency* distribution of the raw score (x) or the raw score deviation (X), and is irrelevant to the derivation of the normal equation of a raw score difference or of a raw score deviation difference. In deriving (3.03) the normal equation of the raw score or raw score deviation as an approximate description of the *statistical* properties of a binomial histogram, our concern is to determine the total frequency of a certain range of discrete score values in accordance with our definition of frequency, i.e. that the sum of all possible score frequencies $(y_x$ or $Y_x)$ is unity, i.e.

$$\sum_{0}^{x=r} y_x = 1 = \sum_{X=-rp}^{X=rq} Y_x . \qquad \cdot \qquad \cdot \qquad \cdot \qquad \cdot \qquad \text{(i)}$$

In this context, both the raw score and the raw score deviation increase by unit steps, i.e. $\Delta x = 1 = \Delta X$, so that

$$\sum_{0}^{r} y_x . \Delta x = 1 = \sum_{-rp}^{rq} Y_x . \Delta X \qquad \cdot \qquad \cdot \qquad \cdot \qquad \cdot \qquad \text{(ii)}$$

Subject to this condition, the area of the histogram is unity, and we may write :

$$\sum_{-rp}^{rq} Y_x = 1 \simeq \int_{-\infty}^{+\infty} Y_x . dX$$

$$\sum_{-rp}^{a} Y_x \simeq \int_{-\infty}^{a+\frac{1}{2}} Y_x . dX.$$

In the above Y_x, the approximate value of the ordinate of the score $(x - M) = X$, is its exact frequency. As the integrand of the probability integral, we actually employ an approximation $Y \simeq Y_x$ such that the value of the complete integral is exactly unity. When (i) holds good, Y_x is therefore approximately equal to the integrand of the integral definitive of the expectation that a score value will lie within a certain range, i.e. the cumulative score frequencies within that range ; but the derivation of (xi) in 4.02 shows that such correspondence between the integrand and the function definitive of the frequency does not always hold good. If we introduce a new symbol $F(U)$ for the integrand of the probability integral of the proportionate score deviation U, we may write as follows the result established at the end of 4.02 :

$$Y_u \simeq \frac{1}{r\sqrt{2\pi V_u}} \; exp \; \frac{-U^2}{2V_u}.$$

$$F(U) = \frac{1}{\sqrt{2\pi V_u}} \cdot exp \; \frac{-U^2}{2V_u} \simeq r \cdot Y_u.$$

The meaning of this *scalar* discrepancy is not far to seek, if we return to the histogram definitive of the exact distribution. The proportionate score and the proportionate score deviation increase by increments equivalent to r^{-1}, i.e.

$$\Delta U = \frac{1}{r}$$

$$\therefore \; Y_u \cdot \Delta U = \frac{1}{r} Y_u \quad . \qquad . \qquad . \qquad . \qquad . \qquad . \quad \text{(iii)}$$

Thus we can express the frequency of obtaining the exact value U as

$$Y_u = r \cdot Y_u \cdot \Delta U \simeq F(U) \cdot \Delta U. \quad . \qquad . \qquad . \qquad . \quad \text{(iv)}$$

For the cumulative frequency of all values of U in the range $U = 0$ to a, when r is large, we may therefore write :

$$\sum_{U=0}^{U=a} r \cdot Y_u \cdot \Delta U \simeq \int_{-\frac{1}{2}}^{a+\frac{1}{2}} F(U) \, dU \quad . \qquad . \qquad . \qquad . \quad \text{(v)}$$

$$E\,(\leqslant a) \simeq \int_{-\infty}^{a+\frac{1}{2}} F(U) \, dU \quad . \qquad . \qquad . \qquad . \quad \text{(vi)}$$

To obtain the appropriate integrand $F(U)$ in the above we merely write

$$F(U)\Delta U \simeq Y_u \qquad . \qquad . \qquad . \qquad . \qquad . \quad \text{(vii)}$$

The reader will note that the infinitesimal product $F(U) \cdot dU$ definitive of the limiting contour of the binomial distribution of the proportionate score deviation has the same form as the corresponding product $F(X) \cdot dX$ w.r.t. raw score deviations, i.e.

$$F(U) \cdot dU = \frac{1}{\sqrt{2\pi V_u}} \cdot exp - (U^2 \div 2V_u)$$

$$F(X) \cdot dX = \frac{1}{\sqrt{2\pi V_x}} \cdot exp - (X^2 \div 2V_x)$$

This product, of which the first factor is the so-called *probability density*, is therefore what all normal distributions, or other distributions of a particular specification, have in common irrespective of the scale of the score system.

If the width of each column of the histogram which describes a *discrete* function y_x is $\Delta x = 1$, the horizontal boundaries of the element of area $y_x \Delta x$ are respectively $(x - \frac{1}{2}\Delta x)$ and $(x + \frac{1}{2}\Delta x)$. With due regard to the fact that there is then only one value of x, namely the integer within the interval so defined, we may express this by saying that y_x is the expectation that the values of the x-score lie in the range $x \pm \frac{1}{2}\Delta x$. If we further subdivide the interval Δx on our horizontal scale into h equal parts of width Δu with h corresponding columns of approximate height $F(u)$ to accommodate discrete values with a fractional increment like that of the proportionate score, the element of area $F(u)\Delta u = y_u$ now approximately defines the expectation that the score u will be in the range $u \pm \frac{1}{2}\Delta u$. Evidently, we can make Δu as small as we like, and can then employ the operations of the integral calculus to make change of scale explicit without going back to first principles.

Having taken cognisance of this convenience, we may now proceed to examine the distribution of the difference (d) of two proportionate scores U_a and U_b respectively referable to samples of a and b items from the same universe. By definition $d = (U_a - U_b)$, so that $U_a = (d + U_b)$. Since we are concerned with proportionate scores, the variances $(V_a$ and $V_b)$ of the parent distributions and the variance (V_d) of the difference distribution have the values:

$$V_a = pq \div a \; ; \; V_b = pq \div b \; ; \; V_d = V_a + V_b$$

When both a and b are large, we then define the probabilities of obtaining particular score values U_a and U_b from one or other sample in accordance with (xvi) of 4.02 and of the scalar convention specified above, *viz.*

$$y_a \simeq \frac{1}{\sqrt{2\pi V_a}} \cdot exp - \frac{U_a{}^2}{2V_a}\Delta U_a$$

and

$$y_b \simeq \frac{1}{\sqrt{2\pi V_b}} \cdot exp - \frac{U_b{}^2}{2V_b} \cdot \Delta U_b \quad . \qquad . \qquad . \quad \text{(viii)}$$

Let us first suppose that our concern with the b-fold sample is only with samples whose proportionate score is the particular value U_b. In other words, we are concerned only with d-scores which we may write in the form $U_a - C$, that is to say, d-scores which increase by an increment $\Delta d = \Delta U_a$. When the value of the a-fold sample score lies in the range $U_a \pm \frac{1}{2}\Delta U_a$, the value of $d = (U_a - U_b)$ therefore lies in the range $d \pm \frac{1}{2}\Delta d$. The expectation (y_a) that U_a will lie within the range $U_a \pm \frac{1}{2}\Delta U_a$ is then the expectation that d will lie within the corresponding range $d \pm \frac{1}{2}\Delta d$ if the b-fold sample score has the particular value U_b. We may write this in the form

$$y_a \simeq \frac{1}{\sqrt{2\pi V_a}} exp \frac{-(d + U_b)^2}{2V_a} \cdot \Delta d.$$

The joint expectation that d will lie in this range, when the b-fold sample score has the fixed value U_b is $y_{ab} = y_a \cdot y_b$. Whence by (viii) above

$$y_{ab} \simeq \frac{1}{2\pi\sqrt{V_a V_b}} exp -\frac{1}{2}\left[\frac{(d + U_b)^2}{V_a} + \frac{U_b{}^2}{V_b}\right]\Delta d . \Delta U_b \quad . \qquad . \quad \text{(ix)}$$

Since our end in view is to define the expectation of the d-score, it is convenient to change this expression as follows :

$$\frac{(d + U_b)^2}{V_a} + \frac{U_b{}^2}{V_b} = \frac{V_b d^2 + 2V_b \cdot U_b d + (V_a + V_b)U_b{}^2}{V_a \cdot V_b}$$

$$= \frac{V_d}{V_a \cdot V_b}\left\{\frac{V_b}{V_d} \cdot d^2 + \frac{2V_b}{V_d} \cdot U_b d + U_b{}^2\right\}$$

$$= \frac{V_d}{V_a \cdot V_b}\left[\left\{\frac{V_b}{V_d}d^2 - \frac{V_b{}^2}{V_d{}^2}d^2\right\} + \left\{\frac{V_b{}^2}{V_d{}^2} \cdot d^2 + \frac{2V_b}{V_d}U_b d + U_b{}^2\right\}\right]$$

$$= \frac{V_d - V_b}{V_a V_d}d^2 + \frac{V_d}{V_a \cdot V_b}\left[U_b + \frac{V_b}{V_d} \cdot d\right]^2$$

Since $(V_d - V_b) = V_a$, the above reduces to

$$\frac{d^2}{V_d} + \frac{V_d}{V_a V_b}\left[U_b + \frac{V_b}{V_d} \cdot d\right]^2$$

Whence (ix) becomes :

$$y_{ab} \simeq \frac{1}{2\pi\sqrt{V_a V_b}} \, exp \, \frac{-d^2}{2V_d} \cdot \Delta d \cdot exp \, \frac{-V_d}{2V_a V_b}\left[U_b + \frac{V_b}{V_d}d\right]^2 \Delta U_b$$

We now recall that this expression defines the expectation of a d-score value in the range $d \pm \frac{1}{2}\Delta d$ when the b-fold sample score has the fixed value U_b ; but our end in view is to define the expectation (y_d) of obtaining such a value when we allow U_b to have any value whatever consistent with the relation $d = (U_a - U_b)$ assumed in the above, and hence also implicitly allow U_a to have any value consistent with the presumption that the d-score lies in the range specified. We therefore perform the summation in the usual way :

$$y_d \simeq \frac{1}{2\pi\sqrt{V_a V_b}} \, exp \, \frac{-d^2}{2V_d} \cdot \Delta d \int_{-\infty}^{\infty} exp \, \frac{-V_d}{2V_a V_b}\left[U_b + \frac{V_b}{V_d} \cdot d\right]^2 dU_b$$

To simplify the integral we then write

$$z = \frac{\sqrt{V_d}}{\sqrt{V_a V_b}} \cdot \left(U_b + \frac{V_b}{V_d} \cdot d\right); \quad dz = \frac{\sqrt{V_d}}{\sqrt{V_a V_b}} \cdot dU_b.$$

$$\therefore \; y_d \simeq \frac{1}{2\pi\sqrt{V_d}} \, exp \, \frac{-d^2}{2V_d} \cdot \Delta d \int_{-\infty}^{\infty} e^{-\frac{1}{2}z^2} dz.$$

We have already determined the value of the integral in this expression, viz.

$$\int_{-\infty}^{\infty} e^{-\frac{1}{2}z^2} dz = 2\int_{0}^{\infty} e^{-\frac{1}{2}z^2} dz = \sqrt{2\pi}.$$

$$\therefore \; y_d \simeq \frac{1}{\sqrt{2\pi V_d}} \cdot exp \, \frac{-d^2}{2V_d} \Delta d \quad . \quad . \quad . \quad . \quad . \quad \text{(x)}$$

In accordance with our assumption that it is permissible to describe the distribution of U_a and U_b by continuous functions as specified in (viii), the integrand of the distribution of the score difference therefore has the normal form anticipated in (xvi) of 4.06 ; but the reader is entitled to suspect that we have sidestepped the difficulty raised at the end of 4.07 by introducing such a postulate. The assumption itself is an approximation, and it is gratuitous to assume that true statements about truly continuous distributions are approximately true of distributions approximately described by continuous functions of the same class. Accordingly, we shall return at a later stage to the rationale of the c-test for a score difference with due regard to the clue disclosed in Fig. 51 ; and we shall there (6.10 below) re-examine the issue by an admittedly approximate method of fitting a continuous curve to the histogram contour of a distribution definitive of discrete score values.

4.09 THE C-TEST FOR A DIFFERENCE

To appreciate the implications of (x) in 4.08, we must now retrace our steps. The *null* hypothesis which we have explored in 4.03 and 4.05 is that two samples in which the proportion of items of a particular class are respectively p_a out of a total of a and p_b out of a total of b do in fact come from the same universe ; and we set up this null hypothesis to examine whether the observed difference $d = (p_a - p_b)$ is an occurrence so infrequent as to *undermine our confidence in the assumption that they do.* Since statistical text-books in circulation cite for a difference of this sort two different c-tests of which only one has any intelligible relation to an admissible null hypothesis, it is important to recall our original intention as stated in 4.03 ; and its implications w.r.t. the use we make of (x) in 4.08.

The former implies that there exists a common universe (or two identical universes) with a parameter p of which p_a and p_b are sample values ; and our best estimate (p_{ab}) of p is obtained by pooling all our material, *viz.* :

$$p_{ab} = \frac{ap_a + bp_b}{a + b} \qquad \qquad \text{(i)}$$

The derivation of (x) in 4.08 also implies this. Only if it is so, can we make any intelligible use of the relation. The equation defines the distribution of the deviation of a variate from its mean value which is zero if (and only if) the two variates have the same mean value p (p. 166). Only if this is so, can d in (x) of 4.08 correspond numerically with an observed difference ; and we can then define V_d as in (xv) of 4.06

$$V_d = pq\left(\frac{1}{a} + \frac{1}{b}\right) = pq\left(\frac{a + b}{ab}\right).$$

Since we do not know the exact value of p, we have to fall back on the best estimate consistent with the null hypothesis, i.e. p_{ab} as defined by (i) above. We therefore write

$$V_d = p_{ab} \cdot q_{ab} \cdot \left(\frac{a + b}{ab}\right) \qquad \qquad \text{(ii)}$$

If we neglect the half-integral correction of (iv) in 3.05, we therefore base our c-test on the ratio

$$h = \frac{d}{\sigma_d} = (p_a - p_b)\sqrt{\frac{ab}{p_{ab} \cdot q_{ab} \cdot (a + b)}} \qquad \qquad \text{(iii)}$$

In the example already used in 4.02 we have

$$p_{ab} = 0.084 ; \quad q_{ab} = 0.916 ; \quad p_{ab} \cdot q_{ab} = 0.077 ;$$
$$p_a = 0.011 ; \quad p_b = 0.122 ; \quad p_a - p_b = 0.111 ;$$
$$a = 279 ; \quad b = 539 ; \quad (a + b) = 818 ;$$
$$ab = 150381.$$

$$\therefore h = 0.111 \sqrt{\frac{150381}{0.077 (818)}} \simeq 5.$$

Thus the single c-test based on (xvi) of 4.07 is more sensitive than either of the two c-tests based on (viii) and (ix) of 4.02 in the sense that it entitles us to reject our null hypothesis more confidently.

EXERCISE 4.09

In each case formulate your *null hypothesis* EXPLICITLY

1. Examine the following figures with reference to (*a*) differential liability of gastric and duodenal ulcers to perforations ; (*b*) differential mortality w.r.t. the two types.

(*a*)	All Cases.	Perforated.	Deaths.
Gastric . .	1397	174	16
Duodenal .	5465	363	17

2. Do the following data for comparable periods with reference to mortality for peptic ulcer in two classes of hospitals indicate a greater risk of death in one or the other ?

	No. of Cases.	Deaths.
Civilian . .	184	16
Military . .	18	9

3. In an investigation on treatments of Impetigo, records were available for new cases and for relapses. To validate comparison between the assessment of the treatments in terms of mean duration of stay the distribution of the two types of cases must be random. Ascertain whether this is so.

	O.	*S.*	*P.*
New cases .	434	356	54
Relapses . .	104	69	14
Total . .	538	425	68

4. Examine the following figures with reference to males and females in military service (1933–34).

Cases of Appendicitis.	♂	♀
With acute imflammation .	191	62
Interval cases . . .	16	13
Normal	15	20
Indefinite or unspecified .	158	203
Total . .	380	298

5. Examine the following data to assess whether there were statistically significant differences (a) with reference to treatment prescribed by the two types of hospitals, and (b) with reference to the incidence of complications in either type of hospital or both together

Treatment of Scarlet Fever, 1943–44.	Military Hospitals.		Civilian Hospitals.	
	No. of Patients.	No. with Complications.	No. of Patients.	No. with Complications.
Serum only . . .	57	9	123	43
Sulphonamides only . .	33	10	61	25
Serum and Sulphonamides	8	4	33	14
No specific treatment .	86	21	308	97

6. Make a similar analysis for the following data for the efficacy of the treatments with reference to late complications.

Treatment of Scarlet Fever, 1943–44.	Military Hospitals.		Civilian Hospitals.	
	No. of Patients.	No. with late Complications.	No. of Patients.	No. with late Complications.
Serum only . . .	57	5	213	19
Sulphonamides only . .	33	4	61	13
Serum and Sulphonamides	8	0	33	4
No specific treatment .	86	13	308	61

7. Interpret the following figures with reference to serum treatment of diphtheria

First dose of Serum.	All Patients.	With any Complications.	With Polyneuritis.
Before end of 2nd day from onset	81	15	5
After end of 2nd day from onset	77	19	13

8. What conclusions with reference to prognosis of polyneuritis following diphtheria do the following figures justify ?

	All Patients.	With Polyneuritis.
With palatal paralysis .	36	16
Without palatal paralysis	209	10

9. Interpret the following information with reference to knee joints.

	All Patients.	With ruptured Cruciate Ligaments.
With excessive anteroposterior mobility at knee joint	39	4
Knee joint mobility normal	463	10

10. What conclusions can you draw from the following figures with reference to treatment of soldiers for internal derangement of the knee (1943).

Type of Hospital.	With Operative Treatment.		Without Operative Treatment.	
	All Cases.	Medically Downgraded.	All Cases.	Medically Downgraded.
Civilian . .	77	37	32	15
Military . .	29	16	30	9

11. Are the differences within the three service categories between syphilitics and a control group of soldiers in the following table significant ?

	Under Service.	Discharged.	Untraceable.
Syphilitics .	534	756	38
Controls . .	544	745	39

12. By the same method test the comparability of the two groups with reference to risk of accident, impetigo, and psychiatric disorders.

	All Cases.	Accidents.	With Impetigo.	With Psychiatric Breakdown.
Syphilitics . .	1328	49	29	68
Controls . .	1328	38	22	76

13. What conclusions do you draw from the following information with reference to soldiers suffering from syphilis ?

	All Cases.	No. above Median
Syphilitics . .	492	165
Controls . .	481	205

14. What conclusions can you draw by a comparison of jaundice incidence among patients respectively treated with whole blood from individual donors and pooled plasma ?

	No. of Patients.	Cases of Jaundice.
Whole blood .	248	2
Pooled .	214	13

15. Examine the following figures for incidence of jaundice among battle casualties :

	With Transfusion.	Without Transfusion.
All patients .	757	2352
Cases of jaundice .	33	12

16. Examine the following figures with reference to serum reaction at the end of a six months' follow up of syphilitics :

Initial Reaction Treatment.		Reaction after Six Months.		
		Negative.	Doubtful.	Positive.
Sero-negative	Short-term Arsenic .	68	2	0
	Penicillin .	64	4	3
Sero-positive	Short-term Arsenic .	105	15	3
	Penicillin .	87	7	11

17. Evaluate the relative efficacy pair by pair of the three treatments of gonorrhœa shown below.

	All Cases.	Relapses on Primary Course.
Penicillin .	217	14
Sulphathiazole .	358	37
Sulphapyridine .	115	8

18. Investigate the following results:

	No. of Cases.	Cures.
Penicillin alone	251	203
Sulphathiazole alone	569	321
Sulphathiazole and simultaneous irrigation with KMnO$_4$.	204	172
Sulphathiazole and simultaneous irrigation with HgOCN .	95	67

19. Examine the relative efficacy of different dosages of sulphapyridine for cases of gonorrhœa treated therewith as below:

	< 15 Grams.	15–20 Grams.	> 20 Grams.
No of cases . .	9	49	130
Complete cures .	5	25	75

20. Assess the following data w.r.t. early and late surgical treatment of *cholecystitis*:

Operation.	No. of Patients.	Deaths.
Immediate . .	206	13
Delayed . .	273	13

4.10 Unbiassed Estimate of the Critical Ratio

A satisfactory figure for the critical ratio c presupposes a reliable estimate of the variance of samples of a given size extracted from the putative parent universe. When an r-fold sample is large we may assume that the observed proportionate score $p_0 = (x \div r)$ will not often differ greatly from the true universe value p. In so far as this is so, we may assume that the estimate $\sigma_0^2 (= p_0 q_0 \div r)$ will rarely differ materially from the true variance $\sigma^2 (= pq \div r)$ of the proportionate score distribution; but σ_0 so defined is not the most satisfactory estimate of σ. The sample statistic s which will give us the most satisfactory estimate is the statistic of which σ itself is the mean value for all r-fold samples, and we shall now see that the mean value of σ_0 is somewhat less than σ.

To define a statistic s^2 in accordance with our definition, we shall write its mean value as $M(s^2)$, so that

$$M(s^2) = \sigma^2 - \frac{pq}{r} \qquad . \qquad . \qquad . \qquad . \qquad . \qquad (i)$$

The value of the estimate σ_0^2 is

$$\frac{p_0 q_0}{r} = \frac{x(r-x)}{r^3}.$$

Its mean value is therefore given by

$$M(\sigma_0^2) = \sum_{x=0}^{x=r} y_x \cdot \frac{x(r-x)}{r^3}$$

$$= \sum_{x=0}^{x=r} \frac{x(r-x)}{r^3} \cdot \frac{r!}{x!\,(r-x)!} \cdot p^x q^{r-x}$$

$$= \frac{1}{r^2} \sum_{x=0}^{x=r} \frac{(r-1)!}{(x-1)!\,(r-x-1)!} p^x q^{r-x}$$

$$= \frac{(r-1)pq}{r^2} \sum_{x=0}^{x=r} \frac{(r-2)!}{(x-1)!\,(r-x-1)!} p^{x-1} \cdot q^{r-x-1}$$

$$= \frac{(r-1)pq}{r^2} \cdot (p+q)^{r-2}$$

$$= \frac{(r-1)pq}{r^2}.$$

$$\therefore M(\sigma_0^2) = \frac{r-1}{r}\sigma^2 \quad \text{and} \quad \sigma^2 = \frac{r}{r-1}M(\sigma_0^2).$$

Thus a statistic s^2 which satisfies (i), i.e. the condition that its mean value is σ^2 is :

$$\frac{r}{r-1}\sigma_0^2 = \frac{r p_0 q_0}{r(r-1)} = \frac{p_0 q_0}{r-1}.$$

In accordance with (ii) we should therefore write the critical ratio for the difference of two samples respectively composed of a and b items as

$$\frac{p_a - p_b}{\sqrt{p_{ab}q_{ab}\left(\dfrac{1}{a-1} + \dfrac{1}{b-1}\right)}}$$

$$= (p_a - p_b)\sqrt{\frac{(a-1)(b-1)}{p_{ab} \cdot q_{ab}\,(a+b-2)}} \qquad . \qquad . \qquad . \qquad . \quad \text{(iii)}$$

The importance of the correction involved in the substitution of $(a-1)$ for a and $(b-1)$ for b in (iii) of 4.09 is open to question inasmuch as : (a) it makes very little difference if the pooled sample is large ; (b) we are entitled to assume a normal distribution of $d = (p_a - p_b)$ only when the pooled sample is in fact large. In any case, we have now at our disposal (vide infra, Chapter 6) the means of exploring the precise magnitude of the error of judgment involved in using an estimate of σ_a or σ_b based on the observed proportionate score p_{ab}.

13

4.11 Two Ways of Detecting a Real Difference

When called upon to pronounce judgment on the efficacy of methods of prophylaxis or treatment, the yardstick we employ may be, and quite commonly is, an attribute which admits of a 2-fold split of the items in the universe of choice.   Thus a human population is amenable to division into those who have had cholera and those who have not had it, and the former are themselves divisible into those who die of an attack and those who do not.   Such a classification may be merely qualitative in the sense that our criterion of what is *A* and what is not *A* is *all or none ;* but it may be a more or less arbitrary threshold specifiable by recourse to a number, as if we ask what proportions of patients respectively treated *ceteris paribus* with and without liver extract for six weeks have an r.b.c. count of over 4 million.   So long as we score the efficacy of a treatment by the *number of individuals in a class,* the criterion we adopt for delimiting the class itself is immaterial to the mathematical statement of the problem.   From the standpoint of experimental technique, it may be better to employ a quantitative Kahn in preference to a qualitative Wasserman as a basis for distinguishing patients as seronegative and seropositive ; but a preference for one or the other involves no issue relevant to statistical procedure if our yardstick of relative efficacy w.r.t. long-term and short-term arsenotherapy is the proportion of seropositive patients at the end of a follow-up.

The method of detecting a real difference dealt with in this chapter is the only method at our disposal when the difference, like the difference between Mendel's green and yellow peas, admits of no quantitative specification in the ordinary sense of the term ; and it is the best line of attack when the effect of a treatment is one we have good reason to associate with a fairly clear-cut threshold.   This by no means exhausts the criteria of efficacy we may adopt when the end in view is to assess the biological effect of an external agency or even of a particular gene complex.   We expect a soporific to exert an appreciable effect on the hours of sleep of almost anyone, if undisturbed ; and the appropriate criterion of its action is some average which takes into account its effect on *every* individual in the group.   In distinguishing groups by an average, our criterion explicitly takes no stock of the number of individuals which belong to one or other of a system of two or more classes.   Nor do we do so, when we distinguish two pure lines of beans by the mean weight of the seed or the mean number of seeds per pod.

The detection of a real difference of the sort last stated will be a subject for treatment in Chapter 7, where we shall have to give more precise definition to the implications of two ways of scoring a sample.   The type of difference to which analysis hitherto undertaken in these pages is relevant belongs to the domain of what statistical text-books usually call sampling of attributes, an expression which suggests a peculiarity referable to what we score.   The essential dichotomy refers to the method of scoring we adopt.   Of two methods of scoring samples distinguished later as *taxonomic* and *representative,* the former alone has so far occupied our attention.   The problem of detecting a difference dealt with in this chapter is how to detect a difference which involves a *taxonomic* score.   How to detect a difference which we refer to as a representative score, e.g. a mean or median value, calls for separate investigation of the distribution laws of averages.

CHAPTER 5

SIGNIFICANCE AND CONFIDENCE

The issue dealt with in the last chapter drew attention to two classes of statistical problems, severally distinguishable as problems of verification and problems of estimation. Of recent years, the implications of this distinction have come sharply into focus ; and the most elementary treatment of probability would be incomplete without an examination of statistical judgments *vis-à-vis* this dichotomy. To undertake such an examination with profit we must now try to be more explicit than heretofore about our criteria of credibility and confidence. Inescapably, we find ourselves in controversial territory, when we do so ; but happily the most controversial issues involve judgments of methodology and semantics rather than of mathematical technique. For instance, the weight we attach to introspection as opposed to observation of the behaviour of others is highly relevant to the acceptability of arguments put forward by one school of theoreticians in support of their views ; and it is not possible to do full justice to such reasoning without raising issues about which men of science still entertain diverse opinions. It is therefore pertinent to cite an example from one of the most valuable contemporary treatises. Expounding a view common among mathematicians, the author states : " if we adopt the axiomatic approach in which probability is a measure of attitudes of mind, it is reasonable to take prior probabilities to be equal when nothing is known to the contrary, for the mind holds them in equal doubt ". Confronted with such an assertion, the biologist who is no mathematician is under no obligation to accept it. One may regard it as an illustration of the fact that good mathematics and good biology are consistent with different philosophies of life.

In asking whether the composition of a sample is or is not consistent with a particular hypothesis, we have hitherto been content to seek an answer to the question : would the choice of such a sample be a rare event, if the hypothesis were a correct specification of the conditions of our choice ? So long as we content ourselves with statements of this kind, there is general agreement among statisticians about what procedure we should follow. If we do follow the prescribed procedure, we have at least disclosed what *numerical* information at our disposal is relevant to any verdict we may legitimately pronounce ; but if we do no more than this, we leave the act of judgment itself to others. In ourselves assuming the right to pass a verdict on a particular issue, we accept the responsibility of defining a criterion for a wider class of decisions, *viz.* how to discriminate between the alternatives that : (a) the hypothesis is true in spite of the rarity of an event whose occasional occurrence is consistent with its requirements ; (b) the rarity assigned to the event by the hypothesis is such as to make the hypothesis unacceptable. Such a distinction presupposes some connexion between : (a) the frequency with which an event occurs if a given hypothesis is relevant to the situation in which it does occur ; (b) the frequency with which an individual can *correctly* infer that the same hypothesis is indeed applicable to such a situation.

To be clear about what judgments we can correctly make on the basis of observation of population samples, it is thus helpful *first* to examine a situation about which we have all data relevant to both issues and *then* to explore the consequences of withholding part of our information. Accordingly, we shall suppose that 5 per cent. of mothers with litters of 8 in a laboratory colony of white rats carry a sex-linked lethal gene. The theoretical sex ratio (*females* to *males*) in such litters is 2 : 1, i.e. p the expectation that an offspring of a carrier mother will be male is $\frac{1}{3}$, instead of $\frac{1}{2}$ when the mother is normal. For simplicity, we shall assume, as is often true of rodents, that there is a high measure of intra-uterine competition for survival among fertilised eggs. In so far as this is so, it is pardonable for the heuristic end in view to simplify the situation by neglecting the influence of the lethal gene on the *total* number of offspring born, and therefore on the chance that a mother with a given number of offspring is herself carrier. Given this information, we then have all requisite data to supply an answer to the following question : how often shall I be right in the long run if I deem every mother of 8 to be a carrier because *all her 8 offspring are females?*

In approaching this problem we may distinguish two sorts of information : (a) *prior* information about the background of the situation, *viz.* that 5 per cent. of mothers of 8 are carriers ; (b) the *additional* information in the foreground of the situation, that the mothers of 8 in the present context have exclusively female offspring. We also have two hypotheses upon which we have to pass judgment :

Hypothesis 1.   The mother is a carrier.

Hypothesis 2.   The mother is normal.

Let us consider the frequency of *events* relevant to our judgments about the situation before discussing the issue explicitly raised in the question above, i.e. the frequency of *correct judgments* about the situation. Our prior information about the *background situation* signifies that the mean numbers of carriers and normal females in samples of 1,000,000 mothers of 8 are respectively 50,000 and 950,000. The additional information that the mothers under discussion have female offspring only has different implications depending on whether *hypothesis* 1 or *hypothesis* 2 is applicable to the situation *as a whole*. If the mother of 8 is a carrier, the theoretical frequency of an exclusively female litter is

$$(\tfrac{2}{3})^8 = 0.0389.$$

If she is normal, the theoretical frequency of the event is

$$(\tfrac{1}{2})^8 = 0.0039.$$

In samples of 50,000 litters of 8 produced by mothers who are carriers, the mean number of exclusively female litters will thus be

$$0.0389 \times 50,000 = 1945.$$

In samples of 950,000 litters of 8 produced by normal mothers, the mean number of exclusively female litters will be

$$0.0039 \times 950,000 = 3705.$$

We thus arrive at the following balance sheet :

Mother of Eight.	*All* Offspring Female.	*Some* Offspring Male.	Total.
Carrier .	1945	48,055	50,000
Normal .	3705	946,295	950,00
Total .	5650	994,350	1,000,000

Among mothers of 8 whose offspring are exclusively females the *long-run* ratio of carriers to normal rats will thus be 1945 : 3705 ; and the expectation that the mother of such a litter will herself be a carrier is

$$\frac{1945}{5650} = \frac{(0\cdot05) \times (0\cdot0389)}{(0\cdot05) \times (0\cdot0389) + (0\cdot95) \times (0\cdot0039)} \qquad \cdot \quad \cdot \quad \cdot \quad \cdot \text{(i)}$$

This means that we should *come to the right conclusion* about 1945 times in 5650, if we regard the production of an exclusively female litter by a mother of 8 as sufficient grounds for deeming her to be a carrier. In other words, we should be about twice as often wrong as right. This is the answer to the question stated above ; and it is an assertion about the expectation that our judgment will be right if we act on the assumption that *hypothesis* 1 is applicable to the situation as a whole. Let us denote this by $E(h_1)$. Accordingly, the expectation of a correct judgment on the alternative assumption that *hypothesis* 2 is correct will be

$$E(h_2) = \frac{3705}{5650} = \frac{(0\cdot95) \times (0\cdot0039)}{(0\cdot05) \times (0\cdot0389) + (0\cdot95) \times (0\cdot0039)} \qquad \cdot \quad \cdot \quad \cdot \text{(ii)}$$

The theoretical frequency of a correct judgment about the background situation is the *prior* expectation that a mother of 8 will be as specified by which hypothesis dictates our decision *in the absence of the additional information* that restricts the range of choice, *viz.* that all her offspring are female. We shall denote by $E(p_1)$, the frequency of a correct judgment, if we deem *every* mother of 8 to be a carrier and by $E(p_2)$ the frequency of a correct judgment, if we deem every mother of 8 to be normal. Thus $E(p_2) = 0.95$ means that 95 per cent. of my judgments will be right in the long run, if I assume that *hypothesis* 2 is applicable to the background situation. The expectation that a mother of 8 will have exclusively female offspring is the theoretical frequency of an *event* which is independent of one's ability to make a correct judgment, but does depend upon the specification of the mother in terms of one or other hypothesis under discussion. We shall denote it by $E(f_1)$ if the first hypothesis is applicable, and by $E(f_2)$ if the second is applicable. In this example $E(f_1) = 0\cdot0389$ and $E(f_2) = 0\cdot0039$. What was originally a balance sheet of information about rats thus becomes a balance sheet of information about the frequency of correct judgments, when we generalise (i) and (ii) by recourse to these symbols, i.e. :

$$E(h_1) = \frac{E(p_1) \times E(f_1)}{E(p_1) \times E(f_1) + E(p_2) \times E(f_2)} \qquad \cdot \quad \cdot \quad \cdot \quad \cdot \text{(iii)}$$

$$E(h_2) = \frac{E(p_2) \times E(f_2)}{E(p_1) \times E(f_1) + E(p_2) \times E(f_2)} \qquad \cdot \quad \cdot \quad \cdot \quad \cdot \text{(iv)}$$

This is the form the famous theorem of Bayes takes, when only two hypotheses are admissible. As applied to situations which offer a choice between *n* hypotheses, we may write it in the form :

$$E(h_m) = \frac{E(p_m) \cdot E(f_m)}{\sum_{x=1}^{x=n} E(p_x) \cdot E(f_x)} \qquad \cdot \quad \cdot \quad \cdot \quad \cdot \text{(v)}$$

It is customary to speak of $E(h_m)$ as the *posterior* probability of *hypothesis m* in contradistinction to $E(p_m)$, its *prior* probability in the absence of additional information which is a necessary basis for assigning a value to $E(f_m)$. The latter we call the *likelihood* of the occurrence on the assumption that the mth hypothesis is correct. These terms do not make explicit the essential content of the balance sheet of Bayes. We may capture it better by referring to $E(h_m)$ as the *operational value* of the hypothesis, to $E(p_m)$ as its *commendability* and to $E(f_m)$ as the theoretical frequency accordingly ascribed to the *occurrence itself*.

Though *Bayes' Theorem* has been the battleground of a long and vigorous controversy, the theorem itself is unexceptionable. It is indeed an elaborate tautology. What is open to exception is not its form or content. The argument is about its usefulness. As it stands, it answers no question until we can assign a value to the prior probabilities $E(p_x)$, i.e. the expectation that one or other hypothesis is applicable to the *background* situation ; and we commonly have to make judgments about events without any exact information of this sort. In any case, the exactitude of the statement presupposes that the number of hypotheses which might apply to the situation is specifiable ; and any such assumption invites Bacon's comment that the operations of nature are more various than the conceptions of man.

In a scholium of his posthumous work (*Phil. Trans. Roy. Soc.*, 1763) Bayes proffered a tentative recipe for action by recourse to the axiom that prior probabilities are equal if we have no available information to the contrary. In accordance with this famous postulate, we should have to put $E(p_1) = \frac{1}{2} = E(p_2)$ in the foregoing example, if we did not know what proportion of females in our colony were carriers. Let us see where this would lead us. We should then write

$$E(h_1) = \frac{\frac{1}{2}(0 \cdot 0389)}{\frac{1}{2}(0 \cdot 0389) + \frac{1}{2}(0 \cdot 0039)} = \frac{389}{428}.$$

In accordance with Bayes' postulate we thus arrive at the conclusion that we should be right about 39 times out of 43 in the long run, if we accepted an exclusively female litter as sufficient grounds for deeming a mother of 8 to be a carrier. In other words, we should be ten times as often right as wrong. Actually, as we have seen above, we should be more often wrong than right ; indeed, nearly twice as often. Needless to say, the school which adheres to Bayes' postulate would not claim that its use necessarily gives us a correct assessment of the value of such a judgment in a particular situation. What they do claim is that we shall make more correct judgments than wrong ones in the long run, if we apply both Bayes' theorem and his axiom consistently to situations in which we lack background information relevant to a precise judgment.

Applying the theorem itself to everyday affairs means acting in accordance with the hypothesis which assigns highest posterior probability to an occurrence evoking an act of judgment ; and to do so it is necessary to assign a numerical value to the prior probabilities of the hypotheses. Applying the axiom signifies that we assume the equality of the latter in the absence of any knowledge about them. In effect this means that we then adopt the hypothesis which assigns to the occurrence the greatest *likelihood* in the technical sense defined above. Such a recipe for action in the practical affairs of life invites objections of two sorts. One is that it appears to imply a model urn of judgments from which we draw with equal frequency in the long run counters which do or do not correctly specify a given situation, and critics of the postulate of Bayes fail to recognise compelling reasons for undertaking so exacting an exploit of the imagination.

A more serious objection directs attention to what useful purpose the balance sheet of Bayes' theorem serves as a *caveat* against over-confidence in the absence of sufficient relevant evidence. The limitation implicit in the postulate itself is an idealisation which is at loggerheads with the theorem so conceived. For it has little relevance to situations in which we commonly make decisions about the relative merits of hypotheses. To be sure, it is not hard to illustrate a precise application of Bayes' theorem and the implications of the postulate by recourse to an urn problem, if the end in view is to provide a text-book with an apposite example ; but the world's work commonly confronts us with situations in which the postulate is inappropriate. The very fact that one takes the trouble to test a hypothesis in real life commonly arises because one has good grounds for believing that it is right or for suspecting that it is wrong. We cannot therefore say that we have *no* relevant prior knowledge at our disposal. Though we have no grounds for assigning a *numerical* value to the prior probabilities which make the balance sheet audit, we commonly have good reasons for believing that they are unequal, and indeed grossly unequal. In such circumstances, we are faking the balance sheet if we ascribe the numerical value 0·5 to the prior probabilities of alternate hypotheses. Our rat colony example emphasises this. If the statement of the problem withheld the relevant information that $E(p_1) = 0.05 = 1 - E(p_2)$, any biologist invited to express a judgment would start off with the knowledge that female carriers of sex-linked lethals are happily uncommon in colonies kept for laboratory use other than research on such genes.

It is indeed important to recognise that we make statistical judgments against a background of information, only part of which—and often a small part—is susceptible of sufficiently precise statement for incorporation in a balance sheet of the frequency of correct verdicts ; and this fact is highly relevant to a realistic appreciation of what meaning we may legitimately attach to the term *statistical significance*. The pivot of criticism which Bayes' postulate has provoked has been the claim that it can contribute to solution of problems involving *inverse* probability, i.e. inferences about a universe of which our knowledge is confined to samples ; but the theorem as propounded by its author is truly relevant only to problems involving *direct* probability, i.e. decisions about the source of samples which may come from one or other universe of known composition. One use it serves is therefore to bring into focus the limitations inherent in any numerical criterion of significance ostensibly concerned with the operational value of putatively admissible hypotheses.

To appreciate what is salutary in Bayes' theorem from this point of view, it is necessary to remove certain ambiguities customary in verbal designations of the expressions it contains. As a balance sheet, it is easy to illustrate by recourse to a fictitious example, and easy to visualise as in Figs. 52–53 ; but the verbal interpretation of the symbols is either ambiguous or cumbersome. An outstanding ambiguity in current statements arises from the customary implication of the terms *prior* and *posterior* as applied to applicability of the specified hypotheses. The *prior* probability is an assessment of the frequency with which the hypothesis applies to a general class of situations referred to above as the *background* of the event. The *posterior* probability refers to its applicability to a particular member of such a general class of situations ; and the additional information which it takes into account changes the scope of the hypothesis. In so far as the scope of a hypothesis is in fact relevant to its specification, the two probabilities do not literally refer to the same hypothesis. While a succinct verbal formula which resolves this dilemma is difficult to devise, it is not impossible to find a form of words which conveys the gist of the matter. What the balance sheet of Bayes discloses is that an exact decision in favour of one or other alternative hypothesis involves *both*

(a) An assessment of the frequency (*likelihood*) each hypothesis assigns to the observed event as it stands.

(b) An assessment of the frequency with which they respectively apply to a class of events sharing every characteristic of the observed one other than the particular numerical specification which *explicitly* provides the raw material for a statistical judgment.

An exact decision here signifies an answer to a question of the form : how often shall I be right if I act on the assumption that such and such a hypothesis is correct ? The dual

$E(p_1) \cdot \frac{2}{4} \cdot \frac{1}{4}$ $E(p_2) \cdot \frac{2}{4} \cdot \frac{1}{4}$ $E(p_3) \cdot \frac{1}{4}$ FIG 52

$p_1 = \frac{1}{4}$ $p_2 = \frac{1}{2}$ $p_3 = \frac{3}{4}$

$E(f_1)$ $E(f_2)$ $E(f_3)$

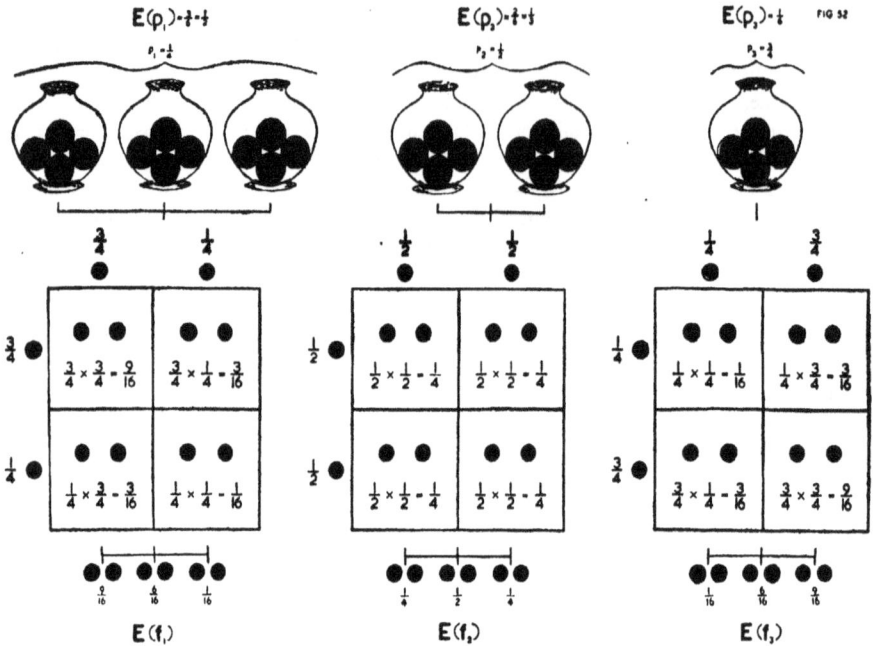

		$E(h_1)$	$E(h_2)$	$E(h_3)$
●	●	$\left(\frac{3}{4}\cdot\frac{9}{16}\right) \div \left(\frac{3}{4}\cdot\frac{9}{16}+\frac{2}{4}\cdot\frac{1}{4}+\frac{1}{4}\cdot\frac{1}{16}\right) = \frac{27}{32}$	$\left(\frac{2}{4}\cdot\frac{1}{4}\right) \div \left(\frac{3}{4}\cdot\frac{9}{16}+\frac{2}{4}\cdot\frac{1}{4}+\frac{1}{4}\cdot\frac{1}{16}\right) = \frac{4}{32}$	$\left(\frac{1}{4}\cdot\frac{1}{16}\right) \div \left(\frac{3}{4}\cdot\frac{9}{16}+\frac{2}{4}\cdot\frac{1}{4}+\frac{1}{4}\cdot\frac{1}{16}\right) = \frac{1}{32}$
●	●	$\left(\frac{3}{4}\cdot\frac{6}{16}\right) \div \left(\frac{3}{4}\cdot\frac{6}{16}+\frac{2}{4}\cdot\frac{1}{2}+\frac{1}{4}\cdot\frac{6}{16}\right) = \frac{9}{20}$	$\left(\frac{2}{4}\cdot\frac{1}{2}\right) \div \left(\frac{3}{4}\cdot\frac{6}{16}+\frac{2}{4}\cdot\frac{1}{2}+\frac{1}{4}\cdot\frac{6}{16}\right) = \frac{8}{20}$	$\left(\frac{1}{4}\cdot\frac{6}{16}\right) \div \left(\frac{3}{4}\cdot\frac{6}{16}+\frac{2}{4}\cdot\frac{1}{2}+\frac{1}{4}\cdot\frac{6}{16}\right) = \frac{3}{20}$
●	●	$\left(\frac{3}{4}\cdot\frac{1}{16}\right) \div \left(\frac{3}{4}\cdot\frac{1}{16}+\frac{2}{4}\cdot\frac{1}{4}+\frac{1}{4}\cdot\frac{9}{16}\right) = \frac{3}{20}$	$\left(\frac{2}{4}\cdot\frac{1}{4}\right) \div \left(\frac{3}{4}\cdot\frac{1}{16}+\frac{2}{4}\cdot\frac{1}{4}+\frac{1}{4}\cdot\frac{9}{16}\right) = \frac{8}{20}$	$\left(\frac{1}{4}\cdot\frac{9}{16}\right) \div \left(\frac{3}{4}\cdot\frac{1}{16}+\frac{2}{4}\cdot\frac{1}{4}+\frac{1}{4}\cdot\frac{9}{16}\right) = \frac{9}{20}$

FIG. 52. Urn Model of Bayes' Balance Sheet (*with Replacement*). We may here choose 2 balls from one of 6 urns containing black (*success*) and red (*failure*) balls : (i) 3 urns each containing 3 red and 1 black ball ($p_1 = \frac{1}{4}$) ; (ii) 2 urns each containing equal numbers of black and red balls ($p_2 = \frac{1}{2}$) ; (iii) 1 urn containing 1 red and 3 black balls ($p_3 = \frac{3}{4}$). The expectation of choosing a ball from (i) is $\frac{1}{4}$, and this is the expectation $E(p_1)$ of choosing a ball from an urn with a chance of $\frac{1}{4}$ in favour of success. That of choosing a ball from an urn with a chance of $\frac{1}{2}$ in favour of success has an expectation $E(p_2) = \frac{1}{3}$, and that of choosing a ball from an urn with a chance of $\frac{3}{4}$ in favour of success has an expectation $E(p_3) = \frac{1}{6}$. The chessboard shows the likelihood $E(f_1)$, etc., of getting 0, 1 or 2 successes in a 2-fold draw from each class of urns, and the staircase model exhibits the contingent probability of each event with due regard to the *prior* expectation of choosing the 2 balls from one class of urns rather than another. The balance sheet at the bottom exhibits the probability $E(h_1)$, etc., of drawing a right conclusion about which class of urns a specified pair of balls may have come from.

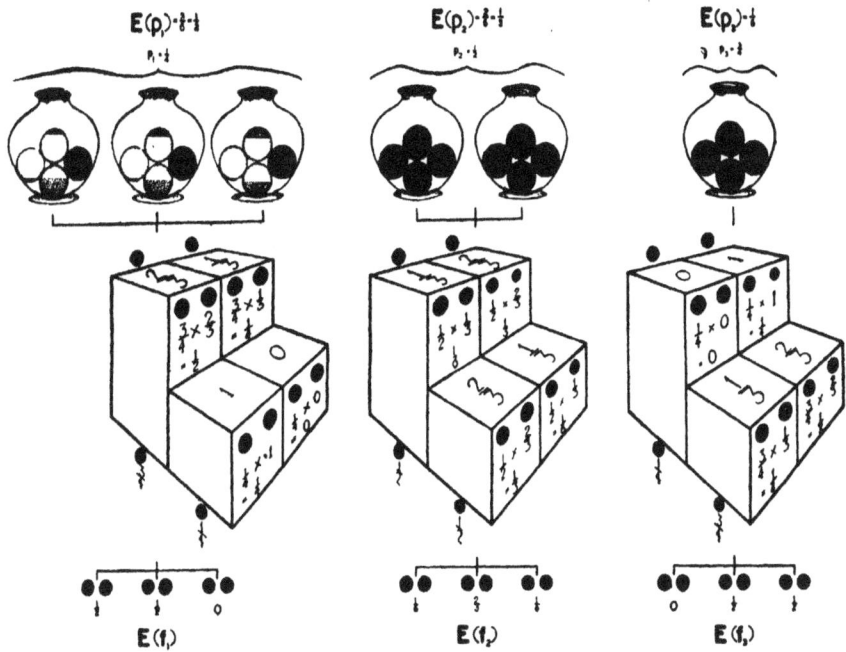

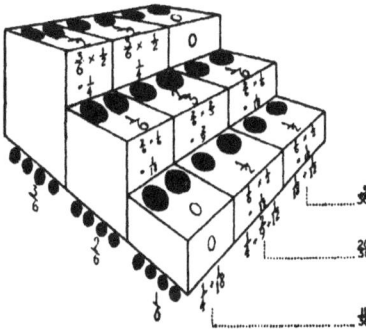

FIG. 53. Urn Model of Bayes' Balance Sheet (*without Replacement*). The set-up is essentially as for Fig. 52 except in so far as it is necessary to use the staircase model for non-replacement to evaluate the likelihood of drawing each class of pairs from a particular class of urns.

requirements of an exact decision in this sense as set forth in the balance sheet of Bayes are not peculiar to judgments about inverse probability. Nor did Bayes set forth his theorem with that end in view. Accordingly, we shall re-examine our criterion of significance *vis-à-vis* the class of judgments we pronounce when the end in view is the assessment of an *a priori* hypothesis as defined in Chapter 3 before proceeding to explore in greater detail the ulterior problem of estimation raised in Chapter 4. We shall therefore need to recall the nature of the assessment which we designate a test of significance.

Before doing so, we may pause to take stock of a distinction between two uses of numbers which enter into the complete balance sheet. We started our enquiry by discussing only the frequency of *events* : (a) how often rat mothers of 8 in our model colony are normal or otherwise ; (b) how often rat mothers of one sort or the other have exclusively female offspring. We end it by allocating the *frequency of the truth of a proposition* identifying a particular mother of 8 exclusively female offspring as a mother of one sort or the other. The two classes of frequency which enter into our balance sheet call attention to a dichotomy in contemporary statistical thought. One school prefers to identify the concept of probability exclusively with the frequency of external events. Another identifies it exclusively with the frequency of correct judgments about them. Between the two extremes a more eclectic view is admissible, *viz.* : (a) frequencies ascribed to events by particular hypotheses do not provide a sufficient basis for assessing the frequency of applying the latter correctly ; (b) valid judgments about the frequency of applying hypotheses correctly have to rely on information derived from observations about the frequency of events which invoke their use.

One property of (v) deserves further comment before we take leave of Bayes' theorem. If each of m hypotheses has equal prior probability we may write $E(p_m) = 1/m$, so that

$$E(h_m) = \frac{E(f_m)}{\sum_1^m E(f_x)} \quad \text{i.e.} \quad E(h_m) \propto E(f_m).$$

This is tantamount to saying that the hypothesis which has highest posterior probability is the one which assigns highest likelihood to the occurrence. If we take as our criterion of the operational value of a hypothesis the condition that it assigns maximum likelihood to the event, we are therefore bringing in Bayes' postulate by the back door. It is pertinent to assert this explicitly, because some opponents of Bayes' postulate have espoused the principle last stated ; and protagonists of the axiom have raised the objection that Bayes' postulate is implicit in the principle. Those who propose the method of maximum likelihood as an escape from the dilemma which the postulate sidesteps assert that the application of Bayes' axiom does not always lead to the same results as the method of maximum likelihood itself ; but the examples adduced by them refer to continuous distributions. It seems to the writer that we should regard the issue at stake as semantic rather than mathematical. The assumption of continuity is a useful fiction for construction of convenient formulæ for computations of sufficient precision for practical purposes ; but we may well ask whether an assumption of this sort is relevant to the logic of sampling of attributes as such. At least in this domain we are concerned only with enumeration of discrete entities. If assumption of continuity leads to paradox, may it not therefore be so because the assumption is essentially, albeit conveniently, paradoxical ?

EXERCISE 5.01

1. I toss a penny ten times in succession. It always turns up a head. If one penny in every million coined has two heads, what is the chance that my penny has two heads ?

2. A purse contains ten coins, each of which is either a florin or a half-crown ; a coin is drawn and found to be a half-crown, what is the chance that this is the only half-crown ?

3. A purse contains ten coins, which are either pennies or florins. If a single coin drawn from the purse is a penny, what is the chance that it is the *only* one ?

4. A year which contained 53 Sundays was not the last year of a century. What are the odds against the truth of the surmise that it was a leap year ?

5. A, B, C entered for a race, and their respective chances of winning were estimated at $\frac{2}{11}$, $\frac{4}{11}$, $\frac{5}{11}$. Circumstances which come to our knowledge raise the chance of A's success to $\frac{1}{4}$. What are then the prospects of B and C ?

6. The dealer removes *one* of a pack of 52 cards. From the remainder of the pack a player draws 2 which are both spades. Assess the chance that the missing card is also a spade.

7. Jones tells the truth three times out of four, Smith four times out of five ; and they agree in asserting that a ball drawn from a bag containing nine balls, each of a different colour, is red. Show that the probability of the truth of their assertion is $\frac{96}{97}$.

8. Snooks makes a true report four times out of five, Wright three times out of five, and Johnson five times out of seven. If Wright and Johnson both report that an experiment failed and Snook reports the reverse, what is the chance that the experiment succeeded ?

5.02 THE CRITICAL RATIO

When a sampling distribution is normal, we can cite the long-run frequency of an event, i.e. its *likelihood* in the sense defined above, in conformity with a particular hypothesis, by recourse to what we have hitherto called the *c*-ratio. For each numerical value (h) of this *critical ratio*, the probability integral assigns the fraction of the total area under the normal curve of unit variance (and unit area) enclosed between the mean ordinate and the positive ordinate (to the right) corresponding thereto. We are at liberty to make use of this information in two ways. To distinguish between them, it will suffice to cite a small-scale type of problem amenable to direct application of the binomial distribution. Let us suppose that a mother of ten has nine girls. Our concern may be to state how often a normal (1 : 1) sex ratio prescribes that the number of boys or girls in a fraternity of ten will exceed or fall short of the expected mean value 5 by as much as 4. In conformity with the symbolism of 3.01, we shall here denote this by $E(\geqslant |4|)$. Alternatively, and more especially if we entertain good reason for suspecting that the genetical constitution of the mother prescribes a sex ratio of 2 : 1, our concern may be to state in conformity with the same assumption : how often the number of girls will be as great as nine. We here denote this as $E(\geqslant + 4)$, if the score specifies the actual number of girls in the family or by $E(\leqslant - 4)$ if the score specifies the number of boys.

Hitherto we have taken no cognisance of the second type of question, which receives scant attention in many statistical treatises ; but the distinction is not trivial from a logical viewpoint. Since we shall return to it at a later stage, it will be convenient to forestall repeated periphrases by recourse to appropriate verbal labels. We shall therefore speak of : (*a*) a specification of the first type which we denote $E(\geqslant | X |)$ as the *modular* likelihood ; (b) the alternative specification which we denote by $E(\geqslant + X)$ or $E(\leqslant - X)$ as the *vector* likelihood of the event in conformity with the selected hypothesis, e.g. a normal sex-ratio in the preceding illustration. When the numbers involved justify recourse to the use of the normal distribution function we proceed to evaluate them as follows :

(i) *Modular likelihood.* The area of the normal curve between 0 and h defines half the value of the expectation that a deviation corresponding to $h = X \div \sigma_x$ will fall inside the range $\pm(X \div \sigma_x)$. If we denote it by $A(0, + h)$

$$E(| X | \geqslant h\sigma) = 1 - 2A(0, + h).$$

Modern tables (e.g. Table 2, p. 439, Vol. 1, 1st edition of Kendall's *Advanced Theory of Statistics*) usually cite for different values of the critical ratio the numerical value of the total area under the normal curve from $- \infty$ up to $+ h$. If we write this as $A(- \infty, + h) = 0.5 + A(0, + h)$,

$$E(| X | \geqslant h\sigma) = 2 - 2A(- \infty, + h) \quad . \quad . \quad . \quad . \quad \text{(i)}$$

(ii) *Vector likelihood.* We are free to choose which of two exclusive classes such as boys and girls we specify explicitly by the score, and we shall here assume that the class to which the score explicitly refers is the class which excites suspicion in virtue of excess over expectation. The vector likelihood is then the area of the curve to the right of the ordinate specified by the critical ratio, i.e.

$$E(X \geqslant h\sigma) = 1 - A(- \infty, + h) = \tfrac{1}{2}E(| X | \geqslant h\sigma) \quad . \quad . \quad . \quad \text{(ii)}$$

The formulæ (i) and (ii) are applicable to any *symmetrical* sampling distribution normal or otherwise.

Example. Kendall's table cites for $h = 3$, $(X = 3\sigma)$, the value 0.99865. In accordance with (i) the theoretical frequency of a deviation *numerically* as great as 3σ regardless of sign is

$$2 - 2(0.99865) = 0.0027.$$

The odds against are therefore $9973 : 27 = 370 : 1$. In accordance with (ii) the theoretical frequency of a *positive* deviation as great as 3σ is

$$1 - 0.99865 = 0.00135.$$

The odds against are therefore about $740 : 1$.

* * * * * * *

Our last numerical example provides an opportunity for being more explicit than previously about an issue which is relevant to what follows. Two numerical values of the critical ratio have a special interest, viz. the 3σ level cited above and the particular value 0.675σ (*vide* 6.02 *infra*) sometimes called the *probable error*. The latter defines a modular likelihood of 0.5, i.e. equal odds for or against the occurrence. For reasons which we shall discuss at a later stage, the particular interest which attaches to the 3σ level resides in the fact that the odds against occurrence increase very rapidly thereafter, as shown by the following :

Critical Ratio	Modular Likelihood	Approximate Adverse Odds
3·0	0·00270	370 : 1
3·1	0·00194	500 : 1
3·2	0·00138	700 : 1
3·3	0·00096	1,400 : 1
3·4	0·00068	1,800 : 1
3·5	0·00046	2,200 : 1
· · ·	· · ·	· · ·
4·0	0·00006	16,000 : 1

At the 2σ level the *adverse* odds are about 20 : 1, and thus increase about 18-fold in the interval between 2σ and 3σ. Between 3σ and 4σ they increase about 45-fold. The 2σ level has nothing to commend it, other than the fact that it is easy to remember the modular likelihood (5 per cent.) and the corresponding adverse odds. What merits any other specificable values of the critical ratio such as the probable error or the 3σ criterion enjoy is an issue which comes into clearer focus if we distinguish between two objectives which a significance test may subserve.

After examining a class of phenomena, we may be led to the conclusion that only one hypothesis in common parlance makes sense of the data. If the requirements of the hypothesis are numerical, we may then want some assurance that the fit is satisfactory. If the numerical divergence between theory and observation falls near the 0.675σ level, we can then say that agreement is satisfactory in the sense that even greater divergence would be somewhat more common than a discrepancy smaller than the observed one. In assessing the fit as satisfactory in this sense, we make no assertion about how often we should err by acting on the assumption that the hypothesis correctly describes the class of situations which invoke its guidance. In fact, we merely record the judgment that the observations we invoke to test its truth do not disappoint the hope which motivated the test. We therefore proceed on the assumption that the hypothesis is a good one, until we encounter new information which compels us to modify or to discard it.

The situation discussed in the last paragraph is one which arises when there is some new synthesis of knowledge, and we may illustrate it by an appropriate example. In one of Mendel's original experiments he obtained a progeny of 5474 round and 1850 wrinkled peas. Hypothesis prescribed a 3 : 1 ratio of round to wrinkled, i.e. an expectation of 0.2500 for wrinkled. The observed proportion was $1850 \div 7324 = 0.2526$. The s.d. of the theoretical distribution calculated in accordance with the formula on p. 149 is approximately 0.0050, and the probable error $(0.675\ \sigma)$ is approximately 0.0033. The deviation $0.2526 - 0.2500$ is 0.0026, and is therefore inside the range which defines equal odds.

Such situations arise when there is a comprehensive synthesis of theoretical knowledge bringing a diversity of phenomena hitherto unco-ordinated within the framework of a single generalisation. It would be difficult to cite an example from the social sciences, and it is an exceptional happening in biology to date. Only in the physical sciences is it a considerable preoccupation of statistical procedure. That this is so, may partly explain why mathematicians in contact with experimental physics commonly espouse the viewpoint which identifies probability exclusively with the frequency of external events. For what is characteristic of situations comparable to the above is that no explicit assessment of the frequency of making a correct judgment enters into the interpretation of the test.

In the rough and tumble of operational research in the domain of biology, psychology or economics, we rarely undertake a test with so unique an aim. More often than otherwise, the end in view when we perform a significance test is to arbitrate on the respective merits of *different* hypotheses, of which (*a*) one has more general applicability to the class of situations with which we are concerned ; (*b*) the other is (or others are) in closer agreement with the numerical data of the total situation at face value. The customary procedure is to accept the latter *in default*, if the likelihood assigned to the event by the former, i.e. the *null hypothesis*, falls short of an arbitrary limit defined by the critical ratio, e.g. 2σ or 3σ, according to taste. In general, the null hypothesis is therefore the *conservative* hypothesis, i.e. the hypothesis which accords most commonly with our prior knowledge of the background situation in the absence of the additional information which the event itself supplies.

So far, we have paid no attention to considerations which are relevant to the choice of such a limit in a particular class of background situations. The following problem will serve as a model. If a man spins a coin ten times and scores nine heads am I to conclude that the penny has a bias ? The implication is that the bias favours heads. So the question involves choosing between : (i) the hypothesis that there is no appreciable bias ($p = \frac{1}{2} = q$) ; (ii) the hypothesis that the penny has a bias ($p > \frac{1}{2}$). If we agree to accept (ii) *by default*, we do not have to raise the issue of estimation, since the hypothesis we erect to *nullify* as a condition of our preference for the other specifies the relevant parameter of the putative universe.

Since the possible bias with which we are here concerned has a specified direction, we may make our assessment of the theoretical frequency of the occurrence in conformity with the null hypothesis in terms of its vector likelihood as defined above. The expectation of scoring more than eight heads in a 10-fold trial is $10(\frac{1}{2})^{10} + (\frac{1}{2})^{10} = \frac{11}{1024}$ or odds of 93 : 1 against the occurrence. *Vis-à-vis* the common practice of regarding a deviation beyond the 2σ level as significant, these are high odds ; but the fact that the odds are high does not suffice to provide an exact answer to our question, if we interpret it in an operational sense, *viz.* shall I be right more often if I act in accordance with hypothesis (ii) that if I act in accordance with hypothesis (i) ?

When faced with such a choice we have seen that an exact answer involves due consideration to the *prior probabilities*. If the prior probabilities were equal we should certainly reject (i), since we can construct an indefinitely large number of hypotheses of type (ii) by postulating different values of p between $\frac{1}{2}$ and 1 ; and any such hypothesis would assign a higher likelihood to the event. Clearly we have no grounds for supposing that all such hypotheses have as much and as little to commend them as the hypothesis which predicates no appreciable bias ; and the situation illustrates the difficulty of applying Bayes' axiom if we concede its validity. Although we have no basis for assigning a numerical value to the prior probabilities involved, the attendant circumstances may be such as to justify the assumption that (i) has a much higher probability than (ii). A variety of factual information not disclosed in the preceding statement would in given circumstances prove to be more or less relevant : whether I have good grounds for believing that coins of the realm in general have no appreciable bias ; whether I am free to examine the penny and reinforce my confidence in the belief that it is in fact a coin of the realm ; whether I have greater or less experience of counterfeit coinage ; whether I know the man to be financially honest ; and what proportion of men in general are endowed with habitual honesty w.r.t. the consequences of spinning a coin.

The student will now appreciate a good reason for exhibiting the eighteenth century museum piece which was the subject-matter of 5.01. The important issue which Bayes' balance sheet brings into focus is that a hypothesis with a high intrinsic likelihood may have a low operational value, if it has very low extrinsic *commendability*. If we reject Bayes' axiom as a mere act of faith which no experience can justify, we are therefore driven to the following conclusions :

(*a*) we should act in accordance with a hypothesis which experience shows to be usually applicable to the relevant class of *background* situations, unless it assigns very high odds against the particular occurrence under consideration ;

(*b*) since we commonly lack exact knowledge about how often the *conservative* hypothesis does in fact apply to relevant background situations, what odds we regard as *high* involves an act of judgment embracing an intimate knowledge of the subject-matter.

There is therefore no universal criterion of what is *high or low* in this context ; but the fact that there is no such universal criterion endows with a peculiar interest the rapidity with which adverse odds increase beyond the 3σ level. If the null hypothesis, which is commonly also the conservative hypothesis, ascribes to an event a critical ratio above 3, an alternative hypothesis of very low commendability (i.e. prior probability) may have higher operational value, if it assigns a relatively high likelihood to the event. Conversely, one may well hesitate to embrace such an alternative hypothesis, if the conservative hypothesis assigns a lower critical ratio to the event ; but this does not mean that we should reject it. All it signifies is that the matter invites further investigation. In any case, there would seem to be only one logical alternative to acceptance of the Bayes postulate—or, what is semantically equivalent, the principle of maximum likelihood. If we reject it we have to recognise a *sliding scale of critical ratios ;* and the operational value of a particular significance level involves an act of judgment which the investigator at home with the materials of the problem cannot afford to relegate to the mathematical specialist who has no first hand familiarity with them.

By the same token, the investigator at home with his materials (and others who share relevant familiarity with the background situation) has then to be the arbiter of what critical ratio justifies more or less confidence in rejecting a null hypothesis. This makes it important to recognise two different criteria which motivate the acceptance of the hypothesis one selects for nullification with a view to adopting an alternative hypothesis by default. One may do so primarily because it is the only plausible hypothesis which is itself susceptible of exact statement or because it is the more conservative alternative of two hypotheses each amenable to exact statement. In practice, the distinction is not clear cut. The fact that a given hypothesis is the only available one susceptible of exact formulation very often signifies that it is the hypothesis in line with common experience of the background situation. On the other hand, the specification of a hypothesis as *conservative* is itself open to more than one interpretation. The investigator may have special sources of information, which point to the view that a generally accepted hypothesis has less applicability to the background situation than current well-informed opinion concedes. If so, a critical ratio which would reinforce the investigator's confidence would not necessarily commend the alternative hypothesis to a critic without disclosure of information which lies outside the scope of statistical analysis. To that extent a statistical judgment can invoke agreement only in so far as it invites the exploration of a middle way.

This implication of the credentials of a statistical judgment at the operational level has a special relevance to the theme we shall next pursue, namely the search for an explicit formulation of the limitations involved in so-called significance tests which involve an act of *estimation*. Our next objective will be to give precise specification to the frequency with which errors of a given magnitude arise in this domain. In accordance with the viewpoint of the foregoing sections, the terms of reference of such a specification must address themselves to the assent of investigators who do not share the same relevant knowledge of the background situation. In short, a satisfactory statement of the relevant data should also be an implicit invitation to agreement with respect to what magnitude of sampling error has a frequency too low to compromise the verdict of the test.

5.03 Proportionate Error of Estimation

In any use we make of the critical ratio within the domain of taxonomic scoring, it is necessary to draw a distinction between two classes of null hypotheses :

(i) the null hypothesis which explicitly postulates the exact value of the expected proportionate score p, or raw score rp, as when we test (Chapter 3) the progeny of a mating to ascertain its conformity to a particular Mendelian ratio ;

(ii) the null hypothesis which compels us to rely on an estimate of the parameter p, as when the end in view is the recognition (Chapter 4) of a real difference.

As regards (i), the only issue at stake is one of *verification*, i.e. : is the null hypothesis acceptable or otherwise ? We then have to interpret the odds assigned by the critical ratio in the

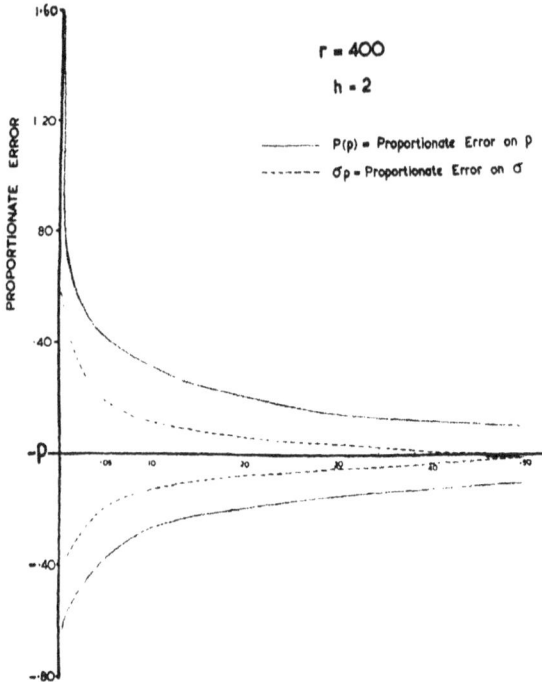

FIG. 54. The error involved in estimating the s.d. of the proportionate score distribution is always smaller than the error involved in estimating the value of the true proportion of successes.

light of considerations discussed in 5.02 with due regard to the possibility that our sample is not very representative of the universe from which it comes. The alternative type (ii) of null hypothesis raises quite another issue. *Ceteris paribus* everything turns on the magnitude of the error involved in assigning an appropriate value to p on the basis of *data derived from sampling alone*. The kernel of the problem of statistical *estimation* is to safeguard ourselves against distortion of judgment attributable to errors from this source.

It is scarcely necessary to labour this distinction. Indeed, every thoughtful student first confronted with the procedure for detecting a real difference adopted in 4.08 must have experienced a misgiving, which we may formulate as follows. Our only source of knowledge concerning the true proportion p in the putative common universe, being the pooled sample of $(a + b)$ items before us, is subject to an unspecifiable sampling error on that account ; and any conclusions we base on the critical ratio of the proportionate score difference with respect to sub-samples of a and b items is likewise subject to an unspecifiable sampling error, since the value assigned to the *s.d.* of the proportionate score difference distribution depends on what value we assign to p itself.

We shall now examine how far it is possible to assess the order of magnitude of the uncertainty arising from this circumstance. At the outset, we can fortify self-confidence by the consideration that the impossibility of specifying the exact value of a parameter does not necessarily signify the impossibility of specifying how often we are liable to make an error of a given order of magnitude in assessing it, and hence how often any judgments based on what value we do assign to it will err beyond specifiable limits. It will clear the ground for a new approach to this issue if we make explicit in our formulæ both what we do know with certainty and what we do not know when we apply c-tests, as in Chapter 4.

Let us suppose that we have before us an estimate p_{ab} of an expected proportionate score based on a sample of $(a + b) = r$ items. We do not actually know the true value p of the proportion of specified items in the putative parent universe. So we do not know the exact value of the standard deviation of the proportionate score distribution in a sample of a items, of b items or of r items. To make this uncertainty explicit, we therefore assume that p differs from p_{ab} by an unspecified error ϵ which we can express as the product of a critical ratio h and of the true standard deviation σ_t of the proportionate score distribution in an r-fold sample. In this way we can investigate : (a) how big an error arises in our estimated standard deviations from a given proportionate error w.r.t. the value p_{ab} assigned to p ; (b) the odds that the error inherent in the estimation of p or of parameters based thereon shall not exceed a specified order of magnitude ; (c) hence the odds for or against making an error of a given magnitude in assigning any critical ratio based on the estimated standard deviation of a score or score difference distribution w.r.t. a sample of given size.

If $p_{ab} = (1 - q_{ab})$ is such an estimate of the unknown true value of the parameter $p = (1 - q)$, and $\epsilon = h\sigma_t$ is the error of estimation w.r.t. p, all we actually know about ϵ and σ_t is the odds that their ratio h will exceed a certain level in a sample of $(a + b) = r$ items. For instance, the odds are roughly 20 : 1 that a deviation will not exceed twice the s.d. (σ_t) of the p distribution, so that the critical ratio h will not exceed 2. We can thus explore the consequences of assigning different values to h within the framework of the following relations which are necessarily true :

$$p = p_{ab} + \epsilon \quad and \quad p_{ab} = p - \epsilon . \quad . \quad . \quad . \quad . \quad \text{(i)}$$

$$q = q_{ab} - \epsilon \quad and \quad q_{ab} = q + \epsilon . \quad . \quad . \quad . \quad . \quad \text{(ii)}$$

$$\sigma_t^2 = \frac{pq}{a + b} \quad . \quad . \quad . \quad . \quad . \quad . \quad . \quad \text{(iii)}$$

$$\epsilon = \frac{h . \sqrt{pq}}{\sqrt{(a + b)}} \quad . \quad . \quad . \quad . \quad . \quad . \quad \text{(iv)}$$

The customary method of testing for the existence of a real difference between proportions of items in two samples as set forth in 4.09 is to set up the null hypothesis that both samples are referable to the same universe or to two identical universes. If so, the expected proportionate score difference is zero. Accordingly, we compute the s.d. of the difference between the proportionate scores of two samples of appropriate size (a and b) on the assumption that the best estimate of the common p implicit in our null hypothesis is the pooled sample value p_{ab}. The unknown true value (σ_d) of the s.d. of the proportionate score *difference* distribution w.r.t. samples of a and b items from the *same* putative universe, and that of our estimate (s_d) * based on p_{ab} are respectively given by

$$\sigma_d^2 = \frac{pq}{a} + \frac{pq}{b} = pq\left(\frac{a + b}{ab}\right). \quad . \quad . \quad . \quad . \quad \text{(v)}$$

$$s_d^2 = p_{ab} \, q_{ab} \cdot \left(\frac{a + b}{ab}\right) = \frac{(p - \epsilon)(q + \epsilon)(a + b)}{ab} \quad . \quad . \quad \text{(vi)}$$

Hence, it follows that

$$\frac{s_d^2}{\sigma_d^2} = \frac{(p - \epsilon)(q + \epsilon)}{pq} = 1 + \frac{(p - q)}{pq}\epsilon - \frac{\epsilon^2}{pq} \quad . \quad . \quad . \quad \text{(vii)}$$

* We here neglect the refinement of 4.09 on the assumption that our pooled sample is large.

14

From (iv) above

$$\frac{s_d^2}{\sigma_d^2} = 1 + \frac{h(p-q)}{\sqrt{pq(a+b)}} - \frac{h^2}{a+b} \qquad \text{(viii)}$$

As they stand, the foregoing equations are merely tautological; but they suffice to give us a new insight into the relation between a proportionate error $P_{(p)}$ in the estimation of p * and the proportionate error $P_{(\sigma)}$ in the estimation of σ_d. For our present purpose, it will suffice to focus attention on the values σ_d may assume when p is in the neighbourhood of 50 per cent. The student may derive profit from exploring the implications of (viii) at the opposite limit, when p is very small, so that $(p-q) \simeq -q$. By definition

$$P(p) \quad \frac{\epsilon}{p} = \frac{h\sqrt{pq}}{p\sqrt{a+b}} = \frac{h\sqrt{q}}{\sqrt{p(a+b)}} \qquad \text{(ix)}$$

$$P(\sigma) = \frac{\sigma_d - s_d}{\sigma_d} = 1 - \frac{s_d}{\sigma_d} \qquad \text{(x)}$$

Whence, from (viii) above,

$$P(\sigma) = 1 - \left[1 + \frac{h(p-q)}{\sqrt{pq(a+b)}} - \frac{h^2}{(a+b)} \right]^{\frac{1}{2}} \qquad \text{(xi)}$$

At the 50 per cent. level, i.e. when $p = \frac{1}{2} = q$, (xi) reduces to

$$P(\sigma) = 1 - \left[1 - \frac{h^2}{(a+b)} \right]^{\frac{1}{2}} \qquad \text{(xii)}$$

In general, the ratio $h^2 : (a+b)$ † will be less than unity, and its square, cube, etc., will therefore be small in comparison with it. If we expand the expression in brackets by the binomial theorem as in Ex. 1.04, p. 29, rejecting all terms after the second as negligible for the reason stated, we have

$$P(\sigma) = 1 - \left[1 - \frac{h^2}{2(a+b)} \right] = \frac{h^2}{2(a+b)} \qquad \text{(xiii)}$$

Subject to the same condition that $p = \frac{1}{2} = q$, (ix) becomes

$$P(p) = \frac{h}{\sqrt{a+b}} .$$

From this it follows that

$$P(\sigma) = \frac{h}{2\sqrt{a+b}} . P(p) \qquad \text{(xiv)}$$

For $h = 2$ when p is in the neighbourhood of 50 per cent. we therefore have

$$P(\sigma) = \frac{1}{\sqrt{r}} . P(p) \qquad \text{(xv)}$$

* For reasons explained below we here consider only values of p less than or equal to $\frac{1}{2}$.

† That we may invoke the normal distribution in this context signifies that we are talking about large samples, so that $(a+b)$ is a large number. On the other hand, values of h greater than 3 can occur very seldom.

The importance of the foregoing investigation resides in what we know about h, i.e. that it suffices to fix the magnitude of ϵ. Though we do not know its actual numerical value, we do know the odds against an error as great as $h\sigma_t$ for any arbitrary value we care to assume w.r.t. h itself. To say that the odds are $20 : 1$ against the occurrence of a p_{ab} deviation exceeding twice the s.d. of the p-distribution for an r-fold sample is to say the odds are $20 : 1$ against an error exceeding the magnitude of a deviation fixed by $h = 2$. The odds against an error exceeding the limit fixed by the critical ratio $h = 4$ are, of course, enormous ; and in that case the proportionate error of estimation involved in assigning a value to σ_d, when p is near the 50 per cent. level, is in the ratio $(2 : \sqrt{r})$ times the proportionate error involved in assigning to p the value p_{ab}. Hence p_{ab} will rarely deviate so far from 50 per cent. as to affect *grossly* the value assigned to the s.d., if the pooled sample of $(a + b)$ items is large. By the same token, errors of estimation w.r.t. p will rarely distort the critical ratio for the sampling difference so grossly as to invalidate the c-test of 4.08.

For illustrative purposes, let us suppose that the pooled sample of $r = 400$ items (Fig. 54). If p were exactly 50 per cent., we should have

$$\sigma_t = \frac{\sqrt{0\cdot5 \times 0\cdot5}}{20} = 0\cdot025.$$

For the critical ratio $h = 3$, the corresponding deviation would be $\pm 0\cdot075$. For the reason stated in 5.02 above, we can thus say that the odds would be (p. 204) about $370 : 1$ against an observed value of p_{ab} lying outside the range 42·5 to 57·5 ; and the proportionate error w.r.t. the estimation of p at either of these limits would be 15 per cent. in accordance with (ix). The corresponding percentage error of the s.d. for the proportionate score difference distribution with respect to a-fold and b-fold samples in accordance with (xiv) would be approximately

$$\frac{3}{2 \cdot \sqrt{400}} \times 15 = 1\cdot125 \text{ per cent.}$$

Evidently therefore, fairly gross errors w.r.t. the estimation of p near the 50 per cent. level in a sample of 400 will have little effect on the critical ratio of the c-test for the significance of the difference $d = (p_a - p_b)$. This example thus suffices to justify two conclusions, *viz.* :

(a) We can give precise specification to our legitimate confidence in a value assigned to a critical ratio on the basis of an estimated parameter, whose exact value is *not* implicit in the null hypothesis under investigation ;

(b) In specifiable circumstances depending only on the order of magnitude of p_{ab} and of $r = (a + b)$, we can assume with confidence that the c-test of 4.08 for the detection of a real difference will *rarely* let us down.

An exact specification of the relevance of both quantities mentioned in (b) to the assurance with which we are entitled to interpret an *estimated* likelihood assigned to an event by our null hypothesis is a matter which will not concern us further, because the method dealt with below provides a more satisfactory way of dealing with the same issue. It suffices to say that (ix) and (xi) disclose all the relevant data.

5.04 CONFIDENCE LIMIT

Till the recent contributions of Neyman and Egon Pearson, it was customary to frame the problem of estimation in its simplest form on the assumption that it is meaningful to ask : what is the *best* estimate of a parameter p consistent with our sample data ? The standpoint we shall now explore seeks no answer to this question which it rejects as unanswerable and in that sense without meaning. Instead, we shall ask : what is the expectation that such a parameter will lie between certain specifiable limits ? To keep our feet on the ground, let us consider the following result of a therapeutic trial. In a sample of 496 individuals subjected to a specified treatment there were 397 (80 per cent.) *cures* and 99 (20 per cent.) failures. Our problem is to state as precisely as possible the prognosis of cure.

As a first approximation to such prognosis, we shall adopt the traditional approach, i.e. assume that sampling errors of sufficient magnitude to distort grossly our assessment of the s.d. of the distribution w.r.t. the proportion of cures will not occur frequently. Accordingly, we shall postulate that the proportion of cures associated with the treatment involved will not commonly differ from the sample value, i.e. 80 per cent., by any quantity which will grossly affect the computation of σ. We therefore write $p \simeq 0.8$, $q \simeq 0.2$, and $(a + b) = r = 496$, so that

$$\sigma^2 \simeq \frac{0.8 \times 0.2}{496},$$

$$\therefore \ \sigma \simeq 0.018.$$

For a deviation of $2\sigma = 0.036$ within the framework of the assumption stated, we may thus say that the observed value of the proportion (p_0) in any sample of 496 would lie within the limits 0.80 ± 0.036, i.e. that any such sample value will lie between 0.764 and 0.836. Since the odds are 20 : 1 against the occurrence of a deviation as great as 2σ, the odds are 20 : 1 against the possibility that the proportion p_0 of cures in a second sample of the same size would lie outside these limits, if our initial assumption is justifiable. In accordance with the same initial assumption, the odds are 370 : 1 against the occurrence of a deviation as great as 3σ, i.e. that p_0 would lie outside the limits $0.80 \pm 3(0.018)$, i.e. outside the range 0.746 to 0.854. In so far as we are entitled to assume that our pooled sample furnishes us with a representative value of p, we are thus entitled to regard a second treatment as *significantly* more efficacious in the statistical sense of the term, if it resulted in 90 per cent. cures in a sample of 496.

So far, our only justification for such an assumption is the hint offered in 5.03, postulating that the p of our calculations lies within limits of error too small to affect materially our judgment about the frequency with which other sample values of the proportion of cures will occur outside a range set by a critical ratio specified accordingly. Relying on this consideration, we have implicitly assumed that our sample value of p is the best estimate we can make of it. In short, we have evaded the obligation to define the basis for a precise answer to the alternative question : what is the expectation that the true value of p lies within particular specified limits ? To do so, we must make a fresh start in accordance with the procedure of 5.03, making *no* assumption about how close to the true value p our r-fold sample value $p_0 (= 0.80)$ lies. Instead, we shall content ourselves with the assertion that the two differ by an *unknown* sampling error $\epsilon = h\sigma$. We shall therefore put

$$p_0 = (p - \epsilon); \ q_0 = (q + \epsilon) \quad . \qquad . \qquad . \qquad . \quad \text{(i)}$$

$$\frac{\epsilon^2}{h^2} = \sigma^2 = \frac{pq}{r}.$$

$$\therefore \ r\epsilon^2 = h^2(p_0 + \epsilon)(q_0 - \epsilon) = h^2[p_0q_0 + (q_0 - p_0)\epsilon - \epsilon^2],$$

$$\therefore \ (r + h^2)\epsilon^2 + (p_0 - q_0)h^2\epsilon - h^2p_0q_0 = 0.$$

$$\therefore \ \epsilon = \frac{-(p_0 - q_0)h^2 \pm \sqrt{(p_0 - q_0)^2 \cdot h^4 + 4h^2 \cdot (r + h^2)p_0q_0}}{2(r + h^2)} \qquad . \qquad . \quad \text{(ii)}$$

The only unknown quantity required for the solution of this quadratic is h the critical ratio of the sampling error ϵ. Let us assume that $h = 2$. Since $(a + b) = r = 496$, $p_0 = 0.80$ and $q_0 = 0.20$

$$\epsilon = \frac{-4(0.6) \pm \sqrt{16(0.8 - 0.2)^2 + 4(4)(500)(0.16)}}{1000}$$

$$= +0.034 \quad or \quad -0.038.$$

Accordingly, we are entitled to say that the expectation is approximately 95 per cent. (20 : 1 odds) that the true value of p will lie within limits set by $+0.034$ and -0.038 in (i). Since $p = (p_0 + \epsilon)$, this means that p lies within the limits $(p_0 + 0.034) = 0.834$ and $(p_0 - 0.038) = 0.762$. If we put $h = 3$ in (ii) we have for samples of the same size

$$\epsilon = -0.059 \quad or \quad +0.048.$$

In this case, we can say that the odds are 370 : 1 that p lies within the limits $(p_0 + 0.048) = 0.848$ and $(p_0 - 0.059) = 0.741$.

It is customary to designate as *fiducial limits* the boundaries fixed by a critical ratio calculated from the value of σ by recourse to the approximation $p_0 \simeq p$, and to refer to the boundaries fixed by the solution of the quadratic (ii) as *confidence limits*. For the example cited, the two sets of boundaries are as follows :

Critical Ratio	Fiducial Limits	Confidence Limits
2	0.764 — 0.836	0.762 — 0.834
3	0.746 — 0.854	0.741 — 0.848

5.05 THE CONFIDENCE SIEVE

We have now at our disposal the means to remove the misgiving which very properly arises from the circumstance emphasised at the beginning of 5.03, *viz.* results from the performance of a c-test for the recognition of a real difference depend on the reliability of the proportionate score assigned to the putative common universe. We shall therefore return to the type specimen of 4.02. The relevant data are as follows :

	Proportion Attacked	Total Number at Risk
Inoculated (Sample A) . .	0.011	279
Untreated (Sample B) . .	0.122	539
Total population . . .	0.084	818

We first apply equation (ii) of 5.04 to ascertain the confidence limits of the pooled estimate $p_{ab} = 0.084 = p - \epsilon$, *viz.* :

$$\epsilon = \frac{-(p_{ab} - q_{ab})h^2 \pm \sqrt{(p_{ab} - q_{ab})^2 . h^4 + 4h^2(r + h^2)p_{ab}q_{ab}}}{2(r + h^2)}.$$

Since
$$q_{ab} = (1 - p_{ab}) = 0.916 ;$$
$$p_{ab} - q_{ab} = -0.832 ;$$
$$p_{ab}q_{ab} = 0.07694 ; \quad (p_{ab} - q_{ab})^2 = 0.6922.$$
$$\epsilon = \frac{h^2(0.832) \pm h\sqrt{(0.6922)h^2 + 4(818 + h^2)0.0769}}{1636 + 2h^2}.$$

If $h = 2$, $\epsilon = +0.02147 \quad or \quad -0.01742.$
If $h = 3$, $\epsilon = +0.03381 \quad or \quad -0.02476.$

Since $p = p_{ab} + \epsilon$, the confidence limits are (to 3 decimals) :

When $h = 2$, $p = 0.084 + 0.021$ and $0.084 - 0.017$.

When $h = 3$, $p = 0.084 + 0.034$ and $0.084 - 0.025$.

Thus the odds are $20 : 1$ that the true value of p will be within the range 0.105 and 0.067 if the null hypothesis is correct, and $370 : 1$ that the true value of p will lie within the range 0.118 and 0.059.

When applying the c-test in the usual way we take 0.084 as our estimate of p, so that $q \simeq 0.916$. The observed proportionate score difference of the two samples is $0.122 - 0.011 = 0.111$, and the estimated variance (s_d^2) of the difference distribution is given by

$$(0.084 \times 0.916) \left\{ \frac{1}{279} + \frac{1}{539} \right\}$$

$$= 0.0769 \left\{ \frac{818}{150381} \right\}$$

$$= (0.0769)(0.00544)$$

$$= 0.000418$$

$$s_d = \sqrt{0.000418} = \pm 0.0204.$$

Hence the estimated critical ratio of the proportionate score difference is

$$0.111 \div 0.0204 \simeq 5.4.$$

If we abandon the presumption that 0.084 is the best estimate of p in the putative common universe of choice in any sense other than the lack of any means of specifying a single alternative which is quite certainly better, it is still possible to investigate the value of the critical ratio for the difference at the limits of a specified confidence range of p values. For the confidence range defined by $h = 2$, we have

Upper limit. $p_1 = 0.105$; $q_1 = 0.895$; $p_1 q_1 = 0.093975$.

Lower limit. $p_2 = 0.067$; $q_1 = 0.933$; $p_2 q_2 = 0.062511$.

These give us two limiting estimates of σ_d, *viz.* :

(i) $\sqrt{(0.093975)(0.00544)} \simeq \pm 0.0226.$

(ii) $\sqrt{(0.062511)(0.00544)} \simeq \pm 0.0184.$

The corresponding critical ratios are

(i) $0.111 \div 0.0226 \simeq 4.9.$

(ii) $0.111 \div 0.0184 \simeq 6.0.$

We may thus derive with 95 per cent. expectation a figure over 4.8 for the critical ratio of the observed difference in conformity with the null hypothesis, a finding which makes the likelihood of the event extremely small. However, we may wish to apply an even more exacting criterion. With odds of $370 : 1$ we may prescribe confidence limits to p as follows :

Upper limit. $p_3 = 0.118$; $q_3 = 0.982$; $p_3 q_3 = 0.11588$.

Lower limit. $p_4 = 0.059$; $q_4 = 0.941$; $p_4 q_4 = 0.0555$.

These give two new limiting estimates of σ_d, *viz.*:

$$\text{(iii)} \quad \sqrt{(0{\cdot}11588)(0{\cdot}00544)} \simeq \pm\, 0{\cdot}0251.$$

$$\text{(iv)} \quad \sqrt{(0{\cdot}00555)(0{\cdot}00544)} \simeq \pm\, 0{\cdot}0174.$$

The corresponding critical ratios are

$$0{\cdot}111 \div 0{\cdot}0250 \simeq 4{\cdot}4,$$

$$0{\cdot}111 \div 0{\cdot}0175 \simeq 6{\cdot}4.$$

Evidently there is no limit to the procedure outlined, if the end in view is to make the test more exacting. In practice, it is possible to explore the error involved in assigning a numerical value to the critical ratio of a score difference by a less laborious procedure, which is somewhat less precise. For reasons outlined in 5.03, the fiducial limits of an estimate usually lie close to the confidence limits at the same critical level ; and it is always safe to assume they do so, if (*a*) the number of items (*r*) in the pooled sample is fairly large ; (*b*) the pooled sample value of the proportionate score is neither very small nor very large. For this example, we have already found the s.d. of the distribution of the pooled sample value for 818 items is

$$\sqrt{(0{\cdot}084)(0{\cdot}916) \div 818} \simeq 0{\cdot}0097.$$

Hence we have the following values for the fiducial and confidence limits of p :

Critical Ratio	Fiducial Limits	Confidence Limits
2	0·064 — 0·103	0·067 — 0·105
3	0·055 — 0·113	0·059 — 0·118

Since the fiducial differ so little from the confidence limits, we should not go far astray if we contented ourselves with a specification of the range within which the critical ratio of a difference lies by using the alternative values σ_d assumes for values of p at either corresponding fiducial boundary. The importance of the *confidence sieve* is that it provides a rationale for the *c*-test when the null hypothesis prescribes no specification of the necessary parameter other than information inherent in the sample itself. In practice, it is rarely necessary to apply it, when the range assigned by the cruder method of fiducial limits is very small. Even if this is not so, recourse to fiducial limits (Fig. 55), which are easier to compute, usually offers a sufficient safe-guard against over-confidence, if we use them with due regard to the fact that they do not necessarily tally closely with the true boundaries of the confidence range and may deviate grossly therefrom. Indeed, the two sets of limits diverge widely, if the sample itself is small, for extreme values of the estimated parameter p. What constitutes a large sample in this context, and when we are to regard an estimated value as extreme in this sense, are matters on which experience of statistical data must dictate the need for more or less caution w.r.t. procedure.

The method of 4·08 which assigns a critical ratio to the actual difference between propor-tionate scores of two samples is not the only way of dealing with the same issue. In 4.01 we provisionally explored an alternative procedure : are the values 0·011 and 0·122 admissible sample values of one and the same universe from which we extract 279 and 539 items respectively ? If we are content to regard 0·084 as the best value we can assign to the putative common universe we derive for sample A

$$\sigma_p \simeq \sqrt{\frac{(0{\cdot}084)(0{\cdot}916)}{279}} = \pm\, 0{\cdot}017.$$

CONFIDENCE AND FIDUCIAL LIMITS OF p

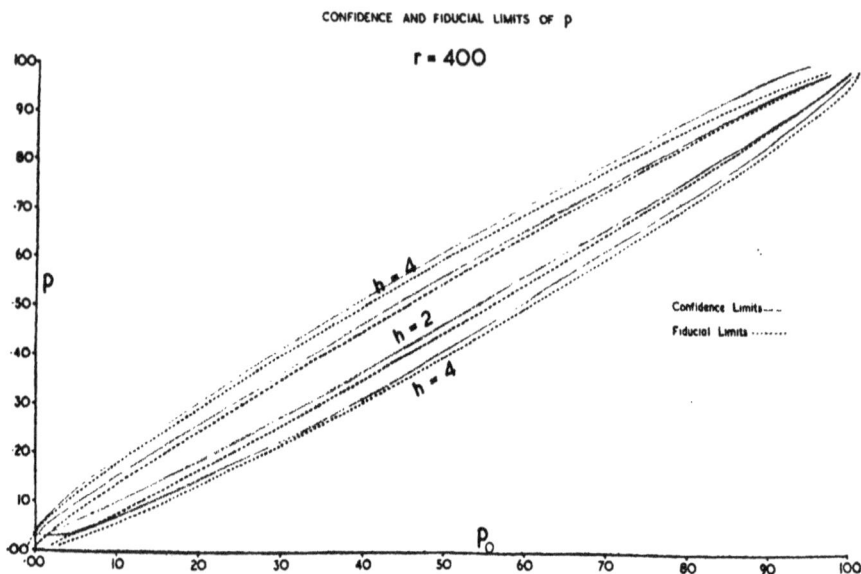

FIG. 55. Except at extreme values, the confidence range within which the true proportion (p) of successes in the universe of choice corresponding to an observed value p_0 tallies closely with the fiducial limits for critical ratios from 2 to 4.

The deviation of the sample value 0·011 from the estimate $p \simeq 0.084$ is 0·073, and the critical ratio is $(0.073) \div (0.017) \simeq 4.3$. At the confidence limits corresponding to $h = 2$, we have two estimates of σ_p :

$$\text{(i)} \quad \sqrt{(0.105)(0.895) \div 279} \simeq \pm 0.018.$$

$$\text{(ii)} \quad \sqrt{(0.067)(0.933) \div 279} \simeq \pm 0.015.$$

The corresponding critical ratios are

$$(0.105 - 0.011) \div 0.018 \simeq 5.2,$$
$$(0.067 - 0.011) \div 0.015 \simeq 3.7.$$

At the confidence limits for $h = 3$, we have as our two estimates of σ_p

$$\text{(i)} \quad \sqrt{(0.118)(0.882) \div 279} \simeq \pm 0.019.$$

$$\text{(ii)} \quad \sqrt{(0.059)(0.941) \div 279} \simeq \pm 0.014.$$

The corresponding critical ratios are

$$(0.118 - 0.011) \div 0.019 \simeq 5.6.$$
$$(0.059 - 0.011) \div 0.014 \simeq 3.4.$$

This result illustrates a good reason for preferring the difference-distribution test. The divergence between critical ratios assigned at corresponding pairs of confidence boundaries by the latter is less than the last set of figures exhibits ; and this is generally so, for a reason set forth in 5.03. The proportionate error involved in the estimation of an s.d. is less than the proportionate error involved in estimating the parameter p requisite for its specification. Whereas a c-test based on the sampling distribution of a difference makes no use of a p estimate except as a means of computing the s.d. in the denominator of the critical ratio, the method last given involves the double disadvantage of employing a direct estimate of p in the numerator of the critical ratio, and hence a figure w.r.t. which the proportionate error of sampling is necessarily greater than that of the s.d. alone. In general the c-test based on the sampling distribution of the difference itself is therefore the more sensitive and the less equivocal. On that account, we speak of it as a more *efficient* statistic.

In stating that the procedure of 4.01–2 is less efficient than the *difference-distribution* test of 4.08, we have assumed that our only concern is to test the validity of the null hypothesis that two samples come from the same universe. In practice, it is also of interest to go a step further. Instead of being content to say what the rejection of the null hypothesis signifies, that one treatment is more efficacious than another, we may wish to know how much improvement will result from substitution of one treatment for another. The difference test does not entitle us to give any precision to an assertion to this effect. On the other hand, the procedure of 4.01–4.02 offers us an alternative formulation of the problem. At a given confidence level we may say that our estimate p_a of proportionate success with one treatment (A) does not *exceed* a limit l_a, and that our higher estimate p_b w.r.t. the alternative treatment (B) does not *fall short* of l_b. Thus we can assign a precise measure of confidence to the assertion that treatment B is $100(l_b - l_a)$ per cent. more efficacious than treatment A.

These considerations invite comment on the increasingly prevalent custom in recent textbooks of prescribing as the correct procedure for testing the reality of a treatment difference a 4-fold table such as that of Exercise 4.02, Nos. 1–3, with a view to performing the *Chi-square* test for 1 degree of freedom. As we shall see in Vol. 2, and as Fisher (1922) was the first to point out, the latter must give exactly the same result as the c-test for a difference as set out in 4.08. Algebraically and numerically the two tests are in fact equivalent ; but the rationale of the Chi-square test relies on much more advanced algebra, having therefore less to commend it on heuristic grounds alone. Aside from this, the practice referred to is exceptionable for two reasons : (a) a 4-fold table of the sort shown in Exercise 4.03, Nos. 1–3, merely summarises the data with a view to performing the test, being of no further use, and fails to exhibit the face-value order of magnitude of the difference involved ; (b) the prescription of the test itself excludes the possibility of assigning a confidence level to the odds for or against the occurrence if the null hypothesis is true.

The Neyman sieve points to a way out of an impasse which arises when the observed proportion of items of a specified class in a small sample is zero. In accordance with the traditional procedure we should then have to regard the best estimate of p as zero with zero range between the fiducial limits. If we put $p_{ab} = 0$ in (ii) of 5.04, we get

$$\epsilon = \frac{h^2}{r + h^2}.$$

At the critical level $h = 2$, the confidence range therefore extends from zero to $4 \div (r + 4)$ and at $h = 3$ from zero to $9 \div (r + 9)$. However, the exact specification of the odds in favour of an error located at either boundary will not usually be such as we can infer from the normal distribution, because the normal distribution gives a good fit for very small values of p only if r itself is very large.

5.06 SIGNIFICANCE AND SAMPLE SIZE

A type of question to which research workers in medicine often seek an answer takes the following form : will samples of such and such a size suffice to demonstrate a significant difference between two treatments ? So stated, the question admits of no singular answer ; but it raises an issue which is of sufficient interest to merit examination.

In this context, as throughout this chapter, our concern is with *taxonomic* score differences, e.g. differences w.r.t. the *proportion* of sampled individuals among whom a given treatment evokes or fails to evoke a particular response. We have seen that we should distinguish between two sorts of answers obtainable from a significant test relevant to such. Only one of them was our concern in Chapter 4, where we sought an answer to the question : does a real difference exist, i.e. is treatment A better than treatment B or *vice versa ?* It is then necessary to remember— and all too easy to forget—that a difference which is highly significant in the statistical sense may be utterly trivial at the operational level *vis-à-vis* the return we get from a much greater outlay of effort or of limited resources requisite for other objectives. For instance, a 0·15 per cent. difference between the relapse rate for syphilis on short-term arsenotherapy and the relapse rate for syphilis treated with a new fungus extract of the penicillin-streptomycin class in favour of the latter might be highly significant if the two treatment samples were both very large ; but a difference with a confidence range of 0·10 per cent. to 0·20 per cent. might be of no practical interest if available supplies of the alternative drug were in much greater demand than arsenicals for other classes of patients. In general, it is of little interest to know that one treatment is better than another, unless the lower limit of the confidence range we impose upon the difference leaves a margin of relative efficacy appropriate to the dictates of considerations which lie outside the proper province of statistics. How big such a margin must be is not an issue in which the statistician should have to adjudicate.

Let us first deal with the more restricted—and more academic—issue raised in Chapter 4 by asking what size of sample is appropriate to the recognition that a real difference exists. What the statistician can say when asked a question of this sort is : the size of your samples must be such and such, if the *proportionate* difference you wish to vindicate is as great as so-and-so. An honest answer in such terms presupposes that there is already agreement about our criterion of significance, i.e. the lower boundary of the confidence range we are prepared to impose on the difference involved ; and this is a matter which invites due consideration of our prior knowledge of the situation in the light of the foregoing exposition of Bayes' theorem. In any case, no answer to the class of questions with which we are here concerned is in fact possible unless we place some limitation on the *relative*, as well as the absolute, size of the two samples which we select in order to test the existence of a real difference. In practice, it is desirable, and evidently economical, to choose samples of comparable size. For what follows, we shall therefore assume that we are talking about *equal* samples, so that $a = r = b$ in the symbolism of 4.07–4.08. Thus $ab = r^2$, $(a + b) = 2r$ and

$$\sigma_d^2 = p_{ab} \cdot q_{ab}\left(\frac{1}{a} + \frac{1}{b}\right) \quad \frac{2p_{ab} \cdot q_{ab}}{r} \quad . \qquad . \qquad . \qquad \text{(i)}$$

$$c^2 = \frac{r(p_a - p_b)^2}{2p_{ab} \cdot q_{ab}} . \qquad . \qquad . \qquad . \qquad . \qquad \text{(ii)}$$

With due regard to the inescapable limitations mentioned above, an evaluation of sample size relevant to the detection of a difference of specified magnitude at a given confidence level, involves the magnitude of the relevant proportionate score p_{ab} of the pooled sample from the putative common universe of the null hypothesis. This is necessarily so, because the standard deviation of the difference distribution, and hence the critical ratio of the observed difference, is a function both of p_{ab} and of the size of the individual samples. Hence, the best answer we can give to our first question calls for separate tabulation w.r.t. *different* confidence levels of sample size requisite to validate an observed difference of a given magnitude in terms of the magnitude of p_{ab} itself (*vide infra*, 5.07). On the assumption that the significance levels included in a table such as Table 1 suffice to justify our assurance *vis-à-vis* a particular situation, it gives us an answer to the question : does a real difference exist ? It tells us nothing about how big the difference is.

As the criterion of significance Table 1 actually specifies *fiducial*, as opposed to confidence limits, which, as Fig. 55 shows, closely coincide with the former except for very small values of p; but the distinction is immaterial to our purpose which is merely exemplary, and with that end in view it does not much matter whether p is a parameter of the putative common universe of the null hypothesis or a pooled sample estimate thereof. The procedure embodied in the build-up of the table is as follows. If we impose a fiducial range of $\pm 3\sigma$ as our criterion of a real difference, so that $c^2 = 9$, (ii) above becomes

$$\frac{r(p_a - p_b)^2}{2p_{ab} \cdot q_{ab}} = 9$$

Within the framework of the assumptions stated, let us now suppose that the parameter p_{ab} of our putative universe has the value 0·25, so that $p_{ab} \cdot q_{ab} = (0·25)(0·75) = 0·1875$. If we now wish to detect a difference $d = (p_a - p_b)$ as large as 3 per cent. (0·03)

$$\frac{r(0·03)^2}{2(0·1875)} = 9.$$

$$\therefore r = \frac{9.2 \cdot (0·1875)}{0·009}$$

$$\therefore r = 3750.$$

Accordingly, we find the entry 3750 against $p = 0·25$ and a difference of 3 per cent. in the upper half of Table I, referring to the 3σ level. In this example, the ratio of the difference $d (= 3$ per cent.) to $p_{ab} (= 25$ per cent.) is 12 : 100, so that the proportionate difference *so defined* is 12 per cent. The relevant entry of our table therefore tells us that equal samples have to be as large as 3750 to validate a proportionate difference of 12 per cent. as real, if our criterion of validity is the 3σ level.

The table shown here does not explicitly answer a question of the type : how large must our sample be to detect a difference at least as great as x per cent. ? We are sometimes able to simplify the statement of the problem, inasmuch as we commonly test a new treatment with a proportionate response score of p_b based on b individuals, against a background of much more extensive information concerning some pre-existing treatment to which we can ascribe a proportionate response score p_a referable to a very large sample. In that case, the confidence range of our estimate p_a will be trivial, and we are entitled to fall back on the method of 4.01–4.02. Accordingly, we postulate that p_a is an exact estimate of p, the corresponding parameter of the putative common universe. If our figures suggest that the new treatment is more efficacious, and we wish to ascertain how large a sample of individuals subjected to it would suffice to validate a difference as large as 3 per cent., we thus assume that the difference $(p_b - p) = 0·03$, p_b being the only unknown quantity.

TABLE 1

Minimum number of Cases requisite to confer Significance at the 3σ level

Value of p_{ab}.	Percentage Difference $d = 100 (p_a - p_b)$.										
	1	2	3	4	5	6	7	8	9	10	15
0·01	1782	—	—	—	—	—	—	—	—	—	—
0·02	3528	882	392	—	—	—	—	—	—	—	—
0·03	5238	1309	582	—	—	—	—	—	—	—	—
0·04	6912	1728	768	432	276	192	141	—	—	—	—
0·05	8550	2137	950	534	342	237	174	134	106	—	—
0·10	16200	4050	1800	1012	648	450	331	253	200	162	72
0·15	22950	5738	2550	1434	918	637	468	359	283	230	102
0·20	28800	7200	3200	1800	1152	800	588	450	356	288	128
0·25	33750	8438	3750	2109	1350	937	689	527	417	338	150
0·30	37800	9450	4200	2363	1512	1050	771	591	467	378	168
0·35	40950	10238	4550	2559	1638	1137	836	640	506	410	182
0·40	43200	10800	4800	2700	1728	1200	882	675	533	432	192
0·45	44550	11138	4950	2784	1782	1237	909	696	550	446	198
0·50	45000	11250	5000	2813	1800	1250	918	703	556	450	200

At the 2σ level the sample value p_b may exceed its mean value by a quantity defined by the fiducial boundary :

$$2\sqrt{p_b q_b} \div b = 2\sqrt{p_b(1 - p_b)} \div b.$$

Instead of asking that our estimate p_b should exceed p, as defined above, by 0·03, we now ask that the quantity defined by the *lower* fiducial limit specified above should satisfy this condition, i.e.

$$\left[p_b - 2\sqrt{p_b(1 - p_b)} \div b\right] - p = 0.03.$$

In doing so, we here assume that the value of b and p_b is such that the fiducial will coincide very closely with the confidence limits. The student should be able to adapt the preceding argument to the requirements of the more refined and logically more satisfactory procedure of 5.05.

The foregoing assumption that $p_a (= p)$ is subject to no appreciable sampling error is not essential. We can, in fact, proceed on the assumption that p_a and p_b are estimates each subject to errors of sampling. Within this more general framework, we may then suppose for illustrative purposes that : (*a*) a difference less than 5 per cent. in favour of a new treatment is of no operational interest ; (*b*) that the pre-existing procedure guarantees 60 per cent. success. An

TABLE 1—*Continued*

Fiducial Limits (3σ and 2σ) on differences of a Specified Magnitude at the 2σ level

Value of p_{ab}.	Percentage Difference $d = 100 (p_a - p_b)$.										
	1	2	3	4	5	6	7	8	9	10	15
0·01	792	—	—	—	—	—	—	—	—	—	—
0·02	1568	392	174	—	—	—	—	—	—	—	—
0·03	2328	582	259	146	93	—	—	—	—	—	—
0·04	3072	768	341	192	123	85	63	—	—	—	—
0·05	3800	950	422	238	152	106	78	59	47	—	—
0·10	7200	1800	800	450	288	200	147	113	89	72	32
0·15	10200	2550	1133	638	408	283	208	159	126	102	45
0·20	12800	3200	1422	800	512	356	261	200	158	128	57
0·25	15000	3750	1667	938	600	417	306	234	185	150	67
0·30	16800	4200	1867	1050	672	467	343	263	207	168	75
0·35	18200	4550	2022	1138	728	506	371	284	225	182	81
0·40	19200	4800	2133	1200	768	533	392	300	237	192	85
0·45	19800	4950	2200	1238	792	550	404	309	244	198	88
0·50	20000	5000	2222	1250	800	556	408	313	247	200	89

actual difference of 0·05, signifies that the new treatment sample confers 65 per cent. success, so that for equal samples $p_{ab}(= 0\cdot625)$ is the mean of 60 per cent. and 65 per cent. If we calculate r for $p = 0\cdot625$ and $d = 0\cdot05$ at the agreed significance level (e.g. 3σ), we know how large r must be in order that an observed difference of 5 per cent. would be indicative of the existence of a real difference *however* small. Evidently therefore r would have to be much greater than this, if the end in view were to validate the existence of a difference *not less than* 5 per cent.

For a given value of r we can assign how small the proportionate value of d, as defined above, must be, if we wish to establish the conclusion last stated. If the samples are of equal size, so that $p_{ab} = \frac{1}{2}(p_a + p_b)$, the proportionate value of d, subject to the understanding that $p_{ab} < 0\cdot5$ is given by

$$d_p = \frac{d}{p_{ab}} = \frac{p_a - p_b}{\frac{1}{2}(p_a + p_b)} \quad . \quad . \quad . \quad . \quad . \quad . \quad \text{(iii)}$$

At the significance level $h\sigma$, our criterion of the existence of a real difference is

$$d > h \cdot \sigma_d.$$

A difference exactly $h\sigma_d$ would on this showing be real, but it might prove to be of trivial magnitude in a long run of trials. To say that a real difference as great as x exists is therefore to say that it exceeds $h\sigma_d$ by that amount. Accordingly, the criterion of the real existence of a difference at least as great as x is

$$d \geqslant h\sigma_d + x \qquad . \qquad . \qquad . \qquad . \qquad . \qquad \text{(iv)}$$

In this equation we know x and h, also

$$\sigma_d^2 = \frac{2p_{ab}(1 - p_{ab})}{a}$$

$$d = d_p \cdot p_{ab}.$$

Hence (iv) becomes

$$d_p \cdot p_{ab} \geqslant x + h\sqrt{2p_{ab}(1 - p_{ab}) \div a} \qquad . \qquad . \qquad . \qquad \text{(v)}$$

For purposes of tabulation, we can specify particular values of d_p in (v). In general, experience permits us to assign a good estimate of p_a, from which p_{ab} is obtainable by recourse to (iii) above.

In this context we assume that p_{ab} is less than 0·5 as in Table 1, and hence less than q_{ab}. Otherwise we define d_p as the ratio of d to q_{ab} the smaller of the two complementary parameters. The reader may ask why some cells in Table 1 have no entries. Since we assume that $a = r = b$, our pooled estimate p_{ab} must be the mean value of p_a and p_b, i.e. $2p_{ab} = (p_a + p_b)$. Since $d = (p_a - p_b)$,

$$2p_{ab} - d = 2p_b.$$

Hence $2p_{ab} - d$ must be positive, and any estimate of r inconsistent with this necessary condition is inadmissible. For instance, when $p_{ab} = 0·01$, no difference as great as, or greater than, 2 per cent. could arise within the scope of the procedure outlined.

The importance of this consideration arises from the circumstance which makes it unnecessary to extend the p values of the table beyond the 50 per cent. level. It is immaterial whether we score the result of a therapeutic trial in terms of cures or failures. If $p_a = 41$ per cent., $q_a = 59$ per cent., and if $p_b = 47$ per cent., $q_b = 53$ per cent., so that

$$(p_b - p_a) = 0·06 = (q_a - q_b).$$

In short, the difference is *numerically* identical, whether referable to the proportion of successes or to the proportion of failures. It is therefore immaterial whether we label successes by p or by $q = (1 - p)$; and what is applicable to values of p in Table 1 is therefore applicable to $1 - p$, e.g. entries for $p = 0·99$ would be the same as for $p = 0·01$.

5.07   SIGNIFICANCE, SYMMETRY AND SKEWNESS

Our foregoing treatment of the confidence range proceeds from the assumption that the sampling distributions involved are approximately normal. It is therefore appropriate at this stage to examine circumstances which justify such an assumption. We have seen that the normal distribution of a taxonomic raw score or proportionate score is a good approximation to the contour of the binomial histogram when r, the size of the sample, is fairly large in comparison with the reciprocal of p. We can express the fact that M, the mean raw score, is rp by saying that r is M times the reciprocal of p. In other words, the absolute value of M is a measure of what we mean by fairly large in this context.

In deriving the normal equation as the limit of the exact binomial expression in 3.03–3.05, we have already given cursory consideration to the geometrical meaning of the statement that M

is fairly large. The closeness of the approximation depends on the approach to symmetry of the histogram contour. The range of the histogram extends from $x = 0$ to $x = r$, that of the normal from $(x - M) = X = -\infty$ to $(x - M) = X = +\infty$, the curve being symmetrical about $X = 0$, i.e. $x = M$. Hence a convenient measure of the approach to symmetry is (Fig. 34) what proportion of the total area of the histogram lies outside the range defined by a raw score $x = 0, X = -M$ to $x = 2M, X = +M$.

We have seen that the Poisson series gives us a good approximation for the ordinates of the binomial histogram, when r, though itself large, is not large compared with the reciprocal of p; and M is then small compared with the maximum value of $x = r$. If M is very small, the Poisson histogram is itself very skew, like the corresponding exact distribution; but the contour of the histogram defined by the Poisson series (Fig. 56) becomes more symmetrical, approaching the normal curve as we increase the value of M. Thus it is possible to get a close-up view of the issue stated above by a back door entrance, if we ask ourselves the question : what fraction of the total area of the Poisson histogram lies outside the range $x = 0$ to $x = 2M$, when we assign a particular value to M and hence to the ratio of r to the reciprocal of p ?

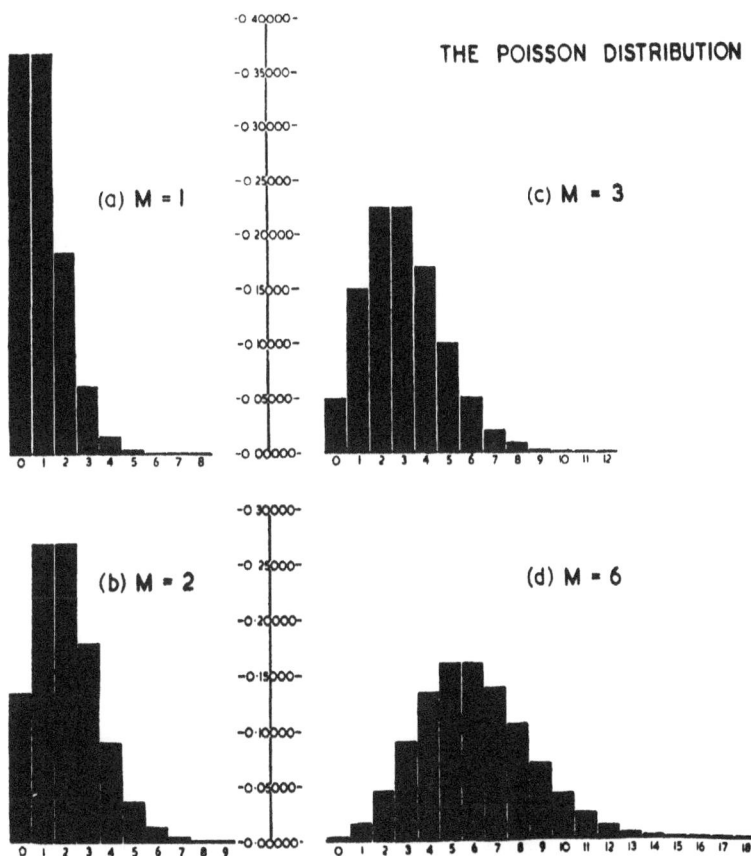

FIG. 56. The Poisson Distribution closely approaches symmetry when the mean value of the raw score distribution is 6.

TABLE 2

Total Area of Poisson Histogram in the Range x = 0 to x = 2M

	Poisson Series.			Binomial ($r = 100$).
X.	M = 1.	M = 2.	M = 3.	M = 3.
− 3	—	—	0·050	0·048
− 2	—	0·135	0·199	0·195
− 1	0·368	0·406	0·423	0·420
0	0·736	0·677	0·647	0·647
1	0·920	0·857	0·815	0·818
2	(0.981)	0·947	0·916	0·919
3	(0·996)	(0·983)	0·966	0·969
4	(0·999)	(0·995)	(0·988)	(0·989)
5	—	(0·999)	(0·996)	(0·997)
6	—	—	(0·999)	(0·999)

The total area of the Poisson histogram is given (p. 136) by the following expression :

$$A_M = e^{-M} \sum_{x=0}^{x=\infty} \left(\frac{M^x}{x!} \right) = 1 \quad . \quad . \quad . \quad . \quad . \quad . \quad . \quad \text{(i)}$$

$$\therefore A_M = e^{-M} \left(1 + M + \frac{M^2}{2!} + \frac{M^3}{3!} + \ldots + \frac{M^{2M}}{(2M)!} + \ldots \right)$$

$$= e^{-M} \left(1 + M + \frac{M^2}{2!} + \frac{M^3}{3!} + \ldots + \frac{M^{2M}}{(2M)!} \right)$$

$$+ e^{-M} \left(\frac{M^{2M+1}}{(2M+1)!} + \frac{M^{2M+2}}{(2M+2)!} + \ldots \right).$$

The area (S_M) defined by the range $x = 0$ to $x = 2M$ is therefore given by

$$S_M = e^{-M} \left(1 + M + \frac{M^2}{2!} + \frac{M^3}{3!} \ldots + \frac{M^{2M}}{(2M)!} \right).$$

We may now paint into this expression particular numerical values of $M (= 4, 5, 6, 7, 10)$ as below :

$$S_4 = e^{-4} \left(1 + 4 + \frac{4^2}{2!} \ldots \frac{4^8}{8!} \right) = 0·9786.$$

$$S_5 = e^{-5} \left(1 + 5 + \frac{5^2}{2!} \ldots \frac{5^{10}}{10!} \right) = 0·9864.$$

$$S_6 = e^{-6}\left(1 + 6 + \frac{6^2}{2!} \cdots \frac{6^{12}}{12!}\right) = 0.9889.$$

$$S_7 = e^{-7}\left(1 + 7 + \frac{7^2}{2!} \cdots \frac{7^{14}}{14!}\right) = 0.9943.$$

$$S_{10} = e^{-10}\left(1 + 10 + \frac{10^2}{2!} \cdots \frac{10^{20}}{20!}\right) = 0.9984.$$

Thus the value $M = 6$ defines the condition that almost exactly 99 per cent. of the area of the Poisson histogram lies in the hump about $X = \pm M$. In other words, less than 1 per cent. of the total area lies in the tail if r is more than 6 times the reciprocal of p and less than 0·2 per cent. if r is as much as ten times as great as the reciprocal of p. The accompanying tables (2 and 3) will give the reader some insight into the discrepancies between Poisson and normal approximations to skew binomial distributions.

Our preceding examination of the area of the binomial histogram within the range $x = 0$ to $x = 2M$ presupposes that the Poisson series gives a good fit for the relevant value of M. We shall now approach the issue more directly. Our task will be to answer the following question : how small is the sum $(1 - S_M)$ of the frequencies of the terms outside the range $x = 0$ to $x = 2M$, when a binomial is very skew ? This sum is

$$(1 - S_M) = \sum_{x = 2M + 1}^{x = r} y_x = \sum_{a = 1}^{a = r - 2M} y_{2M + a} \qquad \cdot \qquad \cdot \qquad \cdot \qquad \cdot \quad \text{(ii)}$$

To answer the question last stated, we shall seek to express the series defined by (ii) as fractions of the value y_x assumes when $x = 2M$. We first note that

$$\frac{y_{b+1}}{y_b} = \frac{r!}{(b+1)!\,(r-b-1)!} p^{b+1} q^{r-b-1} \cdot \frac{b!\,(r-b)!}{r!} p^{-b} q^{b-r},$$

$$y_{b+1} = \frac{(r-b)p}{(b+1)q} \cdot y_b.$$

Similarly we derive

$$y_{b+2} = \frac{(r-b-1)p}{(b+2)q} \cdot y_{b+1} = \frac{(r-b)(r-b-1)p^2}{(b+2)(b+1)q^2} \cdot y_b,$$

$$\therefore \ y_{b+2} = \frac{(r-b)^{(2)} p^2}{(b+2)^{(2)} q^2} \cdot y_b.$$

By iteration therefore

$$y_{b+a} = \frac{(r-b)^{(a)} p^a}{(b+b)^{(a)} q^a} \cdot y_b.$$

When $b = 2M$, we therefore have

$$y_{2M+a} = \frac{(r-2M)^{(a)} p^a}{(2M+a)^{(a)} q^a} \cdot y_{2M} \qquad \cdot \qquad \cdot \qquad \cdot \qquad \cdot \quad \text{(iii)}$$

When p is small so that $q \simeq 1$, the value of the denominator in (iii) is evidently greater than $(2M)^a$. We may write the numerator in the form

$$(r - 2M)p \cdot (r - 2M - 1)p \cdot (r - 2M - 2)p \ \ldots \ \text{to } a \text{ factors}$$
$$= (rp - 2Mp)(rp - 2Mp - p)(rp - 2Mp - 2p) \ \ldots \qquad \textit{ditto}$$
$$= (M - 2Mp)(M - 2Mp - p)(M - 2Mp - 2p) \ \ldots \qquad \textit{ditto}$$

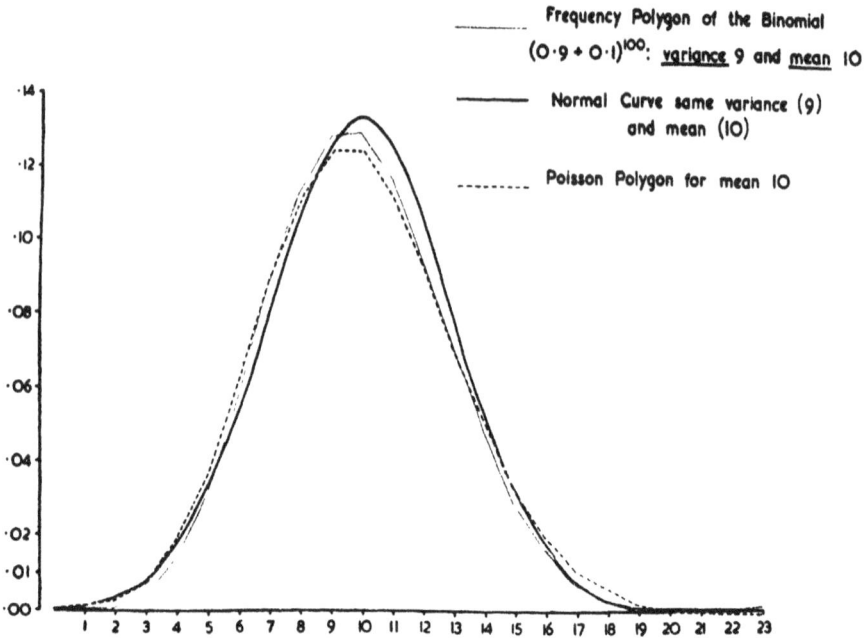

Fig. 57. For mean score values of 10 the Poisson and Normal distribution give about equally good correspondence with the Binomial for a sample of 100.

It is therefore evident that the numerator of (iii) is less than M^a. Hence we may write

$$y_{2M+a} < \frac{M^a}{(2M)^a} y_{2M},$$

$$y_{2M+a} < \frac{1}{2^a} y_{2M}.$$

By substitution in (ii) we therefore derive

$$(1 - S_M) < y_{2M} \sum_{a=1}^{a=r-2M} (\tfrac{1}{2})^a.$$

Now the expression under the summation sign is a geometric progression, of which we know from our schoolbooks that

$$\sum_{a=2}^{a=\infty} (\tfrac{1}{2})^a = 1.$$

Hence we conclude that

$$(1 - S_M) < y_{2M}.$$

A good approximation to y_{2M} is possible when r is large by recourse to Stirling's theorem (p. 48), viz. :

$$y_{2M} \simeq \frac{\sqrt{2\pi} \cdot r^{r+\frac{1}{2}} \cdot e^{-r} \cdot M^{2M} \cdot (r-M)^{r-2M}}{\sqrt{2\pi} \cdot (2M)^{2M+\frac{1}{2}} \cdot e^{-2M} \cdot \sqrt{2\pi} \cdot (r-2M)^{r-2M+\frac{1}{2}} \cdot e^{-r+2M} \cdot r^r},$$

TABLE 3

Raw score frequency for Distributions with a Mean Value of 10 computed in accordance with : (a) the Poisson Series, (b) the Terms of the Binomial for r = 100, (c) the Normal Curve with the same Variance (σ = 3) as the Terms of $(0·9 + 0·1)^{100}$. Note the Cumulative Frequencies cited for the Normal Distribution represent the sum of the Ordinates cited as Individual Score Frequencies (see Table 4).

Frequency of Individual Scores.				Cumulative Frequency.			
Score.	Normal.	Binomial.	Poisson.	Score.	Normal.	Binomial.	Poisson.
0	0·0005	0·0000	0·0000	0	0·0005	0·0000	0·0000
1	0·0015	0·0003	0·0004	1	0·0020	0·0003	0·0004
2	0·0038	0·0016	1·0023	2	0·0058	0·0019	0·0027
3	0·0087	0·0059	0·0076	3	0·0145	0·0078	0·0103
4	0·0180	0·0159	0·0189	4	0·0325	0·0237	0·0292
5	0·0332	0·0339	0·0378	5	0·0657	0·0576	0·0670
6	0·0547	0·0596	0·0631	6	0·1204	0·1172	0·1301
7	0·0807	0·0889	0·0901	7	0·2011	0·2061	0·2202
8	0·1065	0·1148	0·1126	8	0·3076	0·3209	0·3328
9	0·1258	0·1304	0·1251	9	0·4334	0·4513	0·4579
10	0·1330	0·1319	0·1251	10	0·5664	0·5832	0·5830
11	0·1258	0·1199	0·1137	11	0·6922	0·7031	0·6967
12	0·1065	0·0988	0·0948	12	0·7987	0·8019	0·7915
13	0·0807	0·0743	0·0729	13	0·8794	0·8762	0·8644
14	0·0547	0·0513	0·0521	14	0·9341	0·9275	0·9165
15	0·0332	0·0327	0·0347	15	0·9673	0·9602	0·9512
16	0·0180	0·0193	0·0217	16	0·9853	0·9795	0·9729
17	0·0087	0·0106	0·0128	17	0·9940	0·9901	0·9857
18	0·0038	0·0054	0·0071	18	0·9978	0·9955	0·9928
19	0·0015	0·0026	0·0037	19	0·9993	0·9981	0·9965
20	0·0005	0·0012	0·0019	20	0·9998	0·9993	0·9984
21	0·0001	0·0005	0·0009	21	0·9999	0·9998	0·9993
22	0·0000	0·0002	0·0004	22	0·9999	1·0000	0·9997
23	0·0000	0·0001	0·0002	23	0·9999	1·0000	0·9999

TABLE 4

For large values of r, the *ordinates* of the normal curve approximately correspond to those of the corresponding mid-points at the head of the columns of the binomial histogram. Hence their sum corresponds closely to the cumulative frequency specified by summation of successive terms of the binomial. To get the corresponding area of the *normal integral* it is necessary to make the half interval adjustment of p. 114. The figures below show : (a) the result of summing the ordinates of the normal curve (σ = 3) as given in Table 3 ; (b) the corresponding area of the integral with the appropriate half-interval correction.

Score Deviation.	Sum of Ordinates as in Table 3.	Area of the Integral with half-interval Correction.
0	0·5664	0·5662
1	0·6922	0·6915
2	0·7987	0·7977
3	0·8794	0·8783
4	0·9341	0·9332
5	0·9673	0·9666
6	0·9853	0·9849
7	0·9940	0·9938
8	0·9978	0·9977
9	0·9993	0·9992
10	0·9998	0·9998

$$\simeq \frac{r^{\frac{1}{2}} \cdot (r - M)^{r - 2M} \cdot M^{2M}}{\sqrt{\pi} \cdot 2^{2M + 1} \cdot M^{2M + \frac{1}{2}} \cdot (r - 2M)^{r - 2M + \frac{1}{2}}},$$

$$\simeq \sqrt{\frac{r}{\pi M}} \cdot \frac{1}{2^{2M + 1}} \cdot \frac{(r - M)^r}{(r - 2M)^{r - 2M + \frac{1}{2}}},$$

$$\simeq \frac{1}{2^{2M + 1}\sqrt{\pi M}} \cdot \frac{\left(1 - \dfrac{M}{r}\right)^{r - 2M}}{\left(1 - \dfrac{2M}{r}\right)^{r - 2M + \frac{1}{2}}},$$

$$\simeq \frac{1}{2^{M + 1}\sqrt{\pi M}} \cdot \frac{(1 - p)^{r(1 - 2p)}}{(1 - 2p)^{r(1 - 2p) + \frac{1}{2}}}.$$

When p is very small we therefore have

$$y_{2M} \simeq \frac{1}{2^{2M + 1}\sqrt{\pi M}} \cdot \frac{(1 - p)^r}{(1 - 2p)^r}.$$

On the foregoing assumption we may employ (i) of page 46, i.e.

$$(1 - p)^r \simeq e^{-rp} = e^{-M} \quad \text{and} \quad (1 - 2p)^r \simeq e^{-2M}.$$

$$y_{2M} \simeq \frac{e^M}{2^{2M + 1}\sqrt{\pi M}}.$$

Whence by (ii)

$$(1 - S_M) < \frac{e^M}{2^{2M + 1}\sqrt{\pi M}} \qquad \qquad \text{. . . . . (iv)}$$

The last expression does not depend on r, but merely upon $M = rp$, on the assumption that r is large and p is small. If we put $M = 6$ in (iv), we have

$$e^M \simeq 403 \cdot 4; \quad 2^{2M + 1} = 8192; \quad \sqrt{\pi M} \simeq 4 \cdot 34,$$

$$\therefore (1 - S_M) < 0 \cdot 012.$$

If we put $M = 10$ in (iv), we have

$$e^M \simeq 22026; \quad 2^{2M + 1} = 2097152; \quad \sqrt{\pi M} \simeq 5 \cdot 605,$$

$$\therefore (1 - S_M) < 0 \cdot 002.$$

Within the framework of our assumptions that r is very large and p very small, we therefore derive

(i) when $M = 6$

$$\sum_{x=0}^{x=2M} y_x > 0 \cdot 988.$$

(ii) when $M = 10$

$$\sum_{x=0}^{x=2M} y_x > 0 \cdot 998.$$

When p is in fact small $rpq \simeq rp = M$, so that $\sigma = \sqrt{6} \simeq 2 \cdot 45$ when $M = 6$. So the area of the histogram beyond the range $x = 12$, $X = 6 \simeq 2 \cdot 45\sigma$ is about 1 per cent. of the whole. When $M = 10$, less than $0 \cdot 2$ per cent. of the total area lies outside the range $\pm 3\sigma$. For values of M greater than or equal to 10 and high values of r the normal curve will in fact give a better description of a skew binomial than does the Poisson distribution, and for most practical purposes we are on safe ground when we use it.

CHAPTER 6

INTERLUDE ON THE METHOD OF MOMENTS

6.01 THE METHOD OF MOMENTS

IN Chapter 3 we have examined the possibility of finding approximate expressions for the frequency of raw scores, proportional scores and the deviations of either from their mean values in the domain of taxonomical (p. 194) statistics. Our enquiry embraced two types of sampling : (a) *with* replacement in accordance with the expansion of $(p + q)^r$; (b) *without* replacement in accordance with the expansion of $(nq + np)^{(r)} \div n^{(r)}$. In general, the objective of such analysis is to sidestep laborious computation by recourse to tables of appropriate functions. Suitability for tabulation is necessarily a criterion of what is in fact appropriate in this context ; but there is another which is equally relevant to the end in view.

Distributions exactly defined by the terms of either type of binomial referred to above do not include all kinds of sampling distributions on which statistical analysis relies ; but two classes of related functions for which tables exist enter into approximate expressions for those of a very large class which includes them. These functions—the Gamma and the Beta functions —are the central theme of this chapter. In a derivative form we have already made the acquaintance of each. For the score deviation (X) distribution defined by the exact binomial $(p + q)^r$ we obtained (pp. 119-120) in Chapter 3 two approximate expressions from each of which the normal distribution falls out as a special case when r is large. The first of these we developed by transforming into a differential equation an exact difference equation of the form

$$\frac{\Delta y_x}{y_x} = f(x)\, \Delta x.$$

The corresponding differential equation is

$$\frac{dY}{dX} \simeq \frac{rp}{q} \cdot \log\left(1 + \frac{X}{rp}\right) - \frac{X}{q} + K . \qquad . \qquad . \qquad . \qquad \text{(i)}$$

The solution of this is

$$Y = Y_0 \cdot e^{-\frac{X}{q}}\left(1 + \frac{X}{M}\right)^{\frac{M}{q}} \qquad . \qquad . \qquad . \qquad . \qquad \text{(ii)}$$

For an element of area dA, we therefore have

$$dA = Y \cdot dX = Y_0 \cdot e^{-\frac{X}{q}}\left(1 + \frac{X}{M}\right)^{\frac{M}{q}} \cdot dX \qquad . \qquad . \qquad . \qquad \text{(iii)}$$

The replacement of X by a linear function z of X merely involves a shift of origin and change of scale. If we put $X = (qz - M)$

$$dX = q \cdot dz ; \quad \left(1 + \frac{X}{M}\right) = \frac{qz}{M}; \quad \frac{X}{q} = z - \frac{M}{q},$$

$$\therefore\ dA = Y_0 \cdot q \cdot exp\left(-z + \frac{M}{q}\right) \cdot \left(\frac{qz}{M}\right)^{\frac{M}{q}} \cdot dz.$$

To simplify this we shall put $(M \div q) = (a - 1)$,

$$\therefore \ dA = Y_0 \cdot q^a \cdot M^{1-a} \cdot e^{a-1} \cdot e^{-z} \cdot z^{a-1} \cdot dz.$$

If C is a constant equal to $Y_0 \cdot q^a \cdot M^{1-a} \cdot e^{a-1}$

$$dA = C \cdot e^{-z} \cdot z^{a-1} \cdot dz \qquad . \qquad . \qquad . \qquad . \qquad . \qquad \text{(iv)}$$

Now the area under the whole curve is unity. By putting $Y = 0$ in (ii) we find that the limits of the curve are $X = - M$ and $X = \infty$. When $X = \infty$, $z = \infty$, and when $X = - M$, we have $\left(1 + \dfrac{X}{M}\right) = 0 = z$, so that

$$Y_0 \int_{-M}^{\infty} e^{-\frac{X}{q}} \left(1 + \frac{X}{M}\right)^{\frac{M}{q}} dX = 1 = C \int_{0}^{\infty} e^{-z} \cdot z^{a-1} \cdot dz,$$

$$\therefore \ \frac{1}{C} = \int_{0}^{\infty} e^{-z} \cdot z^{a-1} \cdot dz \qquad . \qquad . \qquad . \qquad . \qquad . \qquad \text{(v)}$$

Being a definite integral (v) is a function of a alone, called the *Gamma* function ; and tables of this function are available. It is customary to write it as $\Gamma(a)$, so that (iv) becomes

$$dA \simeq \frac{1}{\Gamma(a)} \cdot e^{-z} \cdot z^{a-1} \cdot dz \ . \qquad . \qquad . \qquad . \qquad . \qquad \text{(vi)}$$

If $Y = f(z)$ and $K^{-1} = \Gamma(a)$, we speak of it as a *Gamma variate* when

$$f(z) = K \cdot e^{-z} \cdot z^{a-1} \qquad . \qquad . \qquad . \qquad . \qquad . \qquad \text{(vii)}$$

A second class of functions mentioned in the first paragraph has emerged in connection with the central difference equation (xviii) in 3.04 corresponding to (i) above and as an approximate description of the hypergeometric series, i.e. successive terms of $(f + s)^{(r)} \div n^{(r)}$, which defines the non-replacement distribution in accordance with (xvi) and (xix) of 3.09. To obtain the latter, we made the assumption that the sample extracted from the universe is a small fraction of it. The removal of this limitation, as we shall later see, suggests a general pattern which includes as special cases all the distributions of Chapter 3, and leads to an empirical method of finding a curve suitable for the description of sampling distributions. In the unfolding of this pattern, we encounter a class of parameters (i.e. constants) which have a special descriptive value ; and our first task will be to define them.

For descriptive purposes it is not enough that a distribution function * should be convenient with a view to tabulation alone. To fulfil the end in view, its parameters should be easily calculable from the distribution. They should also help us to visualise its character ; and hence to choose a descriptive function of the right sort. It is this circumstance which endows the class of parameters called *moments* with special interest.

When we write it as below, our expression for the normal distribution involves two such constants, the *mean* (M) and the *variance* (V) :

$$y = \frac{1}{\sqrt{2\pi V}} \ exp \ \frac{-(x - M)^2}{2V}.$$

The *mean* of the distribution is in fact the weighted arithmetic average of the scores, the *variance* being the weighted average of the squares of the corresponding score deviations. The corre-

---

* Contemporary writers commonly use this term for the integral of the frequency function.

sponding weighted average (V_0) of the squares of the scores themselves has a simple relation to the mean and to the variance, *viz.* :

$$V = V_0 - M^2 \qquad . \qquad . \qquad . \qquad . \qquad . \qquad . \qquad \text{(viii)}$$

The constants M, V_0 and V which suffice to define a particular type of sampling distribution suggest the exploration of a large class of indices, called *moments* by analogy with corresponding functions in mechanics.* We shall henceforth speak respectively of M and V_0 as the first and second *zero moments* of a distribution, and shall denote them by the symbols μ_1 and μ_2. The weighted average of the deviations or first moment *about the mean* is necessarily zero. The average of the square deviations, i.e. the variance, we shall call in this context the *second mean moment*, henceforth denoted m_2. We may set out these definitions thus for a score range from 0 to r :

$$\mu_1 = \sum_0^r y \cdot x = M ; \quad m_1 = \sum_0^r y(x - \mu_1) = \sum_{-\mu_1}^{r-\mu_1} Y \cdot X = 0.$$

$$\mu_2 = \sum_0^r y \cdot x^2 = V_0 ; \quad m_2 = \sum_0^r y(x - \mu_1)^2 = \sum_{-\mu_1}^{r-\mu_1} Y \cdot X^2 = V.$$

It is important to recognise that m_1 is not the same as an index of dispersion called the *mean deviation* (η). The latter is the weighted average deviation regardless of sign, i.e.

$$\sum_{-\mu_1}^{r-\mu_1} Y \cdot | X | = \sum_0^r y \cdot | x - \mu_1 | \qquad . \qquad . \qquad . \qquad . \qquad \text{(ix)}$$

By analogy we can define moments based on higher powers, e.g.

$$\mu_3 = \sum_0^r y \cdot x^3 ; \quad m_3 = \sum_{-\mu_1}^{r-\mu_1} Y \cdot X^3 ; \quad \mu_4 = \sum_0^r y \cdot x^4 ; \quad m_4 = \sum_{-\mu_1}^{r-\mu_1} Y \cdot X^4.$$

More generally, we may write for the kth moments, about zero and about the mean respectively :

$$\mu_k = \sum_0^r y \cdot x^k ; \quad m_k = \sum_{-\mu_1}^{r-\mu_1} Y \cdot X^k \qquad . \qquad . \qquad . \qquad . \qquad \text{(x)}$$

In conformity with this symbolism

$$\mu_0 = \sum_0^r y = 1.$$

When r is indefinitely large and x (hence also X) increases † by unit steps ($\Delta x = 1 = \Delta X$) we may write (x) in the approximate form

$$\mu_k = \int_0^\infty y \cdot x^k \cdot dx ; \quad m_k = \int_{-\infty}^\infty Y \cdot X^k \cdot dX . \qquad . \qquad . \qquad \text{(xi)}$$

* The *moment of force* about a point is the product of the force and the perpendicular distance of its application from the same point. We define the centre of gravity (*centroid*) of a system of particles as the point about which the algebraic sum of their moments w.r.t. gravitational attraction is zero. The *moment of inertia* involves the second power of the distance of such a system of particles. From a formal viewpoint, the first mean moment is thus comparable to the centroid and the variance to the moment of inertia about the centroid.

† Otherwise we have to make the appropriate scalar change of (vii) in 4.08 to obtain the correct form of the integrand. For simplicity we assume throughout this chapter that the scale of x and X is unity. If not, we must replace the frequency y by the integrand $F(x)$ in conformity with the scalar reduction

$$y = F(x)\Delta x.$$

For instance, $F(x) = 3y$ if $\Delta x = 0.8$.

In general the kth zero moment is the mean or expected value of the kth power of a score whose range is positive. We may write it

$$\mu_k = E(x^k) \qquad . \qquad . \qquad . \qquad . \qquad . \qquad \text{(xii)}$$

For our present purpose the special interest of these indices resides in their relevance to the *contour* of a distribution. If a sampling distribution has only one *mode*, i.e. only one maximum (*crest*) value of y, two properties with due regard to its range suffice to indicate its shape : (*a*) whether it is more or less *skew*, i.e. asymmetrical about the mean as origin ; (*b*) whether it is more or less flat. *Odd* moments other than the first have a particular interest with respect to (*a*). The only one we shall use is m_3, the mean cube deviation. Like any higher kth mean moment in which k is an odd number, the third mean moment may be positive, negative or zero. It will be zero if negative and positive cube deviations of equal magnitude are equally frequent, as must be true of a symmetrical curve. It will be positive or negative according as positive are more or less numerous than negative. An excess of one or the other signifies that the distribution is skew.

Even mean moments are necessarily positive, since even powers of negative deviations must themselves be positive. Hence they can tell us nothing about the symmetry of a curve. Their interest resides in what they can tell us about its *flatness*. Thus a high value of the ratio $m_4 : m_2^2$ commonly denoted β_2, signifies a steeper and a low value signifies a flatter contour of a unimodal distribution. This assertion is not very obvious, and the student may find it useful to explore its numerical implications before seeking an algebraic rationale. Accordingly, we give below examples of a 5-class symmetrical universe of score values 1, 2, 3, 4, 5. In virtue of symmetry the mean is 3 and the deviations are therefore -2, -1, 0, $+1$, $+2$. The first is *rectangular*, being as flat as may be, and those that follow are successively steeper, as a freehand sketch of their histograms will suffice to disclose :

(a) $1:1:1:1:1$; $m_2 = \frac{10}{5}$; $m_4 = \frac{34}{5}$; $\beta_2 = 1\cdot7$.

(b) $1:2:3:2:1$; $m_2 = \frac{12}{9}$; $m_4 = \frac{36}{9}$; $\beta_2 = 2\cdot25$.

(c) $1:2:4:2:1$; $m_2 = \frac{12}{10}$; $m_4 = \frac{36}{10}$; $\beta_2 = 2\cdot5$.

(d) $1:4:10:4:1$; $m_2 = \frac{16}{20}$; $m_4 = \frac{40}{20}$; $\beta_2 = 3\cdot125$.

The constants m_4, m_3 and m_2 whose definition involves μ_1, thus suffice to convey a clear picture of a *unimodal* distribution, if we can define its range ; and an expression involving all four constants is therefore adaptable for the description of such distributions. With due regard to both desiderata of a general pattern for sampling distributions which are unimodal, Beta or Gamma functions involving the first four mean moments as their constants therefore have special advantages. Such is the rationale of the *method of moments* for fitting a continuous curve to a discrete frequency distribution. Its exposition calls for an interlude to provide the student with opportunity for revision of, or for first acquaintance with, some relevant applications of infinitesimal calculus. Sections here included with that end in view will also provide occasion to establish conclusions elsewhere stated without proof or without illustration.

Before proceeding further, the student will find it helpful to gain some preliminary practice in the notation of moments. Both for purposes of algebraic manipulation and for computation of mean moments as dispersion indices of empirical distributions, it is often convenient to be able to derive them from zero moments or *vice versa*. One relation of this sort is already familiar, *viz.*

$$V = V_0 - M^2 \quad or \quad m_2 = \mu_2 - \mu_1^2 \qquad . \qquad . \qquad . \qquad . \qquad \text{(xiii)}$$

The method employed in the derivation of this relation is quite general. Thus we have

$$m_3 = \sum_0^r y(x - \mu_1)^3 = \sum_0^r y(x^3 - 3\mu_1 x^2 + 3\mu_1^2 x - \mu_1^3)$$

$$= \sum_0^r y \cdot x^3 - 3\mu_1 \sum_0^r y \cdot x^2 + 3\mu_1^2 \sum_0^r y \cdot x - \mu_1^3 \sum_0^r y$$

$$= \mu_3 - 3\mu_1 \cdot \mu_2 + 3\mu_1^2 \cdot \mu_1 - \mu_1^3$$

$$= \mu_3 - 3\mu_1 \cdot \mu_2 + 2\mu_1^3 \qquad . \qquad . \qquad . \qquad . \qquad . \qquad . \qquad (\text{xiv})$$

By substitution from (xiii) we may write this in the form

$$\mu_3 = m_3 + 3\mu_1 \cdot m_2 + \mu_1^3 \qquad . \qquad . \qquad . \qquad . \qquad . \qquad (\text{xv})$$

In the same way we derive

$$m_4 = \mu_4 - 4\mu_1\mu_3 + 6\mu_1^2 \mu_2 - 3\mu_1^4 \qquad . \qquad . \qquad . \qquad . \qquad (\text{xvi})$$

EXERCISE 6.01

1. Find the first four zero and mean moments of the Poisson distribution of p. 136.

2. The cards of a pack being numbered consecutively from 0 to n, determine by appropriate formulæ for exact summation (Ex. 8 on p. 23) the first four zero moments ($\mu_1 - \mu_4$) and hence the first four mean moments ($m_1 - m_4$) of the player's score w.r.t. choice of *one* card only.

3. For the *rectangular* distribution specified in 2 above, show by integration that for large values of n

(a) m_2 approaches a limit of $\frac{1}{12} n^2$;

(b) m_4 approaches a limit of $\frac{1}{80} n^4$.

4. For the 3-class universe of score values -1, 0 and $+1$ with frequencies $1 : 4 : 1$ determine the value of β_2, and use the chessboard method to find that of β_2 for 2-fold and 3-fold samples.

5. Compare the results of 2 and 3. If $\beta_2 = (m_4 \div m_2^2)$ is an index of flatness of a distribution, what are its exact and limiting values for the rectangular distribution ?

6. In a draw of r cards with replacement, the player who gets x hearts scores 3^x points. Write down a general expression for his mean heart score so defined.

7. If he scores 2^x points, find an expression for his mean score as a finite difference series.

8. If he scores e^{tx} points, develop an expression for his mean score by the exponential series.

9. What relation exists between the coefficients of $(t^n \div n!)$ in the series of 7 above and the moments of the raw score distribution of the player ?

10. By recourse to (viii), (xiv) and (xvi) above find the second, third and fourth zero moments and mean moments of the distribution of the

(i) score deviation difference w.r.t. 4-fold and 3-fold samples from a 2-class universe ;

(ii) proportionate score difference w.r.t. 3-fold and 5-fold samples from a 2-class universe.

6.02 Integration by Series and Integration by Parts

At the outset, the student who hopes to master the elements of curve fitting by moments should be *au fait* with methods of integration. The aim of this section is to give the student who is not in training an opportunity to revise such methods, more particularly with a view to study of the properties of the Gamma and Beta functions. These two functions have a close connection with the factorial numbers which play so prominent a role in the algebra of choice and chance, and it would be redundant to remind the student of this book of the advantages of using the approximate method of the infinitesimal calculus to sidestep the laborious computations entailed by their exact evaluation. We have already made the acquaintance of the approximate formula of Stirling (1.09) for computation of factorials of large numbers; and it will also be fitting to examine its rationale in this context.

There are four principal devices for reducing an integral to a standard pattern available for reference, if not already committed to memory:

(a) Trigonometrical substitution such as

$$x = r \sin \theta ; \quad \sqrt{r^2 - x^2} = r \cos \theta ;$$

(b) resolution into partial fractions, a device to which we have had recourse already for solution of differential equations in Chapter 3;

(c) expansion of the integrand as an infinite power series;

(d) resolution of the integrand into factors with a view to integration by parts.

The last two are specially relevant to the issues dealt with in this chapter.

Integration by Series. By means of Maclaurin's theorem (1.09) it is possible to expand a continuous function of x as an infinite series of powers of x with constant coefficients. We can then integrate term by term in accordance with the following pattern:

$$\int f(x)dx = \int [A_0 + A_1 x + A_2 x^2 + A_3 x^3 \ldots] dx$$

$$= \int A_0 dx + \int A_1 . x . dx + \int A_2 . x^2 . dx + \int A_3 . x^3 . dx \ldots \text{ etc.}$$

$$= A_0 . x + \tfrac{1}{2}A_1 . x^2 + \tfrac{1}{3}A_2 . x^3 + \tfrac{1}{4}A_3 . x^4 \ldots \text{ etc.} \qquad . \qquad . \qquad . \text{ (i)}$$

If the resulting series is rapidly convergent, this device offers a convenient method of computing a numerical result to any required order of precision. As an example, we may consider the normal integral

$$y = \frac{1}{\sqrt{2\pi}} \int e^{-\frac{1}{2}c^2} . dc.$$

In this case, we recall the standard expansion:

$$e^a = 1 + a + \frac{a^2}{2!} + \frac{a^3}{3!} + \frac{a^4}{4!} + \frac{a^5}{5!} \ldots \text{ etc.}$$

$$e^{-\frac{1}{2}c^2} = 1 - \frac{c^2}{2} + \frac{c^4}{8} - \frac{c^6}{48} + \frac{c^8}{384} - \frac{c^{10}}{3840} \ldots \text{ etc.}$$

$$y = \frac{1}{\sqrt{2\pi}} \left[\int dc - \tfrac{1}{2}\int c^2 dc + \tfrac{1}{8}\int c^4 dc - \tfrac{1}{48}\int c^6 dc + \tfrac{1}{384}\int c^8 dc - \tfrac{1}{3840}\int c^{10} . dc \ldots \right]$$

$$= \frac{1}{\sqrt{2\pi}} \left[c - \frac{c^3}{6} + \frac{c^5}{40} - \frac{c^7}{336} + \frac{c^9}{3456} - \frac{c^{11}}{42240} \ldots \right].$$

Evidently, the series is rapidly convergent for values of $c < 1$. As an illustration of its use for computation of a table of the probability integral, we may consider the value $c = 0.675$ ($X = 0.675\sigma$), which bounds almost exactly half the area of the normal curve. Our problem is then to evaluate

$$A = \frac{2}{\sqrt{2\pi}} \int_0^{0.675} e^{-\frac{1}{2}c^2} . dc$$

$$= 0.7979 \ldots \left[c - \frac{c^3}{6} + \frac{c^5}{40} - \frac{c^7}{336} \cdots \right]_0^{0.675}.$$

The first six terms suffice to fix the first four significant figures. For terms inside the brackets we then have

POSITIVE			NEGATIVE		
$c = 0.6750$; $c \div 1$	$= 0.6750$		$c^3 = 0.3075$; $c^3 \div 6$	$= 0.0513$	
$c^5 = 0.1410$; $c^5 \div 40$	$= 0.0035$		$c^7 = 0.0638$; $c^7 \div 336$	$= 0.0002$	
$c^9 = 0.0290$; $c^9 \div 3456$	$= 0.0000$		$c^{11} = 0.0132$; $c^{11} \div 42240$	$= 0.0000$	
TOTALS	0.6785			0.0515	

Hence we may put

$$A = 0.7978 \ldots (0.6785 - 0.0515) = (0.7978)(0.6270)$$
$$= 0.5002$$

Since we have already had occasion to remark on the significance ratio 0.675σ, commonly designated the *probable error* of the distribution, no more need be said about it, except that it is one to remember. Two other examples of series integration are of general interest.

Example 1. It is possible to derive the logarithmic series by consideration of the properties of the integral

$$\int \frac{dx}{1 + x} = \log (1 + x).$$

By direct division we can write

$$\frac{1}{1 + x} = 1 - x + x^2 - x^3 + x^4. \ldots$$

This series is convergent if $x < 1$. Subject to this restriction we can write

$$\log (1 + n) = \int_0^n \frac{dx}{1 + x} = \int_0^n (1 - x + x^2 - x^3 + x^4 \ldots)dx,$$

$$\therefore \ \log (1 + n) = \left[x - \frac{x^2}{2} + \frac{x^3}{3} - \frac{x^4}{4} \ldots \right]_0^n = n - \frac{n^2}{2} + \frac{n^3}{3} - \frac{n^4}{4} + \frac{n^5}{5}, \text{ etc.}$$

Example 2. The breakdown of another standard integral leads to Gregory's series for a numerical value for π:

$$\int_0^1 \frac{dx}{1 + x^2} = \left[\tan^{-1} x \right]_0^1 = \int_0^1 (1 - x^2 + x^4 - x^6 + x^8 \ldots)dx,$$

$$\therefore \ \left[\tan^{-1} x \right]_0^1 = \left[x - \frac{x^3}{3} + \frac{x^5}{5} - \frac{x^7}{7} + \ldots \right]_0^1.$$

Since $\tan (\frac{1}{4}\pi) = 1$ and $\tan 0 = 0$, $\tan^{-1} 1 = \frac{1}{4}\pi$ and $\tan^{-1} 0 = 0$, so that

$$\pi = 4[1 - \frac{1}{3} + \frac{1}{5} - \frac{1}{7} + \frac{1}{9} \ldots].$$

The legitimate use of integration by series presupposes that the series involved are convergent, i.e. that the sum of the terms can never exceed a limiting value, however many terms we include. Its usefulness also presupposes that the derived series converges rapidly, i.e. that the sum of very few initial terms approaches closely to such a limit. The recognition of a convergent series is a less esoteric issue than one might infer from the space devoted to the topic in mathematical treatises. For we make the acquaintance of such a series when we learn to identify as the fraction $\frac{1}{9}$, the unlimited series

$$\frac{1}{10} + \frac{1}{100} + \frac{1}{1000} + \frac{1}{10,000} \ldots \text{ etc.} = 0 \cdot 1111 \ldots \text{ etc.}$$

If every term after some finite number of terms of one series is positive and less than the term of equivalent rank in a series which we know to be convergent, it goes without saying that the sum of all its terms cannot exceed a fixed value. This is evidently true of the series whose terms are reciprocals of the factorials, since successive terms after the tenth $\left(\text{i.e. } \frac{1}{10!} \right)$ decrease by multiplying the denominator by a factor which is greater than 10 itself and hence diminish more rapidly than successive terms of the series $0 \cdot 1$. A power series of x whose coefficients are less than unity must also fall off at a certain level more rapidly than a G.P. whose common ratio is x, when x itself is less than unity. Commonly, therefore, we can apply some such yardstick as a recurring decimal or the reciprocals of the factorials to decide whether a series is or is not convergent. If the signs of the terms of a series alternate and their numerical values decline consistently, we can replace it by one whose terms are all positive by pairing off adjacent terms, and apply the same criterion to the series so constituted, since the fact of pairing off the terms in this way does not affect the summation.

*   *   *   *   *   *   *

Integration by Parts. Instructions for integration by parts as given in many elementary text-books on the infinitesimal calculus leave more than is necessary to the ingenuity of the student not gifted with great facility for manipulating symbols. The following schema sets forth the necessary steps in a way which minimises the difficulties. We start with the standard expression for the differentiation of a product, *viz.* :

$$d(wz) = wdz + zdw \qquad \qquad \text{(ii)}$$

Whence we have

$$wz = \int wdz + \int zdw \qquad \qquad \text{(iii)}$$

If we put

$$dw = y \cdot dx ; \quad w = \int y \cdot dx ; \quad wz = z \int y \cdot dx . \qquad \text{(iv)}$$

From (iv) we have

$$\int z \cdot dw = \int yz \, dx \qquad \qquad \text{(v)}$$

$$\int w \cdot dz = \int dz \left(\int y \cdot dx \right) = \int \frac{dz}{dx} \left(\int y \cdot dx \right) dx \qquad \text{(vi)}$$

By substitution of (iv)-(vi) in (iii) we have a general pattern (vii) below for *memorisation* :

$$z \int y dx = \int \frac{dz}{dx} \left(\int y \cdot dx \right) dx + \int yz \, dx,$$

$$\therefore \int yz \cdot dx = z \int y \cdot dx - \int \left(\frac{dz}{dx} \right) \left[\int y \cdot dx \right] dx \qquad \qquad \text{(vii)}$$

To use (vii) with profit, everything depends on choosing which factor of the product to label as x or y for differentiation and integration respectively.

Example 3.

$$\int x \cdot \cos x \cdot dx = x \int \cos x \, dx - \int \frac{dx}{dx}\left[\int \cos x \, dx\right]dx$$

$$= x \sin x - \int \sin x \, dx$$

$$= x \sin x + \cos x.$$

(*Check* by differentiation.)

Example 4.

$$\int \log x \cdot dx = \int 1 \cdot \log x \, dx$$

$$= \log x \int dx - \int \frac{d \log x}{dx}\left[\int dx\right]dx$$

$$= x \log x - \int \frac{1}{x} \cdot x \cdot dx$$

$$= x \log x - x.$$

(*Check* as before.)

Example 5.

$$\int \sin^2 x \cdot dx = \int \sin x \cdot \sin x \cdot dx = -\sin x \cos x + \int \cos^2 x \, dx$$

$$= -\sin x \cdot \cos x + \int (1 - \sin^2 x) \, dx$$

$$= -\sin x \cos x + x - \int \sin^2 x \, dx,$$

$$\therefore \int \sin^2 x \cdot dx = \tfrac{1}{2}(x - \sin x \cdot \cos x).$$

* * * * * * *

In anticipation of what follows (6.04) we may use the method of integration by parts to obtain a *very coarse* approximation for factorials of large numbers, noting first that

$$\log n! = \log 1 + \log 2 + \log 3 \ldots \log (n-1) + \log n$$

$$= \sum_{r=1}^{r=n} \log r.$$

As a rough and ready estimate, we may treat the expression on the right as an integral in accordance with the method of summation explained in 1.10 with due regard to the correct limits of integration :

$$\sum_{r=1}^{r=n} \log r \simeq \int_{\frac{1}{2}}^{n+\frac{1}{2}} \log r \cdot dr \qquad \ldots \qquad \text{(viii)}$$

Hence, in accordance with Example 4 above,

$$\sum_{r=1}^{r=n} \log r \simeq \left[r \log r - r\right]_{\frac{1}{2}}^{n+\frac{1}{2}}$$

$$= (n + \tfrac{1}{2}) \cdot \log (n + \tfrac{1}{2}) - (n + \tfrac{1}{2}) - \tfrac{1}{2}\log \tfrac{1}{2} + \tfrac{1}{2},$$

$$\therefore \log n! \simeq \log \frac{(n+\frac{1}{2})^{n+\frac{1}{2}}}{(\frac{1}{2})^{\frac{1}{2}}} - n,$$

$$\log . \frac{n!}{(n+\frac{1}{2})^n \sqrt{2n+1}} \simeq -n,$$

$$\therefore \frac{n!}{(n+\frac{1}{2})^n \sqrt{2n+1}} \simeq e^{-n},$$

$$\therefore n! \simeq (n+\tfrac{1}{2})^n \sqrt{2n+1} . e^{-n}.$$

The student will find it instructive to compare results obtained by the above and by Stirling's formula w.r.t. 10! and 12!. The result shown above is *not* very good for a reason explained in 1.10, namely that the function is rapidly increasing throughout the whole range. By this method we can thus get at best a gross approximation for $n!$. So we need not be fastidious about neglecting small terms in the search for a clue to a better one. Thus $\frac{1}{2} \log \frac{1}{2}$ is negligible, when n is large, and we are entitled to try out what happens if we put

$$(n+\tfrac{1}{2}) \log (n+\tfrac{1}{2}) \simeq (n+\tfrac{1}{2}) \log n.$$

We then obtain

$$n! \simeq n^{n+\frac{1}{2}} e^{-n} = \sqrt{n} . n^n . e^{-n}.$$

This result is not even 50 per cent. correct, but it has a highly suggestive arithmetical property which leads to one demonstration of Stirling's theorem cited in 1·09. For large values of n the expression on the extreme right above becomes almost exactly equal to $n!$ if we multiply it by a constant factor which turns out to be $\sqrt{2\pi}$.

EXERCISE 6.02

1. Obtain a series for $\log (1 - x)$ by the method of Example 1 in 6.02.

2. By the same method obtain a series for $\log \dfrac{1+x}{1-x}$ and check the result by recourse to the series for $\log (1 + x)$ and $\log (1 - x)$ on p. 47.

3. Specify power series for $\sin x$ and $\cos x$ by Maclaurin's method (p. 47), and hence show that
$$\int \cos x . dx = \sin x \quad and \quad \int \sin x . dx = - \cos x.$$

4. Evaluate the following :
$$\int x^2 . \cos x . dx ; \quad \int x . \sin x . dx ; \quad \int x^3 . \cos x . dx ;$$
$$\int \cos^2 x . dx ; \quad \int \sin^3 x . dx ; \quad \int_0^{\pi/2} \sin^4 x . dx.$$

5. Find the value of
$$\int x^2 . \log x . dx ; \quad \int (1 - x)^2 . x^3 . dx ; \quad \int e^{-x} . x^2 . dx ; \quad \int e^{-x} . x^3 . dx ;$$
$$\int x . \cos x . dx ; \quad \int x^2 . \sin x . dx ; \quad \int \sin x . \cos x . dx ; \quad \int \sin x . \cos^3 . x . dx.$$

6. Obtain comparable expressions by recourse to (vii) in 6·02 for each pair of the following integrals, and use the combined result to evaluate each:

$$\text{(i)} \int e^x . \cos x . dx \quad and \quad \int e^x . \sin x . dx,$$

$$\text{(ii)} \int e^{ax} . \cos bx . dx \quad and \quad \int e^{ax} . \sin bx . dx.$$

7. Show that

(a) $\int x^m (1-x)^{n-2} . dx = \dfrac{1}{m+1} x^{m+1}(1-x)^{n-2} + \dfrac{n-2}{m+1} \int x^{m-1}(1-x)^{n-3} . dx.$

(b) $\int x^{m+1}(1-x)^{n-3} . dx = \dfrac{1}{m+2} x^{m+2}(1-x)^{n-3} + \dfrac{n-3}{m+2} \int x^{m+2}(1-x)^{n-4} . dx.$

8. Use the results of (7) to show that

$$\int x^{m-1}(1-x)^{n-1} . dx = \frac{1}{m} x^m(1-x)^{n-1} + \frac{n-1}{m(m+1)} x^{m+1}(1-x)^{n-2}$$

$$+ \frac{(n-1)(n-2)}{m(m+1)(m+2)} . x^{m+2}(1-x)^{n-3} + \frac{(n-1)(n-2)(n-3)}{m(m+1)(m+2)} \int x^{m+2}(1-x)^{n-4} dx,$$

and hence that

$$\int x^{m-1}(1-x)^{n-1} . dx = \sum_{r=0}^{r=(p-1)} \frac{(n-1)^{(r)}}{(m+r)^{(r+1)}} x^{m+r}(1-x)^{n-r-1} + \frac{(n-1)^{(p)}}{(m+p-1)^{(p)}} \int x^{m+p-1}(1-x)^{n-p-1} dx.$$

9. Show that

(a) $\int e^{-x} x^{n-2} dx = - x^{n-2} e^{-x} + (n-2) \int e^{-x} . x^{n-3} . dx.$

(b) $\int e^{-x} x^{n-3} dx = - x^{n-3} e^{-x} + (n-3) \int e^{-x} . x^{n-4} . dx.$

10. Use the results of the last example to show that

$$\int e^{-x} . x^{n-1} dx = - x^{n-1} . e^{-x} - (n-1)x^{n-2} . e^{-x} - (n-1)(n-2)x^{n-3} . e^{-x}$$
$$+ (n-1)(n-2)(n-3) \int e^{-x} . x^{n-4} . dx.$$

11. Show that

$$\sec x = 1 + \frac{x^2}{2} + \frac{5x^4}{4} + \frac{61x^6}{720} \cdots$$

$$\tan x = x - \frac{x^3}{3} + \frac{2x^5}{15} - \frac{17x^7}{315} \cdots$$

12. Integrate the following by the method of parts:

$$\int \sin^n x \, dx ; \quad \int \cos^n x \, dx.$$

Also show by recourse to Maclaurin's series that

$$e^{ix} = \cos x + i \sin x,$$

hence that

$$e^{i\pi} = -1.$$

6.03 THE WALLIS PRODUCT FOR π

As an illustration of the use of integration by parts leading to continued products as in Ex. 8-10 of 6.02, a method of deriving an expression for π due to Wallis, a contemporary of Newton, is instructive and also necessary for subsequent examination of Stirling's formula for factorials of large numbers. It involves the evaluation of the definite integral :

$$\int_0^{\pi/2} \sin^m x \,.\, dx = \int_0^{\pi/2} \sin x \,.\, \sin^{m-1} x \,.\, dx$$

$$= \left[\sin^{m-1} x \,.\, \cos x \right]_0^{\pi/2} + \int_0^{\pi/2} (m-1) \sin^{m-2} x \,.\, \cos^2 x \,.\, dx.$$

Since $\sin (0) = 0$, $\sin^{m-1}(0)$ and $\cos \pi/2 = 0$

$$\int_0^{\pi/2} \sin^m x \, dx = \int_0^{\pi/2} (m-1) \sin^{m-2} x \,.\, \cos^2 x \,.\, dx$$

$$= (m-1) \int_0^{\pi/2} \sin^{m-2} x(1 - \sin^2 x)dx$$

$$= (m-1) \int_0^{\pi/2} \sin^{m-2} x \, dx - (m-1) \int_0^{\pi/2} \sin^m x \,.\, dx,$$

$$\therefore \; m \int_0^{\pi/2} \sin^m x \,.\, dx = (m-1) \int_0^{\pi/2} \sin^{m-2} x \,.\, dx,$$

$$\therefore \; \int_0^{\pi/2} \sin^m x \,.\, dx = \frac{m-1}{m} \int_0^{\pi/2} \sin^{m-2} x \,.\, dx.$$

By the same token :

$$\int_0^{\pi/2} \sin^{m-2} x \,.\, dx = \frac{m-3}{m-2} \int_0^{\pi/2} \sin^{m-4} x \,.\, dx,$$

$$\int_0^{\pi/2} \sin^m x \,.\, dx = \frac{(m-1)(m-3)}{m(m-2)} \int_0^{\pi/2} \sin^{m-4} x \,.\, dx.$$

More generally :

$$\int_0^{\pi/2} \sin^m x \,.\, dx = \frac{(m-1)(m-3)(m-5) \ldots (m-2p+3)(m-2p+1)}{m(m-2)(m-4) \ldots (m-2p+4)(m-2p+2)} \int_0^{\pi/2} \sin^{m-2p} x \,.\, dx. \quad (i)$$

If m is an *even* integer, we can replace it by $2n$, and (i) becomes

$$\int_0^{\pi/2} \sin^{2n} x \,.\, dx = \frac{(2n-1)(2n-3)(2n-5) \ldots (2n-2p+3)(2n-2p+1)}{2n(2n-2)(2n-4) \ldots (2n-2p+4)(2n-2p+2)} \int_0^{\pi/2} \sin^{2(n-p)} x \,.\, dx.$$

If we now put $p = n$

$$\int_0^{\pi/2} \sin^{2n} x \, dx = \frac{(2n-1)(2n-3)(2n-5) \ldots 3 \,.\, 1}{2n(2n-2)(2n-4) \ldots 4 \,.\, 2} \int_0^{\pi/2} dx$$

$$= \frac{(2n-1)(2n-3)(2n-5) \ldots 5 \,.\, 3 \,.\, 1}{2n(2n-2)(2n-4) \ldots 6 \,.\, 4 \,.\, 2} \cdot \frac{\pi}{2} \qquad \cdot \quad (ii)$$

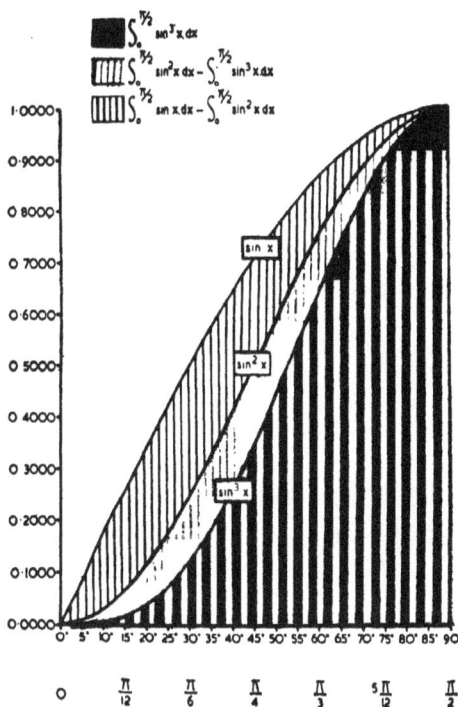

FIG. 58.

For explanation see text.

If m in (i) is an *odd* integer, we may write either

$$\int_0^{\pi/2} \sin^{2n+1} x \, dx = \frac{2n(2n-2)(2n-4) \ldots (2n-2p+4)(2n-2p+2)}{(2n-1)(2n-1)(2n-3) \ldots (2n-2p+5)(2n-2p+3)} \int_0^{\pi/2} \sin^{2(n-p)+1} x \, . \, dx,$$

or :

$$\int_0^{\pi/2} \sin^{2n-1} x \, dx = \frac{(2n-2)(2n-4)(2n-6) \ldots (2n-2p+2)(2n-2p)}{(2n-1)(2n-3)(2n-5) \ldots (2n-2p+3)(2n-2p+1)} \int_0^{\pi/2} \sin^{2(n-p)-1} x \, . \, dx.$$

When $p = n$ we therefore have

$$\int_0^{\pi/2} \sin^{2n+1} x \, dx = \frac{2n(2n-2)(2n-4) \ldots 4 \cdot 2}{(2n+1)(2n-1)(2n-3) \ldots 5 \cdot 3} \qquad . \qquad . \quad \text{(iii)}$$

We must then put $p = (n-1)$ in the second integral above, and

$$\int_0^{\pi/2} \sin^{2n-1} x \, dx = \frac{(2n-2)(2n-4)(2n-6) \ldots 4 \cdot 2}{(2n-1)(2n-3)(2n-5) \ldots 5 \cdot 3} \qquad . \qquad . \quad \text{(iv)}$$

For brevity, it is convenient to denote the integrals defined by (ii), (iii) and (iv) respectively as I_{2n}, I_{2n+1} and I_{2n-1}. Within the range 0 to $\pi/2$, $\sin x$ is less than unity, so that $\sin^m x$ must

16

exceed $\sin^{m+1} x$ for positive values of m, and the area under the curve $y = \sin^{m+1} x$ must be less than the area (Fig. 58) under the curve $y = \sin^m x$. In the new notation therefore

$$I_{2n-1} > I_{2n} > I_{2n+1}$$
$$(I_{2n-1} \div I_{2n+1}) > (I_{2n} \div I_{2n+1}) > 1 \quad . \quad . \quad . \quad . \quad . \quad (v)$$

From (iii) and (iv) we have

$$(I_{2n-1} \div I_{2n+1}) = \frac{2n+1}{2n}.$$

When n is indefinitely large, $(2n+1) \simeq 2n$; and

$$(I_{2n-1} \div I_{2n+1}) \simeq 1.$$

Hence from (v) above

$$Lt(I_{2n} \div I_{2n+1}) = 1.$$

For indefinitely large values of n we therefore have in accordance with (ii) and (iii)

$$\frac{(2n+1)[(2n-1)(2n-3)(2n-5) \ldots 5.3.1]^2}{[2n(2n-2)(2n-4) \ldots 6.4.2]^2} \cdot \frac{\pi}{2} = 1,$$

$$\therefore (n + \tfrac{1}{2})\pi = \left[\frac{2.4.6 \ldots (2n-4)(2n-2)2n}{1.3.5 \ldots (2n-3)(2n-1)} \right]^2.$$

Since our assumption is that n is very large, $(n + \tfrac{1}{2})\pi \simeq n\pi$; and

$$\sqrt{n\pi} = \frac{2.4.6.8 \ldots (2n-4)(2n-2)2n}{1.3.5.7 \ldots (2n-3)(2n-1)} \quad . \quad . \quad . \quad (vi)$$

The expression on the right hand is Wallis' formula which exhibits the relation between π and the limiting ratio of two continued products, the latter related (Ex. 8 in 1.01, p. 12) to $2n!$ and $n!$. Its discovery naturally suggested that π might enter into a convenient expression such as that of Stirling for computing factorials of large numbers.

6.04 Stirling's Formula

Mathematicians of the late seventeenth and early eighteenth century obtained continued product and continued fraction formulæ for π and e, the former, as we have seen, involving expressions which recall factorial numbers. At a time when the calculation of trigonometrical tables and tables of logarithms for navigation and trade endowed the numerical evaluation of π and e at an assignable level of precision with peculiar topical interest, the binomial theorem was also a new tool ; and any clue which might throw light on the properties of the factorial numbers embodied in binomial coefficients excited widespread concern. Stirling's theorem did not indeed hatch out fully fledged. Others beside Stirling in part anticipated it. So no economical demonstration of its validity can now recall the tortuous path which led to the discovery of the formula for which we here cite alternative derivations. Many approximations for computing factorials of large numbers are indeed more or less admissible, but none more precise and equally thrifty.

One method of demonstrating the theorem of Stirling depends on the properties of a continued product function, here denoted G_n, due to Glaisher (Q. J. Math., 1877), and defined as follows :

$$G_n = \left(\frac{4}{2}\right)^3 \cdot \left(\frac{6}{4}\right)^5 \cdot \left(\frac{8}{6}\right)^7 \cdots \left(\frac{2n}{2n-2}\right)^{2n-1} \cdot \left(\frac{2n+2}{2n}\right)^{2n+1} \qquad \cdot \quad \text{(i)}$$

$$\therefore \log G_n = \sum_{r=1}^{r=n} (2r+1) \log\left(1 + \frac{1}{r}\right) \qquad \cdot \quad \text{(ii)}$$

We may rearrange (i) as follows :

$$G_n = \frac{1}{2} \cdot \frac{4^3}{2^2} \cdot \frac{6^5}{4^3 \cdot 4^2} \cdot \frac{8^7}{6^5 \cdot 6^2} \cdot \frac{10^9}{8^7 \cdot 8^2} \cdots \frac{(2n)^{2n-1}}{(2n-2)^{2n-3}(2n-2)^2} \cdot \frac{(2n+2)^{2n+1}}{(2n)^{2n-1}(2n)^2}$$

$$= \frac{1}{2} \cdot \frac{1}{2^2} \cdot \frac{1}{4^2} \cdot \frac{1}{6^2} \cdot \frac{1}{8^2} \cdots \frac{1}{(2n-2)^2} \cdot \frac{1}{(2n)^2} \cdot \frac{(2n+2)^{2n+1}}{1}$$

$$= \frac{1}{(2.4.6.8 \ldots 2n)^2} \cdot \frac{(2n+2)^{2n+1}}{2} \qquad \cdot \quad \text{(iii)}$$

We may write the factor on the right as

$$\tfrac{1}{2}(2n)^{2n+1}\left(1+\frac{1}{n}\right)^{2n+1} = 2^{2n} \cdot n^{2n+1} \cdot \left(1+\frac{1}{n}\right)^{2n} \cdot \left(1+\frac{1}{n}\right).$$

When n is very large (1.08) :

$$\left(1+\frac{1}{n}\right)^{2n} \simeq e^2 ; \quad \left(1+\frac{1}{n}\right) \simeq 1,$$

$$\therefore 2^{2n} \cdot n^{2n+1}\left(1+\frac{1}{n}\right)^{2n}\left(1+\frac{1}{n}\right) \simeq 2^{2n} \cdot n^{2n+1} \cdot e^2.$$

Hence (iii) becomes

$$G_n \simeq \frac{2^{2n} \cdot n^{2n+1} \cdot e^2}{(2.4.6.8 \ldots 2n)^2} \qquad \cdot \quad \text{(iv)}$$

$$\therefore \sqrt{G_n} \simeq \frac{2^n \cdot n^{n+\frac{1}{2}} \cdot e}{2.4.6.8 \ldots 2n} \qquad \cdot \quad \text{(v)}$$

Hence as in Ex. 8 in 1.01 (p. 12) we have

$$\sqrt{G_n} \simeq \frac{n^{n+\frac{1}{2}} \cdot e}{n!}$$

$$n! \simeq \frac{n^{n+\frac{1}{2}} \cdot e}{\sqrt{G_n}} \qquad \cdot \quad \text{(vi)}$$

Similarly, we may write

$$(2n)! = \frac{(2n)^{2n+\frac{1}{2}} \cdot e}{\sqrt{G_{2n}}} = \frac{2^{2n+\frac{1}{2}} \cdot n^{2n+\frac{1}{2}} \cdot e}{\sqrt{G_{2n}}}.$$

By Ex. 8 in 1.01 we also have

$$(2n)! = \frac{(n!)^2 \cdot 2^{2n}}{W}.$$

In this expression $W \simeq \sqrt{n\pi}$ when n is very large, as defined by (vi) of 6.03 above, so that

$$(2n)! \simeq \frac{(n!)^2 \cdot 2^{2n}}{\sqrt{n\pi}},$$

$$\therefore \frac{(n!)^2 \cdot 2^{2n}}{\sqrt{n\pi}} \simeq \frac{2^{2n+\frac{1}{2}} \cdot n^{2n+\frac{1}{2}} \cdot e}{\sqrt{G_{2n}}},$$

$$\therefore (n!)^2 \simeq \frac{\sqrt{2n\pi} \cdot n^{2n+\frac{1}{2}} \cdot e}{\sqrt{G_{2n}}} \quad . \qquad . \qquad . \qquad . \qquad \text{(vii)}$$

We now divide (vii) by (vi), and thus obtain

$$n! \simeq \sqrt{2n\pi} \cdot n^n \cdot \sqrt{G_n \div G_{2n}} \quad . \qquad . \qquad . \qquad . \qquad \text{(viii)}$$

It now remains to show that the ratio $\sqrt{G_n \div G_{2n}} = e^{-n}$ in accordance with Stirling's formula (iii) of 1.09. In (ii) above, we can conveniently substitute as follows :

$$a = \frac{1}{2r+1}; \quad \frac{1+a}{1-a} = \frac{2r+2}{2r} = \left(1 + \frac{1}{r}\right).$$

We can then put

$$(2r+1)\log\left(1 + \frac{1}{r}\right) = \frac{1}{a}\log\frac{1+a}{1-a} = \frac{1}{a}[\log(1+a) - \log(1-a)].$$

From Ex. 2 in 6.02 :

$$\log(1+a) - \log(1-a) = 2a + \tfrac{2}{3}a^3 + \tfrac{2}{5}a^5 + \tfrac{2}{7}a^7 \ldots \text{ etc.},$$

$$\frac{1}{a}[\log(1+a) - \log(1-a)] = 2 + \tfrac{2}{3}a^2 + \tfrac{2}{5}a^4 + \tfrac{2}{7}a^6 \ldots \text{ etc.},$$

$$\therefore (2r+1)\log\left(1 + \frac{1}{r}\right) = 2 + \frac{2}{3} \cdot \frac{1}{(2r+1)^2} + \frac{2}{5} \cdot \frac{1}{(2r+1)^4} + \frac{2}{7} \cdot \frac{1}{(2r+1)^6} \ldots \text{ etc.},$$

$$\therefore \sum_{r=1}^{r=n}(2r+1)\log\left(1 + \frac{1}{r}\right) = 2\sum_{1}^{n}1 + \frac{2}{3}\sum_{1}^{n}\frac{1}{(2r+1)^2} + \frac{2}{5}\sum_{1}^{n}\frac{1}{(2r+1)^4} + \frac{2}{7}\sum_{1}^{n}\frac{1}{(2r+1)^6} \ldots \text{ etc.}$$

By (ii) above the series on the right defines $\log G_n$. We now write for brevity

$$\sum_{r=1}^{r=n} \frac{1}{(2r+1)^x} = s_x \quad . \qquad . \qquad . \qquad . \qquad . \qquad \text{(ix)}$$

With this convention, $s_0 = n$, so that

$$\log G_n = 2n + \tfrac{2}{3}s_2 + \tfrac{2}{5}s_4 + \tfrac{2}{7}s_6 \ldots \text{ etc.} \quad . \qquad . \qquad . \qquad \text{(x)}$$

Now all the series defined by (ix) are very rapidly convergent, if $x > 1$. They therefore arrive at saturation limit after comparatively few terms, attaining a limit *independent of n*, if n is large. Furthermore, successive values of s_x diminish rapidly and the limiting value of successive terms in (x) rapidly approach zero as we increase x. Thus (x) is also a series of terms whose sum closely approaches its upper limiting value, when n is large, so that its value no longer increases in virtue of making n larger. Subject to the restriction that n is large, we can therefore confidently replace the series whose first term in (x) involves s_x by a constant *log K^2 independent of* the numerical value of n, so that

$$\log G_n \simeq 2n + \log K^2,$$

$$\therefore \log \frac{G_n}{K^2} \simeq 2n,$$

$$\therefore G_n \simeq K^2 e^{2n}.$$

Since K^2 does not involve n, we may also write

$$G_{2n} \simeq K^2 e^{4n},$$

$$\therefore G_n \div G_{2n} \simeq e^{-2n},$$

$$\therefore \sqrt{G_n \div G_{2n}} \simeq e^{-n}.$$

On substituting the above value of $\sqrt{G_n \div G_{2n}}$ in (viii), we have Stirling's formula :

$$n! \simeq \sqrt{2n\pi} \, . \, n^n \, . \, e^{-n}.$$

An alternative derivation of the theorem proceeds from (viii) of 6.02, where we have already explored the approximate summation formula :

$$\log n! = \sum_{r=1}^{r=n} \log r \simeq \int_{\frac{1}{2}}^{n+\frac{1}{2}} \log x \, dx \qquad . \qquad . \qquad . \qquad . \qquad . \qquad \text{(xi)}$$

Now we may write

$$\int_{\frac{1}{2}}^{n+\frac{1}{2}} \log x \, dx = \int_{\frac{1}{2}}^{1\frac{1}{2}} \log x \, dx + \int_{1\frac{1}{2}}^{2\frac{1}{2}} \log x \, dx + \ldots \int_{n-\frac{1}{2}}^{n+\frac{1}{2}} \log x \, . \, dx$$

$$= \sum_{r=1}^{r=n} \int_{r-\frac{1}{2}}^{r+\frac{1}{2}} \log x \, dx$$

$$= \sum_{r=1}^{r=n} \left[(r + \tfrac{1}{2}) \log (r + \tfrac{1}{2}) - (r - \tfrac{1}{2}) \log (r - \tfrac{1}{2}) - (r + \tfrac{1}{2}) + (r - \tfrac{1}{2}) \right]$$

$$= \sum_{r=1}^{r=n} \left[\log (r - \tfrac{1}{2}) - (r + \tfrac{1}{2}) \log \left(\frac{r - \frac{1}{2}}{r + \frac{1}{2}} \right) - 1 \right] \qquad . \qquad . \qquad . \qquad \text{(xii)}$$

Since the expression on the right of (xi) is inexact, we may make the error ϵ explicit by setting it out in the form

$$\log n! = \int_{\frac{1}{2}}^{n+\frac{1}{2}} \log x \, dx + \epsilon \qquad . \qquad . \qquad . \qquad . \qquad \text{(xiii)}$$

Whence from (xi) and (xii) :

$$\epsilon = \sum_{r=1}^{r=n} \log r - \sum_{r=1}^{r=n} \left[\log (r - \tfrac{1}{2}) - (r + \tfrac{1}{2}) \log \left(\frac{r - \frac{1}{2}}{r + \frac{1}{2}} \right) - 1 \right]$$

$$= \sum_{r=1}^{r=n} \left[1 + (r + \tfrac{1}{2}) \log \left(\frac{r - \frac{1}{2}}{r + \frac{1}{2}} \right) + \log \left(\frac{r}{r - \frac{1}{2}} \right) \right] \qquad . \qquad . \qquad . \qquad \text{(xiv)}$$

It is thus convenient to express the error ϵ in the form

$$\epsilon = \sum_{r=1}^{r=n} u_r \, . \qquad . \qquad . \qquad . \qquad . \qquad . \qquad \text{(xv)}$$

$$\therefore u_r = 1 + (r + \tfrac{1}{2}) \log \left(\frac{r - \frac{1}{2}}{r + \frac{1}{2}} \right) + \log \left(\frac{r}{r - \frac{1}{2}} \right)$$

$$= 1 + (r + \tfrac{1}{2}) \log \left(1 - \frac{1}{r + \frac{1}{2}} \right) + \log \left(1 + \frac{1}{2r - 1} \right).$$

By recourse to the logarithmic series, we therefore have

$$u_r = 1 + (r + \tfrac{1}{2})\left[-\frac{1}{r + \tfrac{1}{2}} - \frac{1}{2(r + \tfrac{1}{2})^2} \cdots \right] + \left[\frac{1}{2r - 1} - \frac{1}{2(2r - 1)^2} \cdots \right].$$

Evidently, the first term (*unity*) goes out, and we are left with a very quickly converging series, whose sum is of the order r^{-2}. Thus the series ϵ defined by (xiv) is also quickly convergent for high values of n, approaching a limit which is independent of n (the number of terms) when n itself is very large. So we may write $\epsilon = -\log A$ as a constant without serious error in (xiii), whence

$$\log n! \simeq (n + \tfrac{1}{2}) \log (n + \tfrac{1}{2}) - \tfrac{1}{2} \log \tfrac{1}{2} - n - \log A.$$

Since we now assume that n is very large, we may neglect $\tfrac{1}{2} \log \tfrac{1}{2}$ and also write $(n + \tfrac{1}{2}) \log (n + \tfrac{1}{2}) \simeq (n + \tfrac{1}{2}) \log n$, so that

$$\log \left[\frac{n!}{A \cdot n^{n + \tfrac{1}{2}}}\right] \simeq -n.$$

$$\therefore n! \simeq A \cdot n^{n + \tfrac{1}{2}} \cdot e^{-n} . \qquad \qquad \text{(xvi)}$$

To evaluate A we now make use of the result established in Example 8 of 1.01, *viz.* :

$$W = \frac{(n!)^2 \, 2^{2n}}{(2n)!} \simeq \frac{A^2 \cdot n^{2n+1} \cdot e^{-2n} \cdot 2^{2n}}{A \cdot (2n)^{2n + \tfrac{1}{2}} \cdot e^{-2n}}.$$

$$\therefore W \simeq A \sqrt{\tfrac{1}{2}n}.$$

From (vi) in 6.03 $W = \sqrt{n\pi}$, so that

$$A \simeq \sqrt{2\pi}.$$

Hence from (xvi) we again derive :

$$n! \simeq \sqrt{2\pi} \cdot n^{n + \tfrac{1}{2}} \cdot e^{-n}.$$

$$\therefore n! \simeq \sqrt{2n\pi} \cdot n^n \cdot e^{-n}.$$

6.05 THE GAMMA FUNCTION

We have already defined the *Gamma function* as a definite integral by (v) in 6.01, *viz.* :

$$\Gamma(n) = \int_0^\infty e^{-x} \cdot x^{n-1} \cdot dx . \qquad \qquad \text{(i)}$$

An alternative form of (i) more explicitly related to a large class of statistical functions is obtainable by the substitution $x = \tfrac{1}{2}z^2$, so that

$$x^{n-1} = 2^{1-n} \cdot z^{2n-2} \quad \text{and} \quad dx = z \cdot dz.$$

Since $z = 0$ when $x = 0$ and $z = \infty$ when $x = \infty$

$$\int_0^\infty e^{-x} x^{n-1} \, dx = 2^{1-n} \int_0^\infty e^{-\tfrac{1}{2}z^2} \cdot z^{2n-1} \, dz,$$

$$\therefore \Gamma(n) = \frac{1}{2^{n-1}} \int_0^\infty e^{-\tfrac{1}{2}z^2} \cdot z^{2n-1} \, dz . \qquad \qquad \text{(ii)}$$

In conformity with (i) we may also write

$$\Gamma(n + 1) = \int_0^\infty e^{-x} \cdot x^n \cdot dx . \qquad \qquad \text{(iii)}$$

The method of integration by parts resolves the expression on the right of (iii) as follows :

$$\left[- e^{-x} \cdot x^n \right]_0^\infty + n \int_0^\infty e^{-x} \cdot x^{n-1} \cdot dx,$$

$$\therefore \ \Gamma(n+1) = n\Gamma(n) + \left[- e^{-x} \cdot x^n \right]_0^\infty \qquad . \qquad . \qquad . \qquad . \qquad \text{(iv)}$$

Now we may write $e^{-x} \cdot x^n = (e^x x^{-n})^{-1}$ in which

$$e^x x^{-n} = \frac{1}{x^n}\left(1 + x + \frac{x^2}{2!} + \frac{x^3}{3!} + \frac{x^4}{4!} \cdots \right)$$

$$= \left[\frac{1}{x^n} + \frac{1}{x^{n-1}} + \frac{1}{2!\, x^{n-2}} + \frac{1}{3!\, x^{n-3}} \cdots + \frac{1}{(n-1)!\, x} \right] +$$

$$\left[\frac{1}{n!} + \frac{x}{(n+1)!} + \frac{x^2}{(n+2)!} \cdots \text{etc.} \right]$$

$$= \sum_{r=0}^{r=(n-1)} \frac{1}{r!\, x^{n-r}} + \sum_{r=0}^{r=\infty} \frac{x^r}{(n+r)!} \qquad . \qquad . \qquad . \qquad . \qquad . \qquad \text{(v)}$$

From (v), when $x = 0$

$$e^x x^{-n} = \infty \quad and \quad e^{-x} x^n = 0.$$

When $x = \infty$

$$e^x x^{-n} = \infty \quad and \quad e^{-x} x^n = 0,$$

$$\therefore \ \left[e^{-x} x^n \right]_0^\infty = 0.$$

Thus (iv) becomes

$$\Gamma(n+1) = n\Gamma(n) \ . \qquad . \qquad . \qquad . \qquad . \qquad \text{(vi)}$$

In accordance with (i)

$$\Gamma(1) = \int_0^\infty e^{-x}\, dx = 1 \ . \qquad . \qquad \qquad . \qquad . \qquad \text{(vii)}$$

Whence by (v)

$$\Gamma(2) = \Gamma(1+1) = 1\Gamma(1) = 1 \qquad\qquad = 1!$$
$$\Gamma(3) = \Gamma(2+1) = 2\Gamma(2) = 2 \cdot 1 \qquad = 2!$$
$$\Gamma(4) = \Gamma(3+1) = 3\Gamma(3) = 3 \cdot 2 \cdot 1 \qquad = 3!$$
$$\Gamma(5) = \Gamma(4+1) = 4\Gamma(4) = 4 \cdot 3 \cdot 2 \cdot 1 = 4!$$

In general for integral values of n

$$\Gamma(n+1) = n! \quad and \quad \Gamma(n) = (n-1)! \ . \qquad . \qquad . \qquad \text{(viii)}$$

In particular, when $n = 0$,

$$0! = \Gamma(0+1) = \Gamma(1) = 1 \qquad . \qquad . \qquad . \qquad \text{(ix)}$$

When $n = 0$, (viii) becomes

$$\Gamma(0) = (-1)!,$$
$$\therefore \ \Gamma(0) = \pm \infty \qquad . \qquad . \qquad . \qquad . \qquad . \qquad \text{(x)}$$

In accordance with (vi) we may also write

$$\Gamma(n) = (n-1)\Gamma(n-1).$$
$$\Gamma(0) = -1\Gamma(-1).$$
$$\therefore \ \Gamma(-1) = \pm \infty.$$
$$\Gamma(-1) = -2\Gamma(-2).$$
$$\therefore \ \Gamma(-2) = \pm \infty.$$

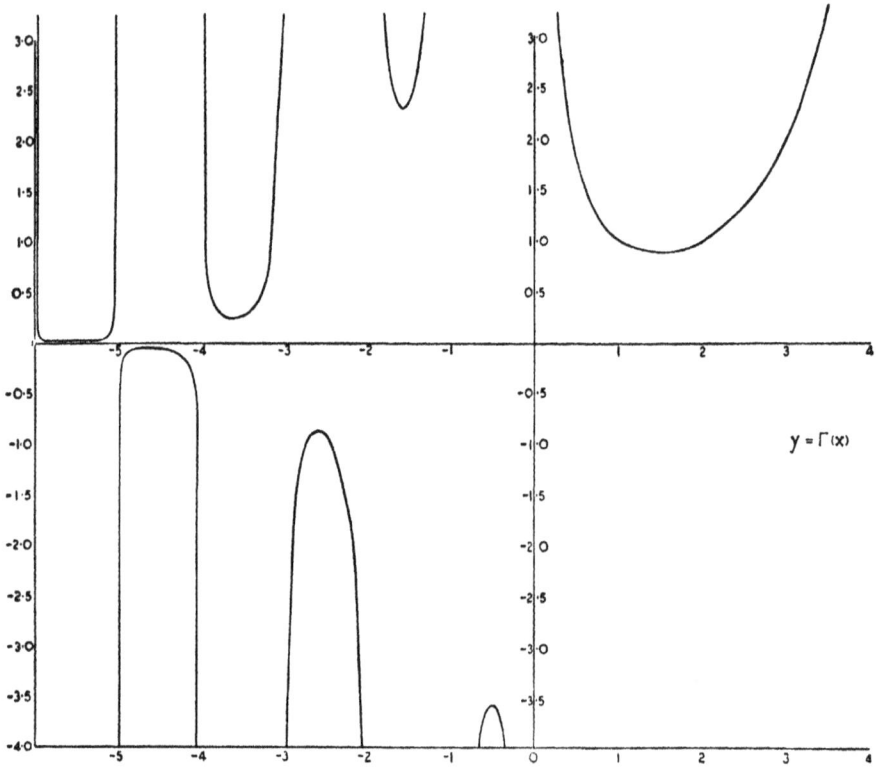

FIG. 59. The Complete Gamma Function.

For integral values of n we therefore have

$$\Gamma(5) = 4! \qquad\qquad \Gamma(0) \ \ = \pm \infty$$
$$\Gamma(4) = 3! \qquad\qquad \Gamma(-1) = \pm \infty$$
$$\Gamma(3) = 2! \qquad\qquad \Gamma(-2) = \pm \infty$$
$$\Gamma(2) = 1! \qquad\qquad \Gamma(-3) = \pm \infty$$
$$\Gamma(1) = 0! \qquad\qquad \Gamma(-4) = \pm \infty$$

For integers positive or negative alike, the numerical value of $\Gamma(n)$ is thus consistent with the algebraic definition of $(n-1)!$ in accordance with (viii). For fractional values of n, it is a fraction related to π, whose evaluation throughout the range $n = 1$ to $n = 2$ suffices for the build up of an exhaustive table by recourse to (vi). One method of obtaining $\Gamma(\frac{1}{2})$, and hence of the Gamma function for $n = \frac{3}{2}, \frac{5}{2}, \frac{7}{2}$, etc., depends on (ii) in which we substitute $n = \frac{1}{2}$, so that

$$\Gamma(\tfrac{1}{2}) = \sqrt{2} \int_0^\infty e^{-\frac{1}{2}z^2} \, . \, dz.$$

By (xxii) of 3.04, the integral on the right is equal to $\sqrt{\tfrac{1}{2}\pi}$, so that

$$\Gamma(\tfrac{1}{2}) = \sqrt{\pi} \simeq 1 \cdot 7722.$$

An alternative derivation in 6·07 (p. 251) depends on the relation of the *Beta* to the *Gamma* function. From the relation specified by (vi)

$$\Gamma(1 \cdot 5) = \tfrac{1}{2}\Gamma(\tfrac{1}{2}) = \tfrac{1}{2}\sqrt{\pi}\,; \quad \Gamma(2 \cdot 5) = 1 \cdot 5\Gamma(1 \cdot 5) = \tfrac{3}{4}\sqrt{\pi}, \text{ etc.}$$

Since $\Gamma(n - 1) = \Gamma(n) \div (n - 1)$, we also have

$$\Gamma(-\tfrac{1}{2}) = \Gamma(\tfrac{1}{2} - 1) = \Gamma(\tfrac{1}{2}) \div (\tfrac{1}{2} - 1) = -2\sqrt{\pi}.$$
$$\Gamma(-1 \cdot 5) = \Gamma(-\tfrac{1}{2} - 1) = \Gamma(-\tfrac{1}{2}) \div (-1 \cdot 5) = +\tfrac{4}{3}\sqrt{\pi}, \text{ etc.}$$

Accordingly, we can draw up a table which discloses the general contour (Fig. 59) of the function as follows :

n	$\Gamma(n)$	n	$\Gamma(n)$
− 4·5	− 0·022	+ 0·5	1·772
− 4·0	± ∞	+ 1·0	1·000
− 3·5	+ 0·270	+ 1·5	0·886
− 3·0	± ∞	+ 2·0	1·000
− 2·5	− 0·945	+ 2·5	1·429
− 2·0	± ∞	+ 3·0	2·000
− 1·5	+ 2·363	+ 3·5	2·658
− 1·0	± ∞	+ 4·0	6·000
− 0·5	− 3·544	+ 4·5	18·61
0·0	± ∞	+ 5·0	24·000

Values of $\Gamma(n)$ for fractional values of n other than odd multiples of 0·5 are obtainable by trigonometrical methods, a clue to which is in 6·07.

The student should be able to deduce the following relation from (i)

$$\int_0^\infty e^{-ax}\, x^{n-1}\, dx = \frac{\Gamma(n)}{a^n} \qquad . \qquad . \qquad . \qquad . \qquad . \qquad \text{(xi)}$$

EXERCISE 6.05

1. Show that $\Gamma(p + n) = (p + n - 1)^{(n)} \cdot \Gamma(p)$.

2. Given the following values of e^{-n}, evaluate $\int_0^n e^{-x} x^2\, dx \div \Gamma(2)$ for integral values of n from 1 to 10, and draw a graph of the function

n	e^{-n}	n	e^{-n}
1	0·36788	6	0·002479
2	0·13534	7	0·000912
3	0·049787	8	0·000336
4	0·018316	9	0·000123
5	0·006738	10	0·000045

3. Evaluate $\int_0^n e^{-x} \cdot x^2 \cdot dx \div \Gamma(3)$ for the above integral values of n and draw a graph.

6.06 MOMENTS AS GAMMA FUNCTIONS

It is sometimes possible to express moments of a distribution as Gamma functions, in which case it is possible to express a score distribution as an incomplete Gamma function. If the distribution is *symmetrical* and k is even, the kth moment about the origin (*mean*) in accordance with (xi) of 6.01 is approximately given by

$$m_k = 2 \int_0^\infty Y \cdot X^k \cdot dX.$$

For the normal distribution $Y = (2\pi V)^{-\frac{1}{2}} exp(-X^2 \div 2V)$, and we arrive at the identity $V = m_2$, if we write the above in the form

$$m_2 = \frac{2}{\sqrt{2\pi V}} \int_0^\infty e^{\frac{-X^2}{2V}} X^2 \cdot dX \qquad . \qquad . \qquad . \qquad . \qquad (i)$$

From (ii) in 6.05 we have

$$\int_0^\infty e^{-\frac{1}{2}z^2} z^{2n-1} \, dz = 2^{n-1} \cdot \Gamma(n) \qquad . \qquad . \qquad . \qquad (ii)$$

$$\therefore \int_0^\infty e^{-\frac{1}{2}z^2} z^2 \, dz = \sqrt{2}\, \Gamma(1 \cdot 5) = \sqrt{\frac{\pi}{2}} \qquad . \qquad . \qquad . \qquad (iii)$$

If we substitute $z = X \div \sqrt{V}$ in (i) we obtain

$$m_2 = \frac{2V^{\frac{3}{2}}}{\sqrt{2\pi V}} \int_0^\infty e^{\frac{-z^2}{2}} \cdot z^2 \cdot dz.$$

Hence from (ii) $m_2 = V.$

The 4th moment of the normal distribution is obtainable in the same way :

$$m_4 = \frac{2}{\sqrt{2\pi V}} \int_0^\infty e^{\frac{-X^2}{2V}} \cdot X^4 \cdot dX$$

$$= \frac{V^2 \sqrt{2}}{\sqrt{\pi}} \int_0^\infty e^{-\frac{1}{2}z^2} \cdot z^4 \cdot dz.$$

The integral is in this case equivalent to (ii) above when $n = 2 \cdot 5$, so that

$$m_4 = \frac{V^2 \sqrt{2}}{\sqrt{\pi}} \cdot 2^{1 \cdot 5} \, \Gamma(2 \cdot 5).$$

$$= \frac{(2V)^2}{\sqrt{\pi}} \cdot \frac{3\sqrt{\pi}}{4}.$$

$$m_4 = 3V^2 \qquad . \qquad . \qquad . \qquad . \qquad . \qquad . \qquad . \qquad (iv)$$

In the same way, we derive the mean deviation regardless of sign, as specified by η in 6.01 (p. 231) :

$$2 \int_0^\infty Y \cdot X \cdot dX.$$

For the normal distribution this is

$$\frac{2}{\sqrt{2\pi V}} \int_0^\infty e^{\frac{-X^2}{2V}} \cdot X \cdot dX$$

$$= \sqrt{\frac{2V}{\pi}} \int_0^\infty e^{-\frac{1}{2}z^2} \cdot z \cdot dz.$$

The integral in this case is equivalent to (ii) above when $n = 1$, so that

$$\int_0^\infty e^{-\frac{1}{2}s^2} z \, dz = \Gamma(1) = 1,$$

$$\therefore \eta = \sqrt{\frac{2V}{\pi}} \simeq \frac{4\sigma}{5} \qquad . \qquad . \qquad . \qquad . \qquad . \qquad (v)$$

Thus the mean deviation of a normal distribution is roughly four-fifths of the standard deviation.

6.07 THE BETA FUNCTION

We have noted alternative forms of the Gamma function defined by (i) and (ii) of 6·05. The Beta function is a definite integral which we may meet in three disguises of which the one most commonly cited is

$$B(m, n) = \int_0^1 x^{m-1} . (1 - x)^{n-1} . dx \qquad . \qquad . \qquad . \qquad . \qquad (i)$$

Let us now make the substitution

$$x = \frac{1}{1 + u}; \quad 1 - x = \frac{u}{1 + u},$$

$$\therefore u = \frac{1 - x}{x}.$$

From this we see that $u = \infty$ when $x = 0$ and $u = 0$ when $x = 1$, also

$$\frac{dx}{du} = \frac{-1}{(1 + u)^2}; \quad dx = \frac{-du}{(1 + u)^2},$$

$$\therefore B(m, n) = -\int_\infty^0 \frac{u^{n-1}}{(1 + u)^{m+n}} . du,$$

$$\therefore B(m, n) = \int_0^\infty \frac{u^{n-1}}{(1 + u)^{m+n}} . du \qquad . \qquad . \qquad . \qquad . \qquad (ii)$$

Alternatively, we may make the substitution $x = \sin^2 a$, so that $x = 1$ when $a = \frac{1}{2}\pi$ and $x = 0$ when $a = 0$. We then have

$$\frac{dx}{da} = \frac{d \sin^2 a}{d \sin a} . \frac{d \sin a}{da} = 2 \sin a . \cos a.$$

$$dx = 2 \sin a \cos a . da.$$

$$x^{m-1} . (1 - x)^{n-1} = \sin^{2m-2} a . (1 - \sin^2 a)^{n-1} = \sin^{2m-2} a . \cos^{2n-2} a,$$

$$\therefore B(m, n) = 2 \int_0^{\frac{\pi}{2}} \sin^{2m-1} a . \cos^{2n-1} a . da \qquad . \qquad . \qquad . \qquad (iii)$$

If we put $m = \frac{1}{2}$, $n = \frac{1}{2}$:

$$B(\tfrac{1}{2}, \tfrac{1}{2}) = 2 \int_0^{\frac{\pi}{2}} da = \pi . \qquad . \qquad . \qquad . \qquad . \qquad (v)$$

The relation between the Beta and Gamma functions is most easy to recognise by recourse to (i). If we integrate by parts

$$\int_0^1 x^{m-1} (1 - x)^{n-1} dx = \left[\frac{(1 - x)^{n-1} x^m}{m} \right]_0^1 + \frac{(n - 1)}{m} \int_0^1 x^m (1 - x)^{n-2} dx$$

$$= \frac{n - 1}{m} \int_0^1 x^m (1 - x)^{n-2} dx.$$

Similarly,

$$\int_0^1 x^m (1-x)^{n-2}\, dx = \frac{n-2}{m+1}\int_0^1 x^{m+1}.(1-x)^{n-3}.\,dx,$$

$$\int_0^1 x^{m+1}(1-x)^{n-3}\, dx = \frac{n-3}{m+2}\int_0^1 x^{m+2}.(1-x)^{n-4}.\,dx,$$

$$\therefore \int_0^1 x^{m-1}(1-x)^{n-1}\, dx = \frac{(n-1)(n-2)(n-3)}{(m+2)(m+1)m}\int_0^1 x^{m+2}.(1-x)^{n-4}.\,dx$$

$$= \frac{(n-1)^{(3)}}{(m+2)^{(3)}}\int_0^1 x^{m+2}(1-x)^{n-4}\,dx.$$

More generally,

$$\int_0^1 x^{m-1}(1-x)^{n-1}\, dx = \frac{(n-1)^{(p)}}{(m+p-1)^{(p)}}\int_0^1 x^{m+p-1}(1-x)^{n-p-1}\,dx \qquad . \qquad . \quad \text{(v)}$$

If n is an integer, and $n = (p+1)$, so that $p = (n-1)$,

$$\int_0^1 x^{m-1}(1-x)^{n-1}\, dx = \frac{(n-1)^{(n-1)}}{(m+n-2)^{(n-1)}}\int_0^1 x^{m+n-2}\,dx$$

$$= \frac{(n-1)^{(n-1)}}{(m+n-1)^{(n)}}.\left[x^{m+n-1}\right]_0^1$$

$$= \frac{(n-1)!}{(m+n-1)(m+n-2)\ldots m}$$

$$= \frac{(n-1)!\,.\,(m-1)!}{(m+n-1)!}$$

$$B(m,\,n) = \frac{\Gamma(n)\,.\,\Gamma(m)}{\Gamma(m+n)} \qquad . \qquad . \qquad . \qquad . \qquad . \qquad . \quad \text{(vi)}$$

When n is not an integer $(1-x)^{n-p-1}$ does not become $(1-x)^0 = 1$ after $p = (n-1)$ operations of the type which led us to (v); and the identity defined by (vi) calls for separate proof; but we can obtain independent confirmation of its applicability to fractional values of m and n by recourse to (iii). Thus we may substitute $m = \frac{1}{2} = n$ in (vi) to obtain by recourse to (iv) a result already established in 6.05, *viz.* :

$$\pi = B(\tfrac{1}{2},\,\tfrac{1}{2}) = \frac{\Gamma(\tfrac{1}{2})\Gamma(\tfrac{1}{2})}{\Gamma(1)}$$

$$\therefore \Gamma(\tfrac{1}{2}) = \sqrt{\pi}.$$

Similarly, we obtain by substituting $m = \frac{3}{2}$ and $n = \frac{1}{2}$ in (iii)

$$B(\tfrac{3}{2},\,\tfrac{1}{2}) = 2\int_0^{\frac{\pi}{2}} \sin^2 x\,.\,dx.$$

From Example 5 in 6.02 we have

$$\int_0^{\frac{\pi}{2}} \sin^2 x\,.\,dx = \left[\frac{x-\sin x\,.\,\cos x}{2}\right]_0^{\frac{\pi}{2}} = \frac{\pi}{4}.$$

$$\therefore B(\tfrac{3}{2},\,\tfrac{1}{2}) = \frac{\pi}{2}.$$

If we make this substitution in (vi), we also get as elsewhere shown

$$\frac{\pi}{2} = \frac{\Gamma(\frac{3}{2}) \cdot \Gamma(\frac{1}{2})}{\Gamma(2)} = \sqrt{\pi} \cdot \Gamma(\frac{3}{2}).$$

$$\therefore \ \Gamma(\frac{3}{2}) = \frac{1}{2}\sqrt{\pi}.$$

We are now in a position to appreciate the meaning of the transformation from (xvi) to (xix) for the hypergeometric distribution in 3.09, *viz.* :

$$\int_m^n k\left(1 - \frac{X}{a}\right)^u \left(1 + \frac{X}{b}\right)^v dX = C \int_{\frac{m+b}{b+a}}^{\frac{n+b}{b+a}} z^v (1-z)^u dz \quad . \quad . \quad . \quad \text{(vii)}$$

We can at once evaluate C from (xx) in 3.09, *viz.* :

$$\frac{1}{C} = \int_0^1 z^v (1-z)^u \, dz.$$

Since $v = (v+1) - 1$ and $u = (u+1) - 1$,

$$\frac{1}{C} = B(1+u, 1+v) \quad . \quad . \quad . \quad . \quad . \quad \text{(viii)}$$

In the particular case when $a = b$ and $u = v$, so that the curve is symmetrical with a range $X = \pm a$

$$Y_0\left(1 - \frac{X^2}{a^2}\right)^u dX = C \cdot z^u (1-z)^u \, dz \quad . \quad . \quad . \quad \text{(ix)}$$

$$C = \frac{1}{B(1+u, 1+u)} \quad . \quad . \quad . \quad . \quad \text{(x)}$$

In (vii) above $k = Y_0$ and

$$C = \frac{k(a+b)^{u+v+1}}{a^u \cdot b^v} = \frac{Y_0 (a+b)^{u+v+1}}{a^u \cdot b^v},$$

$$\therefore \ Y_0 = \frac{a^u \cdot b^v}{(a+b)^{u+v+1} \cdot B(1+u, 1+v)} \quad . \quad . \quad . \quad \text{(xi)}$$

We thus see that the incomplete Beta function describes the contour of more than one class of sampling distributions. When $n = m$ so that $B(n, m) = B(n, n)$, it is possible to express $B(n, n)$ as a simple multiple of $B(\frac{1}{2}, n)$, as follows :

$$B(n, n) = \int_0^1 x^{n-1} (1-x)^{n-1} \, dx$$

$$= \int_0^{\frac{1}{2}} x^{n-1} (1-x)^{n-1} \, dx + \int_{\frac{1}{2}}^1 x^{n-1} (1-x)^{n-1} \, dx$$

If we put $x = (1-y)$ so that $y = (1-x)$ and $dx = -dy$, we have

$$x = \tfrac{1}{2}, y = \tfrac{1}{2} \quad \text{and} \quad x = 1, y = 0,$$

so that

$$\int_{\frac{1}{2}}^1 x^{n-1} (1-x)^{n-1} \, dx = - \int_{\frac{1}{2}}^0 y^{n-1} (1-y)^{n-1} \, dy = \int_0^{\frac{1}{2}} x^{n-1} (1-x)^{n-1} \, dx.$$

$$\therefore \ B(n, n) = 2 \int_0^{\frac{1}{2}} x^{n-1} (1-x)^{n-1} \, dx$$

$$= - 2 \int_{\frac{1}{2}}^0 (x - x^2)^{n-1} dx.$$

If we now put $x = \frac{1}{2}(1 - \sqrt{u})$, so that $dx = -\frac{1}{4}u^{-\frac{1}{2}}du$ and $(x - x^2)^{n-1} = (1 - u)^{n-1} \div 2^{2n-2}$ we then have $u = 0, x = \frac{1}{2}$ and $u = 1, x = 0,$

$$B(n, n) = \frac{1}{2^{2n-1}} \int_0^1 u^{-\frac{1}{2}} (1 - u)^{n-1} \, du,$$

$$B(n, n) = (\tfrac{1}{2})^{2n-1} B(\tfrac{1}{2}, n) \qquad \qquad \qquad \text{(xii)}$$

Whence it also follows that

$$\Gamma(n) \cdot \Gamma(n + \tfrac{1}{2}) = 2^{1 - 2n} \sqrt{\pi} \, \Gamma(2n) \qquad \qquad \text{(xiii)}$$

EXERCISE 6.07

1. Evaluate $\int_0^a x^{m-1} (1 - x)^{n-1} \div B(m, n)$ when $m = 1, n = 2$

for values of a from $0 \cdot 1$ to $0 \cdot 9$ by intervals of $0 \cdot 1$, and draw a graph of the function.

2. Tabulate the *Beta* variate for the values suggested above and graph the result when $m = 2$, $n = 2$.

3. Show that
$$\int_0^\infty e^{-a^2 x^2} \, dx = \frac{1}{2a}\sqrt{\pi}.$$

4. Show that
$$B(m, n) = \frac{n-1}{m} B(m + 1, n - 1).$$

5. Show that
$$\int_{-a}^a (a + x)^{m-1} \cdot (a - x)^{n-1} \cdot dx = (2a)^{m+n-1} \cdot B(m, n).$$

6.08 THE PEARSON SYSTEM

What we commonly call curve-fitting by Karl Pearson's method of moments signifies the specification of the (y) frequency distribution of a score x by an equation involving as its definitive constants (parameters) the moments of the distribution itself. For reasons which we have already seen, the relative values of these suffice to exhibit the two essential features of a curve of the type to which sampling distributions—theoretical or empirical—most commonly conform. Of these, one is skewness, the other steepness. It is convenient to combine the moments relevant to one or the other in an index of which the numerator and denominator each involve the same power of the deviations (X); and we shall name these at the outset. We define the coefficient of skewness (β_1) independently of scale, so that its value is necessarily zero when the curve is symmetrical in which case m_3 of 6.01 is zero :

$$\beta_1 = \frac{(m_3)^2}{(m_2)^3} \qquad \qquad \qquad \text{(i)}$$

The definition of the coefficient of flatness (β_2) is given by

$$\beta_2 = \frac{m_4}{(m_2)^2} \qquad \qquad \qquad \text{(ii)}$$

As pointed out on page 232, a relatively low value of β_2 so defined signifies a relatively flat distribution. For the normal distribution $\beta_1 = 0$, since $m_3 = 0$, and from (iv) of 6.06, we obtain

$$m_4 = 3V^2 = 3m_2^2,$$
$$\therefore \beta_2 = 3 \qquad \qquad \qquad \text{(iii)}$$

The flattest type of distribution is the *rectangular*, defined by equal frequency of every score value like the distribution of single toss scores of a cubical die ($y = \frac{1}{6}$). By definition, the frequency is then the reciprocal of the number of classes defined by particular score values. For the score range $0 - r$, signifying that the number of classes is $(r + 1)$, the kth mean moment therefore takes the form defined by (x) in 6.01 as

$$m_k = \sum_{X = -\frac{1}{2}r}^{X = \frac{1}{2}r} \frac{1}{r+1} X^k.$$

When r is large we may write this as

$$m_k \simeq \frac{1}{r} \int_{\frac{1}{2}r}^{\frac{1}{2}r} X^k \, dX = \frac{1}{r}\left[\frac{X^{k+1}}{k+1}\right]_{\frac{1}{2}r}^{\frac{1}{2}r};$$

Subject to the restriction stated, we thus obtain

$$m_2 \simeq \frac{1}{r}\left[\frac{X^3}{3}\right]_{-\frac{1}{2}r}^{\frac{1}{2}r} = \frac{r^2}{12},$$

$$m_3 \simeq \frac{1}{r}\left[\frac{X^4}{4}\right]_{\frac{1}{2}r}^{\frac{1}{2}r} = 0,$$

$$m_4 \simeq \frac{1}{r}\left[\frac{X^5}{5}\right]_{-\frac{1}{2}r}^{\frac{1}{2}r} = \frac{r^4}{80}.$$

Whence from (i) and (ii)

$$\beta_1 = 0 \quad and \quad \beta_2 = \tfrac{9}{5} \qquad . \qquad . \qquad . \qquad . \qquad . \qquad . \qquad \text{(iv)}$$

We call a unimodal curve *platykurtic*, if the numerical value of β_2 lies between limits defined by (iii) and (iv).

Our problem is now to find a model expression which will assist us to make explicit the geometrical properties of a distribution by recourse to these indices. It must therefore contain *at least* three constants. Clearly also, the curve it represents is likely to be skew, if it is to implicate $\beta_1{}^*$; and we can get the clue we need by retracing our footsteps to the difference equation (ii) which we derived in 3.09 for the hypergeometric distribution. With that end in view, we shall regard it merely as a pattern, and have no interest in the special meaning of the constants it contains. We there based our treatment on the approximate equation

$$\Delta y_x \simeq \frac{y_x \cdot n(M - x)}{C + Cx + x^2} \Delta x.$$

To do so, we made an assumption justified by the end in view, *viz.* to show that the limiting form of the non-replacement distribution is the normal. Had we not made it, we might have made the skewness of the curve when p and q are unequal explicit in the derivation of the original difference equation, i.e. (i) of 3.09. As in that context, we can still use the approximations $(rp - q) \simeq rp$, $(n + 1) \simeq n$ and $C = (nq - r) \simeq (nq - r + 1)$. If we put $na = r$, (i) of 3.09 in which $\Delta x = 1$ then becomes

$$\frac{\Delta y_x}{y_x} \simeq \frac{nM + r - nx}{C + Cx + x^2} = \frac{-n(x - M - a)}{C + Cx + x^2}.$$

---

* If a curve is symmetrical $m_3 = 0$, hence also $\beta_1 = 0$; but the converse is not always true.

We can now write this in the form :

$$\Delta Y_{\bullet} \simeq \frac{Y_{\bullet}(X - a)\, \Delta X}{C_0 + C_1 X + C_2 X^2},$$

$$\therefore \frac{dY}{Y} \simeq \frac{(X - a)dX}{C_0 + C_1 X + C_2 X^2} \qquad . \qquad . \qquad . \qquad . \qquad . \qquad \text{(v)}$$

The last expression contains four constants, and it is not difficult, though a little tedious, to express them uniquely in terms of the first four zero moments which suffice to specify the two Pearson coefficients defined by (i) and (ii) above. It is thus a pattern for a very comprehensive family of functions suitable for describing sampling distributions ; but it will not be necessary for us to develop the more important variants of the pattern from first principles, as set forth in Kendall's treatise (Chapter 6, Vol. I). We have already seen how they may arise in conformity with assumptions more directly relevant to situations they describe ; and it is therefore possible to approach our goal by a more direct and far less laborious route.

In what follows we shall confine our attention to three of the five most important families of curves of the Pearson system, viz. Types I-III. Of these Type III is an incomplete Gamma function, Type I being an incomplete Beta function of which Type II is a limiting case. In its turn, the normal is likewise a limiting case of all three, each of which we have met in preceding pages. The fact that these three families, together with the normal, constitute some of the important functions descriptive of sampling distributions, endows the method of curve fitting by moments with a special claim to consideration in virtue of a peculiar property of their successive zero moments. This arises from the fact that the incomplete Gamma and the incomplete Beta function alike contain a power of x as a factor. If, therefore, we can express the integrand (y) definitive of the probability density (p. 185) of a score x as a function of this sort, we obtain the rth zero moment which is the complete integral of yx^r, by merely stepping up the factor involving definitive of the probability density (p. 185) of a score x as a function of this sort, we obtain the rth zero moment, i.e. the complete integral of yx^r, by merely stepping up the factor involving a power of x in the function without change of form. As the reader will have the opportunity of seeing more clearly below, this means that any zero moment of a Gamma variate is itself expressible as a complete Gamma function and every zero moment of a Beta variate is expressible as a complete Beta function.

The simplicity of the operation of determining the zero moments of one or other will appear at a later stage. To exploit it, we must recall the procedure for transforming zero moments into mean moments, as expounded in 6.01 ; and it will be convenient to set out once more the relevant formulae :

$$m_2 = \mu_2 - \mu_1^2 \qquad . \qquad . \qquad . \qquad . \qquad . \qquad . \qquad . \qquad \text{(vi)}$$

$$m_3 = \mu_3 - 3\mu_1 . \mu_2 + 2\mu_1^3 \qquad . \qquad . \qquad . \qquad . \qquad \text{(vii)}$$

$$m_4 = \mu_4 - 4\mu_1 . \mu_3 + 6\mu_1^2 . \mu_2 - 3\mu_1^4 \qquad . \qquad . \qquad . \qquad \text{(viii)}$$

The complete system of Pearson curves includes several which are merely museum exhibits, but the three first types (I-III) which involve Beta or Gamma variates are of special importance in connexion with the theory of sampling. They are the theme of what follows. Types VI and VII which define important sampling distributions mentioned at the end of Chapters 7 and 10, will be the subject of separate treatment in Vol. II. Since we have already seen how a Gamma or a Beta variate may arise as the limiting form of a sampling distribution, it will not be necessary to traverse the tortuous steps which led Pearson himself to discover the general pattern of his system. It will suffice to show :

(a) how it is possible to express a Gamma or Beta variate in a form which contains no constants other than the mean and the mean moments m_2, m_3, m_4 ;

(b) what numerical values of the Pearson coefficients β_1 and β_2 as defined by (i) and (ii) above are indicative of the suitability of Types I-III for giving a good area fit to a sampling distribution when we know what its moments are.

Type III.—To define a Gamma variate (Pearson's Type III) in the most general way we recall the identity specified by (xi) in 6.05, *viz.* :

$$\int_0^\infty e^{-kx}\, x^{n-1}\, dx = k^{-n} \int_0^\infty e^{-z} z^{n-1}\, dz = k^{-n}\, \Gamma(n) \qquad . \qquad . \qquad (ix)$$

$$\therefore \frac{k^n}{\Gamma(n)} \int_0^\infty e^{-kx}\, x^{n-1}\, dx = 1.$$

The above satisfies the necessary condition that the entire area under the curve of a probability function must be unity. In virtue of this identity, we may therefore assign a probability density y to a score x which has a continuous range of values from 0 to ∞ by the equation definitive of a Gamma variate whose exponent is $(n-1)$ with scalar constant k, namely

$$y = \frac{k^n \cdot e^{-kx} \cdot x^{n-1}}{\Gamma(n)} \qquad . \qquad . \qquad . \qquad . \qquad (x)$$

By definition, the rth zero moment of a distribution so defined is given by

$$\mu_r = \frac{k^n}{\Gamma(n)} \int_0^\infty x^r \cdot e^{-kx} x^{n-1} \cdot dx$$

$$= \frac{k^n}{\Gamma(n)} \int_0^\infty e^{-kx} x^{n+r-1}\, dx.$$

The integral in this expression is a Gamma variate of exponent $(n+r-1)$ with scalar constant k, whence by (ix)

$$\int_0^\infty e^{-kx} x^{n+r-1}\, dx = \frac{\Gamma(n+r)}{k^{n+r}}$$

$$\therefore \mu_r = \frac{k^n\, \Gamma(n+r)}{k^{n+r}\, \Gamma(n)} \qquad . \qquad . \qquad . \qquad . \qquad (xi)$$

In virtue of the fundamental property of the Gamma function

$$\Gamma(n+2) = (n+1)\Gamma(n+1) = n(n+1)\Gamma(n),$$

and in general $\qquad \Gamma(n+r) = (n+r-1)^{(r)}\, \Gamma(n) \qquad . \qquad . \qquad . \qquad (xii)$

Thus (xi) reduces to

$$\mu_r = k^{-r}\, (n+r-1)^{(r)} \qquad . \qquad . \qquad . \qquad (xiii)$$

We thus derive the first four zero moments of the Gamma variate as

$$\mu_1 = n \cdot k^{-1}; \qquad \mu_2 = (n+1)^{(2)} k^{-2}.$$

$$\mu_3 = (n+2)^{(3)} k^{-3}; \qquad \mu_4 = (n+3)^{(4)} k^{-4}.$$

By means of (vi)-(viii) we obtain the mean moments by substitution :

$$m_2 = n \cdot k^{-2}; \qquad m_3 = 2n \cdot k^{-3}; \qquad m_4 = 3n(n+2)k^{-4} \qquad . \qquad . \qquad (xiv)$$

In accordance with (i) and (ii) above, we therefore get

$$\beta_1 = \frac{4}{n} \quad \text{and} \quad \beta_2 = 3 + \frac{6}{n} \qquad . \qquad . \qquad . \qquad (xv)$$

17

THE INCOMPLETE GAMMA FUNCTION

(Pearson's Type III)

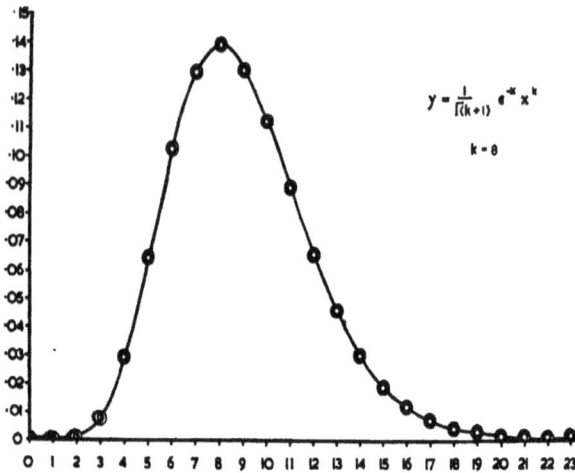

$$y = \frac{1}{\Gamma(k+1)}\, e^{-x}\, x^{k}$$

$$k = 8$$

FIG. 60.

The last relation implies that $\beta_2 = 3(1 + \frac{1}{2}\beta_1)$. When n is very large $\beta_1 \simeq 0$ and $\beta_2 \simeq 3$. The Pearson coefficients of skewness and flatness then differ insensibly from those of the normal distribution. For practical purposes (xv) suffices to define when the Gamma variate is likely to provide a good fit for a unimodal distribution, *viz.* $\beta_1 > 0$ and $\beta_2 > 3 \cdot 0$, and $\beta_2 = 3(1 + \frac{1}{2}\beta_1)$. Thus the Gamma variate describes a skew curve extending from 0 to infinity, being more steep in the neighbourhood of the peak than a truly normal distribution.

Either of the Pearson coefficients of the distribution, as specified by (xv), suffice to fix n in (x); and k has a simple relation to the mean score value $M(= \mu_1)$ in virtue of (xiii) from which we obtain $M = nk^{-1}$. From (xv)

$$n = \frac{4}{\beta_1} \quad and \quad (n-1) = \frac{4 - \beta_1}{\beta_1} \qquad . \qquad . \qquad . \qquad . \qquad \text{(xvi)}$$

Hence we have also

$$k = \frac{4}{M\beta_1} \cdot \qquad . \qquad . \qquad . \qquad . \qquad . \qquad \text{(xvii)}$$

Thus we may transform (x) so as to contain no constants other than the mean, the second mean moment and the third mean moment :

$$y = \frac{\left(\dfrac{4}{M\beta_1}\right)^{\frac{4}{\beta_1}} \cdot e^{-\frac{4x}{M\beta_1}} \cdot x^{\frac{4 - \beta_1}{\beta_1}}}{\Gamma\left(\dfrac{4}{\beta_1}\right)} \qquad . \qquad . \qquad . \qquad . \qquad \text{(xviii)}$$

This is Pearson's Type III when the origin is at $x = 0$. It is customary to cite it in an alternative form by transferring the origin to the *mode* at $x = a$. The mode being the turning point of the curve, we determine as usual by equating the first derivative to zero, i.e.

$$D_x(y) = C\{-ke^{-kx}x^{n-1} + (n-1)e^{-kx}x^{n-2}\} = 0.$$

$$\therefore \ x = \frac{n-1}{k} = a \quad . \qquad . \qquad . \qquad . \qquad . \qquad \text{(xix)}$$

If we now represent the score deviation *from the mode* by X, we have $X = (x - a)$ and $x = (a + X)$ so that

$$x^{n-1} = x^{ka} = (a + X)^{ka}.$$

$$\therefore \ x^{n-1} = a^{ka}\left(1 + \frac{X}{a}\right)^{ka}.$$

We then write (x) in the form

$$Y = \frac{k^n \cdot a^{ka} \cdot e^{-kX}\left(1 + \dfrac{X}{a}\right)^{ka}}{e^{ka}\ \Gamma(n)} \quad . \qquad . \qquad . \qquad . \qquad \text{(xx)}$$

From (xix) we have $ka = (n - 1)$, so that

$$k^n\, a^{ka} = \frac{k^n\, a^{ka+1}}{a} = \frac{(n-1)^n}{a}.$$

For brevity * we may write

$$(n - 1) = p = \frac{4 - \beta_1}{\beta_1} \quad . \qquad . \qquad . \qquad . \qquad \text{(xxi)}$$

By (xvii) and (xix)

$$a = \frac{M(4 - \beta_1)}{4} \quad . \qquad . \qquad . \qquad . \qquad . \qquad \text{(xxii)}$$

Whence by substitution in (xx) we obtain an equation involving three constants respectively defined by (xxi), (xxii) and (xvii) in terms of the moments of the distribution :

$$Y = \frac{p^{p+1} \cdot e^{-kX} \cdot \left(1 + \dfrac{X}{a}\right)^{ka}}{e^p \cdot a \cdot \Gamma(p + 1)} \quad . \qquad . \qquad . \qquad . \qquad \text{(xxiii)}$$

This is the most usually cited form of Type III, the range of the distribution being now from $X = -a$ to $X = \infty$. The reader should with little difficulty be able to derive the analogous expression for a Pearson Type III curve with the *mean* as origin.

   Types I and II.—Just as we introduced a scalar constant in the simple incomplete Gamma function to derive the general form of the Gamma variate or Type III distribution, we shall now modify the Beta function by the scalar substitution $x = az$, so that $x = a$ when $z = 1$, and

$$\int_0^1 z^{j-1} \cdot (1 - z)^{k-1} \cdot dz = B(j, k) = \int_0^a \frac{x^{j-1} \cdot (a - x)^{k-1} \cdot dx}{a^{j+k-1}}.$$

$$\therefore \ \frac{1}{a^{j+k-1} \cdot B(j, k)} \int_0^a x^{j-1} \cdot (a - x)^{k-1} \cdot dx = 1 \quad . \qquad . \qquad . \qquad \text{(xxiv)}$$

* It is customary to use p as in (xxiii) ; but it is (needless to say) not the same as the parameter p of the binomial variate.

We may then define the probability density (y) of a score whose values increase continuously from o to a in the form

$$y = \frac{x^{j-1} \cdot (a - x)^{k-1}}{a^{j+k-1} \cdot B(j, k)} \qquad . \qquad . \qquad . \qquad . \qquad . \qquad \text{(xxv)}$$

This generalised Beta variate is Pearson's Type I when j and k are unequal, and includes as a special case Type II when $j = k$. The constant a has then a simple meaning, as we shall now see.

By our definition of the mean M of any distribution, we can determine that of (xxv) as follows :

$$M = \int_0^a \frac{(x)x^{j-1} \cdot (a - x)^{k-1} \cdot dx}{a^{j+k-1} \cdot B(j, k)}$$

$$= \frac{1}{a^{j+k-1} \cdot B(j, k)} \int_0^a x^j \cdot (a - x)^{k-1} \cdot dx \qquad . \qquad . \qquad . \qquad \text{(xxvi)}$$

From (xxiv) above

$$\int_0^a x^j \cdot (a - x)^{k-1} \cdot dx = \int_0^a x^{j+1-1} \cdot (a - x)^{k-1} \cdot dx = a^{j+k}B(j + 1, k).$$

$$\therefore M = \frac{a \cdot B(j + 1, k)}{B(j, k)} = \frac{a\,\Gamma(j + k)}{\Gamma(j) \cdot \Gamma(k)} \cdot \frac{\Gamma(j + 1)\,\Gamma(k)}{\Gamma(k + j + 1)}.$$

Since $\Gamma(j + 1) = j\,\Gamma(j)$ and $\Gamma(k + j + 1) = (k + j)\,\Gamma(k + j)$,

$$\therefore M = \frac{a \cdot j}{j + k} \qquad . \qquad . \qquad . \qquad . \qquad . \qquad \text{(xxvii)}$$

When $j = k$ in (xxv), we therefore have $M = \frac{1}{2}a$, and

$$y = \frac{[x(2M - x)]^{j-1}}{(2M)^{2j-1}B(j, j)} \qquad . \qquad . \qquad . \qquad . \qquad \text{(xxviii)}$$

If we now transfer the origin to the mean by the usual substitution $X = (x - M)$, $x = (M + X)$,

$$x(2M - x) = (M + X)(M - X)$$

$$= M^2\left(1 - \frac{X^2}{M^2}\right).$$

Hence (xxv) becomes :

$$y = \frac{M^{2j-2}\left(1 - \dfrac{X^2}{M^2}\right)^{j-1}}{(2M)^{2j-1} \cdot B(j, j)}.$$

$$\therefore y = \frac{\left(1 - \dfrac{X^2}{M^2}\right)^{j-1}}{M \cdot 2^{2j-1} \cdot B(j, j)} \qquad . \qquad . \qquad . \qquad . \qquad \text{(xxix)}$$

This is Pearson's Type II, two of the essential properties of which are evident if we put $X = 0$ or $\pm M$. It is symmetrical and extends either side of the origin from $- M$ to $+ M$. It is now necessary to determine only one constant j in terms of the moments with a view to recognising when it is likely to give a satisfactory description of a distribution which is indeed symmetrical. We proceed to get the rth zero moment of the distribution in its most general form as for the mean in (xxvi) above, viz. :

$$\mu_r = \int_0^a \frac{x^{j+r-1} \cdot (1-x)^{k-1} \cdot dx}{a^{j+k-1} \cdot B(j,\, k)}$$

$$= \frac{a^{j+k+r-1} \cdot B(j+r,\, k)}{a^{j+k-1} B(j,\, k)}$$

$$= \frac{a^r \cdot B(j+r,\, k)}{B(j,\, k)}.$$

In virtue of the fundamental property of the Gamma function, i.e. $\Gamma(n+1) = n\Gamma(n)$, this becomes :

$$\mu_r = \frac{a^r \cdot (j+r-1)^{(r)}}{(j+k+r-1)^{(r)}} \qquad . \qquad . \qquad . \qquad . \qquad . \qquad \text{(xxx)}$$

When $j = k$ as in (xxviii) above :

$$\mu_r = \frac{a^r (j+r-1)^{(r)}}{(2j+r-1)^{(r)}}.$$

Hence we obtain

$$\mu_2 = \frac{a^2 j (j+1)}{2j(2j+1)} = \frac{a^2(j+1)}{2(2j+1)}.$$

$$\mu_3 = \frac{a^3(j+2)}{4(2j+1)}.$$

$$\mu_4 = \frac{a^4(j+2)(j+3)}{4(2j+1)(2j+3)}.$$

We can convert these into mean moments as in the derivation of the constants of Type III, and obtain by simple algebra

$$m_2 = \frac{a^2}{4(2j+1)}; \quad m_3 = 0; \quad m_4 = \frac{3a^4}{16(2j+1)(2j+3)}.$$

Hence we have :

$$\beta_1 = 0 \quad and \quad \beta_2 = \frac{3(2j+1)}{(2j+3)} = 3 - \frac{6}{(2j+3)} \qquad . \qquad . \qquad . \qquad \text{(xxxi)}$$

The constant j is therefore determined uniquely by the coefficient of flatness, i.e. in (xxix) above

$$2j = \frac{3(\beta_2 - 1)}{(3 - \beta_2)} \quad and \quad (j-1) = \frac{5\beta_2 - 9}{2(3 - \beta_2)} \qquad . \qquad . \qquad \text{(xxxii)}$$

From (xxxi) we see that $\beta_2 < 3$, since j is positive ; but when j is very large $\beta_2 \simeq 3$, so that both Pearson coefficients of the distribution are almost identical with those of the normal. When $j = 1$, we obtain $\beta_2 = \frac{9}{5}$, and since $(j-1) = 0$, y is constant for all values of x or X. We then call the distribution itself *rectangular*, the definitive Pearson coefficients of such a distribution being therefore $\beta_1 = 0$ and $\beta_2 = \frac{9}{5}$ in agreement with (iv) above.

The procedure for choosing Type II as a likely fit for a unimodal distribution which is symmetrical about the mean M, or very nearly so, is to evaluate β_2 by determining its second and fourth moments. If β_2 is less than 3·0, with due regard to the meaning of j as defined by (xxxii), equation (xxix) then specifies a curve which will commonly describe its contour closely. Type II is important as a descriptive curve for a binomial and also of the hypergeometric distribution (p. 142) when $p = \frac{1}{2} = q$. We shall also meet it again in Chapter 8.

If j and k are unequal, the determination of their separate values in terms of β_1 and β_2 is more tedious. The curve is asymmetrical about the mean which we have determined above, and β_1 does *not* vanish. The second Pearson coefficient is less than 3·0 as for Type II. Thus the generalised Beta variate or Type I distribution defines a skew curve more flat than the normal. As β_1 tends to zero Type I merges into Type II and into the normal as β_2 also approaches the numerical value 3·0.

We can change (xxv) into a form like that of (xxix), if we transfer the origin to a fixed point other than the mean; and it is convenient to choose the value of x at the mode, which is the value of x at which the turning point occurs. Thus the differential coefficient is zero, and

$$D_x[x^{j-1}(a-x)^{k-1}] = 0.$$

$$x = \frac{(j-1)a}{k+j-2} = A \qquad . \qquad . \qquad . \qquad . \qquad . \qquad \text{(xxxiii)}$$

We now write $V = (x - A)$, so that

$$x^{j-1}(a-x)^{k-1} = (A+V)^{j-1} \cdot (\overline{a-A}-V)^{k-1}.$$

In this expression we put $B = (a - A)$, so that

$$B = \frac{(k-1)a}{k+j-2} . \qquad . \qquad . \qquad . \qquad . \qquad \text{(xxxiv)}$$

$$\therefore \ x^{j-1} \cdot (a-x)^{k-1} = A^{j-1} \cdot B^{k-1} \cdot \left(1+\frac{V}{A}\right)^{j-1}\left(1-\frac{V}{B}\right)^{k-1}.$$

By substitution in (xxv) above, we then get the generalised Beta or Type I distribution in its *modal* form as :

$$Y = \frac{A^{j-1} \cdot B^{k-1} \cdot \left(1+\dfrac{V}{A}\right)^{j-1} \cdot \left(1-\dfrac{V}{B}\right)^{k-1}}{a^{j+k-1} \cdot B(j,k)}. \qquad . \qquad . \qquad \text{(xxxv)}$$

The curve then has the range $-A$ to $+B$.

The evaluation of j, k and a in (xxxv), hence likewise of A and B, in terms of the moments is tedious and the relevant expressions are cumbersome. The student who seeks further information concerning the Pearson system will find a full treatment in Elderton's *Frequency Curves and Correlation*, and will find Chapter 6, Vol. I, of Kendall's treatise helpful.

Choice of a Descriptive Curve.—Within the framework of our present discussion which circumscribes most of the fundamental curves descriptive of sampling distributions we may summarise as follows the considerations which guide our choice in the search for an appropriate form :

(*a*) determine the mean, the second, third and fourth zero moments of the distribution, hence also β_1 and β_2 ;

(*b*) if β_1 is nearly zero and β_2 is nearly 3·0, a normal curve may give a satisfactory fit ;

(*c*) if β_1 is nearly zero and β_2 is less than 3·0, Pearson's Type II is likely to be more appropriate than the normal ;

(d) if the distribution is asymmetrical, so that β_1 differs appreciably from zero, the choice depends on the numerical value of β_2;

(e) if β_2 exceeds 3·0 and the mode of the distribution is relatively near the origin, Type III will be suitable only if also $\beta_2 \simeq 3(1 + \frac{1}{2}\beta_1)$ in accordance with (xv);

(f) if β_2 is less than 3·0, Type I will be our first choice for a skew distribution.

These indications presume that the distribution is unimodal. They do not greatly help us to make a decision unless we define *nearly* in this context. Computation shows that there is little to commend Type I or Type III in preference to the normal if $\beta_1 \simeq 0·1$ and β_2 lies within the range 2·85 to 3·15. If β_2 is exactly zero, as is true of the binomial distribution when $p = \frac{1}{2} = q$ the normal should suffice for most practical purposes if β_2 exceeds 2·75. In theoretical statistics a symmetrical distribution which is steeper ($\beta_2 \simeq 3·0$) than the normal is the t-variate of 7·08. This is a particular case of Pearson's Type VII.

If the foregoing introduction to Pearson's system has whetted the appetite of the reader, the following explanatory remarks will suffice to indicate how to explore the properties of Type VI, which itself defines an important sampling distribution, and is also the parent of Type VII. In 6·07 we have seen that we may specify the complete Beta function by an integral of restricted range (i) and by an integral of infinite range (ii). The integrand in the former is definitive of Pearson's Type I and of Type II as a special case. The integrand of (ii) in 6·07 is Pearson's Type VI when the origin is at zero score value. Type VII defines the distribution of the square root of a Type VI score.

We shall have to examine the properties of Types VI and VII in connexion with sampling distributions which are the theme of Vol. II; and shall therefore defer further reference to them. In anticipation of an important class of distributions we shall examine in the same context, the reader may take the opportunity of proving that (x) above defines the distribution of the square ($Q = c^2$) of the critical ratio, when $k = \frac{1}{2} = n$. When $k = \frac{1}{2}$ and $n = \frac{1}{2}f$, we speak of the Type III distribution defined by (x) as that of *Chi Square* for f degrees of freedom.

Types I and III alike describe skew distributions respectively more or less flat near the mode than is the normal; but Type I differs from Type III and from the normal in virtue of the fact that its range is limited. We can also define a Beta variate of unlimited range by reference to the alternative form of the Beta function specified by (ii) in 6·07. This distribution, which plays a leading role in more recent developments of statistical theory, lies outside the scope of significance tests dealt with in this volume. The Pearson coefficients of any of the curves dealt with above have values which may occur in a binomial distribution of the ordinary type, as is apparent by determining the moments of the binomial distribution by one or other method dealt with in 6·09. For the raw score distribution whose definitive binomial is $(q + p)^r$, we obtain

$$m_2 = rpq; \quad m_3 = rpq(q - p); \quad m_4 = 3r^2p^2q^2 + rpq(1 - 6pq).$$

$$\beta_1 = \frac{(q - p)^2}{rpq} \quad and \quad \beta_2 = 3 + \frac{1 - 6pq}{rpq}.$$

For large values of r the two coefficients evidently tend to the same limit as those of the normal. Otherwise $\beta_1 = 0$ when $p = \frac{1}{2} = q$ and $\beta_2 = 3·0$ when $(1 - 6pq) = 0$, so that $p \simeq 0·79$ or 0·21. For values of p within this range $\beta_2 < 3·0$ and β_1 vanishes only when $p = 0·5$. For values of p outside the range stated $\beta_2 > 3·0$ and β_1 is always numerically greater than zero. A caveat is appropriate in this context *vis-à-vis* the derivation in Chapter 3 of continuous curves to describe the binomial distribution approximately when r is large. The functions we there derived by making *ad hoc* approximations suggest the form of the Gamma variate or of the Beta variate; but the constants in (xviii) of 3·04 are *not* identical with those of a Type II distribution with the same moments as the binomial. They merely serve to disclose a pattern suitable for specifying a fitting curve whose moments are in fact identical with those of an exact distribution.

EXERCISE 6.08

1. The mean moments of the Binomial Distribution $(p + q)^r$ are as follows (see 6.09):

$$m_1 = rp; \quad m_2 = rpq; \quad m_3 = rpq(q - p); \quad m_4 = 3r^2p^2q^2 + rpq\,(1 - 6pq).$$

Tabulate numerical values of both β_1 and β_2 for $(\frac{1}{2} + \frac{1}{2})^r$, $(\frac{3}{4} + \frac{1}{4})^r$ and $(\frac{9}{10} + \frac{1}{10})^r$ for $r = 8, 16,$ 20 and 100.

2. Show that the condition $\beta_2 = 3(1 + \frac{1}{2}\beta_1)$ defined by (xv) is inconsistent with the relation $(p + q) = 1$.

3. Fit a Type II distribution to the binomial $(\frac{1}{2} + \frac{1}{2})^{14}$ and compare the result with the normal distribution shown in Table II of 3.04.

4. Write down the first four mean moments of the incomplete Γ function $y_x = e^{-x} x^{n-1} \div \Gamma(n)$; evaluate β_1 and β_2 and examine the shape which the function takes for $n = \frac{1}{2}, 1, 2, 4$.

6.09   MOMENT GENERATING SERIES

The foregoing introduction to Pearson's method of assigning to a sampling distribution a continuous curve from the area of which it is approximately possible to specify the expectation of a score value within a particular range suffices to justify the search for methods of general applicability to the evaluation of the moments of such a distribution. To materialise a class of series of special importance with this end in view, it may help the reader if we adhere to our custom of regarding a statistic as a way of scoring the player's luck.

In a game of hazard, we are at liberty to adopt any system of scoring, if we state the rule in advance; and it is therefore permissible to postulate the rule that the player who gets a raw score of x records the result as 3^x. If an ace ranks as a success, a deal of all four aces would then count as $81 = 3^4$ points. More generally, a system which assigns a^x points in virtue of a raw score x is one which we shall call exponential scoring; and we shall examine what would then be the mean score of the player denoting the mean * as $E(f_x)$ so that for r-fold trials

$$E(f_x) = \sum_{x=0}^{x=r} y_x a^x.$$

It is convenient to express this in terms of the natural logarithm of a. We shall therefore write $t = \log_e a$, so that $a = e^t$; and

$$E(f_x) = \sum_{x=0}^{x=r} y_x e^{tx} \qquad . \qquad . \qquad . \qquad . \qquad . \qquad \text{(i)}$$

When r is indefinitely large,

$$E(f_x) = \int_0^{\infty} y . e^{tx} dx \qquad . \qquad . \qquad . \qquad . \qquad \text{(ii)}$$

---

* The symbol E often used for the mean value of a statistic is not the step-up operation of 1.10.

We now expand e^{tx} by series integration:

$$E(f_x) = \int_0^\infty y\left(1 + tx + \frac{t^2 x^2}{2!} + \frac{t^3 x^3}{3!} + \frac{t^4 x^4}{4!} \ldots\right)dx \quad . \quad . \quad . \quad \text{(iii)}$$

$$= \int_0^\infty y\,dx + t\int_0^\infty y\,x\,dx + \frac{t^2}{2!}\int_0^\infty y\,x^2\,dx$$

$$+ \frac{t^3}{3!}\int_0^\infty y\,x^3\,dx + \frac{t^4}{4!}\int_0^\infty y\,x^4\,dx \ldots \text{etc.}$$

Hence from (x) in 6.01

$$E(f_x) = \mu_0 + \mu_1 \cdot t + \mu_2 \cdot \frac{t^2}{2!} + \mu_3 \cdot \frac{t^3}{3!} + \mu_4 \cdot \frac{t^4}{4!} + \quad . \quad . \quad . \quad \text{(iv)}$$

The expression on the right is a series in which the form of an individual term is $\mu_x \frac{t^x}{x!}$, the xth

zero moment *of the raw score* (x) *distribution* being the coefficient of $\frac{t^x}{x!}$. In the same way, we

may build up a corresponding series by making use of the player's exponential score *deviation*. We shall denote the mean by $E(f_X)$, so that

$$E(f_X) = \int_{-\mu_1}^{r-\mu_1} Y\,e^{tX}\cdot dX = m_0 + m_1 t + m_2\frac{t^2}{2!} + m_3\frac{t^3}{3!}, \text{ etc.} . \quad . \quad . \quad \text{(v)}$$

In the last expression the coefficients of $t^x \div x!$ are successive mean moments of the distribution of raw scores. It is customary to speak of $E(f_x)$ and $E(f_X)$ as moment *generating functions*. If we can discover such a function w.r.t. a distribution of scores (iv) and (v) provide a method of specifying its moments.

The possibility of doing so depends on the fact that it is possible to expand any continuous function as a power series in the independent variable in accordance with Maclaurin's theorem of 1.06. The series we have obtained is in fact a power series of t, so that we may regard the functions defined by (iv) and (v) as functions of t itself. In accordance with the customary symbolism of Maclaurin's theorem, we shall denote by $E(f_x)_0$ and $E(f_X)_0$ respectively the values assumed when t itself is zero by the two required functions which respectively define the mean value of the exponential score, and that of the exponential score deviation. Let us now investigate their values by the method used to establish the same theorem. If we differentiate successively both sides of (iv) we get

$$\frac{d}{dt}E(f_x) = \mu_1 + \mu_2 t + \mu_3 \frac{t^2}{2!} + \mu_4 \frac{t^3}{3!} \ldots$$

$$\frac{d^2}{dt^2}E(f_x) = \mu_2 + \mu_3 t + \mu_4 \frac{t^2}{2!} \ldots$$

$$\frac{d^3}{dt^3}E(f_x) = \mu_3 + \mu_4 t + \mu_5 \frac{t^2}{2!} \ldots$$

In general, therefore,

$$\frac{d^x}{dt^x}E(f_x) = \mu_x + \mu_{x+1}\cdot t + \mu_{x+2}\cdot\frac{t^2}{2!} \ldots$$

When $t = 0$, every term on the right hand side vanishes except μ_x, so that

$$\frac{d^x}{dt^x}E(f_x)_0 = \mu_x \quad . \quad . \quad . \quad . \quad . \quad . \quad \text{(vi)}$$

Similarly we obtain

$$\frac{d^x}{dt^x} E(f_x)_0 = m_x \qquad \qquad \text{. . . . . (vii)}$$

If we can transform $E(f_x)$ or $E(f_X)$ as defined in (i)-(v) into a function suitable for successive differentiation, we can therefore obtain the xth moment or xth mean moment by equating t to zero in the xth derivative of the function. For the definition of the function we have already obtained the necessary expressions, $viz.$ (i) or (ii) for $E(f_x)$ and when r is large (v) for $\overline{E}(f_X)$. The formula for $E(f_X)$ analogous to (i) is

$$E(f_X) = \sum_{-\mu_1}^{r-\mu_1} Y_X \cdot e^{tX}.$$

Whether the finite summation formula or the integral form of the function is suitable for the purpose depends on ease of manipulation. The moments of the binomial raw score distribution are determinable by recourse to the former. Thus (i) becomes

$$E(f_x) = \sum_0^{x=r} y_x e^{tx} = \sum_0^{x=r} r_{(x)} p^x q^{r-x} \cdot e^{tx}$$

$$= \sum_0^{x=r} r_{(x)} (pe^t)^x q^{r-x}.$$

The last expression is the sum of the terms of the binomial $(pe^t + q)^r$, so that we have

$$E(f_x) = (pe^t + q)^r.$$

This is the required generating function for the zero moments. By (vi) when $t = 0$

$$\frac{d^x}{dt^x} (pe^t + q)^r = \mu_x.$$

We shall write this result in the form

$$D_t^x \cdot \left[(pe^t + q)^r \right]_{t=0} = \mu_x \qquad \qquad \text{. . . . . (viii)}$$

By differentiating once we get the familiar result

$$D_t \left[(pe^t + q)^r \right]_{t=0} = \left[r(pe^t + q)^{r-1} \cdot pe^t \right]_{t=0}$$

$$= rp(p + q)^{r-1} = rp = \mu_1.$$

By successive differentiation we get

$$D_t^2 (pe^t + q)^r = rpe^t (pe^t + q)^{r-1} + r(r-1)p^2 e^{2t} (pe^t + q)^{r-2}$$

$$D_t^3 (pe^t + q)^r = rpe^t (pe^t + q)^{r-1} + 3r(r-1)p^2 e^{2t} (pe^t + q)^{r-2}$$
$$+ r(r-1)(r-2) p^3 e^{3t} (pe^t + q)^{r-3}.$$

If we now set $t = 0$ in the above, substituting $(p + q)^m = 1$ for all values of m

$$D^2 \cdot E(f_x)_0 = rp + r(r-1)p^2 = \mu_2 \qquad \qquad \text{. . . . . (ix)}$$

$$D^3 \cdot E(f_x)_0 = rp + 3r(r-1)p^2 + r(r-1)(r-2)p^3 = \mu_3 \qquad \text{. . (x)}$$

From (viii) and (ix) by means of (xiv) in 6.01, we can derive the 3rd mean moment

$$m_3 = \mu_3 - 3\mu_1 \cdot \mu_2 + 2\mu_1^3$$
$$= rp + 3r(r-1)p^2 + r(r-1)(r-2)p^3 - 3rp(rpq + r^2p^2) + 2r^3p^3,$$
$$\therefore m_3 = rpq(q - p) \cdot \qquad \qquad \text{. . . . . (xi)}$$

The moments of the Poisson distribution for rare occurrences (3.07) are obtainable in the same way. By definition the range is infinite, and

$$y_x = \frac{e^{-M} \cdot M^x}{x!}.$$

$$E(f_x) = \sum_0^\infty \cdot \frac{e^{-M} \cdot M^x \cdot e^{tx}}{x!} = e^{-M} \sum_0^\infty \frac{(Me^t)^x}{x!}.$$

From the definition of the exponential function as a power series

$$\sum_0^\infty \frac{b^x}{x!} = e^b,$$

$$\therefore e^{-M} \sum_0^\infty \frac{(Me^t)^x}{x!} = e^{-M} \cdot exp\ (Me^t)$$
$$= exp\ [M(e^t - 1)].$$

This is the required generating function $E(f_x)$ for the zero moments of the raw score (x) distribution. By successive differentiation

$$D_t \cdot E(f_x) = M\ exp\ [M(e^t - 1) + t].$$
$$D_t^2 \cdot E(f_x) = (M^2 e^{2t} + Me^t)\ exp\ [M(e^t - 1)].$$

The use of the moment generating function calls for the sort of ingenuity which comes with practice for those of us who lack genius. So it will be profitable to consider a few examples of generating functions with comparable properties. Many books use the symbol $G(t)$ for a power series of t like (iv) above with the implication that the coefficients of t have some specified meaning such as μ_x in the same equation. This usage does not make explicit that our interest in the generating function resides in what it generates, i.e. in this case μ_x. To bring into the picture both the variable t of which the parent series is an explicit function and the variable x of which the offspring is an explicit function, we shall here denote by $G_t(f_x)$ the parental expression which we expand as a power series in such a form as (iv).

The corresponding function $E(f_x)$ on the left hand of (iv) specifies the player's exponential score, and the function f_x which turns up as a factor of the coefficient of t^x on the right hand side is the xth zero moment of the raw score distribution. In what follows f_x is any function of x. It need have no special relevance to statistics. When we can express the parent series in a form suitable for differentiation in accordance with (vi) or (vii), it is a pure function of t, such as $(q + pe^t)^r$ above. Its relationship to its offspring appears only in its expansion as a power series of which the xth term necessarily has as its coefficient a function of x.

To get this clear, let us examine the expression

$$G_t(f_x) = \log_e (1 + t) \qquad . \qquad . \qquad . \qquad . \qquad \text{(xii)}$$

In the above $\log_e (1 + t)$ is thus the parent function of t which generates f_x the filial function of x. The function (f_x) is our unknown. To find it we have to use what we know about $\log_e (1 + t)$ and about the generating function itself. In order that $G_t(f_x)$ may conform to the definition of a generating function implicit in our previous treatment it must be a function both of t and of f_x, and hence also of x, such that

(a) we can expand it in a power series of which $(f_x \div x!)$ is the coefficient of t^x, i.e.

$$G_t(f_x) = f_0 + f_1 t + f_2 \frac{t^2}{2!} + f_3 \frac{t^3}{3!} + f_4 \frac{t^4}{4!} \text{etc.} \qquad . \qquad . \qquad \text{(xiii)}$$

(b) we can represent it as the sum of a series of products of the form $y_x e^{tx}$, of which the factor y_x is in general a function of x different from f_x, so that

$$G_t(f_x) = \sum_0^r y_x \cdot e^{tx} \simeq \int_0^r y_x e^{tx}\, dx \qquad \cdots \qquad \text{(xiv)}$$

(c) we obtain the value of f_x by equating t to zero in the xth derivative (D_t^x) of $G_t(f_x)$ w.r.t. t itself, i.e.

$$D_t^x \cdot \left[G_t(f_x) \right]_{t=0} = f_x \qquad \cdots \qquad \text{(xv)}$$

What we have done in preceding examples and what we do in general when we want to find f_x the xth moment of a distribution defined by the frequency function y_x, is to make use of (xiv) in virtue of our prior knowledge of what y_x is in order to express $G_t(f_x)$ as a simple function amenable to differentiation. We then obtain f_x by virtue of (xv). This drill is straightforward if the product $y_x e^{tx}$ is amenable to finite summation or integration by parts. In (xii) we have sidestepped the first operation which presupposes prior knowledge of y_x, but know nothing about the form of the function f_x other than what we can infer from (xv) as follows :

$$
\begin{aligned}
D_t^0 \log(1+t) &= \log(1+t) & &= & 0 \text{ when } t=0 &=f_0 \\
D_t^1 \log(1+t) &= (1+t)^{-1} & &= & 1 \quad ditto &=f_1 \\
D_t^2 \log(1+t) &= -(1+t)^{-2} & &= & -1 \quad ditto &=f_2 \\
D_t^3 \log(1+t) &= 2(1+t)^{-3} & &= & 2! \quad ditto &=f_3 \\
D_t^4 \log(1+t) &= -3.2(1+t)^{-4} & &= & -3! \quad ditto &=f_4 \\
D_t^x \log(1+t) &= (-1)^{x+1}(x-1)!\,(1+t)^{-x} & &= & (-1)^{x+1}(x-1)! &=f_x
\end{aligned}
$$

We may thus write $f_0 = 0$, and if $x > 0$

$$f_x = (-1)^{x+1} \cdot (x-1)!$$

This is evidently consistent with (xii) and (xiii) from which (xv) is deducible, as we see if we substitute the numerical values of f_0, f_1, etc., as above in (xiii). We then have

$$G_t(f_x) = 0 + 1 \cdot t - \frac{t^2}{2!} + \frac{2!\, t^3}{3!} - \frac{3!\, t^4}{4!}, \text{ etc.}$$

$$= t - \frac{t^2}{2} + \frac{t^3}{3} - \frac{t^4}{4} \cdots \text{ etc.}$$

$$= \log_e(1+t).$$

Having witnessed the process of parturition specified by (xv), which delivers the function (f_x) of x out of the womb of $G_t(f_x)$, let us now examine how we trace the parent of the abandoned offspring, i.e. discover the function which generates f_x. For illustrative purposes we shall choose as the latter the function definitive of the *natural number* series, i.e. $f_x = x$. Thus $f_0 = 0, f_1 = 1, f_2 = 2, f_3 = 3$, etc., in (xiii), so that we can specify the parent $G_t(f_x) = G_t(x)$ by

$$G_t(x) = \frac{0}{0!}t^0 + \frac{1}{1!}t^1 + \frac{2}{2!}t^2 + \frac{3}{3!}t^3 + \frac{4}{4!}t^4 + \frac{5}{5!}t^5, \text{ etc.}$$

$$= 0 + t + t^2 + \frac{t^3}{2!} + \frac{t^4}{3!} + \frac{t^5}{4!}, \text{ etc.}$$

$$= t\left(1 + t + \frac{t^2}{2!} + \frac{t^3}{3!} + \frac{t^4}{4!}, \text{ etc.}\right).$$

$$\therefore\ G_t(x) = te^t.$$

We can test the consistency of this result by applying the midwife operation :

$$D_t(te^t) = (te^t + e^t) = 1 \quad \textit{when } t = 0 \quad f_1 = 1$$
$$D_t^2(te^t) = (te^t + 2e^t) = 2 \quad \textit{ditto} \quad f_2 = 2$$
$$D_t^3(te^t) = (te^t + 3e^t) = 3 \quad \textit{ditto} \quad f_3 = 3$$
$$D_t^x(te^t) = (te^t + xe^t) = x \quad \textit{ditto} \quad f_x = x$$

The same procedure is adaptable to the discovery of a generating function for any arithmetic progression, defined by $A_x = (A_0 + kx)$. Thus $A_1 = (A_0 + k)$, $A_2 = (A_0 + 2k)$, etc., and in (xiii) we have

$$G_t(A_x) = A_0 + (A_0 + k)t + \frac{(A_0 + 2k)t^2}{2!} + \frac{(A_0 + 3k)t^3}{3!} \text{ etc.}$$

We can write this as

$$G_t(A_x) = \sum_0^\infty \frac{(A_0 + kx)t^x}{x!} = A_0 \sum_0^\infty \frac{t^x}{x!} + k \sum_0^\infty \frac{xt^x}{x!}$$

$$= A_0 \sum_0^\infty \frac{t^x}{x!} + kt \sum_0^\infty \frac{t^{x-1}}{(x-1)!} \qquad . \qquad . \qquad . \qquad \text{(xvi)}$$

In this equation

$$\sum_0^\infty \frac{t^x}{x!} = 1 + t + \frac{t^2}{2!} + \frac{t^3}{3!} + \frac{t^4}{4!} \text{etc.} = e^t ;$$

$$\sum_0^\infty \frac{t^{x-1}}{(x-1)!} = \frac{1}{(-1)!\,t} + \frac{t^0}{0!} + \frac{t}{1!} + \frac{t^2}{2!} + \frac{t^3}{3!} \ldots \text{etc.}$$

$$= 0 + 1 + t + \frac{t^2}{2!} + \frac{t^3}{3!} \ldots \text{etc.}$$

$$= e^t.$$

Hence, from (xvi)

$$G_t(A_x) = (A_0 + kt)e^t.$$

The G.F. of a geometric series whose common ratio is r is easily obtainable from (xiii). Since $f_x = f_0 r^x$

$$G_t(f_x) = f_0 + f_0 rt + \frac{f_0 r^2 t^2}{2!} + \frac{f_0 r^3 t^3}{3!} \ldots \text{etc.}$$

$$= f_0\left(1 + rt + \frac{r^2 t^2}{2!} + \frac{r^3 t^3}{3!} \ldots \text{etc.}\right).$$

$$\therefore \ G_t(f_x) = f_0 e^{rt}.$$

For the simple harmonic series $H_x = (1 + x)^{-1}$, we have

$$H_0 = 1; \ H_1 = \tfrac{1}{2}; \ H_2 = \tfrac{1}{3}; \ H_3 = \tfrac{1}{4}, \text{etc.,}$$

$$\therefore \ G_t(H_x) = 1 + \frac{t}{2!} + \frac{t^2}{3!} + \frac{t^3}{4!} + \frac{t^4}{5!} \ldots \text{etc.,}$$

$$\therefore \ t \cdot G_t(H_x) = t + \frac{t^2}{2!} + \frac{t^3}{3!} + \frac{t^4}{4!} + \frac{t^5}{5!} \ldots \text{etc.,}$$

$$\therefore t . G_t(H_x) + 1 = 1 + t + \frac{t^2}{2!} + \frac{t^3}{3!} + \frac{t^4}{4!} + \frac{t^5}{5!} \text{ etc.,}$$

$$\therefore t . G_t(H_x) + 1 = e^t,$$

$$\therefore G_t(H_x) = \frac{e^t - 1}{t}.$$

As a final example of this class of generating function, we shall trace the parent of the series of factorial numbers, defined by the function $F_x = x!$, so that (xiii) becomes

$$G_t(F_x) = 1 + t + t^2 + t^3 + t^4 \ldots \text{ etc.}$$

The expression on the right is the sum of a G.P. which is convergent only if $|t|$ is less than unity, in which case its value is $(1 - t)^{-1}$ and

$$D_t^x . G_t(F_x) = x!(1 - t)^{-x+1} = x! \text{ when } t = 0.$$

The moment generating functions defined by (i)-(v) are examples of a larger class of operations commonly employed in statistical theory. We shall therefore denote $G_t(f_x)$ of the previous section by $G_t^1(f_x)$ to distinguish it from other generating functions. Its basic properties are

$$G_t^1(f_x) = f_0 + f_1 t + \frac{f_2 t^2}{2!} + \frac{f_3 t^3}{3!} \ldots \text{ etc. .} \qquad \text{(xvii)}$$

$$D_t^x[G_t^1(f_x)]_{t=0} = f_x . \qquad \text{(xviii)}$$

We may define another type of generating function by the relation

$$G_t^2(f_x) = f_0 + f_1 t + f_2 t^2 + f_3 t^3 + f_4 t^4 \ldots \text{ etc. .} \qquad \text{(xix)}$$

In the above f_x, in contradistinction to $(f_x \div x!)$ as in (xvii), is the coefficient of t^x; and we obtain by successive differentiation as before

$$\frac{1}{x!} D_t^x[G_t^2(f_x)]_{t=0} = f_x \qquad \text{(xx)}$$

If we now represent by f_x in (xix) the frequency function y_x of a score x, it becomes

$$G_t^2(y_x) = y_0 + y_1 t + y_2 t^2 + y_3 t^3 + \ldots y_r t^r.$$

Over the range $x = 0$ to $x = r$, we may write this as

$$G_t^2(y_x) = \sum_{x=0}^{x=r} y_x t^x.$$

This is again equivalent to the mean value of the player's exponential score t^x; but the coefficient of t^x in the power series is now the frequency of the raw score x. If the distribution is binomial

$$G_t^2(y_x) = \sum_{x=0}^{x=r} r_{(x)} q^{r-x} p^x t^x = (q + pt)^r.$$

On differentiating $(q + pt)^r$ successively in accordance with (xx), we thus generate the frequencies of the distribution

$$\frac{1}{1!} \cdot \frac{d}{dt}(q + pt)^r = rp(q + pt)^{r-1} = rpq^{r-1} \quad \text{when} \quad t = 0;$$

$$\frac{1}{2!} \cdot \frac{d}{dt}(q + pt)^r = \frac{r(r-1)}{2!} p^2 (q + pt)^{r-2} = r_{(2)} p^2 q^{r-2} \quad \text{when} \quad t = 0;$$

$$\frac{1}{x!} \cdot \frac{d}{dt}(q + pt)^r = \frac{r^{(x)}}{x!} p^x (q + pt)^{r-x} = r_{(x)} p^x q^{r-x} \quad \text{when} \quad t = 0.$$

The considerable class of generating functions employed in statistical treatises are confusing; and the same author may make use of more than one type. Thus Aitken's *moment* generating function accords with (xxvii) and (xxviii) above; and his *probability* generating function based on (xix) and (xx) defines a frequency as illustrated by the last example. If we seek for a function which will generate zero moments in accordance with (xix) it must have the form

$$G(\mu_k) = \mu_0 + t\mu_1 + t^2\mu_2 + t^3\mu_3$$
$$= 1 + t\Sigma yx + t^2\Sigma yx^2 + t^3\Sigma yx^3, \text{ etc.}$$
$$= \Sigma y_x(1 + tx + t^2x^2 + t^3x^3 \ldots).$$

This is recognisable as a quotient

$$G(\mu_k) = \Sigma y_x \frac{1}{1 - tx}.$$

If we can express the function on the right in a form suitable for differentiation, we therefore obtain μ_k by setting $t = 0$ in the kth derivative with respect to t and multiplying the result by $k!$ as indicated in (xx).

By implication in (xiv), but without explicit statement, we have here defined a generating power series of either type as *either* a series of a limited number of terms reducible to the definitive function $G_i(f_x)$ by finite summation, *or* as an infinite series which we can represent by a definite integral. It goes without saying that the midwife operation is equally applicable to either, the only difference involved being that f_x has an upper limit set by the range of x.

EXERCISE 6.09

1. By successive use of the chessboard device examine the meaning of a *frequency* generating function w.r.t. the a-fold sample from a 4-class rectangular universe.

2. Evaluate the moments of the exponential distribution $y_x = ce^{-cx}$ for $c > 0$ and x from 0 to ∞. Hint: take $|t| < C$.

3. Rig up a generating function for the mean moments of the normal distribution. Hint: put

$$exp\left(tX - \frac{X^2}{2V}\right) = exp\left[-(Vt - X)^2 \div 2V + \frac{Vt^2}{2}\right].$$

6.10 MOMENTS OF A DIFFERENCE DISTRIBUTION

If z is some function of A and B, two variates, i.e. scores of two sets each with a specifiable frequency distribution, we may speak of it as a score function which is itself a variate, if we can also discover a law of its own distribution. So far the only score function we have dealt with is that of the difference $(d = A - B)$ of two binomial variates in the taxonomical domain. In Vol. II we shall introduce an economical notation to describe the frequency of a score function of two variates from considerations of a grid lay-out comparable to those of Figs. 46, 47 and 78 with or without the condition of statistical independence, i.e. obedience to the product rule of 2.06 as there assumed. In this context, however, we are ready to approach a dilemma we had to face in 4.05 and 4.08 from a new viewpoint. We there saw that the chimney landscape (Fig. 50) of the histogram for the proportionate score difference distribution w.r.t. co-prime samples assumes a contour (Fig. 51) suggestive of a normal fit if we coarsen our scale by grouping adjacent

d-score values in equal intervals. Indeed, computation shows that a normal fit is satisfactory for the end in view, i.e. summation of frequencies within a given range, even when the size of neither sample is large.

If the exact value of the frequency of any single admissible d-score is irrelevant to the end in view, a continuous curve which fits the condensed histogram is for practical purposes what we seek ; and we can explore the properties of such a curve by recourse to the method of moments without violating the assumption that the exact distributions of the parent scores u_a and $u_f = (u_a + d)$ respectively tally with successive terms of the binomials $(q + p)^a$ and $(q + p)^b$. To do so, we need first to establish a procedure which enables us to specify the moments of the distribution of the difference between any two independent variates in terms of their own moments.

Here we should remind ourselves that the difference between two proportionate scores of samples from the same universe is also the difference (p. 166) between the corresponding proportionate score deviations from their common mean. So it will be simpler, if we confine our discussion to two sets of sample score deviations (A and B) whose mean values are zero, and we have already seen that the mean of the d-score of sample pairs is then zero by (vii) and (viii) of 4.04. If we use the convention $M(x^r)$ for the mean value of the rth power of any score x whose mean value is zero, our definition of the rth mean moment, $m_r(x)$ of the x-score distribution, then implies that $m_r(x) = M(x^r)$, i.e.

$$m_2(d) = M(A^2 - 2AB + B^2) \ . \qquad . \qquad . \qquad . \qquad . \qquad \text{(i)}$$
$$m_3(d) = M(A^3 - 3A^2B + 3AB^2 - B^3) \qquad . \qquad . \qquad . \qquad \text{(ii)}$$
$$m_4(d) = M(A^4 - 4A^3B + 6A^2B^2 - 4AB^3 + B^4) \qquad . \qquad . \qquad \text{(iii)}$$

From the definition of the mean and by recourse to Figs. 46, 47 and 78, the reader will see that we can write the above in the form

$$m_2(d) = M(A^2) - 2M(AB) + M(B^2) = m_2(A) - 2M(AB) + m_2(B).$$

But we have already established that $m_2(d) = V_d = V_a + V_b = m_2(A) + m_2(B)$ when the two variates are independent. Subject to this condition, $M(AB) = 0$, as is deducible from the chessboard lay-out when the two sets of sample scores are in fact independent in virtue of the relation exhibited in Fig. 78, viz. $M(AB) = M_a \cdot M_b$, since $M_a = 0 = M_b$ if we measure the scores as deviations from their mean. The rule which Fig. 78 demonstrates is a particular case of a more general relation w.r.t. products of independent variates, viz.

$$M(A^n \cdot B^m) = M(A^n) \cdot M(B^m) \qquad . \qquad . \qquad . \qquad \text{(iv)}$$

The truth of this is evident from a consideration of the grid set-up in Fig. 78, if we replace A_0, A_1, etc., by A_0^n, A_1^n as the column border scores and B_0, B_1, etc., by B_0^m, B_1^m, etc., as the row border scores. If we denote by A_c^n with frequency u_c and B_r^m with frequency v_r, the border scores of column c and row r, we may write for the weighted sum of the products in the rth row :

$$u_0 v_r A_0^n B_r^m + u_1 v_r A_1^n B_r^m + u_2 v_r A_2^n B_r^m \ . \ . \ . \ \text{etc.}$$
$$= v_r B_r^m (u_0 A_0^n + u_1 A_1^n + u_2 A_2^n \ . \ . \ . \ \text{etc.})$$
$$= v_r B_r^m \cdot M(A^n).$$

When we sum the row totals weighted by the row score frequencies, we have :

$$= M(A^n) v_0 B_0^m + M(A^n) v_1 B_1^m + M(A^n) v_2 B_2^m \ . \ . \ . \ \text{etc.}$$
$$= M(A^n) \{ v_0 B_0^m + v_1 B_1^m + v_2 B_2^m \ . \ . \ . \ \text{etc.} \}$$
$$= M(A^n) \cdot M(B^m).$$

When A and B are score deviations, (iv) then becomes :

$$M(A^n B^m) = m_n(A)m_m(B) \qquad . \qquad . \qquad . \qquad . \qquad (v)$$

Since the first mean moment is zero it also follows that

$$M(A^n . B) = 0 = M(A . B^m).$$

Thus (ii) becomes :

$$m_3(d) = M(A^3) - M(B^3) = m_3(A) - m_3(B) \qquad . \qquad . \qquad . \qquad (vi)$$

Likewise (iii) becomes :

$$m_4(d) = m_4(A) + 6V_a . V_b + m_4(B) \qquad . \qquad . \qquad . \qquad . \qquad (vii)$$

We can now define as follows the Pearson coefficients for the distribution of the difference of two independent variates measured from the mean

$$\beta_1 = \frac{[m_3(A) - m_3(B)]^2}{(V_a + V_b)^3} \qquad . \qquad . \qquad . \qquad . \qquad . \qquad (viii)$$

$$\beta_2 = \frac{m_4(A) + 6V_a . V_b + m_4(B)}{(V_a + V_b)^2} \qquad . \qquad . \qquad . \qquad . \qquad (ix)$$

The moments of the distribution of a score deviation are the same as those of the corresponding score since the mean value of X in (viii) of page 231 is zero, and the kth moment of the proportionate score distribution of an r-fold sample is obtainable from that of the raw score if we multiply the latter by r^{-k} in virtue of the substitution $u = (x \div r)$ in (x) of 6.01. Thus the moments of the proportionate score deviation distribution of a binomial variate of degree r are on the right below

	raw score and raw score deviation	proportionate score and proportionate score deviation
m_2	rpq	$\dfrac{pq}{r}$
m_3	$rpq(q - p)$	$\dfrac{pq(q - p)}{r^2}$
m_4	$3r^2p^2q^2 + rpq(1 - 6pq)$	$\dfrac{3rp^2q^2 + pq(1 - 6pq)}{r^3}.$

Whence if A and B are proportionate score deviations respectively referable to a-fold and b-fold samples :

$$m_2(d) = \frac{a + b}{ab} pq.$$

$$m_3(d) = \frac{a^2 - b^2}{a^2 . b^2} pq(q - p).$$

$$m_4(d) = \frac{3p^2q^2(a + b)^2}{a^2 . b^2} + \frac{(a^3 + b^3)pq(1 - 6pq)}{a^3 . b^3}.$$

Thus we obtain for the distribution of the proportionate score difference w.r.t. a-fold and b-fold samples :

$$\beta_1 = \frac{(a - b)^2}{ab(a + b)} . \frac{(q - p)^2}{pq} \qquad . \qquad . \qquad . \qquad . \qquad . \qquad (x)$$

$$\beta_2 = 3 + \frac{a^2 - ab + b^2}{ab(a + b)} . \frac{(1 - 6pq)}{pq} \qquad . \qquad . \qquad . \qquad (xi)$$

18

Evidently β_1 becomes zero when either $a = b$ or $p = q$; and the factor $(1 - 6pq)$ in β_2 becomes zero in the neighbourhood of $p = 0.79$ and 0.21, so that β_2 is then exactly 3. The factor

$$(1 - 6pq) \div pq$$

is a minimum when $p = \frac{1}{2} = q$ and is then negative, its value being -2. It is positive when $p > 0.79$ or < 0.21. If the parent samples are of equal size and p has the value near 0.79 or 0.21 when $(1 - 6pq)$ vanishes, the Pearson coefficients of the difference distribution are then exactly equivalent to those of the normal. How closely they tally otherwise depends on the relative size of a and b, the relative magnitude of p and q, and the absolute size (a or b) of one or other sample. To explore the measure of agreement and in what circumstances we might expect a Type I, Type II or Type III curve to give a good fit, it is convenient to change (x) and (xi) by the substitution $a = kb$ and $q = mp$. We then have

$$\beta_1 = \frac{(k - 1)^2 (m - 1)^2}{mk(k + 1)b} \quad and \quad \beta_2 = 3 + \frac{(k^2 - k + 1)(1 - 6mp^2)}{km(k + 1)bp^2} \quad . \quad . \quad \text{(xii)}$$

As an exercise the reader may profitably investigate the last equations by assigning numerical values to k, m and b, viz. :

(i) $k = 1, 2 \ldots 5$ signifying that one sample is equal to the other, or twice . . . five times as great as the smaller of the two ;

(ii) $m = 1, 2, 4, 9$ signifying that $p = \frac{1}{2}, \frac{1}{3}, \frac{1}{5}, \frac{1}{10}$;

(iii) the smaller sample consists of $b = 16, 25, 50$ items.

EXERCISE 6.10

MISCELLANEOUS

1. For the normal distribution, show that $m_6 = 15V^3$ and $m_8 = 105V^4$.

2. Show that Pearson's Type III has no mode unless $n = 1$ consistent with the understanding that x is positive.

3. Repeat the foregoing development to obtain β_1 and β_2 for the raw score difference.

4. Use the foregoing procedure (6.10) to show that β_1 and β_2 w.r.t. the raw score difference of a-fold and b-fold samples from the same universe approach their normal values more quickly than do those of the distribution of the $(a + b)$-fold sample.

5. In 6.08 we used μ_k to determine m_k for Pearson's Types I-III. Could we appropriately use the right-hand expression of (xi) in 6.01 by recourse to their equations with means as origin ? If not, why not ?

CHAPTER 7

THE RECOGNITION OF A MEAN DIFFERENCE
(FOR LARGE SAMPLES)

7.01 TWO METHODS OF SCORING

WHAT has gone before is an attempt to materialise a calculus which began at the gaming table. To date, our contact with practical affairs has been with the class of problems which statisticians commonly refer to as the sampling of attributes. We now enter a new domain of statistical problems ; but we shall still adhere to the view that a statistic is amenable to visualisation in so far as we interpret it as a particular method of scoring the player's luck. The customary distinction between sampling of attributes and sampling of measurements draws attention to a very real distinction between two classes of statistics ; but the form of words blurs what is the most essential difference. It suggests an antithesis between quality and quantity or between discrete enumeration and a metrical continuum. What is really important is to discriminate between two methods of scoring a *sample* in contradistinction to scoring an *individual member* of a sample. For want of terms in general currency we here distinguish them as *taxonomic* and *representative*.

Hitherto, we have classified *sample* scores as raw scores, score deviations, proportionate scores, square score deviations, or—more exotically—*exponential* scores (p. 265). In doing so, we have merely used numbers referable in one or other way to *how many members of a specified class a sample contains*. Scoring of this sort is the essence of what statistical works call the *sampling of attributes ;* but an attribute chosen to label a class as such need not be qualitative in the all-or-nothing sense, like the class which consists only of picture cards or the class which consists only of clubs. All cards with seven pips or all cards with less than three pips, regardless of suit, constitute such a class in its own right, as do all new-born babies weighing 5½ lb. or less, all owners of American cars purchased at cost price under two thousand dollars, all hospital patients with less than 3,000,000 erythrocytes per c.mm. of blood, all farms of under 200 acres, all civil servants with a salary exceeding £1000 per annum, and all university lectures lasting longer than 50 minutes. Thus the pivotal peculiarity of this method of scoring has nothing to do with whether we employ a qualitative or quantitative epithet to label the individual member of a sample as one of a particular class. It resides in the fact that we score the sample by *enumeration* of individuals assigned thereto ; and we here speak of such scoring as taxonomic because the score itself is a number which specifies the size of a *class*. If the class is itself definable in terms of a numerical specification of individuals assigned to it, such numerical specification has nothing to do with the sample score value. It is merely a ticket which entitles the holder to rank as a member of a class or excludes one from doing so.

Commonly, our universe of taxonomic scoring has only two classes ; and any member of it then has one of only two possible scores, being unity or zero according as we deem the result of choosing it a *success* or otherwise. Thus we may define a successful draw of one card from a pack without picture cards as an ace ; and if so the score we attach to a card with 2 pips, 3 pips . . . or any greater number of pips up to 10 is zero. If we define a successful 1-fold draw as a card with over 7 pips, the score of cards with 8, 9, or 10 pips is unity, and that of cards with 1, 2 . . . 7 pips is zero. The score of a 3-fold sample would thus be 3, if it consisted of any of the following combinations

(8) (8) (8) ;   (8) (8) (9) ;   (8) (8) (10) ;   (8) (9) (9) ;   (8) (9) (10) ;
(8) (10) (10) ;   (9) (9) (9) ;   (9) (9) (10) ;   (9) (10) (10) ;   (10) (10) (10).

On the other hand, we should score as 2 the combination (8) (7) (9) with the same total number (24) of pips and the same mean (8) as (8) (8) (8). Though we use the number of pips on the cards to define how we score a successful draw, the taxonomic sample score has therefore no essential connexion with either the total or the mean number of pips on the cards which make up the sample ; and any card whose denomination is 7 or less pips makes *no* numerical contribution to the sample score.

In contradistinction to taxonomic scoring so defined, what we here call *representative* scoring attaches as a numerical specification to a sample a representative figure to which *every* individual member of the sample makes its separate contribution in virtue of its own score. Such a representative figure may be a sum, an arithmetic mean or other average such as the *median*, i.e. a count or measurement characteristic of the mid-member of the sample of items arranged in order of merit. *Ipso facto*, representative scoring explicitly assigns a numerical score to any item of a sample and to any item of the universe. Any class within the latter, or any class within a sample, consists of items each with the same individual score or consecutive range of individual scores. If the representative score definitive of an r-fold sample is the sum (s) of its constituent individual scores or is their mean ($s \div r$), that of the sample of one item (*unit sample*) is therefore itself an individual score.

If the representative score of an r-fold toss of a cubical die is the *sum* of the scores of the r individual tosses, the possible 16 score values of a 3-fold toss are 3, 4, 5 . . . 17, 18. If the representative score is the *mean*, the corresponding 16 values are 1, 1·3, 1·6 . . . 5·6, 6, the limiting values of the distribution of mean scores being always the same as the limiting values of the individual scores, in this case 1, 2 . . . 6, of the items which make up the universe. The members of the class of 3-fold samples defined by a mean score of 3·3 from the 6-class universe of such a die are severally distinguishable in virtue of their individual scores as any of the following combinations whose score-sum is 10 :

(1) (3) (6) ;   (1) (4) (5) ;   (2) (2) (6) ;   (2) (3) (5) ;   (2) (4) (4) ;   (3) (3) (4).

That we can so score the result of tossing a die is not an intrinsic peculiarity of die models, nor is it the only way in which we can do so. We employ the taxonomic procedure to score a 10-fold toss of a coin, when we classify the result by the number of heads uppermost. In the same way we can assign a score to the result of a 10-fold toss of an ordinary cubical die by *enumerating* whether 0, 1, 2 . . . 10 sixes fall face uppermost. This would be a taxonomic sample score comparable with scoring a single toss of ten cubical dice with red, orange, yellow, green, blue and violet faces by the number which fall blue face upward ; and the mere fact that we distinguish individual faces of the coloured die by a qualitative specification does not exclude the possibility of scoring by the alternative method, the result of tossing it. If the colours are monochromatic, we can assign a characteristic wavelength to each, and we can then score the 10-fold sample by reference to the mean, median or other index representative of such individual scores.

NOTE. It may dispose of a difficulty which will occur to some readers, if we take stock of a model situation to which a clear-cut distinction between the two types of scoring does *not* apply. If a flat die like a penny has one pip on one face and no pips on the other, a face score of 1 may be taken to signify 1 success. For the *unit sample* (single toss) 0 or 1 successes correspond to score sums of 0 and 1 with frequencies 0·5 and 0·5, and to proportionate scores or mean scores of 0 and 1. Likewise, score-sums of and 1 and 2 successes in 2-fold tosses, and corresponding mean scores of 0 and ½ and 1, have frequencies 0·25, 0·5 and 0·25 respectively. The same correspondence applies to samples of any size ; but only because : (a) the universe is a binary universe ; (b) the same numerical symbols 0 and 1 respectively label the possible taxonomic scores of a single toss and the possible individual score values of the unit sample in the universe of representative scoring. In fact, the correspondence would break down if two faces of the die respectively carried one and two pips. Individual scores of 1 and 2 would then correspond to 0 and 1 successes or *vice versa* according as two or one pips label a success as such.

Needless to say, we have to rely exclusively on the taxonomic method of scoring a result, if it involves the type of all-or-nothing phenomenon which admits no *numerical* assessment of an individual member of a sample. This is so when we classify individual human beings as normal and tuberculous or as red-haired and otherwise. It is also so when we classify peas as green and yellow or as round and wrinkled ; but it is not so when the focus of our interest is what makes pea plants or human beings tall or short. We are then at liberty to define our samples in either of two ways. For taxonomical scoring of samples we may define a tall man as a man, let us say, 5 ft. 8 in. or over, and a short man as a man under 5 ft. 8 in. We then specify a sample by how many of either class it contains or what proportion of the sample is assignable to one of the two classes. We adopt the method of representative scoring, if we assign a mean or median height, e.g. 5 ft. 5 in. or 5 ft. 9 in., as the sample score. In the practice of physical anthropology, blood group maps invariably refer to taxonomic scores and cephalic index maps commonly specify a geographical sample by a representative score, i.e. the mean value of the population.

In the domain of useful activities, statistical problems often leave open the choice of scoring to the discretion of the investigator, as when we wish to investigate the effect of two diets, A and B, on the growth of a child, or on the red blood cell count of a patient. We use the taxonomic method of scoring the result, if we state the problem in the form : what proportion of infants on diets A and B respectively attain or fail to attain a weight of 3 stone, or what proportion of patients in each treatment group have more or less that 4,000,000 red cells per c.mm. of blood at the end of the period of observation ? We use the representative method of scoring if we state the problem in the form : what are the mean weights of infants or what are the median red blood cell counts of patients after a specified period of treatment with diets A and B ?

Either way, we adopt a quantitative criterion to specify the class to which we assign any member of a sample, and whether the criterion is enumerative (i.e. necessarily *discrete*) or metrical (i.e. presumptively *continuous*) is not peculiar to one or other method of scoring the sample. The essential difference does not depend on whether we count objects or apply a scale to measure them. It resides solely in *what we count*. When we adopt the taxonomic method, we count individual items of the sample. When we adopt the representative method we count blood cells, scale divisions or some other numerically specifiable attribute of each individual, and specify the sample by a figure which is representative of all such counts. Thus the taxonomic method takes no account of *individual variation within the classes which define its operations*.

Whether this is an advantage or otherwise, is not a mathematical issue. When we set out to detect a biological difference, our concern may be to assess the influence of an agency which presumably exerts a clear-cut effect only on a particular genotype, or it may be to assess one which exerts a detectable effect without discrimination on all members of a sample. If our objective is to bring into focus a threshold effect, i.e. a difference of the first sort, the taxonomical method is likely to prove the more suitable. Otherwise, it is likely to be less sensitive than the representative method. Where one method provides an unequivocal answer to the question which prompts us to invoke it, the other may fail to do so. Which is more appropriate depends on the question itself : and it is often instructive to compare the results of scoring both ways.

7.02 The Unit Sampling Distribution

Hitherto, we have examined the implications of detecting a real difference (Chapters 4 and 5) only when the method of scoring is taxonomic. It has been convenient to do so because the specification of an appropriate card pack or other model for the distributions involved raises no difficulties of a sort we have now to meet. Our new difficulties do not arise from the fact that the mathematical problems of representative scoring are intrinsically more sophisticated than those we have dealt with hitherto. They arise from an inescapable modicum of uncertainty about *what we can legitimately postulate as a basis of mathematical analysis.*

To define the distribution of the score of an r-fold sample or the score difference with respect to different samples from one and the same universe or from identical universes, we have to make certain assumptions about the universe or universes involved. What we shall henceforth call the *unit sampling distribution* summarises all the information relevant to our purpose. The unit sampling distribution is a specification of the proportion of items of each class in the parent universe. As its name implies, it also specifies the long-run frequencies with which we extract different classes of *one*-fold samples. An essential difference between taxonomic and representative scoring arises because the former necessarily entails definite information about the unit sampling distribution, whereas the latter does not.

When we adopt the taxonomic method of scoring we usually assign a binary class structure to the universe, so that 1-fold samples can be of only two kinds, respectively defined by a score $x = 0$ or $x = 1$ with corresponding frequencies $y = (1 - p)$ and $y = p$. The exact numerical value of the fraction p may be implicit in our null hypothesis, as is true of Mendel's Laws, or it may be a matter of conjecture. Either way, the *algebraic form* of the unit sampling distribution is fixed by the fact that the class-structure of the universe itself is binary; and the precise numerical value p assumes in a particular problem is irrelevant to the algebra of the card pack model, except in so far as we rely on approximations whose validity depends on the assumption that p is fairly large or very small.

Given a two-class universe specified by the unit sampling distribution $x = 0$, $y = (1 - p)$ and $x = 1$, $y = p$, the chessboard device prescribes the appropriate binomial distribution law for raw scores, proportionate scores and score deviations of independent samples from an indefinitely large or a finite universe, and suffices to clarify in what circumstances the normal equation gives a satisfactory approximation thereto. Given the distribution law for sample scores, the random distribution law (Chapter 4) for the difference between scores of paired samples from the same two-class universe follows as a logical superstructure, which needs no additional empirical information to sustain it. The only empirical issue which enters into our calculations is the appropriate numerical value of the fraction p. An assessment of the risk of error in assigning such a value is possible, as we have seen in Chapter 5.

When the method of scoring is representative, the logical issues raised by the detection of a real difference are much less autonomous. In general, such scoring presupposes a universe in which the number of classes is very large. If every item has a count (e.g. income) or measurement (e.g. height) peculiar to itself, we may then be able to assign to every item a *rank score* which is unique in virtue of *order of merit ;* and all such scores have *equal frequency*. We then say that the unit sampling distribution is *rectangular*, as is that of the cubical die which assigns a frequency of one-sixth to scores of 1 to 6 inclusive. Otherwise, we have no information about its algebraic characteristics apart from what we gain from experience of large samples. To that extent, every problem involving representative scoring other than scoring by rank is a problem *sui generis*. No logical necessity prescribes the algebraic pattern of a unit sampling distribution with respect to stature, weights, incomes, body temperature, red-cell count, metabolic rates,

THE RECOGNITION OF A MEAN DIFFERENCE

plasma volumes or length of life. In this context, examination marks or arbitrary numerical grades which do not correspond with divisions on a universally recognised scale are on all fours with scale and dial measurements of which we customarily postulate continuity of distribution. Measurements in general enjoy no special uniqueness on that account. It is arguable that Time, the father of Newton's fluxions, is the only metric of which we are entitled to postulate continuity in the most exacting sense of the term.

At this point, the prospect would indeed be bleak if we had to rely on logic alone. Happily for the statistician, nature follows the Fabian path of the middle way. Experience of large samples of an immense variety of counts and measurements shows that they are apt to cluster round a fashionable value, from which large deviations are increasingly less frequent. In short, many unit sampling distributions of natural phenomena amenable to numerical specification tally very closely with the Gamma or the Beta function of which the normal is a limiting case, and many sorts of variation are amenable to a unit sampling distribution law of the normal type by recourse to methods of scoring, e.g. intelligence quotients, devised with that end in view. While it is therefore true that detection of a real difference based on representative scoring pre-supposes the validity of an empirical assumption which is not always legitimate, it is also true that the investigator of natural or social phenomena is often able to decide whether it is applicable to a particular problem, if sufficiently at home with his (or her) material.

It is none the less important to recognise that experience alone is the arbiter ; and no satisfactory statistical treatment of measurements can support its weight on a foundation of logical necessity alone. Oddly enough, the assumption of a normal unit sampling distribution first established itself as a fashion in a domain in which its credentials are most open to criticism. From arbitrary assumptions about the distribution of instrumental errors, manifestly false in certain circumstances and at best a very gross approximation to the exigencies of laboratory practice, Gauss deduced a normal law of distribution without recourse to the binomial expansion. Ever since his time, it has been a fashion to assess the significance of differences attributable to errors of observation by recourse to the probability integral or other method involving the assumption of an indefinitely large number of different values which errors involved in one and the same experimental operation may assume. How grossly the conditions of experimental procedure falsify the assumption that errors of observation necessarily conform to any such law would scarcely merit comment, if the practice were no longer widespread. Since it is still common to assess the significance of differences which entail errors of observation in accordance with the Gaussian law, the issue is of sufficient importance to call for illustrative comment in this context.

We shall suppose that we are comparing the calcium content of two samples of blood by a volumetric micro-method of estimation. Our null hypothesis is that any difference of the mean values of a series of titrations of each of the two samples is attributable to instrumental error alone. We reject the null hypothesis, if we can show that the expectation of the observed difference is very low ; and to do so we have to know the distribution law of the mean difference. Whether we are entitled to assume that the law is normal is an issue which admits of no dubiety, since we know all the possible sources of instrumental error over which an experienced labora-tory worker has no control. The size of the last drop which falls from the orifice of the burette uniquely determines which scale division defines the end-point of the reaction in the titrating flask, and the size of the drop which the orifice of a pipette admits uniquely defines the quantity of fluid therein. What the investigator cannot control may result in readings which range over one, two or three, or perhaps four scale divisions, but an indefinitely large number of deter-minations of the same sample from the same source can ring the changes only on a small number of the infinite range of scale divisions implicit in a normal unit sampling distribution of errors.

While the foregoing illustration emphasises the need for common sense in the choice of statistical techniques, and the pitfalls which beset such choice in the absence of a clear appreciation of assumptions inherent in their mathematical credentials, we need not throw out the baby with the bath water. If statistical procedures which invoke comparison of averages and other representative scores presuppose the validity of assumptions which the mathematician has no proper authority to impose on the initial statement of the problem, there are in fact many situations in which experience does justify particular postulates prerequisite to mathematical analysis. In particular, many assemblages of individual counts and measurements in nature admit of sufficient variety and range to invite description by recourse to a normal unit sampling distribution without serious error ; and the goodness of fit of such a distribution often justifies the conjecture that assumption of normality will not lead us far astray. Such an assumption can never be exact in so far as it presupposes an unlimited range of individual variation ; but error arising from the circumstance that we encounter adult men as short as 2 inches or as tall as 20 feet only in *Gulliver's Travels* may be negligible, because the area in the tails of the Gaussian (i.e. normal) curve is a negligible fraction of the total area.

We can carry over much of the mathematical apparatus appropriate to one domain of scoring into the other domain ; but we can fully appreciate when it is proper to do so only if we clearly recognise where one domain ends and the other begins. Because it is possible to do so, statistical theory employs terms, such as random variable, normal variate and the like, to describe distributions in virtue of mathematical properties having more or less relevance to *both* types of scoring ; and it will be profitable to clarify the use of such terminology in this context. In Chapter 3 we have seen that the r-fold sample score of *very large* samples of items taken from any two-class universe with replacement is approximately normal, and experience may also show that samples of *one* item taken from a universe of a very large number of classes each with an individual score assigned to it may also be approximately normal. To this extent, we may speak of a score of either sort as a *normal variate*, an expression applicable to any set of scores with which we associate a normal distribution regardless of whether it is that of an individual item like that of the unit sample in the representative domain or of a large collection of individual items each of which can have a taxonomic score of only zero or unity.

Thus a normal variate is simply a score to which we assign a normal frequency distribution. As such, it is any quantity which we can regard with propriety as the independent variable of the normal function, and as such has properties inherent in the nature of the latter regardless of the status of the score itself in the world of practical affairs. One such property set forth in 4.08 and 6.10 is that the difference between two such scores of independently selected a-fold and b-fold samples is itself normally distributed with variance $V_d = (V_a + V_b)$. Translated into the language of everyday affairs, this may be approximately true of situations (*inter alia*) as unlike as : (a) the distribution of the difference between scores of pairs of very large samples of specified size independently selected from a single two-class universe ; (b) the distribution of the difference between pairs of single samples from the same universe with a very large number of classes.

7.03   THE RECTANGULAR UNIVERSE OF THE CUBICAL DIE

The simplest issue of practical statistics in the representative domain is the detection of a difference between two score averages, as when we ask the following questions :

(a) in terms of hours of sleep gained, is the dextro-rotary form of a drug a more efficacious soporific than the lævo-rotary form ?

(b) are beans of one pure line on the whole heavier than beans of another ?

SCORE SUM DISTRIBUTION FOR 2-FOLD SAMPLES

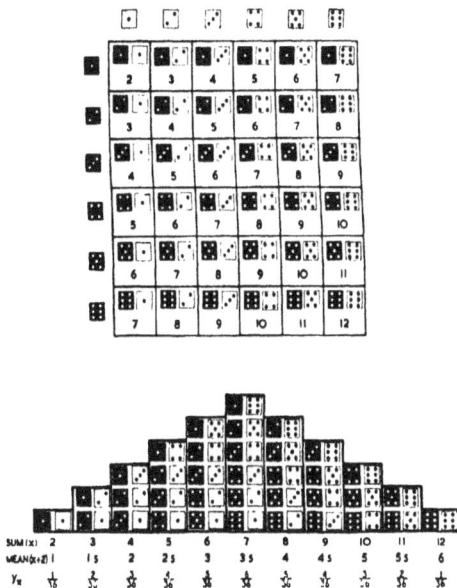

FIG. 61. Sampling from a Rectangular Universe *with* replacement. Derivation of the mean score and score sum distribution of the 2-fold sample of the 6-class universe of the cubical die.

(c) is the eosinophil white cell count above the normal in patients suffering from bilharziosis ?

In all such situations, we may appropriately score the two groups of individuals subject to comparison by reference to an average count or measurement ; and in this context, we shall assume that the particular average we employ is the arithmetic mean. In general, we do not expect to find that all individuals of either group we are comparing will have identical individual scores, i.e. hours of sleep, weights or blood counts, as the case may be ; and we shall therefore expect to find that mean scores of different samples of a particular specification, e.g. Bilharzia patients or beans of a particular pure line, will rarely be exactly equal. Our null hypothesis is the assertion that the difference between the means of two groups subject to comparison arises merely from the circumstance that the two groups are different samples of the same universe. Accordingly, the question to which we seek an answer is : *would a difference between the mean score of two samples from the same universe as large as the observed difference between the mean scores of our control and experimental groups be an exceedingly rare occurrence?*

We can answer a question of this sort only if we are able to define the distribution law of the mean score of independent samples from one and the same universe, i.e. the frequencies of particular mean values of samples extracted from a single universe. Nothing we have dealt with so far offers an answer. To clarify the question itself, and the logical approach to a satisfactory answer, we shall first explore the mean score distribution law of a model which is already

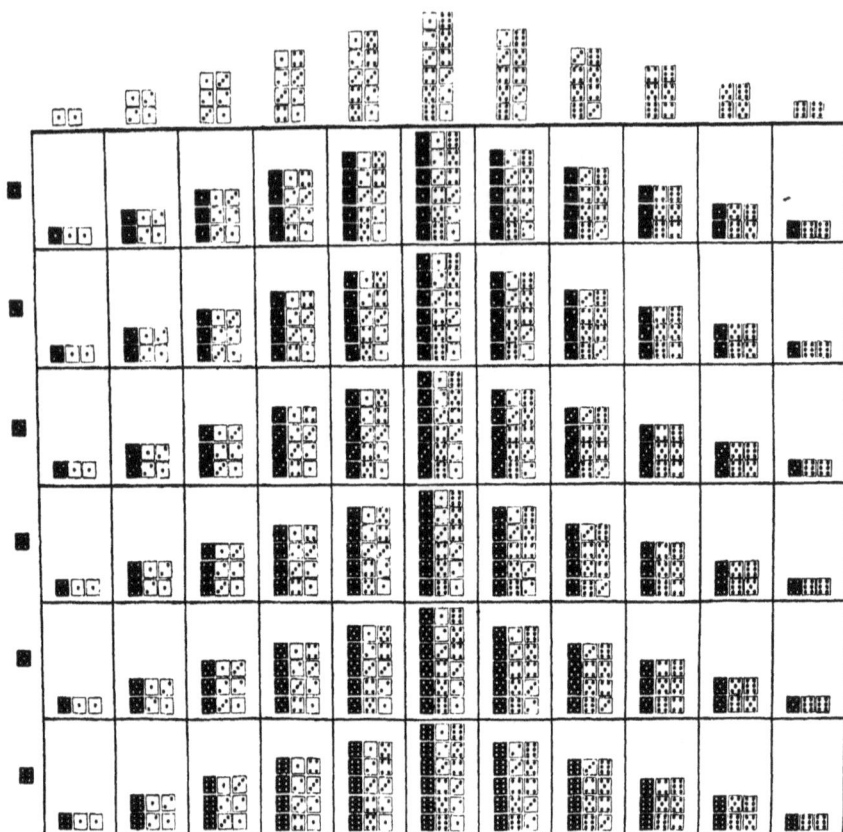

FIG. 62. Sampling from a Rectangular Universe *with* replacement. Derivation of the mean score and score sum distribution of the 3-fold sample of the 6-class universe of the cubical die.

familiar, albeit of little relevance to problems of real life. In doing so, we adhere strictly to the laws of electivity (Chapter 2) which dictate our procedure when the method of scoring is taxonomic. The product rule prescribes the frequency of a particular set of individual scores ; and the addition rule prescribes the total frequency of sets whose sum or mean value conforms to a particular specification *w.r.t.* range.

The chessboard device permits us to set out each permutation of the double toss in Fig. 61, and hence the long-run frequency of any double score sequence. Two such sequences (double-ace or double six) are unique in that the score sum (2 and 12 respectively) or mean score (1 and 6 respectively) associated with them cannot result from any other arrangement. Any other score sum from 3–11 inclusive can result from at least two arrangements. For instance, a score sum of 5 in a 2-fold toss can result from the following score sequences : 1, 4 and 4, 1 ; 2, 3 and 3, 2. Thus 4 out of 36 permutations of the 2-fold toss signify a score sum of 5 with a corresponding mean score of 2·5 ($= 5 \div 2$) and 1 out of 36 signifies a score sum of 12 with a mean of 6. Accordingly, the frequencies of mean scores 2·5 and 6 are respectively 0·$\dot{1}$ ($= 4 \div 36$) and 0·02$\dot{7}$ ($= 1 \div 36$).

As in Figs. 62–63, we set out the 3-fold toss by recourse to the chessboard with due regard to the fact that every 3-fold score sequence is deducible from the result of every possible 2-fold sequence and the result of every possible single toss which can follow it. Having clearly visualised the lay-out in terms of our basic definitions and rules of electivity, we can proceed thereafter to deduce the distribution of 4-fold, 5-fold, etc., tosses by the more economical procedures embodied in the schema of Table 1 for the 3-fold and Table 2 for the 4-fold toss.

For even values of r, it is most economical to lay out the results of successive $\frac{1}{2}r$-fold tosses as in Table 2, and for odd values to obtain the $(2r + 1)$-fold toss distribution by combining the results of r and of $(r + 1)$ tosses, recording score sums and *relative* frequencies separately as in Table 1. To obtain the frequency of a given mean score value we then proceed as follows. In a 4-fold toss, a mean score of 1·5 corresponds to a score sum of 6 in the lower half of Table 2. The *relative* frequencies obtained by applying the product rule in cells of the corresponding diagonal of the upper half of the table are 3, 4, 3. By the addition rule the total frequency of a mean score of 1·5 is $(3 + 4 + 3) \div 6^4 = 0·008$.

We can investigate the law of the distribution of a *mean score difference* w.r.t. samples of the same size or of different sizes by essentially the same method as set out for pairs of 2-fold tosses in Table 3. The student will find it helpful to make up exercises for solution by recourse to it, like those of Exercise 7.03; and to investigate the pattern of the difference distribution. That of the distribution of the mean score in r-fold samples from the rectangular universe of

SCORE SUM DISTRIBUTION FOR 3-FOLD SAMPLES

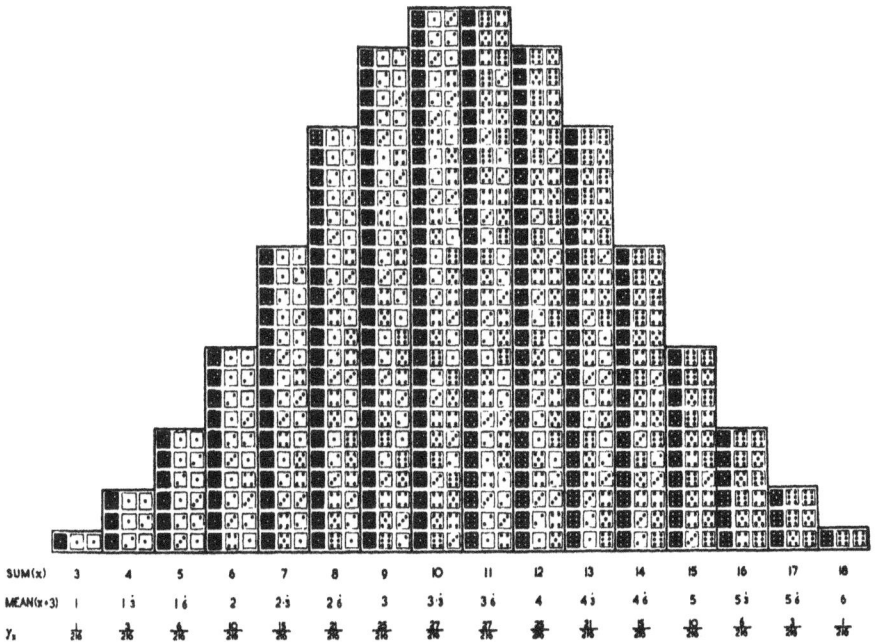

SUM(x)	3	4	5	6	7	8	9	10	11	12	13	14	15	16	17	18
MEAN(x÷3)	1	1⅓	1⅔	2	2⅓	2⅔	3	3⅓	3⅔	4	4⅓	4⅔	5	5⅓	5⅔	6
Yₛ	1/216	3/216	6/216	10/216	15/216	21/216	25/216	27/216	27/216	25/216	21/216	15/216	10/216	6/216	3/216	1/216

FIG. 63. Sampling from a Rectangular Universe with replacement. The histogram of the 3-fold sample based on the data of Fig. 62.

TABLE 1

The 3-fold Toss

Score Sums (x), Mean Scores (M) and Frequencies (y)

Result of First Two Tosses

Result of Third Toss		Score sum	2	3	4	5	6	7	8	9	10	11	12
Score sum	Frequency	Frequency	$\frac{1}{36}$	$\frac{2}{36}$	$\frac{3}{36}$	$\frac{4}{36}$	$\frac{5}{36}$	$\frac{6}{36}$	$\frac{5}{36}$	$\frac{4}{36}$	$\frac{3}{36}$	$\frac{2}{36}$	$\frac{1}{36}$
1	⚀		$x=3$ $M=1$ $y=\frac{1}{216}$	$x=4$ $M=1\cdot3$ $y=\frac{2}{216}$	$x=5$ $M=1\cdot6$ $y=\frac{3}{216}$	$x=6$ $M=2$ $y=\frac{4}{216}$	$x=7$ $M=2\cdot3$ $y=\frac{5}{216}$	$x=8$ $M=2\cdot6$ $y=\frac{6}{216}$	$x=9$ $M=3$ $y=\frac{5}{216}$	$x=10$ $M=3\cdot3$ $y=\frac{4}{216}$	$x=11$ $M=3\cdot6$ $y=\frac{3}{216}$	$x=12$ $M=4$ $y=\frac{2}{216}$	$x=13$ $M=4\cdot3$ $y=\frac{1}{216}$
2	⚁		$x=4$ $M=1\cdot3$ $y=\frac{1}{216}$	$x=5$ $M=1\cdot6$ $y=\frac{2}{216}$	$x=6$ $M=2$ $y=\frac{3}{216}$	$x=7$ $M=2\cdot3$ $y=\frac{4}{216}$	$x=8$ $M=2\cdot6$ $y=\frac{5}{216}$	$x=9$ $M=3$ $y=\frac{6}{216}$	$x=10$ $M=3\cdot3$ $y=\frac{5}{216}$	$x=11$ $M=3\cdot6$ $y=\frac{4}{226}$	$x=12$ $M=4$ $y=\frac{3}{216}$	$x=13$ $M=4\cdot3$ $y=\frac{2}{216}$	$x=14$ $M=4\cdot6$ $y=\frac{1}{216}$
3	⚂		$x=5$ $M=1\cdot6$ $y=\frac{1}{216}$	$x=6$ $M=2$ $y=\frac{2}{216}$	$x=7$ $M=2\cdot3$ $y=\frac{3}{216}$	$x=8$ $M=2\cdot6$ $y=\frac{4}{216}$	$x=9$ $M=3$ $y=\frac{5}{216}$	$x=10$ $M=3\cdot3$ $y=\frac{6}{216}$	$x=11$ $M=3\cdot6$ $y=\frac{5}{216}$	$x=12$ $M=4$ $y=\frac{4}{216}$	$x=13$ $M=4\cdot3$ $y=\frac{3}{216}$	$x=14$ $M=4\cdot6$ $y=\frac{2}{216}$	$x=15$ $M=5$ $y=\frac{1}{216}$
4	⚃		$x=6$ $M=2$ $y=\frac{1}{216}$	$x=7$ $M=2\cdot3$ $y=\frac{2}{216}$	$x=8$ $M=2\cdot6$ $y=\frac{3}{216}$	$x=9$ $M=3$ $y=\frac{4}{216}$	$x=10$ $M=3\cdot3$ $y=\frac{5}{216}$	$x=11$ $M=3\cdot6$ $y=\frac{6}{216}$	$x=12$ $M=4$ $y=\frac{5}{216}$	$x=13$ $M=4\cdot3$ $y=\frac{4}{216}$	$x=14$ $M=4\cdot6$ $y=\frac{3}{216}$	$x=15$ $M=5$ $y=\frac{2}{216}$	$x=16$ $M=5\cdot3$ $y=\frac{1}{216}$
5	⚄		$x=7$ $M=2\cdot3$ $y=\frac{1}{216}$	$x=8$ $M=2\cdot6$ $y=\frac{2}{216}$	$x=9$ $M=3$ $y=\frac{3}{216}$	$x=10$ $M=3\cdot3$ $y=\frac{4}{216}$	$x=11$ $M=3\cdot6$ $y=\frac{5}{216}$	$x=12$ $M=4$ $y=\frac{6}{216}$	$x=13$ $M=4\cdot3$ $y=\frac{5}{216}$	$x=14$ $M=4\cdot6$ $y=\frac{4}{216}$	$x=15$ $M=5$ $y=\frac{3}{216}$	$x=16$ $M=5\cdot3$ $y=\frac{2}{216}$	$x=17$ $M=5\cdot6$ $y=\frac{1}{216}$
6	⚅		$x=8$ $M=2\cdot6$ $y=\frac{1}{216}$	$x=9$ $M=3$ $y=\frac{2}{216}$	$x=10$ $M=3\cdot3$ $y=\frac{3}{216}$	$x=11$ $M=3\cdot6$ $y=\frac{4}{216}$	$x=12$ $M=4$ $y=\frac{5}{216}$	$x=13$ $M=4\cdot3$ $y=\frac{6}{216}$	$x=14$ $M=4\cdot6$ $y=\frac{5}{216}$	$x=15$ $M=5$ $y=\frac{4}{216}$	$x=16$ $M=5\cdot3$ $y=\frac{3}{216}$	$x=17$ $M=5\cdot6$ $y=\frac{2}{216}$	$x=18$ $M=6$ $y=\frac{1}{216}$

TABLE 2

Derivation of the 4-fold Toss Distribution

(a) *Relative Frequencies* ($= y \times 6^4$)

2-Fold Toss

2-Fold Toss												
Score sum		2	3	4	5	6	7	8	9	10	11	12
	Frequency (× 36)	1	2	3	4	5	6	5	4	3	2	1
2	1	1	2	3	4	5	6	5	4	3	2	1
3	2	2	4	6	8	10	12	10	8	6	4	2
4	3	3	6	9	12	15	18	15	12	9	6	3
5	4	4	8	12	16	20	24	20	16	12	8	4
6	5	5	10	15	20	25	30	25	20	15	10	5
7	6	6	12	18	24	30	36	30	24	18	12	6
8	5	5	10	15	20	25	30	25	20	15	10	5
9	4	4	8	12	16	20	24	20	16	12	8	4
10	3	3	6	9	12	15	18	15	12	9	6	3
11	2	2	4	6	8	10	12	10	8	6	4	2
12	1	1	2	3	4	5	6	5	4	3	2	1

(b) *Score sums* (divide each entry by 4 to get corresponding *mean* score)

2	4	5	6	7	8	9	10	11	12	13	14
3	5	6	7	8	9	10	11	12	13	14	15
4	6	7	8	9	10	11	12	13	14	15	16
5	7	8	9	10	11	12	13	14	15	16	17
6	8	9	10	11	12	13	14	15	16	17	18
7	9	10	11	12	13	14	15	16	17	18	19
8	10	11	12	13	14	15	16	17	18	19	20
9	11	12	13	14	15	16	17	18	19	20	21
10	12	13	14	15	16	17	18	19	20	21	22
11	13	14	15	16	17	18	19	20	21	22	23
12	14	15	16	17	18	19	20	21	22	23	24

the cubical die emerges plainly from the numerical results set out in Tables 4–5 by successive application of the method embodied in Tables 1–2.

We may state the rule which emerges from inspection of the figures in Tables 4–5 as follows. If $y_{x(r)}$ is the frequency of a score sum x in an r-fold sample and hence of a mean score $(x \div r)$, we write the corresponding frequency of the score sum x in a sample of $(r - 1)$ items and hence that of a mean score $(x \div \overline{r - 1})$ in the form $y_{x(r-1)}$. The tables show that every term of the series $y_{x(r)}$ is obtainable by recourse to the following rule: sum the terms of the series $y_{x(r-1)}$ for the 6 preceding values of x and divide the result by 6, i.e.

$$y_{x(r)} = \frac{1}{6} \sum_{p=x-6}^{p=x-1} y_{p(r-1)} \quad . \quad . \quad . \quad . \quad . \quad . \quad (i)$$

For instance

$$y_{6(2)} = \frac{1}{6} \sum_{p=0}^{p=5} y_{p(1)} = \frac{1}{36} (1 + 1 + 1 + 1 + 1 + 0) = \frac{5}{36};$$

$$y_{11(3)} = \frac{1}{6} \sum_{p=5}^{p=10} y_{p(2)} = \frac{1}{216} (3 + 4 + 5 + 6 + 5 + 4) = \frac{27}{216}.$$

TABLE 3

Mean Score Difference Distribution for two Double Tosses of a Cubical Die

Each cell shows the score difference d (top) and its frequency (bottom). The row and column "Mean score" values each have the associated Frequency $(\times 6^2)$ shown.

Mean score	Freq $(\times 6^2)$	1 (1)	1.5 (2)	2.0 (3)	2.5 (4)	3.0 (5)	3.5 (6)	4.0 (5)	4.5 (4)	5.0 (3)	5.5 (2)	6.0 (1)
1	1	$d=0$; 1	$d=0.5$; 2	$d=1.0$; 3	$d=1.5$; 4	$d=2.0$; 5	$d=2.5$; 6	$d=3.0$; 5	$d=3.5$; 4	$d=4.0$; 3	$d=4.5$; 2	$d=5.0$; 1
1.5	2	$d=-0.5$; 2	$d=0$; 4	$d=0.5$; 6	$d=1.0$; 8	$d=1.5$; 10	$d=2.0$; 12	$d=2.5$; 10	$d=3.0$; 8	$d=3.5$; 6	$d=4.0$; 4	$d=4.5$; 2
2.0	3	$d=-1.0$; 3	$d=-0.5$; 6	$d=0$; 9	$d=0.5$; 12	$d=1.0$; 15	$d=1.5$; 18	$d=2.0$; 15	$d=2.5$; 12	$d=3.0$; 9	$d=3.5$; 6	$d=4.0$; 3
2.5	4	$d=-1.5$; 4	$d=-1.0$; 8	$d=-0.5$; 12	$d=0$; 16	$d=0.5$; 20	$d=1.0$; 24	$d=1.5$; 20	$d=2.0$; 16	$d=2.5$; 12	$d=3.0$; 8	$d=3.5$; 4
3.0	5	$d=-2.0$; 5	$d=-1.5$; 10	$d=-1.0$; 15	$d=-0.5$; 20	$d=0$; 25	$d=0.5$; 30	$d=1.0$; 25	$d=1.5$; 20	$d=2.0$; 15	$d=2.5$; 10	$d=3.0$; 5
3.5	6	$d=-2.5$; 6	$d=-2.0$; 12	$d=-1.5$; 18	$d=-1.0$; 24	$d=-0.5$; 30	$d=0$; 36	$d=0.5$; 30	$d=1.0$; 24	$d=1.5$; 18	$d=2.0$; 12	$d=2.5$; 6
4.0	5	$d=-3.0$; 5	$d=-2.5$; 10	$d=-2.0$; 15	$d=-1.5$; 20	$d=-1.0$; 25	$d=-0.5$; 30	$d=0$; 25	$d=0.5$; 20	$d=1.0$; 15	$d=1.5$; 10	$d=2.0$; 5
4.5	4	$d=-3.5$; 4	$d=-3.0$; 8	$d=-2.5$; 12	$d=-2.0$; 16	$d=-1.5$; 20	$d=-1.0$; 24	$d=-0.5$; 20	$d=0$; 16	$d=0.5$; 12	$d=1.0$; 8	$d=1.5$; 4
5.0	3	$d=-4.0$; 3	$d=-3.5$; 6	$d=-3.0$; 9	$d=-2.5$; 12	$d=-2.0$; 15	$d=-1.5$; 18	$d=-1.0$; 15	$d=-0.5$; 12	$d=0$; 9	$d=0.5$; 6	$d=1.0$; 3
5.5	2	$d=-4.5$; 2	$d=-4.0$; 4	$d=-3.5$; 6	$d=-3.0$; 8	$d=-2.5$; 10	$d=-2.0$; 12	$d=-1.5$; 10	$d=-1.0$; 8	$d=-0.5$; 6	$d=0$; 4	$d=0.5$; 2
6.0	1	$d=-5.0$; 1	$d=-4.5$; 2	$d=-4.0$; 3	$d=-3.5$; 4	$d=-3.0$; 5	$d=-2.5$; 6	$d=-2.0$; 5	$d=-1.5$; 4	$d=-1.0$; 3	$d=-0.5$; 2	$d=0$; 1

TABLE 4

Mean Score Distribution of the Cubical Die

Score Sum (x).	Unit Sample. Mean x ÷ 1.	Unit Sample. Relative Frequency y × 6	2-Fold Sample. Mean x ÷ 2.	2-Fold Sample. Relative Frequency y' × 6²	3-Fold Sample. Mean Score x ÷ 3.	3-Fold Sample. Relative Frequency y × 6³
1	1	1	0	0	0	0
2	2	1	1	1 + 0 + 0 + 0 + 0 + 0 = 1	0	0
3	3	1	1·5	1 + 1 + 0 + 0 + 0 + 0 = 2	1	1 + 0 + 0 + 0 + 0 + 0 = 1
4	4	1	2	1 + 1 + 1 + 0 + 0 + 0 = 3	1·3	1 + 2 + 0 + 0 + 0 + 0 = 3
5	5	1	2·5	1 + 1 + 1 + 1 + 0 + 0 = 4	1·6	1 + 2 + 3 + 0 + 0 + 0 = 6
6	6	1	3	1 + 1 + 1 + 1 + 1 + 0 = 5	2	1 + 2 + 3 + 4 + 0 + 0 = 10
7			3·5	1 + 1 + 1 + 1 + 1 + 1 = 6	2·3	1 + 2 + 3 + 4 + 5 + 0 = 15
8			4	0 + 1 + 1 + 1 + 1 + 1 = 5	2·6	1 + 2 + 3 + 4 + 5 + 6 = 21
9			4·5	0 + 0 + 1 + 1 + 1 + 1 = 4	3	2 + 3 + 4 + 5 + 6 + 5 = 25
10			5	0 + 0 + 0 + 1 + 1 + 1 = 3	3·3	3 + 4 + 5 + 6 + 5 + 4 = 27
11			5·5	0 + 0 + 0 + 0 + 1 + 1 = 2	3·6	4 + 5 + 6 + 5 + 4 + 3 = 27
12			6	0 + 0 + 0 + 0 + 0 + 1 = 1	4	5 + 6 + 5 + 4 + 3 + 2 = 25
13					4·3	6 + 5 + 4 + 3 + 2 + 1 = 21
14					4·6	5 + 4 + 3 + 2 + 1 + 0 = 15
15					5	4 + 3 + 2 + 1 + 0 + 0 = 10
16					5·3	3 + 2 + 1 + 0 + 0 + 0 = 6
17					5·6	2 + 1 + 0 + 0 + 0 + 0 = 3
18					6	1 + 0 + 0 + 0 + 0 + 0 = 1

The corresponding rule for a die of n faces would be

$$y_{x(r)} = \frac{1}{n} \sum_{x-n}^{x-1} y_{p(r-1)} \quad . \quad . \quad . \quad . \quad . \quad . \quad \text{(ii)}$$

Starting with the data for the 6-fold sample in Table 5, by successive application of this rule we can quickly build up the distribution of the mean score for samples of 7, 8, 9, and so on indefinitely. The picture which emerges is one of a succession of histograms (Fig. 63) more and more closely approaching the normal type. The number of terms and the range of the distribution of the score sum and mean score of an r-fold sample from the 6-fold rectangular universe of the cubical die is evident from the following

Sample of	No. of Terms	Range of Sum	Range of Mean
1	6 = 6 − 0	1 . . . 6	1 . . . 6
2	11 = 2(6) − 1	2 . . . 12	1 . . . 6
3	16 = 3(6) − 2	3 . . . 18	1 . . . 6
4	21 = 4(6) − 3	4 . . . 24	1 . . . 6
5	26 = 5(6) − 4	5 . . . 30	1 . . . 6
6	31 = 6(6) − 5	6 . . . 36	1 . . . 6

The rule for the range of the score sum is r to $6r$, and the rule for the number of terms is

$$t_r = r(6) - (r - 1) = 5r + 1.$$

TABLE 5

Mean Score Distributions of the Cubical Die

Score Sum (x)	Mean Score (x÷4)	4-Fold Samples. Relative Frequency (y × 6^4)	Mean Score (x÷5)	5-Fold Samples. Relative Frequency (y × 6^5)	Mean Score (x÷6)	6-Fold Samples. Relative Frequency (y × 6^6)
3						
4	1	1 = 1				
5	1·25	1 + 3 = 4	1	1 = 1		
6	1·5	1 + 3 + 6 = 10	1·2	1 + 4 = 5	1	1 = 1
7	1·75	1 + 3 + 6 + 10 = 20	1·4	1 + 4 + 10 = 15	1·16	1 + 5 = 6
8	2	1 + 3 + 6 + 10 + 15 = 35	1·6	1 + 4 + 10 + 20 = 35	1·3	1 + 5 + 15 = 21
9	2·25	1 + 3 + 6 + 10 + 15 + 21 = 56	1·8	1 + 4 + 10 + 20 + 35 = 70	1·5	1 + 5 + 15 + 35 = 56
10	2·5	3 + 6 + 10 + 15 + 21 + 25 = 80	2	1 + 4 + 10 + 20 + 35 + 56 = 126	1·6	1 + 5 + 15 + 35 + 70 = 126
11	2·75	6 + 10 + 15 + 21 + 25 + 27 = 104	2·2	4 + 10 + 20 + 35 + 56 + 80 = 205	1·83	1 + 5 + 15 + 35 + 70 + 126 = 252
12	3	10 + 15 + 21 + 25 + 27 + 27 = 125	2·4	10 + 20 + 35 + 56 + 80 + 104 = 305	2	5 + 15 + 35 + 70 + 126 + 205 = 456
13	3·25	15 + 21 + 25 + 27 + 27 + 25 = 140	2·6	20 + 35 + 56 + 80 + 104 + 125 = 420	2·16	15 + 35 + 70 + 126 + 205 + 305 = 756
14	3·5	21 + 25 + 27 + 27 + 25 + 21 = 146	2·8	35 + 56 + 80 + 104 + 125 + 140 = 540	2·3	35 + 70 + 126 + 205 + 305 + 420 = 1161
15	3·75	25 + 27 + 27 + 25 + 21 + 15 = 140	3	56 + 80 + 104 + 125 + 140 + 146 = 651	2·5	70 + 126 + 205 + 305 + 420 + 540 = 1666
16	4	27 + 27 + 25 + 21 + 15 + 10 = 125	3·2	80 + 104 + 125 + 140 + 146 + 140 = 735	2·6	126 + 205 + 305 + 420 + 540 + 651 = 2247
17	4·25	27 + 25 + 21 + 15 + 10 + 6 = 104	3·4	104 + 125 + 140 + 146 + 140 + 125 = 780	2·83	205 + 305 + 420 + 540 + 651 + 735 = 2856
18	4·5	25 + 21 + 15 + 10 + 6 + 3 = 80	3·6	125 + 140 + 146 + 140 + 125 + 104 = 780	3	305 + 420 + 540 + 651 + 735 + 780 = 3431
19	4·75	21 + 15 + 10 + 6 + 3 + 1 = 56	3·8	140 + 146 + 140 + 125 + 104 + 80 = 735	3·16	420 + 540 + 651 + 735 + 780 + 780 = 3906
20	5	15 + 10 + 6 + 3 + 1 = 35	4	146 + 140 + 125 + 104 + 80 + 56 = 651	3·3	540 + 651 + 735 + 780 + 780 + 735 = 4221
21	5·25	10 + 6 + 3 + 1 = 20	4·2	140 + 125 + 104 + 80 + 56 + 35 = 540	3·5	651 + 735 + 780 + 780 + 735 + 651 = 4332
22	5·5	6 + 3 + 1 = 10	4·4	125 + 104 + 80 + 56 + 35 + 20 = 420	3·6	735 + 780 + 780 + 735 + 651 + 540 = 4221
23	5·75	3 + 1 = 4	4·6	104 + 80 + 56 + 35 + 20 + 10 = 305	3·83	780 + 780 + 735 + 651 + 540 + 420 = 3906
24	6	1 = 1	4·8	80 + 56 + 35 + 20 + 10 + 4 = 205	4	780 + 735 + 651 + 540 + 420 + 305 = 3431
25			5	56 + 35 + 20 + 10 + 4 + 1 = 126	4·16	735 + 651 + 540 + 420 + 305 + 205 = 2856
26			5·2	35 + 20 + 10 + 4 + 1 = 70	4·3	651 + 540 + 420 + 305 + 205 + 126 = 2247
27			5·4	20 + 10 + 4 + 1 = 35	4·5	540 + 420 + 305 + 205 + 126 + 70 = 1666
28			5·6	10 + 4 + 1 = 15	4·6	420 + 305 + 205 + 126 + 70 + 35 = 1161
29			5·8	4 + 1 = 5	4·83	305 + 205 + 126 + 70 + 35 + 15 = 756
30			6	1 = 1	5	205 + 126 + 70 + 35 + 15 + 5 = 456
31					5·16	126 + 70 + 35 + 15 + 5 + 1 = 252
32					5·3	70 + 35 + 15 + 5 + 1 = 126
33					5·5	35 + 15 + 5 + 1 = 56
34					5·6	15 + 5 + 1 = 21
35					5·83	5 + 1 = 6
36					6	1 = 1

More generally for an n-fold universe the range of the mean is from 1 to n, the range of the score sum is from n to rn and the number of terms is given by

$$t_r = (n-1)r + 1.$$

A formal proof that the mean score distribution of a large sample from a rectangular universe is approximately normal, and hence that mean difference distributions w.r.t. large samples are also approximately normal would necessitate an algebraic excursion which we may defer to Vol. II ; but the matter is susceptible of investigation by an empirical method, as we shall now see.

To determine the normal curve which gives a good fit to a distribution it is necessary to know the numerical values of only two of its parameters, namely the mean and the variance. Hence we can ascertain whether a normal curve gives a good fit for the distribution of the score mean of r-fold samples if we know (M) the mean value of the r-fold sample mean scores and $V(M)$ the variance of their distribution. The former is simply the mean of the unit sampling distribution. For the cubical die, this is $3\cdot5 = \frac{1}{2}(6+1)$; and in general for the rectangular universe of n items it is $\frac{1}{2}(n+1)$, when the series range from 1 to n. If choice is random, the variance of the mean score of samples of r items from *any* universe has a simple relation to the variance of the unit sampling distribution, and is therefore deducible for the rectangular universe as a particular case. To exhibit this relation we shall employ the following symbols :

$V(u)$ for the variance of the *unit sampling* distribution.

$V_r(s)$ for the variance of the distribution of the score sum w.r.t. r-fold samples.

$V_r(M)$ for the variance of the distribution of the mean score w.r.t. r-fold samples.

$M(s)$ for the mean score sum of an r-fold sample.

Now the variance of the distribution of a score sum or a score difference w.r.t. pairs of samples is the sum of the variances of the two sample distributions (p. 175), so that

$$V_2(s) = V(u) + V(u) = 2V(u) ;$$
$$V_4(s) = V_2(s) + V_2(s) = 2V_2(s) = 4V(u).$$

And in general

$$V_r(s) = rV(u) \qquad . \qquad . \qquad . \qquad . \qquad . \qquad . \qquad (iii)$$

By definition :

$$M(s) = rM \qquad . \qquad . \qquad . \qquad . \qquad . \qquad . \qquad . \qquad (iv)$$

$$V_r(M) = \Sigma y \left(\frac{s}{r}\right)^2 - M^2 = \frac{1}{r^2}[\Sigma y . s^2 - r^2 M^2] \qquad . \qquad . \qquad (v)$$

$$V_r(s) = \Sigma y . s^2 - r^2 M^2 \qquad . \qquad . \qquad . \qquad . \qquad . \qquad (vi)$$

From (v) and (vi) :

$$V_r(M) = \frac{1}{r^2} . V_r(s).$$

Hence from (iii) :

$$V_r(M) = \frac{1}{r} V(u) \qquad . \qquad . \qquad . \qquad . \qquad . \qquad (vii)$$

The last formula is of fundamental importance. It remains to evaluate $V(u)$ for the particular case under consideration. If the individual scores of the n items each of frequency $\frac{1}{n}$ in a rectangular universe are 1 to n,

19

$$V(u) = \frac{1}{n}\sum_1^n x^2 - M^2$$
$$= \frac{(n+1)(2n+1)}{6} - \frac{(n+1)^2}{4} = \frac{(n^2-1)}{12} \qquad \cdot \quad \cdot \quad \cdot \quad \text{(viii)}$$

Hence from (vii) and (viii) above the variance of the mean score distribution is given by

$$V_r(M) = \frac{n^2-1}{12r} \qquad \cdot \quad \cdot \quad \cdot \quad \cdot \quad \cdot \quad \text{(ix)}$$

For 2-fold samples of a 6-fold universe, i.e. double tosses of a die, (ix) gives $(35 \div 24) \simeq 1 \cdot 46$. We may check this by recourse to Table 1 :

$$V_2(M) + (3 \cdot 5)^2 = \tfrac{1}{36}\{1(1)^2 + 2(1 \cdot 5)^2 + 3(2)^2 + 4(2 \cdot 5)^2 + 5(3)^2$$
$$+ 6(3 \cdot 5)^2 + 5(4)^2 + 4(4 \cdot 5)^2 + 3(5)^2 + 2(5 \cdot 5)^2 + 1(6)^2\},$$
$$\therefore\ V_2(M) + 12 \cdot 25 \simeq 13 \cdot 71,$$
$$\therefore\ V_2(M) \simeq 13 \cdot 71 - 12 \cdot 25 = 1 \cdot 46.$$

By the same token the variance of the distribution of the mean score of 6-fold tosses of an ordinary cubical die will be $35 \div 72$, so that we may write as the s.d. of the distribution $\sqrt{35 \div 72} \simeq 0 \cdot 697$. For purposes of reference to the table of the normal integral, the *standard* score corresponding to a mean score of 3 is therefore $3 \div 0 \cdot 697$. Table 5 sets out the exact distribution of the mean score of the 6-fold toss and the student should now be able to make a table by the procedure employed already in 3·04–3·05 and 5·07 setting forth the exact *cumulative* values of the mean standardised scores and the corresponding area of the normal integral, as given in the tables, with due regard to the half-interval correction. It is surprising to find how closely the normal law describes the distribution of the mean score of even small samples from a universe whose definitive unit sample distribution is itself far from normal.

EXERCISE 7.03

1. The four faces of a tetrahedral die respectively carry 1, 2, 3 and 4 pips. Determine the exact sampling distribution of the difference between the mean score of pairs of 2-fold tosses.

2. For the same die, make a table like Table 5 to show the mean score distribution of 2-fold, 3-fold . . . 6-fold tosses.

3. By repeated use of equation (ii) of 7.03 determine the distribution of the mean score of a 10-fold toss.

4. Test the approach to normality of the mean score distribution w.r.t. 4-, 6-, 8- and 10-fold tosses.

7.04 RANK SCORE MEAN DIFFERENCE

Sampling from the rectangular universe of the cubical die is *de facto* (Chapter 2, p. 59) sampling *with replacement*. An appropriate model of sampling without replacement from the rectangular universe is the simultaneous extraction of a sample of given size from a pack of cards numbered consecutively. For instance, we may draw two or more cards at a time from a pack made up of the ace, 2, 3, 4, 5 and 6 of clubs, as in Figs. 64-65.

Whereas the die model has no particular relevance to practical statistics, the corresponding card pack model is the prototype of sampling when the method of scoring is *ordinal*. If we can grade the responses of a group of individuals to one or other of two methods of treatment, we are at liberty to explore the difference between the mean grade (i.e. *rank score*) of the two sub-samples as a criterion of the relative efficacy of the two procedures. If each of n individuals is distinguishable by order of merit and hence has a unique rank, each ordinal score has the same frequency (n^{-1}), and the unit sampling distribution is necessarily rectangular. By the same token, the fact that we are able to assign a rank x to a particular individual of one sub-sample signifies that no individual of the alternative sub-sample can have the rank x. In picking from a pack of cards of only one suit a card which has x pips, we have in fact deprived ourselves of the possibility of finding such a card in the residual pack. Thus rank scoring of two sub-samples signifies sampling from the rectangular universe *without* replacement.

Situations which admit of comparison by recourse to the rank score mean difference are often such as are also amenable to treatment by the Spearman method (Chapter 8), which, as we shall see, is more *efficient*, i.e. sensitive. For instance, we might set out in the following way the data of a fictitious example cited later in 8.02 :

	Scripture.		Pocket Money.	
	Marks.	Rank.	Allowance.	Rank.
A. Upper half of class.				
	95	1	2d.	6
	80	2	3d.	5
	60	3	6d.	4
			Mean of rank score	5·0
B. Lower half of class.				
	55	4	9d.	3
	45	5	1/-	2
	40	6	2/6	1
			Mean of rank score	2·0

This disposition of the data divides them into two w.r.t. success in attaining a certain mark (60) for the examination in scriptural knowledge. We are then able to see whether the mean rank score for pocket allowance of Group A is greater or less than the mean rank score for pocket money of Group B. Without making any assumptions about the nature of the connexion (see 8.01) between schoolboy finance and proficiency in Bible lore, we expect to find such a difference ; and we shall be satisfied to regard its existence as indicative of such a connexion, only if we have the assurance that a comparable result would very rarely occur in a situation consistent with equipartition of opportunity for association.

We thus have before us two sets of figures consisting of the first six integers corresponding to the number of pips on the six cards of our club-suit pack model. One set of figures merely serves to classify our data in two groups of 3 items. The question which we have to ask ourselves concerns only the mean values of the alternative scores of the two groups so distinguished. For our present purpose, we may regard them as 3-fold samples from our card-pack model ; and the likelihood of the observed result is the likelihood of getting a mean score difference numerically not less than $5·0 - 2·0 = 3$ as a result of dividing the pack of six clubs into two equal sub-samples.

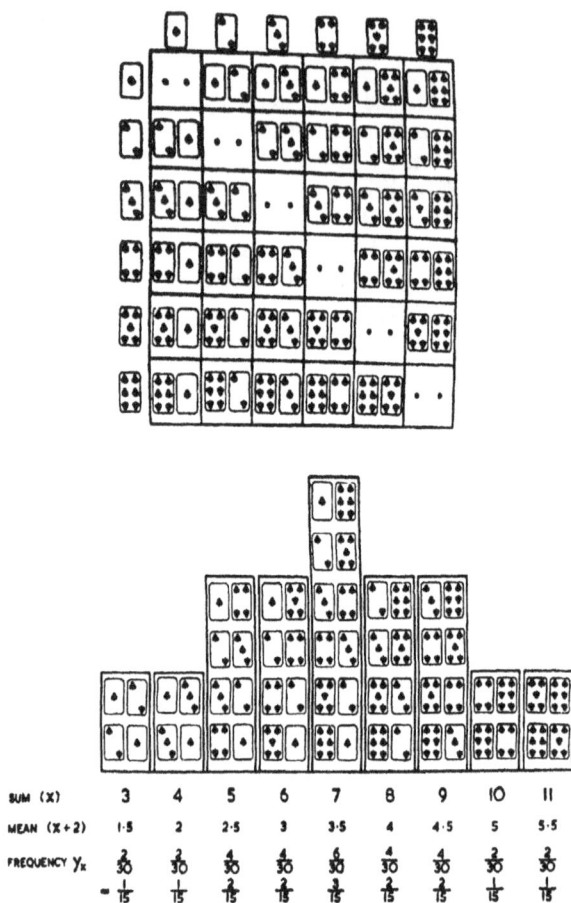

SUM (X)	3	4	5	6	7	8	9	10	11
MEAN (X÷2)	1·5	2	2·5	3	3·5	4	4·5	5	5·5
FREQUENCY y_x	$\frac{2}{30}$	$\frac{2}{30}$	$\frac{4}{30}$	$\frac{6}{30}$	$\frac{6}{30}$	$\frac{6}{30}$	$\frac{4}{30}$	$\frac{2}{30}$	$\frac{2}{30}$
	$=\frac{1}{15}$	$\frac{1}{15}$	$\frac{2}{15}$	$\frac{3}{15}$	$\frac{3}{15}$	$\frac{3}{15}$	$\frac{2}{15}$	$\frac{1}{15}$	$\frac{1}{15}$

FIG. 64. Sampling from a Rectangular Universe of 6 classes *without* replacement. Derivation of the mean score and score sum distribution for the 2-fold sample.

Fig. 64 shows the derivation of the score sum for 2-fold samples of the card pack model from the rectangular unit sampling distribution, and hence the corresponding mean score distribution. The next one (Fig. 65) shows the derivation of the mean score distribution for a 3-fold draw. This provides us with all the necessary data for deducing the distribution of the mean score difference w.r.t. pairs of 3-fold samples. To do so we cannot rely on the chessboard lay-out of Table 2 in 7.03, because the structure of a second sample is not independent of the first, when choice is restrictive. It is therefore necessary to take into account the composition of the residual universe resulting from the removal of a particular sample; and we can summarise the operations involved in the appropriate staircase model as in Table 6 which is self-explanatory.

From Table 6 we see that a mean score difference as great as ± 3 pertains to 12 ($= 6 + 6$) out of 120 possible paired score sequences of three cards. Hence the expectation of a mean score difference numerically as great as ± 3 is 0·1; and the odds are only 9 : 1 against the

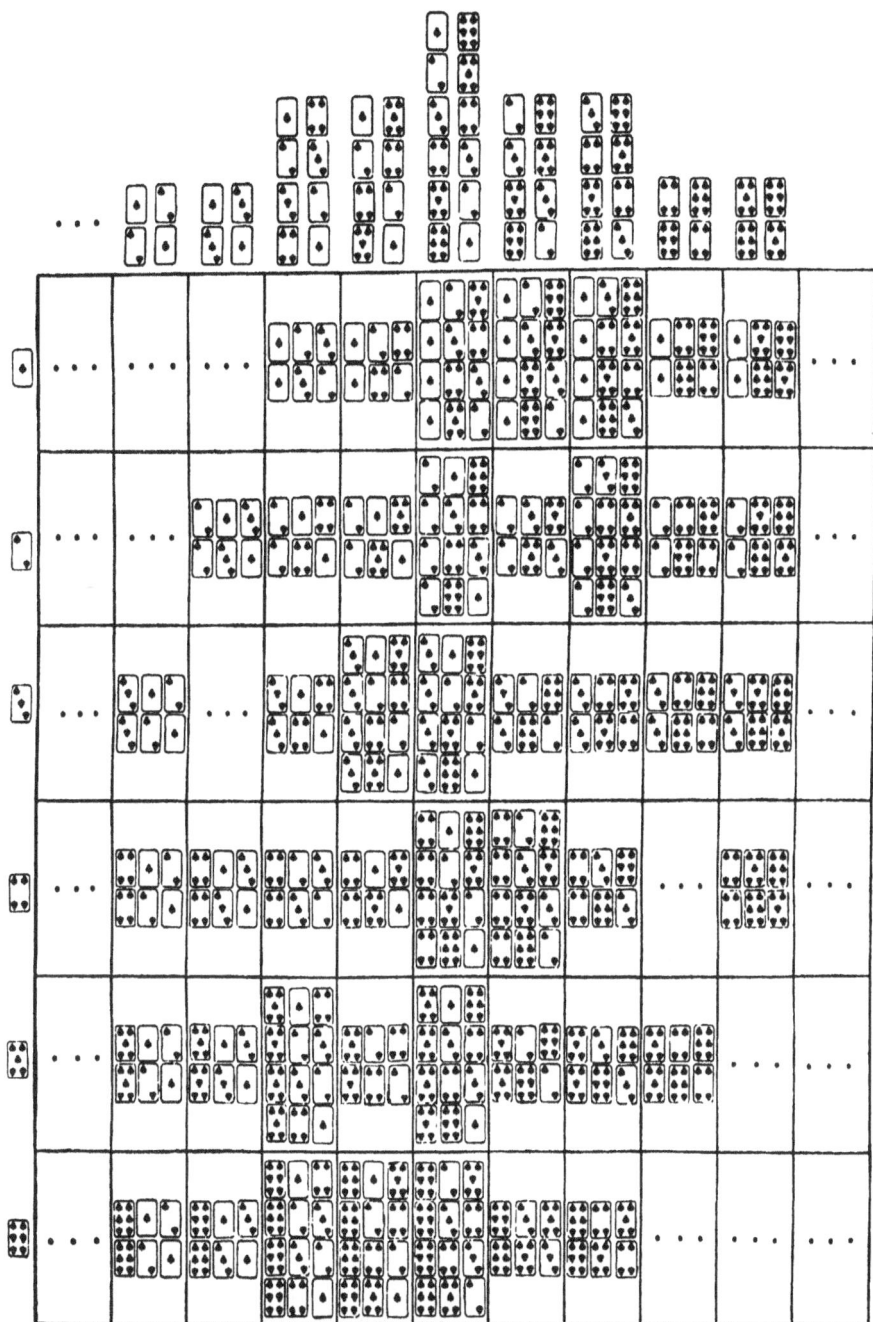

FIG. 65. Sampling from a Rectangular Universe of 6 classes *without* replacement. The build-up of the 3-fold sample from the distribution of Fig. 64.

occurrence. If we took into account all the information at our disposal the likelihood of the result would be far less. There is in fact a one-to-one reverse order correspondence of our two sets of 6 rank scores ; and only one out of 6! ways of arranging one set can correspond exactly to any particular way of arranging the other. Accordingly, the likelihood which the Spearman method (8.02 *below*) assigns to the event is $1 \div 6!$, or odds of 719 : 1 against.

TABLE 6

Non-Replacement Distribution of Mean Score Difference w.r.t. 3-fold Samples of the Rectangular Universe of 6 Items

Initial Sample.			Residual Sample.			Mean Score Difference.	
Mean Score.	Combinations.	No. of Permutations.	Mean Score.	Combinations.	No. of Permutations.	*d.*	No. of Permutations.
2·0	123	6	5·0	456	6	3·0	6
2·3	124	6	4·6	356	6	2·3	6
2·6	125, 134	12	4·3	346, 256	12	1·6	12
3·0	126, 135, 234	18	4·0	345, 246, 156	18	1·0	18
3·3	136, 145, 235	18	3·6	245, 236, 146	18	0·3	18
3·6	146, 236, 245	18	3·3	235, 145, 136	18	− 0·3	18
4·0	156, 246, 345	18	3·0	234, 135, 126	18	− 1·0	18
4·3	256, 346	12	2·6	134, 125	12	− 1·6	12
4·6	356	6	2·3	124	6	− 2·3	6
5·0	456	6	2·0	123	6	− 3·0	6

This discrepancy between the likelihoods which the Spearman and the difference method respectively assign to the event again (p. 217) brings into focus what statisticians mean when they distinguish between more or less *efficient* statistics. One index or statistic, e.g. Spearman's coefficient, is more efficient than another, e.g. the mean rank score difference of the last example, if it gives more weight to relevant information contained in the data. To say that the method which the last example illustrates is, in fact, an inefficient statistic for the reason stated above is not equivalent to saying that it is fallacious. It merely signifies that it fails to take stock of every circumstance which defines what is unique about an observation ; and in failing to do so assigns to it a higher likelihood than a method which takes stock of other information, in this case the fact that the two sets of rank scores correspond exactly in reverse order. Thus a less efficient statistic may make us hesitant to reject a null hypothesis, when an efficient statistic would encourage us to do so ; but the converse is not true. Inasmuch as a null hypothesis is in general the conservative alternative, an inefficient statistic errs on the side of conservative judgment.

7.05 Sampling in a Normal Universe

By recourse to a familiar model, we have now clarified the meaning of a unit sampling distribution, the distribution of a score sum, the distribution of a mean score and the distribution of the difference between the mean score of pairs of samples. To determine each of the distributions last named, we have to start out with some information about the universe itself. When the method of scoring is representative we are rarely in a position to postulate the dis-

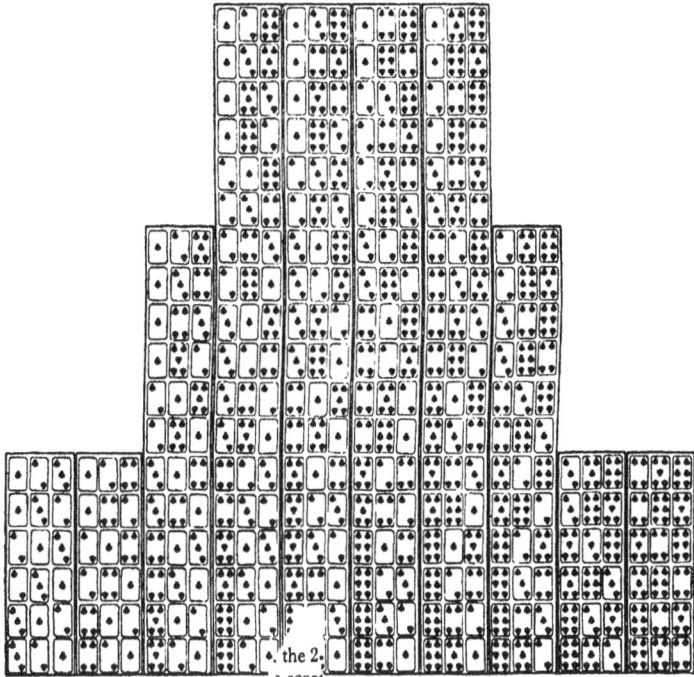

SUM (X)	6	7	8	9	10	11	12	13	14	15
MEAN (X ÷ 3)	2	2·3	2·6	3	3·3	3·6	4	4·3	4·6	5
FREQUENCY (y_x)	$\frac{6}{120}$	$\frac{6}{120}$	$\frac{12}{120}$	$\frac{18}{120}$	$\frac{18}{120}$	$\frac{18}{120}$	$\frac{18}{120}$	$\frac{12}{120}$	$\frac{6}{120}$	$\frac{6}{120}$
=	$\frac{1}{20}$	$\frac{1}{20}$	$\frac{2}{20}$	$\frac{3}{20}$	$\frac{3}{20}$	$\frac{3}{20}$	$\frac{3}{20}$	$\frac{2}{20}$	$\frac{1}{20}$	$\frac{1}{20}$

Fic. 66. Sampling from a Rectangular Universe of 6 classes *without* replacement. Histogram of the mean score and score sum of the 3-fold sample based on the chessboard of Fig. 65.

tribution law of the unit sample without recourse to some empirical information. A histogram or other graph which exhibits the relative frequency of individuals w.r.t. a particular individual score, e.g. heights, weights, incomes, in a large population provides such information, since the electivity of a class in a sample of one individual is by definition its class frequency in the parent universe. Since it often happens that the contour of such a unit sampling distribution, i.e. class frequency distribution of a large population, coincides closely with a normal curve or with a skew type of curve related to the normal, the distribution of means and mean differences of samples from a normal universe has a peculiar interest.

It is not easy to visualise a model with the algebraic properties of a normal universe ; but we now know that the binomial series is the parent of a large family of functions of which the normal is a limiting case. If we can construct a model with a binomial unit-sampling distribution, we have therefore the basis for a more catholic approach to the task of defining the sampling distribution of representative scores than the assumption of strict normality permits. Sampling from a binomial universe in this context signifies sampling from a universe of which :

(a) each item has one of $(n+1)$ scores making up a set which it is possible to arrange consecutively with equal increments and hence, by change of scale, as the numerical sequence S, $(S+1)$, $(S+2)$ . . . $(S+n)$;

(b) subject to the usual convention $(p+q)=1$, successive frequencies of scores S, $(S+1)$ . . . $(S+n)$ then tally with successive terms of the expansion of $(p+q)^n$, so that the frequency of a score $(S+x)$ is

$$\frac{n!}{x!\,(n-x)!}p^x q^{n-x}.$$

It is not difficult to visualise sampling from an indefinitely large universe so defined, as illustrated by the derivation (Figs. 67–69) of the mean score difference distribution when the unit sampling distribution of the die, e.g. a flat circular disc of two faces only, is given by the terms of the binomial $(\frac{1}{2}+\frac{1}{2})^1$. To explore the implications of sampling from an infinite universe containing more than two classes our model will be a tetrahedral die (Figs. 70–72) of which two faces respectively carry 1 and 3 pips, the other two faces each 2 pips. Thus the unit sampling distribution is given by the terms of $(\frac{1}{2}+\frac{1}{2})^2$ in accordance with the following schema :

Score	.	.	.	1	2	3
Frequency	.	.	.	$\frac{1}{4}$	$\frac{1}{2}$	$\frac{1}{4}$

By successive application of the chessboard we obtain the following results :

(a) 2-fold samples

Score sum	.	.	2	3	4	5	6
Mean score	.	.	1	1·5	2	2·5	3
Frequency	.	.	$\frac{1}{16}$	$\frac{4}{16}$	$\frac{6}{16}$	$\frac{4}{16}$	$\frac{1}{16}$

(b) 3-fold samples

Score sum	.	.	3	4	5	6	7	8	9
Mean score	.	.	1	1·3	1·6	2	2·3	2·6	3
Frequency	.	.	$\frac{1}{64}$	$\frac{6}{64}$	$\frac{15}{64}$	$\frac{20}{64}$	$\frac{15}{64}$	$\frac{6}{64}$	$\frac{1}{64}$

Thus the frequency distributions for 2-fold and 3-fold samples are respectively terms of the expansion of $(\frac{1}{2}+\frac{1}{2})^4$ and $(\frac{1}{2}+\frac{1}{2})^6$. This suggests the following rule : if the unit sampling score distribution corresponds to successive terms of the expansion of $(p+q)^n$, the distribution of the mean score of an r-fold sample corresponds to successive terms of the expansion of $(p+q)^{rn}$. That this is true when $n=1$ we have already seen by recourse to the flat circular die of Figs. 67–69, the model for our universe which is both rectangular and binomial. That the rule holds good when p and q are unequal, we can test by recourse to a tetrahedral die (Fig. 73) with one face bearing 1 pip and three faces bearing 2 pips. The unit sampling distribution law is then given by $(\frac{1}{4}+\frac{3}{4})^1$. For 2-fold samples we have

Score sum	.	.	2	3	4
Mean score	.	.	1	1·5	2
Frequency	.	.	$\frac{1}{16}$	$\frac{6}{16}$	$\frac{9}{16}$
			$(\frac{1}{4})^2$	$2(\frac{1}{4})(\frac{3}{4})$	$(\frac{3}{4})^2$

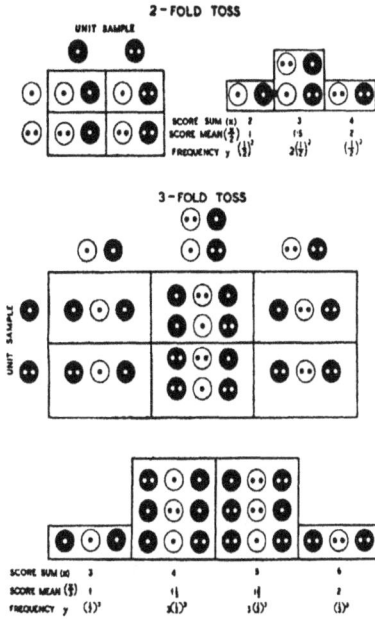

FIG. 67. The Binomial and Rectangular Universe of the 2-face die. The unit sampling distribution is given by the terms of the binomial $(\frac{1}{2} + \frac{1}{2})^1$, that of the mean score or score sum of the 2-fold toss by the terms of the binomial $(\frac{1}{2} + \frac{1}{2})^2$ and that of the mean score and score sum of the 3-fold toss by the terms of the binomial $(\frac{1}{2} + \frac{1}{2})^3$.

In more general terms than the foregoing, we may state a theorem suggested by the foregoing examples as follows :

If the binomial $(\mathbf{p} + \mathbf{q})^a$ *is definitive of an* s-*fold sample score from an indefinitely large universe and that of a* t-*fold sample score is* $(\mathbf{p} + \mathbf{q})^b$, *the distribution of the sum of the score of an* s-*fold and of a* t-*fold sample, i.e. that of the score sum of the* (s + t)-*fold sample, tallies with successive terms of the binomial* $(\mathbf{p} + \mathbf{q})^{a+b}$.

It is implicit in the statement of this theorem that

(i) the scores increase by equal steps and hence by unit steps after appropriate change of scale ;

(ii) unless $a = b$, the score range $(A$ to $\overline{A + a})$ of the s-fold sample will not be the same as the score range $(B$ to $\overline{B + b})$ of the t-fold sample.

With due regard to these considerations, we can derive the theorem stated above from a chessboard lay-out in conformity with the multiplicative and additive operations which respectively specify the electivities of concomitant independent and alternative events. The schema of Table 7 is a grid of this sort, exhibiting simultaneous extraction of

(i) any one of $(a + 1)$ different s-fold samples from a universe with sample scores A, $(A + 1), \ldots (A + a)$;

(ii) any one of $(b + 1)$ different t-fold samples from the same universe with sample scores $B, (B + 1) \ldots (B + b)$.

4 - FOLD TOSS

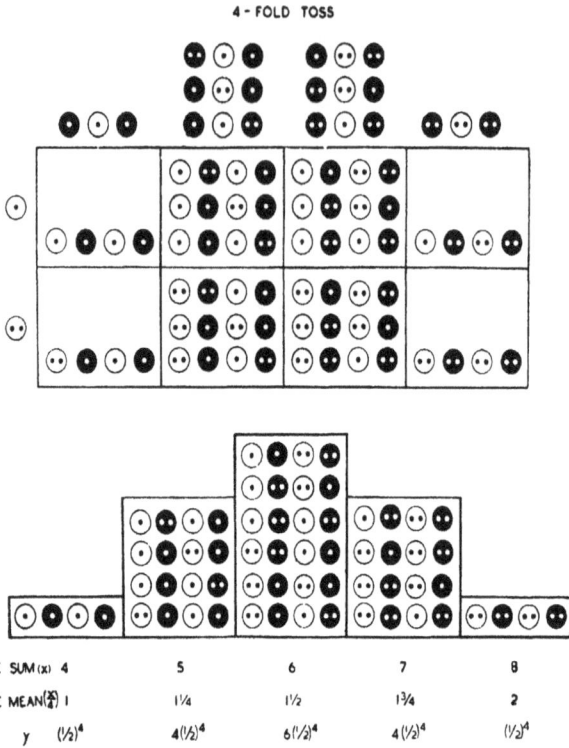

SCORE SUM (x)	4	5	6	7	8
SCORE MEAN ($\frac{x}{4}$)	1	1¼	1½	1¾	2
y	$(\frac{1}{2})^4$	$4(\frac{1}{2})^4$	$6(\frac{1}{2})^4$	$4(\frac{1}{2})^4$	$(\frac{1}{2})^4$

FIG. 68. The build-up of the 4-fold toss mean score and score sum distribution $(\frac{1}{2} + \frac{1}{2})^4$ for the die model of Fig. 67 by combining the results of a single and a 3-fold trial.

TABLE 7

$(A + B) = C; \quad (a + b) = c$

Frequency		$a_{(0)}q^a$ A	$a_{(1)}q^{a-1}p$ $A + 1$	$a_{(2)}q^{a-2}p^2$ $A + 2$	$a_{(3)}q^{a-3}p^3$ $A + 3$	$a_{(4)}q^{a-4}p^4 \ldots$ $A + 4$
	Score					
$b_{(0)}q^b$	B	$a_{(0)}b_{(0)}q^c$ C	$a_{(1)}b_{(0)}q^{c-1}p$ $C + 1$	$a_{(2)}b_{(0)}q^{c-2}p^2$ $C + 2$	$a_{(3)}b_{(0)}q^{c-3}p^3$ $C + 3$	$a_{(4)}b_{(0)}q^{c-4}p^4 \ldots$ $C + 4$
$b_{(1)}q^{b-1}p$	$B + 1$	$a_{(0)}b_{(1)}q^{c-1}p$ $C + 1$	$a_{(1)}b_{(1)}q^{c-2}p^2$ $C + 2$	$a_{(2)}b_{(1)}q^{c-3}p^3$ $C + 3$	$a_{(3)}b_{(1)}q^{c-4}p^4$ $C + 4$	—
$b_{(2)}q^{b-2}p^2$	$B + 2$	$a_{(0)}b_{(2)}q^{c-2}p^2$ $C + 2$	$a_{(1)}b_{(2)}q^{c-3}p^3$ $C + 3$	$a_{(2)}b_{(2)}q^{c-4}p^4$ $C + 4$	—	—
$b_{(3)}q^{b-3}p^3$	$B + 3$	$a_{(0)}b_{(3)}q^{c-3}p^3$ $C + 3$	$a_{(1)}b_{(3)}q^{c-4}p^4$ $C + 4$	—	—	—
$b_{(4)}q^{b-4}p^4$	$B + 4$	$a_{(0)}b_{(4)}q^{c-4}p^4$ $C + 4$	—	—	—	—
—		—	—	—	—	—

If we put $(a + b) = c$, the range of the score sums is then from $(A + B) = C$ to $(A + B + c) = (C + c)$. A glance at the schema shows that the frequency of a score sum $(C + 4)$ is given by

$$(a_{(4)} + a_{(3)}b + a_{(2)}b_{(2)} + ab_{(3)} + b_{(4)}) \cdot p^4 q^{c-4}$$

$$= \sum_{x=0}^{x=4} a_{(x)} \cdot b_{(4-x)} \cdot p^4 q^{c-4}$$

$$= p^4 q^{c-4} \sum_{x=0}^{x=4} a_{(x)} \cdot b_{(4-x)}.$$

We can now recall Vandermonde's theorem, $viz.$

$$(a + b)^{(r)} = \sum_{x=0}^{x=r} \frac{r!}{x!(r-x)!} a^{(x)} b^{(r-x)},$$

$$\therefore \frac{(a + b)^{(r)}}{r!} = \sum_{x=0}^{x=r} \frac{a^{(x)}}{x!} \cdot \frac{b^{(r-x)}}{(r-x)!}$$

$$= \sum_{x=0}^{x=r} a_{(x)} b_{(r-x)}.$$

In virtue of Vandermonde's theorem, the frequency of a score sum $(C + 4)$ as defined above is therefore

$$\frac{(a + b)^{(4)}}{4!} \cdot p^4 q^{c-4} = \frac{c^{(4)}}{4!} \cdot p^4 q^{c-4}.$$

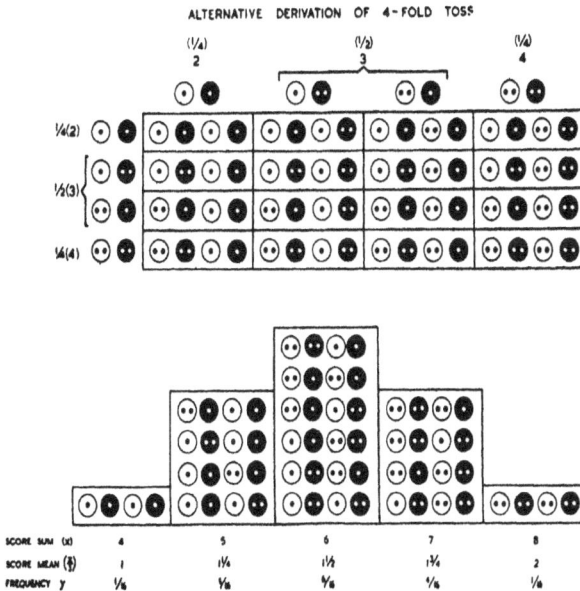

ALTERNATIVE DERIVATION OF 4-FOLD TOSS

FIG. 69. Alternative build-up of the 4-fold toss for the die model of Fig. 67 by combining the results of two 2-fold trials.

More generally, and by the same token, the frequency of a score sum $(C + x)$ is

$$\frac{(a + b)^{(x)}}{x!} \cdot p^x q^{b-x} = \frac{c^{(x)}}{x!} \cdot p^x q^{b-x}.$$

This is the general term of $(p+q)^c = (p+q)^{a+b}$, whence the sum of the raw scores of two unit samples, i.e. that of the score sum of a 2-fold sample, from an infinite binomial universe of $(n + 1)$ score classes is $(p + q)^{n+n} = (p + q)^{2n}$. Hence also the binomial $(p + q)^{2n+n} = (p + q)^{3n}$ defines the distribution of the sum of the scores of 2-fold and unit samples, i.e. that of the score sum of the 3-fold sample ; and more generally $(p + q)^{rn}$ is the binomial definitive of the distribution of the score-sum, hence also the mean score, of r-fold samples taken from a universe whose unit sampling distribution tallies with successive terms of the binomial $(p + q)^n$. By an elementary and now familiar property of the binomial distribution the variance of the score sum of the r-fold sample is therefore given by

$$V(S) = rnpq = r \cdot V(u) \qquad . \qquad . \qquad . \qquad . \qquad . \qquad \text{(ii)}$$

Now the mean score of an r-fold sample of which the corresponding score sum is S is given by $(S \div r)$. Hence we may write for the variance of the distribution of the mean score

$$V(M) = \sum y \left(\frac{S}{r}\right)^2 - \left[\sum y \frac{S}{r}\right]^2$$

$$= \frac{1}{r^2}\left\{\sum y \cdot S^2 - \left(\sum y \cdot S\right)^2\right\}.$$

$$\therefore \ V(M) = \frac{1}{r^2} V(S).$$

Hence from (ii)

$$V(M) = \frac{rV(u)}{r^2} = \frac{V(u)}{r} \qquad . \qquad . \qquad . \qquad . \qquad . \qquad . \qquad \text{(iii)}$$

This agrees with the result (vii) obtained above (7.03) by recourse to more general considerations. We thus arrive at the following conclusion w.r.t. sampling in a binomial universe : *if the unit sampling distribution of a score which increases by unit steps accords with the terms of a binomial distribution of variance* $V(u)$ *the distribution of the mean score of* r-*fold samples accords with the terms of a binomial distribution of variance* $V(u) \div r$. Now we have already shown that : (*a*) the normal curve tallies closely with the contour of the *binomial* histogram when r is large ; (*b*) that this is so, even when p and q are grossly unequal, if rp numerically exceeds a certain limit. As practical folks, we need not therefore soar into the empyrean of the mathematical continuum in quest of a rigorous demonstration that the sum of two normal variates is itself a normal variate, and hence that the distribution of the mean score of samples from a normal universe is itself normal. With due regard to the meaning of the parameter V, i.e. $V = rV(u)$ if x below is the score sum and $V = V(u) \div r$ if x is the mean score, we may thus describe the distribution of a score sum or score mean from an approximately normal universe by a function of which the *integrand* * has the normal form :

$$F(x) = \frac{1}{\sqrt{2\pi V}} \ exp \ \frac{-(x - M_x)^2}{2V}.$$

* See explanatory remarks at the beginning of 4.08 w.r.t. the appropriate scalar change for obtaining the corresponding *frequency* function definitive of the ordinate of the distribution in the interval $x \pm \frac{1}{2}\Delta x$.

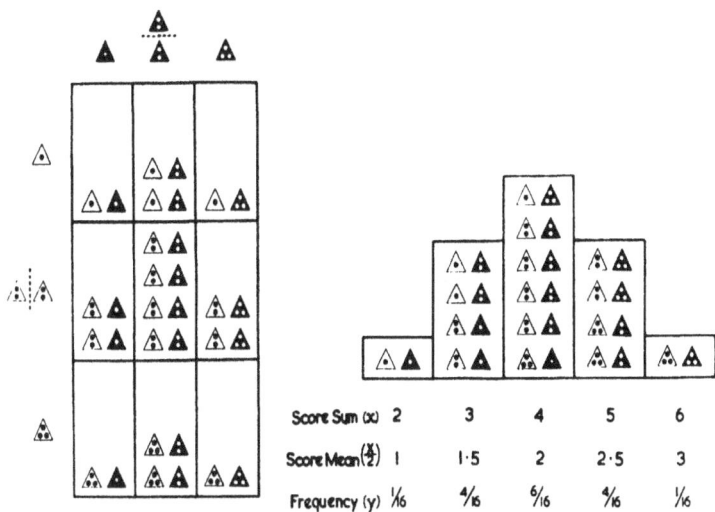

FIG. 70. Sampling from a Symmetrical Binomial Universe of more than two classes. The model is a tetrahedral die with 1 pip on one face, 2 pips on each of two faces, and 3 pips on the remaining face. The binomial $(\frac{1}{2} + \frac{1}{2})^2$ defines the unit sampling distribution. The terms of $(\frac{1}{2} + \frac{1}{2})^4$ define the distributions of the mean score and score sums for 2-fold tosses.

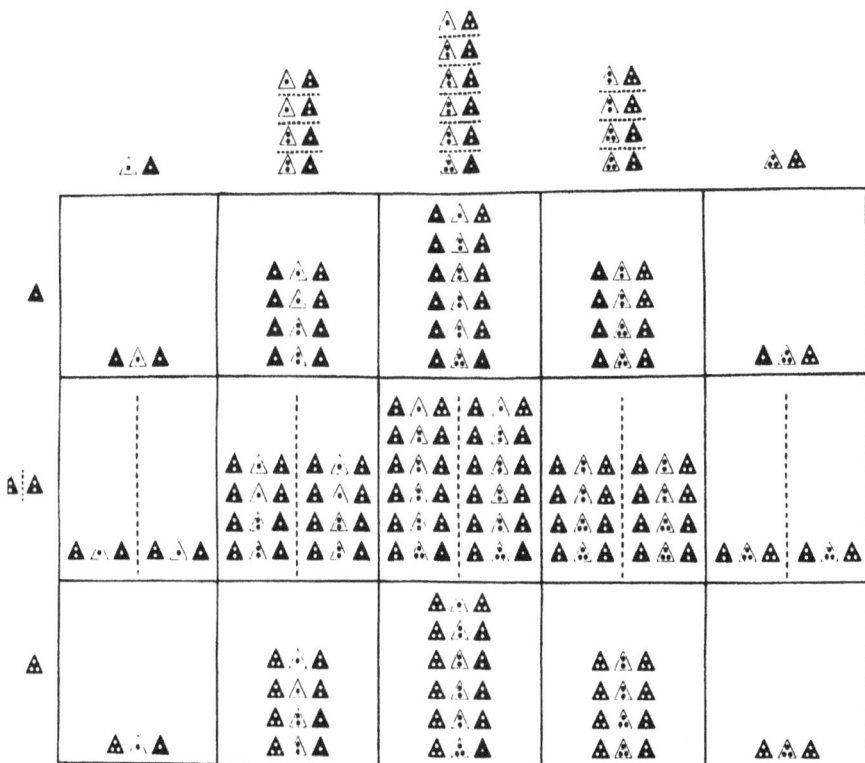

FIG. 71. Derivation of the mean score distribution for the 3-fold toss of the tetrahedral die of Fig. 70.

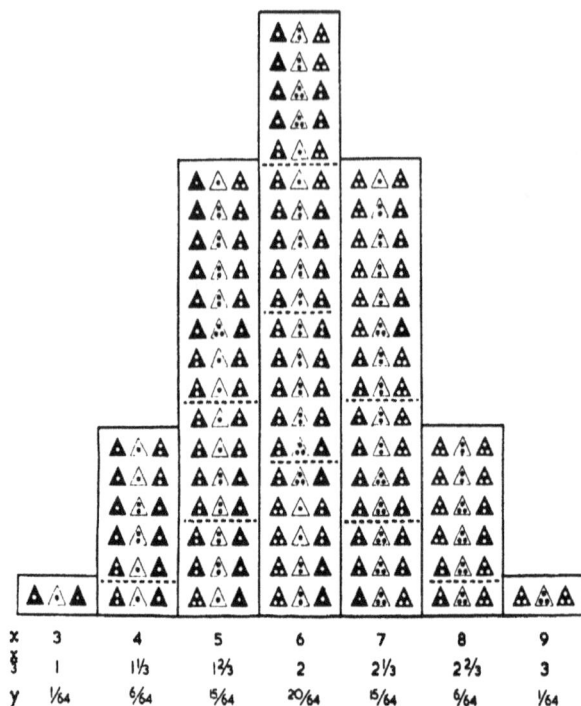

x	3	4	5	6	7	8	9
$\frac{x}{3}$	1	1⅓	1⅔	2	2⅓	2⅔	3
y	1/64	6/64	15/64	20/64	15/64	6/64	1/64

FIG. 72.   Histogram of the mean score and score sum distributions for the 3-fold toss of Fig. 71.

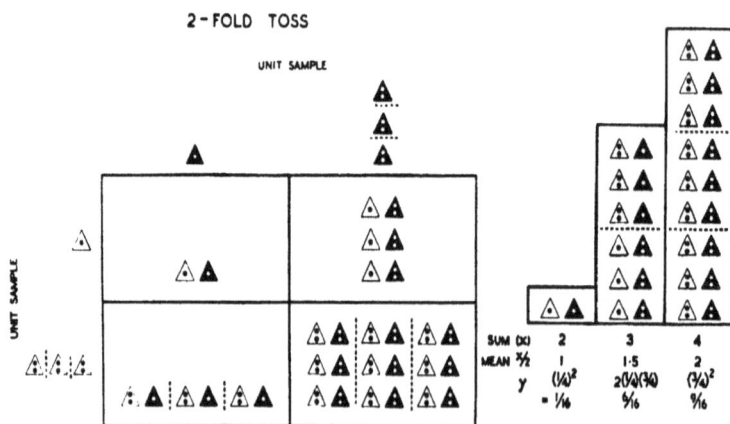

FIG. 73.   Sampling from a Skew Binomial Universe.   The model is a tetrahedral die of which one face carries 1 pip and the other three faces each carry 2 pips.   The unit-sampling distribution accords with the binomial $(\frac{1}{4} + \frac{3}{4})^1$, that of the mean score or score sum of the 2-fold toss with the terms of $(\frac{1}{4} + \frac{3}{4})^2$.

In what follows, we shall confine ourselves to its implications *vis-à-vis* sampling in the normal universe. In a sufficiently large r-fold sample of such a universe we may assume that the possible range of individual scores and their frequencies tally with those of the universe itself, i.e. with the unit sampling distribution. In so far as such a sample is truly representative, the variance of its observed frequency distribution is therefore approximately that of the unit sampling distribution of mean score M, i.e. $V(u)$. Accordingly, we may take as a *provisional* (see p. 304) estimate of the variance $V(u)$ definitive of the distribution of a score x in the parent universe

$$\frac{1}{r}\sum(x - M)^2.$$

Since the variance of the mean score distribution is $V(u) \div r$, our provisional estimate of it will therefore be

$$\frac{1}{r^2}\sum(x - M)^2.$$

Now the distribution of the difference between two normally distributed scores is itself normal. With due regard to the error inherent in any such provisional estimate, we therefore have at our disposal the desiderata for a c-test on a difference between the score means of two groups. Before proceeding to set forth the procedure in detail, one result of our examination of the properties of the tetrahedral die model deserves further comment. We have seen (3.04) that the contour of a skew binomial distribution approaches the normal as we increase the value of the exponent m in $(p + q)^m$. If the unit sampling distribution is a skew type which tallies closely with the binomial $(p + q)^m$, the mean score distribution of the r-fold sample defined by $(p + q)^{rm}$ will be necessarily more symmetrical and more like a normal distribution in virtue of the fact that rm is greater than m, being very much greater if r itself is large. From this consideration, we see that the distribution of the mean of large samples may be approximately normal, even if the unit sampling distribution is somewhat skew.

7.06 THE C-TEST FOR A REPRESENTATIVE SCORE DIFFERENCE

By and large, it is safe to assume that the distribution of scores in a very large sample will tally closely with that of the parent universe, and hence that the variance of the distribution of individual scores within the sample will rarely deviate greatly from the variance of the distribution of individual scores from the putative parent universe, i.e. from the unit sample distribution. If the mean score of every sample were in fact the same as that (M_u) of the parent universe, the mean variance of all possible samples would be the variance of the distribution of the individual scores (x_r). Actually, a particular r-fold sample has its own mean (M_s) and its variance (V_s) given by

$$V_s = \frac{\sum(x_r - M_s)^2}{r} \qquad . \qquad . \qquad . \qquad . \qquad . \qquad \text{(i)}$$

The summation here extends over every individual sample value of the score. If any particular sample score (x_r) occurs with frequency y_r, we may write this in the form

$$V_s = \sum y_r(x_r - M_s)^2 . \qquad . \qquad . \qquad . \qquad . \qquad \text{(ii)}$$

As we shall now see, the mean value of V_s in (i) above is in fact somewhat smaller than $V(u)$ Hence a mean score standardised by recourse to the relation $\sigma_m^2 = V_s \div r$ would be unduly large more often than otherwise. We can give due weight to this circumstance as in 4.10 by defining a statistic whose mean value is identical with the true variance ($\sigma_m^2 = V(u) \div r$) of the mean score distribution. For the sample statistic whose *mean value* is the variance of the score distribution in the parent universe, we shall here use the symbol V_t. Hence if y_s is the expectation of extracting one such sample

$$V(u) = \Sigma\, y_s\, .\, V_t = M(V_t). \qquad . \qquad . \qquad . \qquad . \qquad .\text{~(iii)}$$

Since the statistic which satisfies this criterion is the second moment of the score distribution about the true mean of the parent universe,

$$
\begin{aligned}
V_t &= \Sigma\, y_r\, (x_r - M_u)^2 \\
&= \Sigma\, y_r\, ([x_r - M_s]^2 + [M_s - M_u])^2 \\
&= \Sigma\, y_r\, (x_r - M_s)^2 + 2(M_s - M_u)\, \Sigma\, y_r\, (x_r - M_s) + (M_s - M_u)^2\, \Sigma\, y_x.
\end{aligned}
$$

Now the sum of all the deviations $(x_r - M_s)$ of the sample scores is zero. Hence by (ii)

$$V_t = V_s + (M_s - M_u)^2 \qquad . \qquad . \qquad . \qquad . \qquad .\text{~(iv)}$$

For all possible r-fold samples, we have in accordance with (iii) and (iv)

$$M(V_t) = \Sigma\, y_s V_s + \Sigma y_s\, (M_s - M_u)^2.$$

The last expression on the extreme right is the variance of the distribution of the mean score of all r-fold samples, so that

$$\Sigma\, y_s\, (M_s - M_u)^2 = V(M) = \frac{V(u)}{r}.$$

Hence from (iii)

$$V(u) = \Sigma\, y_s V_s + \frac{V(u)}{r},$$

$$\therefore \Sigma\, y_s V_s = \frac{r-1}{r}\, V(u).$$

Whence from (iii)

$$\Sigma\, y_s V_s = \frac{r-1}{r}\, M(V_t) = \frac{r-1}{r}\, \Sigma\, y_s V_t.$$

From (i) above, it is consistent with this identity to write

$$V_t = \frac{r\,.\,V_s}{r-1} = \frac{\Sigma\, (x_s - M_s)^2}{r-1} \qquad . \qquad . \qquad . \qquad .\text{~(v)}$$

Since V_t is the unbiassed estimate of $V(u)$ we have as our corresponding estimate (s^2) of the variance of the distribution of the mean score

$$s^2 = \frac{V_t}{r} = \frac{\Sigma\, (x_s - M_s)^2}{r(r-1)}$$

$$\therefore s = \sqrt{\frac{\Sigma\, (x_s - M_s)^2}{r(r-1)}} \qquad . \qquad . \qquad . \qquad . \qquad .\text{~(vi)}$$

It is customary to speak of s in (vi) as the *standard error* of the mean score distribution. We have already seen in 4.08 that a difference between sample values w.r.t. normally distributed score differences is itself normally distributed; and the reader should be able to adapt the argument to the situation here discussed (see also 7.07 below). If the sample variances are respectively s_a^2 and s_b^2, that of the score difference distribution is given by (x) of 4.06 as

$$s_d^2 = s_a^2 + s_b^2 \qquad . \qquad . \qquad . \qquad . \qquad \text{(vii)}$$

If the mean scores M_a and M_b of two samples respectively a-fold and b-fold each from the same universe are normally distributed, the integrand of the distribution of the difference $M_d = (M_a - M_b)$ is therefore given approximately by

$$Y_d = \frac{1}{s_d \sqrt{2\pi}} \, exp \, \frac{-M_d^2}{2s_d^2} \qquad . \qquad . \qquad . \qquad . \qquad \text{(viii)}$$

The corresponding *standard score* or critical ratio is

$$c = \frac{M_d}{s_d} = \frac{M_a - M_b}{(s_a^2 + s_b^2)^{\frac{1}{2}}} \qquad . \qquad . \qquad . \qquad . \qquad \text{(ix)}$$

We are now in a position to use the table of the probability integral in the customary way to evaluate the odds against or in favour of the occurrence of a mean difference as great as the c value derived from a set of observations in conformity with the null hypothesis that our a-fold and b-fold samples come from one and the same universe. To explore the null hypothesis so defined by this method, we have to make the best of a bad job, in so far as we do not actually know the true value of the universe mean. Consequently, we must rely on an estimate (M_s which the pooled assemblage of $(a + b)$ score values x_{ab} supplies, i.e.

$$M_s = \frac{\sum x_{ab}}{(a+b)} \qquad . \qquad . \qquad . \qquad . \qquad \text{(x)}$$

In accordance with the value of M_s defined by (x), we thus take as our estimate of the variance of the universe score distribution as prescribed by (v):

$$V(u) \simeq \frac{\sum(x_{ab} - M_s)^2}{(a + b - 1)} \qquad . \qquad . \qquad . \qquad . \qquad \text{(xi)}$$

It follows from (vi) that the variances of the distributions of mean scores in samples of a and b items respectively taken from the putative parent universe are given by

$$s_a^2 = \frac{\sum(x_{ab} - M_s)^2}{a(a + b - 1)}; \qquad s_b^2 = \frac{\sum(x_{ab} - M_s)^2}{b(a + b - 1)}.$$

Whence from (vii)

$$s_d^2 = \frac{(a + b)}{(a + b - 1)ab} \cdot \sum(x_{ab} - M_s)^2 \qquad . \qquad . \qquad . \qquad \text{(xii)}$$

Since our assumption is that the samples are large, so that $\sqrt{(a + b) \div (a + b - 1)}$ does not differ materially from unity, we may evaluate s_d in (ix) as

$$s_d \simeq \sqrt{\frac{\sum(x_{ab} - M_s)^2}{ab}} \qquad . \qquad . \qquad . \qquad . \qquad \text{(xiii)}$$

For convenience in calculation we adapt (ii) and (iii) of 3.06 as follows :

$$nV = nV_0 - nM_s^2$$
$$\therefore \sum(x_{ab} - M_s)^2 = \sum x_{ab}^2 - (a + b)M_s^2$$
$$s_d^2 \simeq \frac{\sum x_{ab}^2 - (a + b)M_s^2}{ab} \qquad . \qquad . \qquad . \qquad . \qquad \text{(xiv)}$$

$$* \qquad * \qquad * \qquad * \qquad * \qquad * \qquad *$$

Example. Dahlberg (*Statistical Methods for Medical Students*) cites figures (centigrams per 100 c.c.) for blood cholesterin of a control group of 25 normal women and a group of 30 female schizophrenics as in the attached table.   From the totals (Table 8) we have

M_a (control group mean)	$383 \div 25 = 15\cdot32$
M_b (patients' mean)	$477 \div 30 = 15\cdot90$
D_m (group mean difference)	$15\cdot90 - 15.32 = 0\cdot58$
M_s (mean of pooled sample)	$(383 + 477) \div 55 = 15\cdot636$

The last quantity is our estimate of the mean of the putative common universe of the null hypothesis.   For the sum of the squares we have

$$\Sigma x^2 = 5931 + 7927 = 13858$$
$$\Sigma(x - M_s)^2 = 13858 - 55\,(15\cdot636)^2 = 411$$

TABLE 8

Cholesterin in Blood of Female Patients suffering from Dementia Præcox and Control Group (milligrams per 10 c.c.)

(i) Control Group (25 Women).		(ii) Patients (30 Women).	
Individual Values (x_a).	x_a^2.	Individual Values (x_b).	x_b^2.
17	289	13	169
13	169	11	121
15	225	12	144
15	225	12	144
16	256	14	196
15	225	12	144
14	196	11	121
15	225	18	324
14	196	22	484
17	289	21	441
15	225	22	484
17	289	19	361
18	324	19	361
13	169	22	484
16	256	21	441
15	225	18	324
15	225	15	225
15	225	14	196
14	196	14	196
14	196	13	169
16	256	17	289
13	169	15	225
18	324	17	289
14	196	14	196
19	361	18	324
		17	289
		13	169
		15	225
		14	196
		14	196
Totals 383	5931	477	7927

If we use (xiii) as a sufficiently good approximation for the variance (s_d^2) of the difference between sample means of the putative common universe :

$$s_d^2 = \frac{411}{30 \times 25} = 0.548,$$

$$s_d = \sqrt{0.548} \simeq 0.74.$$

For the critical ratio of the difference between the sample means we thus have

$$c = \frac{D_m}{s_d} \simeq \frac{0.58}{0.74} \simeq 0.8.$$

By this criterion we have therefore no reason for rejecting the null hypothesis.

NOTE ON STATISTICAL ESTIMATES

Some statistical authors (Kendall, Weatherburn *inter alia*) employ an economical notation for the derivation of a statistical estimate of a universe parameter in terms of an empirical statistic based on the information a sample supplies, e.g. the variance of the distribution of observed scores in a simple r-fold sample, defined by

$$V_s = \frac{\Sigma(x - M_s)^2}{r}.$$

To avoid the inconvenience of unwieldy summations in which it is necessary to distinguish y_s the frequency of a sample of given composition from y_x the frequency of a particular score in our sample under observation, we can denote by $E(. . .)$ the operation of extracting the mean value of the sample means. If we denote by S_e the statistic whose mean value referable to an indefinitely large number of samples is the unknown universe parameter S_u, and distinguish our empirical statistic by S_0, we thus write

$$E(S_e) = S_u.$$

To solve this in terms of S_0, the corresponding parameter of the sample under observation, we first seek its mean value, $E(S_0)$. To determine the statistic (V_e) whose mean value is V_u in terms of the mean square deviation (V_s) of the sample scores, we therefore proceed as follows :

$$E(V_s) = E\left[\frac{1}{r}\Sigma(x - M_s)^2\right] \quad \text{or} \quad r \cdot F(V_s) = E[\Sigma(x - M_s)^2].$$

In this expression M_s is the observed mean of the r-fold sample under observation. We can write $(x - M_s) = (x - M_u) - (M_s - M_u)$ and $(x - M_u) = (x - M_s) + (M_s - M_u)$, so that

$$(x - M_s)^2 = (x - M_u)^2 - 2(M_s - M_u)(x - M_u) + (M_s - M_u)^2$$
$$= (x - M_u)^2 - 2(M_s - M_u)(x - M_s) - 2(M_s - M_u)^2 + (M_s - M_u)^2$$
$$= (x - M_u)^2 - 2(M_s - M_u)(x - M_s) - (M_s - M_u)^2.$$

If we sum all items within a sample, the total of all the deviations $(x - M_s)$ from the sample mean (M_s) is zero, so that

$$\Sigma(x - M_s)^2 = \Sigma(x - M_u)^2 - \Sigma(M_s - M_u)^2.$$
$$\therefore r \cdot E(V_s) = E[\Sigma(x - M_u)^2] - E[\Sigma(M_s - M_u)^2].$$

Each sum of square deviations within the brackets consists of r items, its expected, i.e. long-run mean, value being therefore r times that of the corresponding square deviation itself, so that

$$E(V_s) = E(x - M_u)^2 - E(M_s - M_u)^2.$$

The expressions on the right are respectively mean square deviations of the score from the universe mean and the mean square deviation of r-fold sample means from the universe mean. Thus the first is V_u and the second is $(V_u \div r)$, so that

$$E(V_s) = V_u - \left(\frac{V_u}{r}\right) = \frac{r-1}{r} . V_u,$$

$$\therefore E(V_s) = \frac{r-1}{r} E(V_s).$$

Hence, as above, we derive

$$V_s = \frac{r}{r-1} . V_s = \frac{\Sigma(x - M_s)^2}{r-1}.$$

EXERCISE 7.06

For the following series of individual measurements, assess the significance of sex differences w.r.t. hand length, cephalic length, neck girth, abdominal circumference, bispinous width and bisacromial width, at each of the age-levels specified. See remarks on Ex. 7.07, p. 319.

1. *Hand length.*

Hand Length (Cm.).	9-9½ Years.		13-13½ Years.		16½-17 Years.	
	Boys.	Girls.	Boys.	Girls.	Boys.	Girls.
14·0						
14·5	3	1				
15·0	6	4				
15·5	9	8	3			
16·0	3	4	5	2		
16·5	2	4	6	4		1
17·0	2	2	15	3		
17·5		2	7	10		2
18·0			11	7	1	
18·5			16		2	4
19·0			2	5	2	6
19·5			1	3	2	4
20·0					1	1
20·5						
21·0					2	
21·5					1	
22·0					1	
Total .	25	25	50	50	12	18

2. Cephalic length.

Cephalic Length (Cm.).	9-9¼ Years.		13-13½ Years.		16½-17 Years.	
	Boys.	Girls.	Boys.	Girls.	Boys.	Girls.
16·0		1				
16·5				1		1
17·0	4	7	5	3		
17·5	2	8	4	10		
18·0	12	6	20	17		12
18·5	1	3	3	15		3
19·0	6		17	4	2	2
19·5			1			
20·0					10	
Total .	25	25	50	50	12	18

3. Neck girth.

Neck Girth (Cm.).	9-9¼ Years.		13-13½ Years.		16½-17 Years.	
	Boys.	Girls.	Boys.	Girls.	Boys.	Girls.
23·0						
23·5		2	1			
24·0						
24·5	1	1				
25·0	3	7				
25·5	1	1		2		
26·0	4	5	2	4		
26·5	3	1	1	1		
27·0	4	3	5	5		
27·5	4	1	3	5		
28·0	1	2	8	10		1
28·5			4	6		1
29·0	2		7	6		2
29·5			4	5		3
30·0	1	1	9	2		5
30·5			3	2		3
31·0		1	2	1		1
31·5						1
32·0			1			1
32·5						
33·0					3	
33·5					1	
34·0					2	
34·5					1	
35·0						
35·5					2	
36·0					2	
36·5						
37·0						
37·5						
38·0					1	
Total .	24	25	49	50	12	18

4. *Abdominal circumference.*

9-9½ Years.		13-13½ Years.				16½-17 Years.	
Boys.	Girls.	Boys.		Girls.		Boys.	Girls.
48·0	46·25	52·5	58·0	48·0	61·0	51·5	56·0
49·5	48·0	52·5	58·0	50·0	61·0	61·5	56·5
50·0	48·5	54·0	58·5	54·0	61·5	62·5	59·0
50·5	48·5	54·0	58·5	54·0	61·5	63·0	61·0
51·0	48·5	54·0	59·0	55·0	61·5	63·5	61·5
51·0	48·5	54·5	59·0	55·0	61·75	64·0	62·5
51·25	48·5	55·0	59·0	55·5	62·0	68·5	62·5
51·5	48·5	55·0	59·5	56·0	62·0	69·0	62·5
51·5	49·0	55·0	60·5	56·5	62·0	71·0	62·5
51·5	49·0	55·0	60·5	56·5	62·0	72·5	63·0
52·0	50·0	55·0	60·5	57·0	62·5	72·5	65·0
52·0	50·25	55·5	60·5	57·0	62·5	75·5	66·5
52·5	50·5	55·5	61·0	57·0	63·0		67·0
52·5	50·5	55·5	61·5	58·0	63·0		68·0
52·5	51·0	55·5	61·5	58·0	63·0		68·0
52·5	51·0	56·0	61·5	58·5	64·75		71·0
53·0	51·0	56·5	61·5	59·0	64·0		77·0
53·5	51·5	56·5	62·5	59·0	66·0		77·5
54·25	51·5	56·5	63·5	59·0	67·0		
54·5	53·0	57·0	64·0	59·0	69·0		
55·0	54·0	57·0	64·0	59·0	69·0		
55·75	54·0	57·5	65·0	59·5	69·5		
57·0	59·5	58·0	65·0	60·0	72·0		
57·0	60·0	58·0	67·5	60·5	72·5		
59·5	67·5	58·0	71·0	61·0	74·0		

5. *Bispinous width.*

Bispinous Width (Cm.).	9-9½ Years.		13-13½ Years.		16½-17 Years.	
	Boys.	Girls.	Boys.	Girls.	Boys.	Girls.
6·5		2				
7·0	4	8	3	2		
7·5	6	4	4	3		
8·0	8	9	12	13	2	1
8·5	6	1	16	16	2	1
9·0		1	15	12	7	5
9·5	1			4		4
10·0					1	2
10·5						2
11·0						2
11·5						1
Total .	25	25	50	50	12	18

6. *Bisacromial width.*

Bisacromial Width (Cm.).	9-9½ Years.		13-13½ Years.		16½-17 Years.	
	Boys.	Girls.	Boys.	Girls.	Boys.	Girls.
20·0	1					
20·5						
21·0						
21·5						
22·0						
22·5						
23·0	1	1				
23·5	1					
24·0	7	1				
24·5		1				
25·0	4	2	6			
25·5		1				
26·0	4	4	9			
26·5						
27·0	4	2	9	1		
27·5			1			
28·0	1	7	11			
28·5						
29·0		4	7	2		
29·5	1	1		1		
30·0	1		7	6	2	
30·5		1		1		
31·0				6	1	
31·5				1		
32·0				10	3	1
32·5				5		
33·0				9	3	1
33·5				4		
34·0				2		2
34·5						2
35·0				1	3	5
35·5				1		
36·0						3
36·5						2
37·0						1
37·5						
38·0						1
Total .	25	25	50	50	12	18

7.07 The Treatment of Paired Differences

In biological research such as therapeutic trials and in many classes of sociological problems, we commonly contrast the effect of circumstances on groups in either one of which no particular individual necessarily shares any seemingly relevant peculiarity with a member of the other ; but laboratory enquiry often presents us with the opportunity of *pairing* observations by exposing each individual of a group to both of two procedures, by subjecting to two treatments one or other of a pair of individuals themselves subject to similar external conditions such as time of day, available illumination, humidity, etc., or by exposing to one or other treatment members of a pair selected on account of individual similarities w.r.t. body weight, age, blood relationship (e.g. *litter mates*) and so forth. When we design an experiment in this way, we do so in general because we have reason to suspect that the removal of some source of variation, which would otherwise obscure the difference we wish to test, will help us to expose the existence of the latter, if it is indeed a real one.

The very fact that we do so means that the experiment does *not* conform to the postulates inherent in the argument we have used to justify the appropriate test of the null hypothesis of 7.06. The assumption implicit therein is that each possible sample mean has equal opportunity of association with any other, since we derived the sampling distribution of a group mean difference from the chessboard set-up ; but the circumstance that a member of one sample shares some common characteristic with some member of the other signifies that the two samples are indeed more alike than two samples taken from the same universe need be. Accordingly, we must explore a new approach to the problem. Instead of asking whether the numerical value of the difference between mean scores of two groups is significantly greater than zero, we ask *whether the numerical value of the mean of the differences between the scores of members of a pair differs significantly from zero.*

The logic of the paired difference problem raises for the student difficulties deserving of more attention than they commonly receive. It will therefore be profitable to make explicit the nature of the null hypothesis at each step in the argument. The null hypothesis that there is *no* real difference between alternative scores of one and the same pair signifies that our first concern is with a pair of 1-fold samples from one and the same universe. On the other hand, design of the experiment signifies that the universe from which we take one pair of unit samples is not necessarily the same as the universe from which we take another. Indeed, we pair off our observations because we have reason to believe the contrary. None the less, the mean value of the score difference w.r.t. 2 unit samples from any one such universe is necessarily zero, and the mean value of the sum of such differences referable to an assemblage of universes is itself zero on that account.

We can make no headway towards devising a satisfactory test for our null hypothesis unless we can justifiably rely on additional information about the distribution of our single samples from one and the same universe ; and in this context we shall assume that each universe is a normal universe. We may then formulate our initial assumptions in the following terms :

(*a*) each pair of observations constitutes an act of choice from a particular universe not necessarily the same as any other such universe ;

(*b*) within this framework our null hypothesis may be that each member of a pair of observations is a sample of one item from one and the same universe of normally distributed scores;

(*c*) in virtue of the act of choice involved in extracting from any such normal universe a pair of items with score values x_a and x_b, we assign to the selected pair what we here call a *d*-score, being the difference (D) between the two scores x_a and x_b.

A c-test—or indeed the more sophisticated t-test mentioned below—of the null hypothesis then involves two propositions :

(i) a normal distribution of the d-score from one and the same universe ;
(ii) a normal distribution of the mean of the d-scores from the assemblage of universes which furnish the set of paired observations.

The student may find it easier to follow the ensuing argument, if we first materialise the foregoing assumptions by invoking a model set-up for four pairs of observations such as the following :

(a) We have four urns, each of which contains counters labelled with 0, 1, 2 to 50 pips in proportions specified by the terms of a binomial of the form $(p + q)^{50}$, the precise values of p for the several urns being 0·5, 0·4, 0·3, 0·2 ;

(b) We choose from each urn two samples each of a *single* counter, recording the difference (D) between the number of pips (x_a and x_b) on the two members of the pair as the d-score of the urn.

In this set-up we have four universes each approximately normal in conformity with our findings in 5·08 ; but the means * and variances of the four normal unit sampling distributions are different, the latter being

$$50(0·5)(0·5) ; \quad 50(0·4)(0·6) ; \quad 50(0·3)(0·7) ; \quad 50(0·2)(0·8).$$

It is in this sense that we shall regard the universes from which we extract different pairs of 1-fold samples as being both different from one another and also approximately normal.

With this example to keep our feet on the ground, our first task will be to establish (i) above. This is formally like the problem dealt with in 4·08, but simpler inasmuch as the samples, being unit samples, are of the same size. Being samples from the same universe, they have the same mean (M_u) ; and the variance (V_u) of the distribution of each is the same. It also follows that the difference

$$(x_a - x_b) = D = (X_a - X_b) \quad . \quad . \quad . \quad . \quad . \quad \text{(i)}$$

The variance (V_d) of the distribution of D being the sum of the variances of the distributions of the two samples, we may write

$$V_d = V_u + V_u = 2V_u \quad . \quad . \quad . \quad . \quad . \quad \text{(ii)}$$

If we assume for simplicity that the scores with which we are here concerned increase by unit steps, so that the scale of the d-score distribution is likewise unity, we may simplify the argument of 4.08 by recourse to a chessboard lay-out, denoting by y_c the frequency of the score difference d in a particular cell c with corresponding border scores X_a of frequency y_a and $X_b = (X_a - d)$ of frequency y_b. With no pretension to rigour, we may then regard a normal frequency equation as a good approximation to the *exact* value of y_a or of y_b with due recognition of the fact that we are sampling from different universes with definitive parameters V_a and V_b. On this understanding, the postulate of statistical independence implies the product relation $y_c = y_a \cdot y_b$, so that

$$y_c \simeq \frac{1}{\sqrt{2\pi V_u}} \cdot exp\left(\frac{-X_a{}^2}{2V_u}\right) \cdot \frac{1}{\sqrt{2\pi V_u}} \, exp\left(\frac{-(X_a - D)^2}{2V_u}\right)$$

$$\simeq \frac{1}{\pi . V_d} \, exp - \left[\frac{X_a{}^2 + (X_a - D)^2}{2V_u}\right] \quad . \quad . \quad . \quad . \quad \text{(iii)}$$

* We here state our null hypothesis in the most general terms. When our concern is with *before-after* treatment estimations, the appropriate assumption may be as indicated below, *viz.* that the sampling variance arises solely from instrumental error. If so, we predicate that the unit sampling distributions of our universes have different *means* but identical variances.

We may write

$$\frac{X_a^2 + (X_a - D)^2}{2V_u} = \frac{2X_a^2 - 2X_aD}{2V_u} + \frac{D^2}{2V_u} = \frac{X_a^2 - X_aD}{V_u} + \frac{D^2}{2V_u}$$

$$\cdots = \frac{X_a^2 - X_aD + (\tfrac{1}{2}D)^2}{V_u} - \frac{D^2}{4V_u} + \frac{D^2}{2V_u}$$

$$= \frac{(X_a - \tfrac{1}{2}D)^2}{V_u} + \frac{D^2}{2V_d}.$$

Hence from (iii)

$$y_c \simeq \frac{1}{\pi V_d} \cdot exp\left(\frac{-D^2}{2V_d}\right) exp \frac{-(X_a - \tfrac{1}{2}D)^2}{V_u}.$$

The frequency of the particular d-score D is the total frequency of *all* cells therewith, obtainable as in 4.08 by summation for all values of X_a, so that

$$y_d \simeq \frac{1}{\pi V_d} exp\left(\frac{-D^2}{2V_d}\right) \int_{-\infty}^{\infty} exp \frac{-(X_a - \tfrac{1}{2}D)^2}{V_u} \cdot dX_a \quad . \qquad . \qquad . \quad \text{(iv)}$$

To reduce the integral in (iv), we write

$$\frac{z}{\sqrt{2}} = \frac{X_a - \tfrac{1}{2}D}{\sqrt{V_u}}; \quad dX_a = \sqrt{\frac{V_u}{2}} \cdot dz = \frac{\sqrt{V_d}}{2} \cdot dz,$$

$$\therefore \int_{-\infty}^{+\infty} exp \frac{-(X_a - \tfrac{1}{2}D)^2}{V_u} \cdot dX_a = \sqrt{V_d} \int_{0}^{\infty} e^{-\frac{1}{2}z^2} \cdot dz = \sqrt{\frac{\pi V_d}{2}}.$$

Hence from (iv) above

$$y_d \simeq \frac{1}{\sqrt{2\pi V_d}} exp \frac{-D^2}{2V_d}.$$

Thus D has a normal distribution with *zero mean* and variance V_d. We thus know the *true mean* of the sampling distribution of d-scores from the same universe. If we had before us m such d-scores from one and the same universe, we could therefore write as our un-biassed estimate * of V_d

$$\frac{\Sigma (D - 0)^2}{m} = \frac{\Sigma D^2}{m}.$$

In fact, we have only one such observation, i.e. $m = 1$, and our estimate is

$$V_d \simeq D^2 \quad . \qquad . \qquad . \qquad . \qquad . \qquad . \quad \text{(v)}$$

We have now to establish our second proposition that the mean value of an r-fold sample of such normally distributed d-scores singly derived from the r universes of our r paired observations is itself normal. We first consider the distribution of the sum of two d-scores from different universes, the treatment being essentially as above with due regard to the possibility that the unit sampling distributions of the two d-scores (D_1 and D_2) may not have the same variance.

For reasons explained in 4.08, we write for the variance of the distribution of the 2-fold sum $V(S_2)$

$$V(S_2) = V(D_1) + V(D_2).$$

---

* What follows refers only to the assumption that we propose to perform a c-test. The ratio defined by (vi) is not the t-ratio of 7.08. For reasons we shall examine in Vol. II, we must define s_m in the denominator of the t-ratio, so that it is independent of the mean, and we can fulfil this condition only if we define it in accordance with (vi) of 7·06. Recourse to the c-test is legitimate only if the sample is large.

By the method of 4.08 we derive a normal distribution of the sum about zero mean and with variance as above. We may repeat the process for the sum of 3, 4 and so on d-scores, the distribution of which is normal with variance

$$V(S_r) = \sum V_d.$$

Whence from (v) above

$$V(S_r) \simeq \sum D^2.$$

Since the r-fold d-score sum is normally distributed, that of the mean is normally distributed with estimated variance

$$V_m = \frac{V(S_r)}{r^2} \simeq \frac{\sum D^2}{r^2}.$$

If we denote our estimate of the standard deviation of the distribution of the mean d-score by $s_m = \sqrt{V_m}$ and the observed mean of our d-scores by $M(D)$

$$s_m = \frac{\sqrt{\sum D^2}}{r}$$

For the purpose of an appropriate c-test, we therefore write

$$c = \frac{M(D)}{s_m} = \frac{r \cdot M(D)}{\sqrt{\sum D^2}} = \frac{\sum D}{\sqrt{\sum D^2}} \qquad . \qquad . \qquad . \qquad . \qquad \text{(vi)}$$

Example. Columns (i) and (ii) of Table 9 set out hæmoglobin determinations of blood from the forefinger of 39 pairs of individuals (*a*) before, and (*b*) after constriction of the vessels by a tourniquet for a fixed interval of time. From column (iii) we have

$$M(D) = \sum D \div 39 = 52 \div 39 = 1.33.$$

From column (iv)

$$\sum D^2 = 296$$

$$s_m = \sqrt{296} \div 39.$$

The critical ratio in accordance with equation (vi) is

$$\frac{39(1.33)}{\sqrt{296}} = 3.1.$$

If we are entitled to make all the assumptions we have in fact made in carrying out the last test, we are thus forced to the conclusion that the occurrence of such a mean difference between *B* (*after* constriction) and *A* (*before* ditto) scores of columns (i) and (ii) in Table 9 would be very rare unless constriction did in fact exert an effect. To that extent we are justified in rejecting the null hypothesis if: (*a*) we have good reason to suppose that attendant circumstances conform to the theoretical requirements of the test; (*b*) the sign of the mean difference recorded is indicative of the sort of change which would not surprise us. In this context we shall not concern ourselves with the issue last stated, i.e. the prior probability that the alternative hypothesis is correct. If our concern is to establish the reality of an *increase*, we must in any case pay due regard to the issue raised on page 204, because the question to which we seek an answer —and this is commonly so when our concern is to establish a real difference—is not whether a difference of *any* sort exists. What we want to know is how often we might expect a positive difference $(B - A) = 1.33$, if the null hypothesis is valid.

TABLE 9

(i).	(ii).	(iii).	(iv).
Percentage Hæmoglobin.		Increase (D).	D^2.
Before Constriction (x_a).	After Constriction. (x_b).		
64	66	2	4
72	72	0	0
58	58	0	0
70	70	0	0
58	58	0	0
88	88	0	0
74	76	2	4
84	86	2	4
70	68	−2	4
62	62	0	0
66	70	4	16
88	90	2	4
88	90	2	4
88	90	2	4
86	86	0	0
86	86	0	0
70	72	2	4
86	86	0	0
85	85	0	0
65	64	−1	1
78	78	0	0
83	84	1	1
78	80	2	4
49	49	0	0
86	85	−1	1
57	54	−3	9
85	85	0	0
86	92	6	36
79	80	1	1
80	82	2	4
81	80	−1	1
86	85	−1	1
88	91	3	9
86	91	5	25
82	89	7	49
89	98	9	81
83	87	4	16
89	89	0	0
82	85	3	9
Totals 3035	3087	52	296

That the conclusion, suggested by the mean difference form of the c-test, is in this instance open to some doubt is evident from inspection of the figures in column (iii). We may classify the differences as follows :

Negative	Zero	Positive	Total
6	14	19	39

It thus appears that only half the observed differences are in fact positive, a result which would point to : (a) equal probability of getting or of not getting a positive difference, if confirmed by a long series of trials ; (b) only 3 to 1 odds in favour of a positive result where a difference is in fact detectable.

Another feature of the array of numbers in column (iii) is suggestive, if we compare them with those of columns (i) and (ii). The entire range of figures in column (iii) includes only 11 different values. Since the only integer missing between the limits -3 and 9 is 8, we may infer that the universe of d scores is a universe of at least 13 classes ; but we have no information to suggest that it contains more than 19 and as yet insufficient reason to assume that the distribution of individual d-scores from a universe of 13–20 classes will closely approach an assumed *normal* distribution.

Herein lies an important difference between the test under discussion and that of 7.06. When we compare responses of groups of individuals, we give full play to intrinsic variation of our biological materials w.r.t. the make-up of the individual ; and we may well have good grounds for believing that the sampling distribution arising from this circumstance is approximately normal. In any case, we may assume with confidence a universe of many classes ; but it is not always justifiable to make such an assumption when we pair off observations as in the last example. If we make such paired observations on one and the same individual, we then eliminate *individual variation* more or less completely ; and if the treatment involved is ineffective, our main or sole source of residual variation is *instrumental*. The null hypothesis thus signifies that we are left with errors of observation alone.

If the outcome is sufficiently clear-cut to exonerate us from recourse to a test of significance such a procedure is highly commendable as an experimental technique bringing what is relevant to our end into sharper focus. Otherwise, it carries with it a penalty to which we should always be alert. For the number of scale divisions consonant with competent craftsmanship in successive performance of a simple chemical operation may be so few as to cast doubt on the assumption of approximate normality w.r.t. sampling from the universe of the single pair. From this point of view, not all experiments involving pairing of differences are comparable. Pairing of individuals w.r.t. build, immediate environment, age, ancestry, season and so forth has the advantage of removing a gross source of variation whose presence might conceal a clear-cut difference liable to escape recognition by group comparison. To do so does not eliminate other sources of individual variation which may well assure a sampling distribution implicit in the credentials of the c-test ; and the use of the c-test to assess the mean value of the pair difference is then legitimate. A new issue arises when we pair off measurements separated by a short interval of time on one and the same individual. It then behoves us to scrutinise what distribution of errors of observation is in fact consistent with the set-up.

In short, the Gaussian dogma referred to in 7.02 has too long dominated the treatment of classes of experimental data to which its fundamental postulates have little or no relevance. The assumption of an approximately normal distribution of instrumental errors may be sometimes appropriate to the end result of an intricate sequence of manipulations, and possibly relevant to the type of astronomical observations which occupied the centre of the stage in the days of Laplace and Gauss himself. In the domain of simple repetitions of volumetric analyses of blood or urine, it is manifestly inappropriate ; and the propriety of assessing either the significance of a difference between the mean values of two series of such determinations or the significance of the mean value of a series of pair-differences which putatively involve no other sources of variation calls for a lively appreciation of how many numerical values are consistent with the proper execution of the experimental technique.   Such considerations prompt us to approach any statistical issue in the domain of representative scoring by more than one avenue, when that is possible.

While alert to such a pitfall, we should recall the issue raised at the end of 7.05, where we have seen reason to believe that a normal law of *mean* score distribution may very closely describe a situation in which the unit sampling distribution is by no means normal.   Let us suppose that competent craftsmanship prescribes an instrumental error of only ± 1 scale divisions in a volumetric determination.   It is unlikely that our three possible score values $+ 1$, 0 and $- 1$ will occur with equal frequency.   For illustrative purposes we may suppose : (*a*) that the investigator scores either $+ 1$ or $- 1$ with equal frequency about half as often as 0 ; (*b*) that the investigator is liable to overshoot the mark and scores $+ 1$, 0 and $- 1$ with frequencies in the ratio $16 : 8 : 1$.   If (*a*) is true, the unit sampling distribution tallies closely with terms of the binomial $(0 \cdot 5 + 0 \cdot 5)^2$, and that of the mean of 10-fold samples with that of the binomial $(0 \cdot 5 + 0 \cdot 5)^{20}$. If (*b*) is true, the unit sampling distribution would tally with the terms of the binomial $(0 \cdot 2 + 0 \cdot 8)^2$, and that of the mean of 25-fold samples with the terms of $(0 \cdot 2 + 0 \cdot 8)^{50}$.   Now we have already seen (5.08) that the normal distribution closely fits a binomial if $rp = 10$, as is true of each of these two mean score distributions.

The examples cited emphasise that the assumption of an approximately normal law to describe the distribution of the mean value of successive determinations of a test score may often be justifiable when the possible range of such test scores is in fact very restricted, and when the assumption of normality with respect to the distribution of the test scores themselves, i.e. errors of determination, would be grossly incorrect.   If so, we can justifiably postulate as our null hypothesis that the distribution of differences between the mean of one set of determinations and that of another is approximately normal ; but this does not signify that the distribution of differences between *single* pairs of determinations will be nearly normal.   By recourse to the chessboard, the reader will readily see that the single-pair difference distribution with respect to a determination involving a scale division range of ± 1 as above would have the range ± 2. Thus we should in fact have a 5-class universe of samples defined in accordance with the assumptions stated in (*a*) above by the terms of $(\frac{1}{2} + \frac{1}{2})^4$.

When using the method of paired differences in circumstances which eliminate all variation except experimental errors in the absence of a real difference, much depends on whether the *d*-score is : (*a*) the difference between a *single* determination before and a *single* determination after treatment ; (*b*) the difference between the mean of *repeated* determinations before and repeated determinations after treatment.   In the last resort, the chessboard method set forth in this chapter places at the disposal of the investigator who cares to plot the sampling distribution of his errors of measurement, a method of ascertaining how far the distribution of the mean value of samples of a given size conforms to the statistical test which invokes the normal law.

EXERCISE 7.07

(Significance of paired differences.)

These examples do not involve large numbers, and the student who has access to tables of the *t*-test can with advantage compare the results of testing them by recourse both to the probability integral and to the *t*-function.

1. Egg production of a number of hens in first and second years. (From Turner and Kempster, *Am. J. Physiol.*, 149.)

Hen.	First Year Egg Production.	Second Year Egg Production.
1	145	134
2	152	126
3	194	136
4	206	186
5	145	147
6	201	158
7	172	117
8	188	155
9	209	183
10	165	186

2. *Ditto* hens fed on 10 gms. thyroprotein per 100 lbs. feed. (*Ibid.*)

Hen.	First Year Egg Production.	Second Year Egg Production.
1	112	109
2	218	119
3	192	132
4	151	85
5	152	112
6	193	157
7	206	128
8	152	135
9	145	129
10	214	160
11	168	150
12	225	144
13	216	36

3. Effect of obstruction upon serum urea in rabbits. (From Herrin, *Am. J. Physiol.*, 149.)

Rabbit No.	Serum Urea (mgm./100 c.c.).	
	Normal.	Intestinal Obstruction.
1	44·1	129·0
2	45·0	149·2
3	39·7	284·2
4	42·0	237·0
5	51·7	147·0
6	37·5	144·7
7	41·1	108·0
8	39·0	78·0
9	87·1	174.1

4. *Ditto* upon urea excretion. (*Ibid.*)

Rabbit No.	Urea Excretion (gm./hr.).	
	Normal.	Intestinal Obstruction.
1	0·0761	0·0311
2	0·0841	0·0272
3	0·0618	0·0204
4	0·0589	0·0116
5	0·1437	0·0987
6	0·1136	0·0136
7	0·1837	0·0812
8	0·1620	0·0946
9	0·1165	0·0288

5. Additional hours of sleep gained by the use of two tested drugs. (Fisher, *Statistical Methods for Research Workers.*)

Patient No.	Additional Sleep.	
	Drug A.	Drug B.
1	+ 0·7	+ 1·9
2	− 1·6	+ 0·8
3	− 0·2	+ 1·1
4	− 1·2	+ 0·1
5	− 0·1	− 0·1
6	+ 3·4	+ 4·4
7	+ 3·7	+ 5·5
8	+ 0·8	+ 1·6
9	0·0	+ 4·6
10	+ 2·0	+ 3·4

6. Effect of immersing body in water upon the vital capacity. (Hamilton and Mayo, *Am. J. Physiol.*, 141.)

Subject No.	Vital Capacity.	
	Out of Water.	In Water.
1	5160	5330
2	4160	3960
3	5320	4750
4	3980	3570
5	5090	4760
6	3750	3440
7	5030	4670
8	4270	3820
9	4540	4190
10	4500	4260
11	4760	4260
12	4670	4180
13	4490	3990
14	4920	4700
15	5280	5100
16	4510	4390
17	6230	5750
18	5020	7710
19	5290	4600
20	5930	5530

7. Hæmoglobin response of milk-anaemic rats with iron added to food. (Smith and Medlicott, *Am. J. Physiol.*, 141.)

Rat No.	Hæmoglobin (gms./100 ml. blood).	
	Initial.	After 4 Weeks with Iron (0·5 mg./day).
1	3·4	4·9
2	3·0	2·3
3	3·0	3·1
4	3·4	2·1
5	3·7	2·6
6	4·0	3·8
7	2·9	5·8
8	2·9	7·9
9	3·1	3·6
10	2·8	4·1
11	2·8	3·8
12	2·4	3·3

8. Scores in judging poetry (Abbot-Trabue Test) before and after teaching of technical analysis. (Peters and van Voorhis, *Statistical Procedures.*)

Pair.	Initial Score.	End Score.
1	6	7
2	3	6
3	5	7
4	5	5
5	6	4
6	8	9
7	5	7
8	3	5
9	3	6
10	3	6
11	1	9
12	4	2
13	6	7
14	6	5
15	3	6
16	4	7
17	2	2
18	5	7
19	5	4
20	4	6
21	6	8
22	3	4
23	6	5
24	2	5
25	0	2

9. *Ditto* after some interval of time, without special teaching.   (*Ibid.*)

Pair.	Initial Score.	End Score.
1	4	4
2	7	6
3	7	7
4	8	7
5	6	5
6	7	5
7	2	5
8	6	5
9	7	8
10	6	6
11	3	4
12	3	9
13	5	3
14	4	4
15	5	3
16	3	9
17	4	5
18	4	3
19	3	1
20	6	6
21	6	8
22	2	5
23	3	3
24	5	3
25	1	3

10. Weight at birth and on 10th day for 11 infants. (Davenport, *Body Build and its Inheritance*.)

Subject.	Weight (kg.).	
	At Birth.	Tenth Day.
1	2·126	1·928
2	3·472	3·572
3	2·963	2·927
4	3·558	3·494
5	3·005	3·147
6	3·019	2·977
7	3·175	3·026
8	2·835	2·722
9	3·657	3·536
10	3·232	2·941
11	3·303	2·984

7.08 THE LIMITATIONS OF THE C-TEST

In the domain of the representative score, a normally distributed c-ratio is the quotient of a statistic such as the normally distributed deviation of a mean d-score or of a difference between group means and the standard deviation of the distribution of the same statistic in an indefinitely large number of samples from the parent universe. It is possible to specify the latter exactly if we know the variance of the unit sampling distribution, i.e. the score distribution of the putative common universe ; but in fact our knowledge of the latter is restricted to our single pooled sample ; and the figure we use as the standard deviation or standard error of the distribution of the group mean difference, being based on an estimate of the variance of the unit sampling distribution, is itself an estimate. In so far as it is a good estimate, we are entitled to believe that our observational c-ratio will be normally distributed like the exact value of c ; and it is reasonable to entertain the hope that such an assumption will not often let us down when we are dealing with *large* samples. Otherwise, the hazard is not one which we can dismiss lightly.

It will clarify the distinction between an *empirical* c-ratio which we can actually determine and a *theoretical* c-ratio tabulated for the normal distribution, if we here go back to our models. Let us consider the mean-score distribution of the 4-fold toss of a flat circular die with no pip on one face and a single pip on the other. The binomial $(\frac{1}{2} + \frac{1}{2})^4$ defines the unit sampling distribution of our universe ; and the variance of the score distribution of the universe, i.e. the variance (V_u) of the unit sampling distribution, is $(\frac{1}{2})^2 = 0.25$. The variance (V_m) of the distribution of the mean score of 4-fold samples is accordingly $(0.25 \div 4) = 0.0625$. To keep our feet on the ground, we may check this by recourse to the mean score distribution, exhibiting, with corresponding frequencies in the ratio $1 : 4 : 6 : 4 : 1$, score sums, mean scores, the deviation of the latter from their own mean (M_0) which is also the mean score (0.5) of the unit sampling distribution :

Score sum	0	1	2	3	4
Mean score (M_s) . . .	0	0·25	0·5	0·75	1·0
Mean score deviation $(M_s\text{-}M_0)$.	− 0·5	− 0·25	0	+ 0·25	+ 0·5
Frequency	0·0625	0·2500	0·3750	0·2500	0·0625

From these figures our weighted mean square deviation of the mean score distribution is

$$(0{\cdot}0625)\,(-\,0{\cdot}5)^2 \,+ (0{\cdot}25)\,(-\,0{\cdot}25)^2 + (0{\cdot}25)\,(0{\cdot}25)^2 \,+ (0{\cdot}0625)\,(0{\cdot}5)^2 = 0{\cdot}0625 = V_m.$$

The theoretical c-ratio of the mean score distribution is the ratio of the mean score deviation to the standard deviation ($\sigma_m = \sqrt{V_m}$). The latter is $\sqrt{0{\cdot}0625} = 0{\cdot}25$. Hence we have the following binomial distribution of the c-ratio :

Mean score deviation (M_s-M_0) .	$-0{\cdot}50$	$-0{\cdot}25$	0	$+0{\cdot}25$	$+0{\cdot}50$
c-Ratio (M-M_0) $\div 0{\cdot}25$ . .	-2	-1	0	1	2
Frequency	$0{\cdot}0625$	$0{\cdot}2500$	$0{\cdot}3750$	$0{\cdot}2500$	$0{\cdot}0625$

Let us now investigate what would happen in the long run if we always estimated the standard deviation of the mean score distribution in accordance with (iii) of 7.07. In other words, we regard each 4-fold toss as more or less representative of the sampling distribution of the score in the parent universe. In that event a 4-fold toss with a score sum of 3 is a universe of four items belonging to two classes with scores (x) of 0 or 1 in the ratio 1 : 3. The mean score (M_s) of the sample is 0·75, so that the sum of the square deviations ($x - M_s$) is given by

$$3(1 - 0{\cdot}75)^2 + 1(0 - 0{\cdot}75)^2 = 0{\cdot}75.$$

Our estimate of the variance of the distribution of the mean in accordance with (vi) of 7.06 is, therefore,

$$\frac{\Sigma(x - M_s)^2}{r(r\,-\,1)} = \frac{0{\cdot}75}{(4)\,.\,(3)} = 0{\cdot}0625.$$

Hence an estimate (s_m) of the standard deviation of the mean based on a sample of this composition would be $\sqrt{0{\cdot}0625} = 0{\cdot}25$. If the score sum is 2, our sample universe is made up of two scores of 1 and two scores of zero, so that the sum of the square deviations is

$$2(0{\cdot}5)^2 + 2(-\,0{\cdot}5)^2 = 1.$$

Hence $V_m = \frac{1}{12}$ and $s_m = \dfrac{1}{2\sqrt{3}} = 0{\cdot}288. \ldots$ If the score sum is either 0 or 4, the sum of square deviations is necessarily zero.

Proceeding in this way we obtain the following results :

Score sum	0	1	2	3	4
Mean score (M_s) . . .	0	$0{\cdot}25$	$0{\cdot}50$	$0{\cdot}75$	$1{\cdot}0$
Mean score deviation (M_s-M_0) .	$-0{\cdot}50$	$-0{\cdot}25$	0	$+0{\cdot}25$	$+0{\cdot}50$
Estimated standard deviation s_m of the distribution of the mean score	0	$0{\cdot}25$	$0{\cdot}288$	$0{\cdot}25$	0
Frequency	$0{\cdot}0625$	$0{\cdot}2500$	$0{\cdot}3750$	$0{\cdot}2500$	$0{\cdot}0625$

In contradistinction to a theoretical c-ratio based on the theoretical value of σ_m which is independent of the composition of a particular sample, we shall now denote our empirical estimate of it by t, the ratio of the deviation of the sample mean M_s to the estimated standard deviation s_m computed from the data of an individual sample. In the foregoing table successive values of t are $-\infty$, $-1, 0, 1$ and ∞. We may thus contrast the distribution of t and c for the 4-fold sample of our 2-face die as follows :

c-Ratio	-2	-1	0	$+1$	$+2$
Frequency . . .	$0{\cdot}0625$	$0{\cdot}2500$	$0{\cdot}3750$	$0{\cdot}2500$	$0{\cdot}0625$
t-Ratio	$-\infty$	-1	0	$+1$	$+\infty$

Comparison of the figures above shows that equally spaced values of the c-ratio are referable to frequencies defined by terms of a binomial expansion; but this is not true of the tail ends of the distribution of the t-ratio. Hence the use of the c-test table, i.e. the normal distribution, to assess the significance of an *empirical* c-ratio, i.e. what we here call t, is justifiable only if the size of the sample is sufficiently great to ensure a t-distribution which tallies closely with the normal. In what circumstances it does so is therefore an issue of considerable importance.

Under the pseudonym *Student*, W. S. Gossett published in 1908 an examination of the distribution of the empirical c-ratio (t); and tables of this function are now available in standard treatises such as those of Fisher and of Kendall. Its derivation entails mathematical difficulties we have not surmounted in previous chapters; and we shall not deal with it in this volume. It must here suffice to state that frequencies of corresponding values of t and c are very close if the size of the sample is over 20. For many purposes therefore the older c-test is quite adequate. In any case, the small sample t-distribution postulates a normal distribution of the mean score; and a clear understanding of the less exacting procedure set forth in 7.05–7.07 is therefore a necessary prerequisite to appreciation of its logical credentials.

Like the normal, the t-distribution is a continuous function with an infinite range. Such a distribution can give a good fit to theoretical expectation only if the number of score classes is considerable; and this is customarily true within the domain of representative scores, since the number of mean-score classes of a sample greater than 1 must always be greater than the number of raw score classes in the universe. For instance, a model unit sampling distribution of $(a + 1)$ classes defined by $(p + q)^a$ generates a distribution of mean scores w.r.t. r-fold samples defined by $(p + q)^{ra}$ with $(ra + 1)$ classes. In the domain of taxonomic scores, our customary concern is with a two-class universe, i.e. with a unit sampling distribution defined by $(p + q)^1$, admitting the possibility of only 11 classes of 10-fold samples. It follows that the t-test or *Student*-test, though a small-sample test for differences in the domain of representative scoring, is not a small-sample test appropriate to the assessment of proportionate score differences of the type dealt with in Chapter 4.

In this connexion, we may recall a conclusion there sufficiently illustrated by our investigation of the proportionate score difference distribution in the taxonomic domain. There are numberless pitfalls, if we pay no regard to a distinction between what we may call the *superfinite* universe (i.e. universe of a large number of classes), which we deal with in practical affairs, and the infinite universe of pure mathematics. In particular, derivation of the t distribution presumes statistical independence w.r.t. the square of the deviation of a sample mean and the estimate of the variance of its distribution. In the domain of the mathematical infinite this assumption is justifiable. It is far from true of extraction of 2-fold samples from a universe of 21 score classes, i.e. a universe of which the binomial definitive of the unit sample is $(\frac{1}{2} + \frac{1}{2})^{20}$, though we have seen that the normal curve is a very good descriptive device for the unit sampling distribution of such a universe, and *a fortiori* for the 41 score class distribution of 2-fold samples extracted therefrom.

CHAPTER 8

CORRELATION AND INDEPENDENCE

OUR approach to the applications of the theory of probability has hitherto raised issues of two sorts :

    (a) whether the composition of a given sample is consistent with the assumption that it is a sample from a universe exactly specified by the requirements of a particular (e.g. Mendel's) hypothesis ;

    (b) whether the composition of a given sample is consistent with the assumption that it is a sample from a putative universe of which our only other source of precise information is the composition of another sample taken from it.

Questions of either sort arise in the experimental sciences and in the sciences which are essentially taxonomical. The method with which we shall acquaint ourselves in this chapter is one which has little relevance to the former ; and it is difficult to assess its usefulness justly without due regard to what is peculiar to the latter. It is therefore fitting to start with a recognition of what distinguishes enquiries of one sort from enquiries of the other.

In its most elementary form a scientific law states that an event B happens when an event A happens. Such an assertion may signify either of two sorts of association, distinguishable for lack of accepted terms as *consequence* (B follows A) and *concurrence* (B and A go together). In this context, consequence signifies a stimulus-response relation, as when : (i) the occurrence B is increase of the density of a gas and the occurrence A is external application of pressure to it ; (ii) the occurrence B is recovery from a disease and occurrence A is the injection of a drug. Situations of this sort are the main concern of experimental science whether its subject-matter is living or otherwise ; but it is wrong to suppose that stimulus-response relations, such as Hooke's Law of tension-length or Schafer's discovery of the action of adrenalin on the blood pressure, make up the total subject-matter of scientific law. Before it is possible to ask fruitful questions it is necessary to classify data ; and the province of a large part of science, especially science concerned with living things, is classification with a view to reliable diagnosis or fruitful prognosis. That B happens when A happens thus includes statements of the sort : nearly all people with blue eyes have fair hair. An assertion that the two attributes go together in this sense entails no direct temporal relation between them. If it points to an antecedent, it points to an antecedent common to both *concurrent* events ; and its verification is therefore a challenge to explore the nature of the common antecedent by recourse to *other* sources of information.

At the operational level, a relation of consequence answers the question : *what must we do to bring about the occurrence B ?* A relation of concurrence answers the question : *where should we seek to find B with least effort ?* Where the criterion of both occurrences is all-or-nothing as *blue—not blue, fair—otherwise*, we can deal with concurrent association by recourse to methods already discussed. Whether there is or is not a significant association between blue eyes and fair hair reduces to the issue : *does the proportion of fair-haired persons among people with blue eyes differ significantly from the proportion of fair-haired persons among people with eyes of another colour ?* From one point of view, this is strictly comparable with the end in view when the question is : do tall girls menstruate earlier ? From another, the question last stated raises a new issue. If both occurrences are referable to a *more-or-less* scale, as are height and age at menarche, it is possible to state the problem in terms which are the theme of the ensuing pages. In such a situation we have before us a set of paired observations. Each pair is referable to one individual girl ; and we can regard each member of a pair as the co-ordinate of a point on a height-age graph. If the whole set of points appears to cluster round a hypothetical line in a way suggestive of a trend we may then ask ourselves : *could such clustering arise by chance alone?* To answer such a question in accordance with the practice of this book, it is necessary to set up an appropriate statistical model, as we shall do later. First, it is worth while to notice that the graphical approach to problems of the class last cited has an altogether misleading air of similarity to analytical methods adopted to establish physical laws.

A little reflection on the procedure involved in establishing a simple physical principle such as Hooke's Law suffices to dispel this illusion. The law of the spring (*ut tensio sic vis*) is apposite, because the graphical relation between length and tension is very simple, being approximately linear over a wide range and in any case *monotonic*, i.e. having no turning point. By making paired observations on one and the same spring at fixed temperature in one and the same place, it is possible to plot a graph which shows the consequential relation between the suspended weight and the extension evoked by it. A spring of different thickness or a spring of different material would also behave in accordance with a straight line law, but with a different linear constant (*elastic modulus*). By comparison of sets of successive paired measurements on the *same* spring of one thickness or of one material with sets of successive paired measurements on another spring of different thickness and of the same material, or of the same thickness but of other material, it is possible to prepare tables from which the elastic modulus for a spring of given dimensions and material is calculable. At every stage in the discovery of the law and its subsequent elaboration, the possibility of prescribing how to control the behaviour of a single spring in this way thus presupposes the possibility of assembling a set of successive paired observations on a single spring, complete as a set in the sense that they cover the range within which interpolation is legitimate.

Let us now suppose that we have no knowledge of Hooke's Law, and that we have the opportunity of making one paired (*length-weight*) observation on each of a set of different springs, each seemingly alike. If we find that the points on a graph embodying our experience of such a *population of springs* cluster around a line straight or otherwise, it will certainly confirm our conviction that the springs have much in common, and by so doing establish a criterion of their taxonomic community ; but it will scarcely furnish sufficient basis for the formulation of a law governing the behaviour of an individual spring. In the approach we are about to explore, it is therefore important to be alert at all times to the danger of a false assumption that functional relations established by statistical methods in the domain of concurrence necessarily admit of any legitimate conclusions in the domain of consequence.

8.02 SCORING BY RANK

If we arrange in ascending or descending order any set of individual scores, such as blood sugar level, income or basal metabolic rate, we can assign to each member of the group a secondary score, referable to order in the sequence. On any such derivative scale of *rank* scores, the interval between that of one individual and that of his or her successor is necessarily equal ; and we have already seen that scoring by rank has a peculiar advantage in that it entails a specification of the unit sampling distribution. That is to say, scoring referable to rank signifies sampling from a rectangular universe. For our present purpose, the rank score has an ulterior interest which we may express tersely by saying that it makes data marching in the same or in opposite directions march also in step.

Before making this advantage of the rank score more explicit, let us be clear about the meaning of a term already used. In the *positive* range, we speak of any of the following functions as *monotonic* : (i) $y = 5x + 2$; (ii) $xy = 10$; (iii) $y = \log_{10} x$; (iv) $y = 10^{-x}$. Within the range stated, the increment of y for a positive increment of x is either always positive, as is true of (i) and (iii), or always negative, as is true of (ii) and (iv). This is not true of a periodic function such as $y = \sin x$, which has an infinite number of turning points

$$\left(x = \frac{\pi}{2}, \frac{3\pi}{2}, \text{ etc.} \right).$$

Nor is it true of a parabola such as $y = 2 + 6x - x^2$, since the value of this function increases in the range $x = -\infty$ to $x = 3$, and decreases thereafter.

When discussing whether two scores referable to the same individual such as height and age at menarche go together, we shall assume that any functional relation we hope to establish between them is monotonic in the sense which defines what (i)-(iv) above share in contradistinction to functions with a turning point. *Ceteris paribus*, this means *either* that increase of A entails increase of B *or* that increase of A entails a decrease of B. If we represent paired scores of individuals as points on a graph, a wide range of functional relations is consistent with this restriction, but the functional relation between corresponding rank scores is unique. A single and *fictitious* example * suffices to show that this is so. We shall suppose that we have before us the examination marks (crude scores) of six girls in two subjects as follows :

				Mary	Lucy	Gertrude	Enid	Anne	Muriel
A	French	.	.	53	29	40	82	45	60
B	Music	.	.	51	18	30	90	42	59

We now set out the rank score corresponding to each of the crude scores

Rank	.	.	.	1	2	3	4	5	6
A	French mark	.	82	60	53	45	40	29
B	Music mark	.	90	59	51	42	30	18

Since the crude B score consistently falls as we decrease the A score, the relation between the two sets is monotonic ; but corresponding decrements are not in fixed proportion. Therefore the relationship is not linear. On the other hand, the fact that a monotonic relation holds good signifies also that the rank score with respect to A is always the same as the rank score with respect to B. Hence a graph of the B score as a function of the A score in the rank scale would be the linear relation $B = A$. Let us now consider a second fictitious example of the same type. We see below results of a scriptural knowledge examination and the weekly pocket allowances of six boys :

Pupils . . .	James	Aaron	William	Ian	Neil	Harry
Scripture exam. mark	55	40	60	95	45	80
Weekly pocket money	9d.	2s. 6d.	6d.	2d.	1s.	3d.

* The use of fictitious examples throughout this chapter is deliberate in the hope that it will encourage the student to gain more insight by making up illustrative examples referable to suppositious situations. It is the author's belief that number play of this sort is often more instructive and provocative of self-criticism than working over oft-cited recorded data which do not disclose conclusions of great intrinsic entertainment value.

We may arrange the two sets of figures in order of merit as follows, counting the highest as the figure of unit rank :

Rank	. . .	1	2	3	4	5	6
Scripture mark (A)	.	95	80	60	55	45	40
Pocket money (B)	.	2s. 6d.	1s.	9d.	6d.	3d.	2d.

In terms of rank, the boys thus score as follows :

Pupils	. . .	James	Aaron	William	Ian	Neil	Harry
Rank A	. . .	4	6	3	1	5	2
Rank B	. . .	3	1	4	6	2	5

The suggestion conveyed by these figures is that there is some association between Biblical proficiency and monetary deficiency. To get the figures into clearer focus, we set up one row —it does not matter which—in descending order ; and place beneath each item the corresponding rank score of the other row, thus :

Crude scores	. .	Ian	Harry	William	James	Neil	Aaron
A	. . .	95	80	60	55	45	40
B	. . .	2	3	6	9	12	30
(in pence)							

Rank Scores :

A	. . .	1	2	3	4	5	6
B	. . .	6	5	4	3	2	1

We now see that the crude B scores consistently rise as the A ones fall, but not by increments in fixed proportion. Thus the crude B scores are a monotonic function of the A scores, but not linear. On the other hand, the rank scores correspond exactly in *reverse order ;* and if we plot the rank score B as a function of the rank score A, the relation is truly linear, being $B = (7 - A)$.

In the light of these examples, we may now state three general conclusions, with reference to any crude paired score A and B referable to one and the same individual :

(*a*) if there is perfect *positive correspondence* between crude scores A and B in the sense that B is an increasing monotonic function of A, the two sets of derivative rank scores are in exact *direct* order ;

(*b*) if there is perfect *negative* correspondence between crude scores A and B in the sense that B is a decreasing monotonic function of A, the two sets of derivative rank scores are in exact *reverse* order ;

(*c*) perfect correspondence of either sort signifies an exact linear relation between the rank scores, viz. $B = A$, if positive, and $B = (n + 1 - A)$, if negative.*

Such considerations suggest a method of analysis appropriate to questions of the type illustrated above, viz. : is it more or less likely that a boy who has a large pocket allowance will do well in a scriptural knowledge examination ? Before exploring this suggestion, we should recognise that an affirmative answer to the question does not entitle us to say that a higher all-round pocket allowance would raise the general standard of proficiency in scriptural knowledge. The association might merely signify that parents who encourage Bible study in

* If the relation is linear, $B = mA + A_0$. Since $m = (B_2 - B_1) \div (A_2 - A_1) = -1$, $B = A_0 - A$. When $A = 1$, $B = n$ and $A_0 = n + 1$. Hence $B = (n + 1 - A)$.

the home are more indulgent in matters of finance, that homes which provide opportunities for scrutinising the sacred text are more prosperous, or many other possibilities which need not trouble us. No statistical technique is an adequate substitute for common sense and experience of the materials to which we apply it.

With this reservation, let us now return to the figures of our fictitious class of six boys. Taken at their face value, the conclusion to which they point scarcely needs stating ; and we should not hesitate to draw it, if confronted with equally perfect rank correspondence of a class of sixty pupils. Since the sample is small, the cautious investigator will, however, need some assurance that the result is not a fluke. Fortunately, it is easy to refer this question to a card-pack model. We imagine that : (a) we have before us two incomplete packs (A and B), each of six cards, the ace, 2, 3, 4, 5 and 6 of clubs ; (b) we shuffle each pack and lay side by side the cards of each face upwards in a row ; (c) we then find that the A and B score of cards which occupy the same position in the two rows correspond like the paired examination marks of the above example. Since we call an occurrence a fluke if a comparable occurrence would not be a rare event in a game of chance, a statistical answer to the question stated above is an answer to the question : how often would the result last stated occur in the long run ?

The solution of the problem is elementary. Only one of all possible linear arrangements of all the cards of one pack can correspond exactly to a *particular* arrangement of the cards of the alternate pack. In this case, the number of all possible arrangements is $6! = 720$. If the electivity of such a card event tallies with its long-run frequency, the betting odds are therefore 719 : 1 against the occurrence of the event under discussion in a single trial. The conclusion to which our fictitious figures relating to scripture marks and pocket money lead us is therefore that the result is unlikely to be representative ; and the odds against the supposition that a similar result derived from a class of ten pupils is a fluke would be overwhelming.

If we were content to base judgments of this sort only on one-to-one correspondence, there would be nothing more to say about the problem ; but the result of such an enquiry might not be so clear-cut. By and large, the best examinees might still be the most underpaid, if there were no such *exact* reverse order correspondence. For instance, we might get a result such as the following :

| A | . | . | . | 1 | 2 | 3 | 4 | 5 | 6 |
| B | . | . | . | 5 | 6 | 4 | 2 | 3 | 1 |

In a situation such as this, we might content ourselves with regarding the issue as a two-class difference, either in terms of rank or of actual scores. If the former, we should have

| | A | under 4 | over 3 |
| Mean B | | 5 | 2 |

This reduces the problem to the recognition of a real difference in the domain of representative scoring, and raises no issue which is novel. It has the disadvantage of being an inefficient statistic (*vide* 7.03) and another disadvantage which we shall sidestep in what follows, i.e. it involves recourse to *estimation*.

The method of correlation we are about to explore approaches the issue with due regard to the rank relationship of each pair of scores. When there is perfect positive correspondence, the difference between the rank of any item in the A series and that of the corresponding item in the B series is zero ; and the sum of all such differences is zero. When there is perfect negative correspondence the sum of the differences, *regardless of sign*, is as great as it can be. *Mutatis mutandis* similar remarks apply to any power of the differences. Since the sign of the difference is irrelevant if the power is an *even* number, it is convenient to use the *square* difference ;

and this has a special merit which will emerge later on. When correspondence (or as we shall now say, *correlation*) is perfect, the sum of the square difference will thus be *either* zero (*positive* correlation) *or* a maximum (*negative* correlation) to be defined later. Somewhere between the two must lie a mean value which is that of an indefinitely large number of *independent* sets.

In this context, independence signifies no association between the scores of members of the same pair of a set, as is true of two well-shuffled card packs turned face uppermost in order side by side If our null hypothesis is that the two sets of scores are independent, our algebraic problem will be to define this mean value of the sum and then to specify the frequency distribution of the sum itself in an indefinitely large collection of samples. When we can do so, we can also specify the odds against a departure from the mean value as great as or greater than the observed sum. If they are low, we are not justified in assuming that the figures point to a monotonic relation between the paired scores. A verdict of this sort involves no problem of estimation. Our null hypothesis does not postulate an unknown universe of which we have a sample before us. It postulates a particular card pack model of which all relevant particulars are inherent in the statement of the problem.

If true correspondence between two sets of scores does in fact exist, a different issue arises. We have then to ascertain whether the structure of different samples of paired scores is consistent with the assumption that they come from a common universe of which such correspondence is an essential characteristic. From an elementary point of view, we shall examine the implications of a null hypothesis of this sort in the next chapter of this volume ; but our concern in this one is to explore the implications of a convenient index of correlation only in so far as its numerical value is or is not indicative of correspondence as here defined between one set of scores and another. In doing so, we must be alert to the sort of correspondence which correlation signifies. The null hypothesis that the distribution of two sets of scores is independent predicates that high values of A do not go with high values (positive correlation) or with low values (negative correlation) of B. Hence the conclusion that the distribution of two sets of scores is not inconsistent with the possibility of independent assortment is another way of saying that we have no reason to infer any correlation between them ; but we shall later see (p. 349) that the absence of any correlation does *not* necessarily imply true statistical independence.

8.03 SPEARMAN'S RANK ASSOCIATION INDEX

For economy of space we may visualise the rank of the individual card of the two packs, as in Figs. 74–76, by the pips on the face of a die ; and we can get to grips with the problem most easily by first asking : what must the sum of the square differences be, when there is *perfect reverse* order correspondence. The results shown in Fig. 74 are as follows :

No. of Cards in each Pack (n)	Sum of the Squared Differences ($\sum d^2$)	No of Cards in each Pack (n)	Sum of the Squared Differences ($\sum d^2$)
1	0	4	20
2	2	5	40
3	8	6	70

To discover a rule we apply the method of the vanishing triangle, denoting the sum of the square differences by u_n :

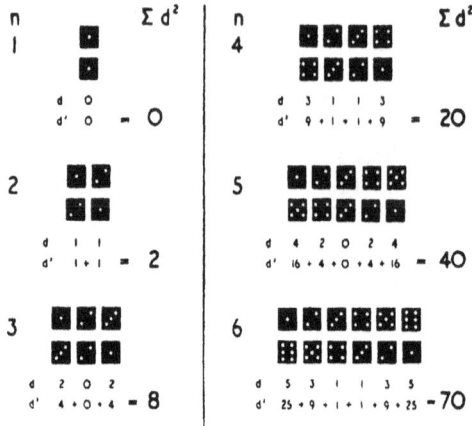

FIG. 74. Representation of the Total Rank Difference when there is perfect *reverse* order correspondence.

n	.	1	2	3	4	5	6
u_n	.	0	2	8	20	40	70
Δu_n	.		2	6	12	20	30
$\Delta^2 u_n$	.			4	6	8	10
$\Delta^3 u_n$	.				2	2	2
$\Delta^4 u_n$	.					0	0
$\Delta^5 u_n$	.						0

Since we start with a term of rank 1, we make the appropriate adjustment of (vii) in 1·05, *viz.*:

$$u_n = (1 + \Delta)^{n-1} . u_1$$

$$= u_1 + (n-1)\Delta u_1 + \frac{(n-1)(n-2)}{2!}\Delta^2 u_1 + \frac{(n-1)(n-2)(n-3)}{3!}\Delta^3 u_1 \ldots \text{etc.}$$

Since the fourth and subsequent differences vanish it is not necessary to add any other terms, so that

$$u_n = 0 + (n-1)(2) + \frac{(n-1)(n-2)}{2!}(4) + \frac{(n-1)(n-2)(n-3)}{3!}(2) \quad (2)$$

$$= (n-1)\left\{2 + 2(n-2) + \frac{(n-2)(n-3)}{3}\right\}$$

$$= \frac{n(n^2-1)}{3}.$$

This is the *maximum* value of the sum of the square difference, and we shall denote it thus :

$$R_d^2 = \frac{n^3 - n}{3} \qquad . \qquad . \qquad . \qquad . \qquad . \quad (i)$$

The minimum value of $\sum d^2$ is, of course, zero. For a set of n observations each assigned two rank values, we thus have

$$\sum d^2 = 0 \text{ when there is perfect } \textit{direct} \text{ order correspondence.}$$

$$\sum d^2 = R_d^2 = \frac{n^3 - n}{3} \text{ when there is perfect } \textit{reverse} \text{ order correspondence.}$$

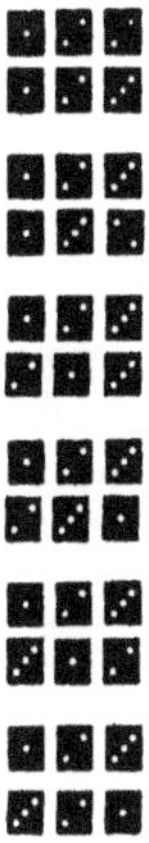

	$\Sigma(d)$	$\dfrac{\Sigma(d)}{n}$	$\Sigma_d{}^2$	$\dfrac{\Sigma_d{}^2}{n}$
	$0+0+0$ = 0	0	$0+0+0$ = 0	0
	$0+1+1$ = 2	$\frac{2}{3}$	$0+1+1$ = 2	$\frac{2}{3}$
	$1+1+0$ = 2	$\frac{2}{3}$	$1+1+0$ = 2	$\frac{2}{3}$
	$1+1+2$ = 4	$\frac{4}{3}$	$1+1+4$ = 6	2
	$2+1+1$ = 4	$\frac{4}{3}$	$4+1+1$ = 6	2
	$2+0+2$ = 4	$\frac{4}{3}$	$4+0+4$ = 8	$\frac{8}{3}$

$\Sigma|d|$	0	2	4
$\dfrac{\Sigma|d|}{n}$	$\frac{0}{2}$	$\frac{2}{3}$	$\frac{4}{3}$
γ	$\frac{1}{6}$	$\frac{2}{6}$	$\frac{3}{6}$

Σd^2	0	2	4	6	8
$\dfrac{\Sigma d^2}{n}$	$\frac{0}{2}$	$\frac{2}{3}$	$\frac{4}{3}$	$\frac{6}{3}$	$\frac{8}{3}$
γ	$\frac{1}{6}$	$\frac{2}{6}$	$\frac{2}{6}$	$\frac{2}{6}$	$\frac{1}{6}$

FIG. 75. The sum of the Rank Difference referable to all possible permutations of three pairs, the order of one set being fixed.

Somewhere between these limits lies a *mean* sum (S_m^2) which signifies no correspondence at all. If the number of double-ranked items is n, S_m^2 is the mean of the sum of the square differences for the $n!$ permutations of the n ranks of one series w.r.t. a fixed order of those of the other. To determine S_m^2 we can proceed as in Figs. 75-76. If the sample consists of one item, the two ranks are necessarily the same, the rank difference zero and $S_m^2 = 0$. If there are two items there are two arrangements, *viz.* :

$$
\begin{array}{cccc}
1 & 2 & 1 & 2 \\
1 & 2 & 2 & 1 \\
\end{array}
$$

$$
\begin{array}{ccccc}
d^2 & 0 & 0 & 1^2 & 1^2 \\
\Sigma d^2 & & 0 & & 2 \\
\end{array}
$$

The mean value of Σd^2 when $n = 2$ is therefore

$$\frac{0+2}{2!} = 1.$$

From Fig. 75 we see that the mean value of Σd^2 when $n = 3$ is

$$\frac{0 + 2(2) + 2(6) + 8}{3!} = 4.$$

$$\Sigma \, d^2$$

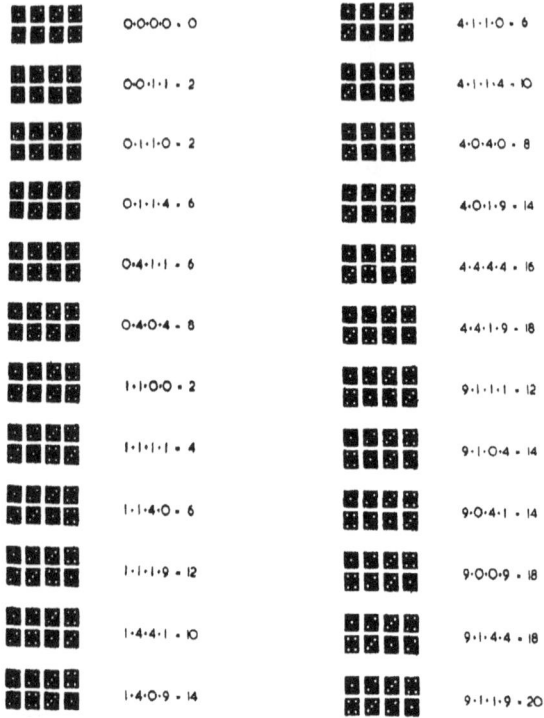

0·0·0·0 · 0		4·1·1·0 · 6
0·0·1·1 · 2		4·1·1·4 · 10
0·1·1·0 · 2		4·0·4·0 · 8
0·1·1·4 · 6		4·0·1·9 · 14
0·4·1·1 · 6		4·4·4·4 · 16
0·4·0·4 · 8		4·4·1·9 · 18
1·1·0·0 · 2		9·1·1·1 · 12
1·1·1·1 · 4		9·1·0·4 · 14
1·1·4·0 · 6		9·0·4·1 · 14
1·1·1·9 · 12		9·0·0·9 · 18
1·4·4·1 · 10		9·1·4·4 · 18
1·4·0·9 · 14		9·1·1·9 · 20

Σd^2	0	2	4	6	8	10	12	14	16	18	20
$\frac{\Sigma d^2}{4}$	0	$\frac{1}{2}$	1	$\frac{3}{2}$	2	$\frac{5}{2}$	3	$\frac{7}{2}$	4	$\frac{9}{2}$	5
Y_4	$\frac{1}{24}$	$\frac{3}{24}$	$\frac{1}{24}$	$\frac{4}{24}$	$\frac{2}{24}$	$\frac{2}{24}$	$\frac{2}{24}$	$\frac{4}{24}$	$\frac{1}{24}$	$\frac{3}{24}$	$\frac{1}{24}$

FIG. 76. The sum of the Rank Differences referable to each of the 4! permutations of four pairs, the order of set being fixed.

From Fig. 76 we also see that the mean of Σd^2 when $n = 4$ is

$$\frac{1(0) + 3(2) + 1(4) + 4(6) + 2(8) + 2(10) + 2(12) + 4(14) + 1(16) + 3(18) + 1(20)}{4!} = 1$$

By the same procedure we find that the mean value of $\Sigma d^2 = 20$ when $n = 5$ (Table 1). can thus make a vanishing triangle :

n	.	. 1		2		3		4		5
u_n	.	. 0		1		4		10		20
Δu_n	.		1		3		6		10	
$\Delta^2 u_n$	.			2		3		4		
$\Delta^3 u_n$	.				1		1			
$\Delta^4 u_n$	.					0				

Accordingly

$$u_n = S_m^2 = 0 + 1(n-1) + \frac{2(n-1)(n-2)}{2!} + \frac{(n-1)(n-2)(n-3)}{3!}$$

$$S_m^2 = \frac{n^3 - n}{6} \quad . \quad . \quad . \quad . \quad . \quad . \quad . \quad \text{(ii)}$$

Hence, from (i)

$$S_m^2 = \tfrac{1}{2}R_d^2; \quad and \quad R_d^2 = 2S_m^2 \quad . \quad . \quad . \quad . \quad \text{(iii)}$$

The last relation suggests a very convenient *summarising index* of correspondence :

$$\rho = 1 - \frac{\sum d^2}{S_m^2} \quad . \quad . \quad . \quad . \quad . \quad \text{(iv)}$$

$$= 1 - \frac{6\sum d^2}{n^3 - n} \quad . \quad . \quad . \quad . \quad . \quad \text{(v)}$$

When correspondence is perfect and positive, $\sum d^2 = 0$, so that

$$\rho = +1 \quad . \quad . \quad . \quad . \quad . \quad . \quad \text{(vi)}$$

When correspondence is perfect and negative, $\sum d^2 = 2S_m^2$ by (iii) above, so that

$$\rho = 1 - \frac{2S_m^2}{S_m^2} = -1. \quad . \quad . \quad . \quad . \quad \text{(vii)}$$

When there is statistical independence the theoretical expectation of $\sum d^2$ is its card pack model mean value, i.e. S_m^2 and

$$\rho = 1 - \frac{S_m^2}{S_m^2} = 0. \quad . \quad . \quad . \quad . \quad \text{(viii)}$$

Example. The two sets of rank scores on page 330 will suffice to illustrate the use of (v), *viz.* :

A	.	.	1	2	3	4	5	6
B	.	.	5	6	4	2	3	1
d	.	.	4	4	1	-2	-2	-5
d^2	.	.	16	16	1	4	4	25

The total value of the square difference is

$$2(16) + 1 + 2(4) + 25 = 66.$$

If there were perfect correspondence in reverse order, the total corresponding to $\rho = -1$ would be

$$\frac{6^3 - 6}{3} = 70.$$

From (v) the actual value of ρ is

$$1 - \frac{6(66)}{6^3 - 6} = -0.886.$$

If our null hypothesis is that there is *no* correspondence, the expected value of *Spearman's coefficient* defined by (v) is *zero* by (viii) above. We therefore reject the null hypothesis, if the value of ρ computed from our data differs significantly from zero ; but we have still to establish a criterion of significance with this end in view. In short, we need to know the sampling distribution of ρ in conformity with the assumption that its mean value is zero.

Before exploring the properties of such a distribution, however, it will be well to take stock of what the numerical value of ρ signifies. For samples of the same size a *numerically* high value of ρ indicates a greater departure from expectation in accordance with the null hypothesis stated above, hence a less frequent occurrence than if no association exists. *Ceteris paribus*, a high numerical value of ρ therefore signifies justification for greater *confidence in the existence of association between the two attributes to which we assign rank scores*. This is not necessarily the same as saying that ρ is a measure of greater or less *correspondence* between the attributes.

The convenience of $\sum d^2$ as a criterion of correspondence depends partly on a formal identity we shall come to later (8.04–5), but it depends also on the fact that its maximum is a simple multiple of its mean value. Hence it is easy to devise a summarising index which is symmetrical, having the values ± 1 for perfect association (positive or negative) and zero for no association at all. By the methods used above the student will easily verify the statement that the mean value (S_m) of $\sum |d|$, the numerical sum of the differences themselves regardless of sign, is given by

$$S_m = \frac{n^2 - 1}{3} \qquad \qquad \text{(ix)}$$

An exact formula for the maximum value of $\sum |d|$ depends on whether n is even or odd. If n is odd, the maximum (R_d) is given by

$$R_d = \frac{n^2 - 1}{2} = \frac{3}{2} S_m \qquad \qquad \text{(x)}$$

For even values of n, $R_d = \frac{1}{2} n^2$. Hence equation (x) is a good approximation for *even* values of n, if n is also large. Spearman has proposed an index (R) based on $\sum |d|$ as the criterion of association, but otherwise comparable to ρ, *viz.* :

$$R = 1 - \frac{\sum |d|}{S_m} = 1 - \frac{3 \sum |d|}{n^2 - 1} \qquad \qquad \text{(xi)}$$

When there is perfect positive correlation $\sum |d| = 0$ and $R = +1$. When there is no correspondence $\sum |d| = S_m$ and $R = 0$. When there is perfect negative correlation

$$\sum |d| = R_d' = \tfrac{1}{2}(n^2 - 1),$$

so that $R = -0.5$. Thus the " footrule " coefficient R is not a symmetrical index of positive and negative correlation, and the negative limit stated is exact only if n is odd.

★ ★ ★ ★ Strictly speaking, that of the sum of the rank score differences or of their squares is a non-replacement distribution, since no item from one and the same of the two n-fold parent universes can be present more than once in the n-fold sample of paired scores. Accordingly, we have derived the mean value of its square without recourse to the chessboard visualisation of sampling with replacement. Actually, the mean difference and that of its square is the same when we do replace each 2-fold sample before taking another. Fig. 77 exhibits the lay-out for the 6-fold rectangular universe $(n = 6)$ of the cubical die. If d_m is the mean value of the difference between the scores of a single 2-fold toss, the mean value of the n-fold sum of such differences is $n \cdot d_m$ and that of the square differences is $n \cdot d_m^2$. From the diagonal entries of the chessboard, we derive the following pattern for the n-fold sum of all the differences :

$$n(0) + 2[(n-1)(1) + (n-2)(2) + (n-3)(3) \ldots 2(n-2) + 1(n-1)].$$

★ ★ ★ ★ Omit on first reading.

SCORE DIFFERENCE DISTRIBUTION FOR 2-FOLD SAMPLES

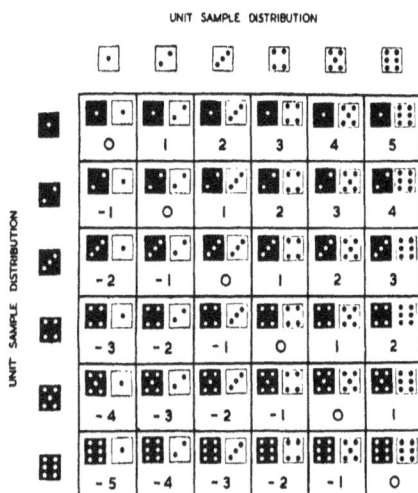

FIG. 77. Chessboard lay-out for all possible pairs of differences between two rank scores.

For the square differences we have

$$n(0^2) + 2[(n-1)1^2 + (n-2)2^2 + (n-3)3^2 \ldots 2(n-2)^2 + 1(n-1)^2] = 2\sum_{1}^{n-1}(n-r)r^2.$$

Since there are n^2 cells in the chessboard

$$d_m^2 = \frac{2}{n^2} \cdot \sum_{1}^{n-1}(n-r)r^2,$$

$$\therefore S_m^2 = n \cdot d_m^2 = \frac{2}{n} \cdot \sum_{1}^{n-1}(n-r)r^2,$$

$$\therefore S_m^2 = 2\sum_{1}^{n-1}r^2 - \frac{2}{n}\sum_{1}^{n-1}r^3.$$

By the method given on page 18 and on page 23 as an exercise

$$\sum_{1}^{n}r^2 = \frac{n(n+1)(2n+1)}{6},$$

$$\therefore 2\sum_{1}^{n-1}r^2 = \frac{n(n-1)(2n-1)}{3}.$$

$$\sum_{1}^{n}r^3 = \frac{n^2(n+1)^2}{4}.$$

$$\frac{2}{n}\sum_{1}^{n-1}r^3 = \frac{n(n-1)^2}{2},$$

$$\therefore S_m^2 = \frac{n(n-1)}{6}(4n-2-3n+3) = \frac{n(n-1)(n+1)}{6} = \frac{n^3-n}{6}.$$

22

We can also derive a formula for R_d^2, the maximum value of $\sum d^2$, by recourse to figurate summation formulæ. When there is perfect correspondence of rank scores B and A in reverse order, we have seen (p. 329) that $B = (n + 1 - A)$. We can therefore write

$$(B - A)^2 = (n + 1 - 2A)^2 = (n + 1)^2 - 4(n + 1)A + 4A^2,$$

$$\therefore \sum_1^n (B - A)^2 = (n + 1)^2 \sum_1^n 1 - 4(n + 1) \sum_1^n A + 4 \sum_1^n A^2$$

$$= n(n + 1)^2 - 4(n + 1) \cdot \tfrac{1}{2}n(n + 1) + 4 \cdot \frac{n(n + 1)(2n + 1)}{6} = \frac{n^3 - n}{3}. \star\star\star\star$$

EXERCISE 8.03

Calculate for each pair of the following sets of examination results :

(a) Spearman's rank coefficient based on the sum of the square differences.
(b) Spearman's footrule coefficient based on the modular sum of the differences.

Pupil.	English Literature.	Geography.	French.	Physics.	Economics.
A	72	55	86	54	55
B	65	68	68	52	63
C	62	64	73	68	78
D	58	60	80	61	69
E	55	35	59	58	42
F	50	42	55	72	61
G	46	81	62	42	59
H	45	29	54	12	50
I	39	51	63	36	73
J	38	76	57	26	26
K	36	56	61	70	49
L	34	71	46	21	32
M	33	17	43	79	29
N	24	49	42	37	42
O	22	37	48	27	19
P	19	42	29	34	4
Q	16	28	37	42	30
R	15	45	18	18	51
S	12	32	23	28	22
T	6	10	15	30	24

8.04 DISTRIBUTION OF SPEARMAN'S COEFFICIENT FOR SMALL SAMPLES

What we have derived in 8.03 is a convenient summarising index of correspondence, the numerical value of which signifies little for our present purpose, unless we can state the odds for or against getting as great a value of ρ in accordance with the null hypothesis that there is *no* correspondence between the attributes concerned. To do so, we must be able to specify the frequency distribution of all possible numerical values of ρ within the framework of the assumption that its mean value is zero. We have therefore to enumerate values of the sum of the square differences for all possible permutations of one set of ranks w.r.t. a fixed order of the other.

TABLE 1

Frequencies of Square Rank Difference Sum for Samples of 1–5

Σd^2	1.	2.	3.	4.	5.	$[\Sigma d^2]'$	Σd^2	1.	2.	3.	4.	5.	$[\Sigma d^2]'$
0	1	$\frac{1}{2}$	$\frac{1}{6}$	$\frac{1}{24}$	$\frac{1}{120}$	0	22	—	—	—	—	$\frac{10}{120}$	484
2		$\frac{1}{2}$	$\frac{2}{6}$	$\frac{3}{24}$	$\frac{4}{120}$	4	24	—	—	—	—	$\frac{6}{120}$	576
4	—	—	$\frac{0}{6}$	$\frac{1}{24}$	$\frac{3}{120}$	16	26	—	—	—	—	$\frac{10}{120}$	676
6	—	—	$\frac{2}{6}$	$\frac{2}{24}$	$\frac{6}{120}$	36	28	—	—	—	—	$\frac{8}{120}$	784
8	—	—	$\frac{1}{6}$	$\frac{2}{24}$	$\frac{7}{120}$	64	30	—	—	—	—	$\frac{8}{120}$	900
10	—	—	—	$\frac{2}{24}$	$\frac{6}{120}$	100	32	—	—	—	—	$\frac{7}{120}$	1024
12	—	—	—	$\frac{2}{24}$	$\frac{4}{120}$	144	34	—	—	—	—	$\frac{6}{120}$	1156
14	—	—	—	$\frac{4}{24}$	$\frac{10}{120}$	196	36	—	—	—	—	$\frac{3}{120}$	1296
16	—	—	—	$\frac{1}{24}$	$\frac{6}{120}$	256	38	—	—	—	—	$\frac{4}{120}$	1444
18	—	—	—	$\frac{3}{24}$	$\frac{10}{120}$	324	40	—	—	—	—	$\frac{1}{120}$	1600
20	—	—	—	$\frac{1}{24}$	$\frac{6}{120}$	400							

Figs. 75 and 76 respectively lay out all possible permutations of three rank score pairs and four rank score pairs with the corresponding values of Σd^2, viz.: $0, 2, 4 \ldots \frac{n^3 - n}{3}$. When $n = 3$,

$$\frac{n^3 - n}{3} = \frac{27 - 3}{3} = 8,$$

$$S_m^2 = \frac{27 - 3}{6} = 4,$$

$$\therefore \rho = 1 - \frac{\Sigma d^2}{4} \qquad \qquad \text{. . . . . . (i)}$$

If $n = 3$, we derive from Fig. 75 the following values of ρ with frequencies in the ratio $1 : 2 : 0 : 2 : 1$

$$(1 - \tfrac{0}{4}),\ (1 - \tfrac{2}{4}),\ (1 - \tfrac{4}{4}),\ (1 - \tfrac{6}{4}),\ (1 - \tfrac{8}{4}).$$

When $n = 4$,

$$\frac{n^3 - n}{3} = \frac{64 - 4}{3} = 20,$$

$$S_m^2 = \frac{64 - 4}{6} = 10,$$

$$\therefore \rho = 1 - \frac{\Sigma d^2}{10} \qquad \qquad \text{. . . . . . (ii)}$$

By recourse to Figs. 75 and 76 we can thus set out the sampling distributions of ρ for $n = 3$ and $n = 4$ as below :

(a) When $n = 3$

Σd^2	0	2	6	8
y	$\frac{1}{6}$	$\frac{1}{3}$	$\frac{1}{3}$	$\frac{1}{6}$
ρ	1·0	0·5	− 0·5	− 1·0

(b) When $n = 4$

Σd^2	0	2	4	6	8	10	12	14	16	18	20
y	$\frac{1}{24}$	$\frac{3}{24}$	$\frac{1}{24}$	$\frac{4}{24}$	$\frac{2}{24}$	$\frac{2}{24}$	$\frac{2}{24}$	$\frac{4}{24}$	$\frac{1}{24}$	$\frac{3}{24}$	$\frac{1}{24}$
ρ	1·0	0·8	0·6	0·4	0·2	0	−0·2	−0·4	−0·6	−0·8	−1·0

The last distribution has an oscillatory character ; and the two distributions conform to no simple algebraical rule connecting ρ or Σd^2 with y in terms of n. Kendall, who has made the most exhaustive investigation of the sampling distribution of ρ has tabulated them for small samples ($n = 2$ to $n = 8$) as above. By recourse to his tabulation we can test the probability of getting a result numerically as great as the one cited on page 335, viz. $\rho = -0·886$ ($\Sigma d^2 = 66$) for six rank score pairs. For an equal positive value of ρ ($= +0·886$), $\Sigma d^2 = 4$. The sum Σd^2 is always an even * number, and the next even numbers respectively below 66 and above 4 are 64 and 6. When $n = 6$ and $\Sigma d^2 = 64$, $\rho = -0·829$. When $n = 6$ and $\Sigma d^2 = 6$, $\rho = +0·829$. Our problem is thus to find the expectation that ρ will lie outside the range $\pm 0·829$. From Kendall's table of Σd^2 (16.1, p. 396, Vol. I of his treatise), we see that the expectation for Σd^2 in the range 6 to 64, i.e. of ρ in the range $\pm 0·829$, is 696 ÷ 720. The expectation of a result numerically as great as $\pm 0·886$ is therefore 24 ÷ 720 $= \frac{1}{30}$. The odds in this case are therefore 29 : 1 against the null hypothesis, if we use the *modular* likelihood (p. 204) of the event as our criterion of significance. We are at liberty to state the issue in the less exacting *vector* form : what are the odds against getting a value of Σd^2 as great as 66, i.e. a value of ρ greater or less than $-0·829$. The expectation is then 12 ÷ 720 and the odds against are 59 : 1. Either way, we should have some reason to assume the existence of negative association.

The distribution cited above for $n = 4$ is distinctly bimodal and oscillatory, recalling the distribution of the proportionate score difference for *co-prime* samples (p. 172). For larger samples the oscillations become less conspicuous, and the dip at the mean becomes less striking. When $n = 8$ the still jagged contour of the histogram is suggestive of a normal distribution. For small samples (*up to* 8), Kendall's tables give an exact distribution of Σd^2 (and hence of ρ as explained above) based on enumeration of all possible permutations of one series of rank scores w.r.t. a fixed order of the other as in Figs. 75 and 76 ; but the labour entailed in extending such tables to include high values of n would be stupendous. To use this test, it is therefore necessary to discover an approximate formula for the frequency of a particular value of ρ or of a range of such values. For reasons set forth in Chapter 6, we proceed first to examine the moments of the distribution of ρ.

Since the distribution of ρ is symmetrical about its mean value ($\rho = 0$) and that of Σd^2 is also symmetrical about its mean value (S_m^2), the odd moments about the mean are all zero. We need therefore concern ourselves only with the variance and fourth moment of the distribution of ρ and of Σd^2 of which ρ is a linear function. Since we are primarily concerned with the distribution of ρ, it will simplify our task if we first establish a general theorem which enables us to determine $V(\rho)$ if we know $V(\Sigma d^2)$. This theorem (8.05) is applicable to any situation in which one variate is a linear function of another.

---

* Since $\Sigma A^2 = \Sigma B^2 = \sum_1^n r^2$, $\Sigma A^2 + \Sigma B^2 = 2\Sigma A^2$. Now $d^2 = (A - B)^2 = A^2 + B^2 - 2AB$.

Hence
$$\Sigma d^2 = \Sigma A^2 + \Sigma B^2 - 2\Sigma AB,$$
$$\therefore \Sigma d^2 = 2(\Sigma A^2 - \Sigma AB).$$

8.05 Variance of a Linear Function of a Variate

In what follows K and C are constants, and p, q are variates connected by the relation

$$p = K + Cq \qquad . \qquad . \qquad . \qquad . \qquad . \qquad . \text{(i)}$$

We denote the mean values of p and q by $M(p)$, $M(q)$, the squares of the mean values by $M^2(p)$, $M^2(q)$, their variances by $V(p)$, $V(q)$, and their second moments about the origin by $V_0(p)$, $V_0(q)$. In accordance with (iii) of 3.06, we may therefore write

$$
\begin{aligned}
V(p) &= \Sigma y(K + Cq)^2 - [\Sigma y(K + Cq)]^2 \\
&= \Sigma y(K^2 + C^2q^2 + 2KCq) - (K\Sigma y + C\Sigma yq)^2 \\
&= K^2\Sigma y + C^2\Sigma y \cdot q^2 + 2KC\Sigma y \cdot q - [K + C \cdot M(q)]^2 \\
&= K^2 + C^2V_0(q) + 2KC \cdot M(q) - [K + C \cdot M(q)]^2 \\
&= C^2V_0(q) - C^2M^2(q) \\
&= C^2[V_0(q) - M^2(q)] \\
&= C^2 \cdot V(q) \qquad . \qquad . \qquad . \qquad . \qquad . \qquad . \qquad . \qquad . \qquad . \text{(ii)}
\end{aligned}
$$

From the definition of ρ in (v) of 8.03, we may write it in the form $\rho = K + C(\Sigma d^2)$, in which $K = 1$ and

$$C = \frac{-1}{S_m^2} = \frac{-6}{n^3 - n}.$$

$$\therefore \; V(\rho) = \frac{1}{(S_m^2)^2} \cdot V(\Sigma d^2) \qquad . \qquad . \qquad . \qquad . \qquad . \text{(iii)}$$

In accordance with the definitions on page 132, we have

$$V(\Sigma d^2) = \Sigma y \cdot (\Sigma d^2)^2 - (S_m^2)^2 \qquad . \qquad . \qquad . \qquad . \text{(iv)}$$

The extreme right-hand column of Table 1 shows the weighted values of $(\Sigma d^2)^2$ corresponding to each weighted value of Σd^2 from 0 to its maximum $2S_m^2$. By weighting each item in this column by its appropriate frequency (y) for a given value of n and adding the results, we get the value of $V_0(\Sigma d^2)$ in the above. For instance, when $n = 3$, we have

Σd^2.	$(\Sigma d^2)^2$.	y.	$y(\Sigma d^2)$.	$y(\Sigma d^2)^2$.
0	0	$\frac{1}{6}$	0	0
2	4	$\frac{2}{6}$	$\frac{2}{3}$	$\frac{4}{3}$
4	16	0	0	0
6	36	$\frac{2}{6}$	2	12
8	64	$\frac{1}{6}$	$\frac{4}{3}$	$\frac{3\,2}{3}$
		Total	4	24
			$= M(\Sigma d^2)$	$= V_0(\Sigma d^2)$
			$= S_m^2$	

In accordance with (iv)

$$V(\Sigma d^2) = 24 - 4^2 = 8.$$

In accordance with (iii) above

$$V(\rho) = \frac{1}{4^2} \cdot 8 = \frac{1}{2}.$$

By this method we may compile a table as follows :

$n.$	$S_m^2.$	$V_0(\Sigma d^2).$	$V(\Sigma d^2).$	$V(\rho).$
1	0	0	0	0
2	1	2	1	1
3	4	24	8	$\frac{1}{2}$
4	10	$\frac{400}{3}$	$\frac{100}{3}$	$\frac{1}{3}$
5	20	500	100	$\frac{1}{4}$

In the same way, we find

n	6	7	8
$V(\rho)$	$\frac{1}{5}$	$\frac{1}{6}$	$\frac{1}{7}$

And, in general, for values of $n > 1$

$$V(\rho) = \frac{1}{n-1} \qquad \qquad \text{(v)}$$

An exact expression for the fourth moment is obtainable by adapting the foregoing procedure, *viz.* :

$$m_4 = \frac{3(25n^3 - 38n^2 - 35n + 72)}{25n(n+1)(n-1)^3}.$$

When n is fairly large, this is approximately

$$\frac{3}{n^2-1} \qquad \qquad \text{(vi)}$$

The value of β_1 (Pearson's coefficient of *skewness*) is zero, since $m_3 = 0$ and that of β_2 (Pearson's coefficient of *flatness*) from (v) and (vi) is approximately

$$\frac{m_4}{m_2^2} \simeq \frac{3(n-1)^2}{(n-1)(n+1)} = \frac{3(n-1)}{(n+1)} \qquad \qquad \text{(vii)}$$

For a symmetrical distribution which has a limited range on either side of the mean, the type of Pearson curve likely to provide a good fit is Type II, of which the general expression is of the form (xxix) in (6.08) :

$$Y = \frac{1}{a \cdot 2^{2z+1} \cdot B(1+z, 1+z)} \cdot \left(1 - \frac{x^2}{a^2}\right)^z \qquad \qquad \text{(viii)}$$

The value of the constants a and z in the above is given by (xxiv)-(xxvii) of 6.08 as

$$a^2 = \frac{2\beta_2 m_2}{3 - \beta_2}; \quad z = \frac{5\beta_2 - 9}{2(3 - \beta_2)} \qquad \qquad \text{(ix)}$$

From (vii) we have

$$2\beta_2 \cdot m_2 = \frac{2 \cdot 3(n-1)}{(n+1)(n-1)} = \frac{6}{n+1} \qquad \qquad \text{(x)}$$

$$3 - \beta_2 = 3 - \frac{3(n-1)}{(n+1)} = \frac{6}{n+1} \qquad \qquad \text{(xi)}$$

Hence from (ix) $a = +1$, and hence from (vii) also

$$5\beta_2 - 9 = \frac{6(n-4)}{n+1} \qquad \qquad \text{(xii)}$$

Hence from (ix), (xi) and (xii)

$$z = \frac{6(n-4)}{n+1} \times \frac{n+1}{2.6} = \frac{n-4}{2} \qquad \text{(xiii)}$$

$$\therefore 1 + z = \frac{n-2}{2} \quad and \quad 2z + 1 = n - 3. \qquad \text{(xiv)}$$

On substituting these values in (viii) we obtain *

$$Y = \frac{1}{2^{n-3} . B\left(\dfrac{n-2}{2}, \dfrac{n-2}{2}\right)} \cdot (1 - \rho^2)^{\frac{n-4}{2}} \qquad \text{(xv)}$$

And for the total area enclosed by the curve, we have

$$A = \frac{2}{2^{n-3} . B\left(\dfrac{n-2}{2}, \dfrac{n-2}{2}\right)} \int_0^1 (1 - \rho^2)^{\frac{n-4}{2}} . d\rho = 1 . \qquad \text{(xvi)}$$

Kendall has shown that this integral, with due regard to the correct limits of integration, gives a very good approximation to the expectation that a value of $\sum d^2$ or the corresponding value of ρ will lie within a specified range even for values of n as small as 10 ; and he has prepared tables for dealing with small samples by recourse to it. When n is very large Pearson's Type II merges into the normal distribution ; but a significance test based on the assumption of normality by the use of (v) to provide the appropriate value of $V(\rho)$ is not very satisfactory unless n is at least 40.

EXERCISE 8.05

1. Investigate the sampling distribution of Spearman's footrule coefficient, and prepare a table like Table 1 for samples of 2 to 5 pairs inclusive.

2. Find a formula for the variance of the footrule coefficient on the assumption that its mean value is zero.

8.06 THE COVARIANCE CRITERION

The approach of 8.03–8.05 is applicable to two variates, if we can assign to each of the n terms a rank which is one of the integers $1 \ldots n$ inclusive. It is then possible to define an index of correspondence involving the ratio between an observed sum of square differences and its theoretical mean value. It is also possible to express this ratio in a form which suggests a different criterion consistent with statistical independence. The usefulness of the alternative expression which involves only the variance of the score distributions and the mean product

---

* We may use the substitution of (xii) in 6.07, *viz.* $B(\frac{1}{2}, z + 1) = 2^{2z+1} B(z + 1, z + 1)$, so that (xv) is reducible to the simpler form :

$$Y = \frac{1}{B\left(\frac{1}{2}, \dfrac{n-2}{2}\right)} \cdot (1 - \rho^2)^{\frac{n-4}{2}}.$$

of corresponding score deviations does not reside in the particular method of scoring on which we have hitherto relied.

For pairs of rank scores the means of each series are identical and the variances are also identical, viz. :

$$M_a = \frac{(n+1)}{2} = M_b.$$

$$V_a = \sum_{r=1}^{r=n} \frac{1}{n} \cdot r^2 - \left[\frac{(n+1)}{2}\right]^2 = V_b,$$

$$\therefore V_a = \frac{n^2 - 1}{12} = V_b.$$

Since $V_a = V_b$, $\sigma_a^2 = \sigma_b^2$ and

$$V_a = \sigma_a \cdot \sigma_b = V_b,$$

$$\therefore \sigma_a \sigma_b = \frac{n^2 - 1}{12}.$$

Hence from (ii) in 8.03

$$S_m^2 = 2n\sigma_a\sigma_b.$$

We may therefore write (iv) of 8.03 in the form

$$\rho = 1 - \frac{\Sigma d^2}{2n \cdot \sigma_a \cdot \sigma_b}$$

$$= \frac{2n\sigma_a\sigma_b - \Sigma d^2}{2n \cdot \sigma_a\sigma_b}. \qquad . \qquad . \qquad . \qquad . \qquad \text{(i)}$$

If we denote the rank of any score of the A series as A and the rank of its pair in the B series by B

$$\Sigma d^2 = \Sigma(A - B)^2$$
$$= \Sigma(\overline{A - M_a} - \overline{B - M_b})^2$$
$$= \Sigma(A - M_a)^2 + \Sigma(B - M_b)^2 - 2\Sigma(A - M_a)(B - M_b)$$
$$= nV_a + nV_b - 2\Sigma(A - M_a)(B - M_b)$$
$$= 2n\sigma_a\sigma_b - 2\Sigma(A - M_a)(B - M_b) \qquad . \qquad . \qquad . \qquad . \qquad \text{(ii)}$$

By substitution of (ii) in (i) we get

$$\rho = \frac{\Sigma(A - M_a)(B - M_b)}{n \, \sigma_a\sigma_b} \qquad . \qquad . \qquad . \qquad . \qquad \text{(iii)}$$

We call the mean value of the product of corresponding score deviations their *covariance*, written $Cov(A, B)$, i.e. for n pairs

$$Cov(A, B) = \frac{\Sigma(A - M_a)(B - M_b)}{n} \qquad . \qquad . \qquad . \qquad . \qquad \text{(iv)}$$

In this formula the frequency (y) of any score is the same, being $1/n$. More generally, for any frequency distribution

$$Cov(A, B) = \Sigma y(A - M_a)(B - M_b). \qquad . \qquad . \qquad . \qquad . \qquad \text{(v)}$$

By substitution in (iii)

$$\rho = \frac{Cov(A, B)}{\sigma_a \cdot \sigma_b} \qquad . \qquad . \qquad . \qquad . \qquad . \qquad \text{(vi)}$$

Since $\rho = 0$ when there is no association, a criterion of associative neutrality (i.e. statistical independence) for rank score pairs is that the covariance of the distribution is zero. This is not a property peculiar to rectangular distributions, such as that of rank scores. We shall now see that it is a law of the chessboard device. Indeed, it is a universal property of a joint distribution of two sets of scores which are statistically independent, though the converse (p. 349) is not necessarily true.

The same lay-out which we have elsewhere used to illustrate the double summation involved in determining the mean of the distribution of a difference (or sum) and its variance will serve to exhibit the steps of the argument. In Fig. 78 we have labelled our A and B scores respectively from A_o to A_a and B_o to B_b, denoting the general B-score by B_l with frequency w_l. Since the student should get familiar with the meaning of the operations involved, no apology is necessary, if we modify our symbols by: (a) denoting as A_1 and B_1 the lowest score values, (b) employing B_j and w_j in place of B_l and w_l in what follows. Thus our two variates are : (i) A_1, $A_2 \ldots A_k \ldots A_a$ with mean M_a, frequency distribution $v_1, v_2 \ldots v_k \ldots v_a$ with variance V_a; (ii) $B_1, B_2 \ldots B_j \ldots B_b$ with mean M_b, frequency distribution $w_1, w_2 \ldots w_j \ldots w_b$ with variance V_b. The covariance is given by the weighted mean product

$$Cov\,(A, B) = \sum_{k=1}^{k=a} \sum_{j=1}^{j=b} v_k w_j\,(A_k - M_a)\,(B_j - M_b) \qquad . \qquad . \qquad . \quad \text{(vii)}$$

$$= \sum_{k=1}^{k=a} \sum_{j=1}^{j=b} v_k w_j A_k B_j + M_a M_b \sum_{k=1}^{k=a} \sum_{j=1}^{j=b} v_k w_j - M_a \sum_{k=1}^{k=a} \sum_{j=1}^{j=b} v_k w_j B_j - M_b \sum_{k=1}^{k=a} \sum_{j=1}^{j=b} v_k w_j A_k.$$

If we carry out the summation first by rows and then by columns or *vice versa*, as explained in page 163, we get

$$\sum_{k=1}^{k=a} \sum_{j=1}^{j=b} v_k w_j = \sum_{k=1}^{k=a} v_k \sum_{j=1}^{j=b} w_j = 1$$

$$\sum_{k=1}^{k=a} \sum_{j=1}^{j=b} v_k w_j B_j = \sum_{k=1}^{k=a} v_k \sum_{j=1}^{j=b} w_j B_j = M_b \sum_{k=1}^{k=a} v_k = M_b$$

$$\sum_{k=1}^{k=a} \sum_{j=1}^{j=b} v_k w_j A_k = \sum_{k=1}^{k=a} v_k A_k \sum_{j=1}^{j=b} w_j = \sum_{k=1}^{k=a} v_k A_k = M_a$$

$$\therefore Cov\,(A, B) = \sum_{k=1}^{k=a} \sum_{j=1}^{j=b} v_k w_j A_k B_j - M_a M_b \qquad . \qquad . \qquad . \qquad . \quad \text{(viii)}$$

By the same procedure

$$\sum_{k=1}^{k=a} \sum_{j=1}^{j=b} v_k w_j A_k B_j = \sum_{k=1}^{k=a} v_k A_k \sum_{j=1}^{j=b} w_j B_j$$

$$= M_b \sum_{k=1}^{k=a} v_k A_k$$

$$= M_b \,.\, M_a.$$

Hence by substitution in (viii)

$$Cov\,(A, B) = 0.$$

For an empirical distribution w.r.t. which y is the frequency of corresponding values of A and B, we may write

$$\Sigma y\,.\,(A - M_a)(B - M_b) = \Sigma y\,.\,AB + M_a M_b\,\Sigma y - M_b\,\Sigma y\,.\,A - M_a\,\Sigma y\,.\,B$$

$$= \Sigma y\,.\,AB + M_a M_b - M_a M_b - M_a M_b,$$

$$\therefore Cov\,(A, B) = \Sigma y\,.\,AB - M_a M_b \qquad . \qquad . \qquad . \qquad . \qquad . \qquad . \quad \text{(ix)}$$

Fig. 78. PRODUCT OF TWO INDEPENDENT VARIATES

$$\sum_{k=0}^{k=a} \sum_{l=0}^{l=b} v_k w_l A_k B_l$$

FREQUENCY		v_0	v_1	. . .	v_k	. . .	v_a	TOTAL
	VARIATE	A_0	A_1	. .	A_k	.	A_a	
w_0	B_0	$w_0 B_0 v_0 A_0$	$w_0 B_0 v_1 A_1$	. . .	$w_0 B_0 v_k A_k$	. . .	$w_0 B_0 v_a A_a$	$w_0 B_0 \sum_{k=0}^{k=a} v_k A_k = w_0 B_0 M_a$
w_1	B_1	$w_1 B_1 v_0 A_0$	$w_1 B_1 v_1 A_1$	. .	$w_1 B_1 v_k A_k$	. .	$w_1 B_1 v_a A_a$	$w_1 B_1 \sum_{k=0}^{k=a} v_k A_k = w_1 B_1 M_a$
		. . .						
. .					. . .			. .
. .								. .
.	. .		.	. .	.	.	. .
. .		. . .		. .		.	. .
w_l	B_l	$w_l B_l v_0 A_0$	$w_l B_l v_1 A_1$	. .	$w_l B_l v_k A_k$	.	$w_l B_l v_a A_a$	$w_l B_l \sum_{k=0}^{k=a} v_k A_k = w_l B_l M_a$
.		. . .					.	. . .
. .								
. .			.				.	
. .	. .	. . .	.	. .			.	. .
w_b	B_b	$w_b B_b v_0 A_0$	$w_b B_b v_1 A_1$	.	$w_b B_b v_k A_k$	. .	$w_b B_b v_a A_a$	$w_b B_b \sum_{k=0}^{k=a} v_k A_k = w_b B_b M_a$
	TOTAL	$v_0 A_0 \sum_{l=0}^{l=b} w_l B_l$ $= v_0 A_0 M_b$	$v_1 A_1 \sum_{l=0}^{l=b} w_l B_l$ $= v_1 A_1 M_b$	. . .	$v_k A_k \sum_{l=0}^{l=b} w_l B_l$ $= v_k A_k M_b$	. . .	$v_a A_a \sum_{l=0}^{l=b} w_l B_l$ $= a_a A_a M_b$	$M_a \sum_{l=0}^{l=b} w_l B_l = M_a M_b$ $M_b \sum_{k=0}^{k=a} v_k A_k = M_a M_b$

FIG. 78. Chessboard summation of the covariance of the joint distribution of two independent scores.

If there are n pairs of values, we may write this in the form

$$Cov\,(A,\,B) = \frac{\Sigma AB}{n} - M_a M_b.$$

By the analogous formula for the variance

$$\sigma_a^2 = \frac{\Sigma A^2}{n} - M_a^2.$$

$$\sigma_b^2 = \frac{\Sigma B^2}{n} - M_b^2.$$

Whence from (vi) above we obtain the more convenient formula for computation

$$\rho = \frac{\Sigma AB - n\,M_a M_b}{\sqrt{(\Sigma A^2 - nM_a^2)(\Sigma B^2 - nM_b^2)}} \qquad . \qquad . \qquad . \qquad . \qquad \text{(x)}$$

The lay-out for numerical work is as in Table 2.

TABLE 2

Items	A	A^2	B	B^2	AB
	A_1	A_1^2	B_1	B_1^2	$A_1 B_1$
	A_2	A_2^2	B_2	B_2^2	$A_2 B_2$
	—	—	—	—	—
	—	—	—	—	—
	A_n	A_n^2	B_n	B_n^2	$A_n B_n$
Totals	ΣA	ΣA^2	ΣB	ΣB^2	ΣAB
	$M_a \quad \frac{1}{n}\Sigma A$			$M_b \quad \frac{1}{n}\Sigma B$	

Equations (vi) and (x) define an alternative method of computing the rank correlation coefficient in accordance with the above numerical schema. If A and B are rank scores as elsewhere defined, it therefore follows that the covariance formula (vi) above has the properties of (iv) in 8.03 in so far as it assumes the values 0, 1 and -1 respectively for independence, complete positive association and complete negative association. For the example on page 330 we have

$$\Sigma A = 1 + 2 + 3 + 4 + 5 + 6 = \Sigma B,$$
$$\therefore\ \Sigma A = 21 = \Sigma B.$$
$$M_a = 21 \div 6 = 3\cdot5 = M_b.$$
$$\Sigma A^2 = 1 + 4 + 9 + 16 + 25 + 36 = \Sigma B^2 = 91.$$
$$\Sigma AB = (1)(5) + (2)(6) + (3)(4) + (4)(2) + (5)(3) + (6)(1)$$
$$= 5 + 12 + 12 + 8 + 15 + 6 = 58.$$

Hence from (x)

$$\rho = \frac{58 - 6(3{\cdot}5)^2}{\sqrt{[91 - 6(3{\cdot}5)^2][91 - 6(3{\cdot}5)^2]}} = -0{\cdot}886.$$

This result agrees with the value of ρ derived (p. 335) by recourse to (iv) and (v) in 8.03. As a computing device (vi) and (x) above have no special advantage; but the method of deriving them does not presume that the scores involved are specifically referable to rank as such. The value of ρ is zero when there is no correspondence between two sets of scores, rank or otherwise; and its possible outside values are ± 1. What are its limiting values when the score is not a rank remains to be seen.

The limiting values of (iv) in 8.03 when correspondence is perfect depend on the assumption that the two sets of scores stand in a linear relation to one another. This is necessarily true of rank score correspondence, because successive rank scores of either series increase by equal increments; but a linear relation between two series of scores which do not increase by equal steps may also exist. Let us therefore examine the meaning of (vi) above on the assumption that the score A is an exact linear function of B. If so, A may increase with B or decrease as B increases, and we may express this by recourse to a linear equation in which h and c are arbitrary constants, *viz.* :

$$A = h + cB \text{ (perfect \textit{positive} linear correspondence).}$$

$$A = h - cB \text{ (perfect \textit{negative} linear correspondence).}$$

When linear correspondence is perfect and positive

$$M_a = \Sigma y(h + cB) = h \,\Sigma y + c\Sigma y \,.\, B$$
$$= h + cM_b \qquad \qquad \qquad \text{(xi)}$$
$$V_a = \Sigma y(h + cB)^2 - M_a^2$$
$$= h^2\Sigma y + c^2\Sigma y \,.\, B^2 + 2hc\,\Sigma y \,.\, B - (h^2 + c^2M_b^2 + 2hcM_b)$$
$$= h^2 + c^2(V_b + M_b^2) + 2hcM_b - h^2 - c^2M_b^2 - 2hcM_b.$$
$$\therefore\ V_a = c^2 \,.\, V_b \,. \qquad \qquad \qquad \text{(xii)}$$
$$\therefore\ \sigma_a\sigma_b = c \,.\, V_b \,. \qquad \qquad \qquad \text{(xiii)}$$

Likewise

$$\Sigma y \,.\, AB = \Sigma y \,.\, hB + \Sigma y \,.\, cB^2$$
$$= hM_b + c(V_b + M_b^2).$$

Hence by (ix)

$$\text{Cov } (A, B) = h \,.\, M_b + c \,.\, V_b + c \,.\, M_b^2 - M_a M_b.$$

And by (xi) above

$$M_a M_b = h \,.\, M_b + c \,.\, M_b^2.$$
$$\therefore\ \text{Cov } (A, B) = c \,.\, V_b.$$

Whence from (xiii)

$$\frac{\text{Cov } (A, B)}{\sigma_a\sigma_b} = \frac{c \,.\, V_b}{c \,.\, V_b} = 1 \qquad \qquad \text{(xiv)}$$

Similarly, if $A = h - cB$

$$\frac{\text{Cov } (A, B)}{\sigma_a\sigma_b} = -1 \,. \qquad \qquad \text{(xv)}$$

Thus the number defined by (vi) above is an index of correspondence w.r.t. any two sets of scores inasmuch as it has the value

(i) *zero* when there is no correspondence ;

(ii) $+ 1$ when there is perfect positive linear correspondence ;

(iii) $- 1$ when there is perfect negative linear correspondence.

The meaning of the zero criterion (i) calls for special comment amplifying the *caveat* in the concluding sentence of 8.02. Hitherto we have assumed the possibility of assigning to each individual a unique A score and a unique B score in virtue of the method of assessment by rank ; but the *product-moment* defined by (vi) or (x) removes this limitation, being applicable to a system of scoring which admits a wide range of B scores consistent with a particular value of the A score and *vice versa*. When this is so, our focus of interest is no longer whether high individual values of B go with high (or low) individual values of A. Our criterion of correspondence is whether the mean value of the set of B scores associated with a high value of the A scores is high or low, or whether the mean value of a set of A scores associated with a high value of the B scores is high or low. To say that A and B are statistically independent signifies that neither the one nor the other is true ; but the absence of any indication of the sort of correspondence implied in the preceding statement does not necessarily signify that the distribution of the two sets of scores satisfies all the requirements of statistical independence.

To say that two sets of scores are independent signifies that we can lay out their joint distribution by recourse to the chessboard device in accordance with the product rule. *Obedience to the product rule*, as Aitken succinctly remarks, is in fact the quintessence of statistical independence ; and a glance at the many examples of the use of the chessboard device in previous chapters suffices to clarify what such obedience entails. We multiply every border score laid out at the row margins by the same column border score to derive the frequency terms of a particular column ; and we multiply every border score laid out at the head of the columns by one and the same row border score to get the frequency terms of a particular row. Thus the relative frequencies of B scores associated with a particular A score are the same whatever its value, and the relative frequencies of A scores associated with a particular B score are the same whatever its value. *Inter alia* this signifies that the mean A score associated with any B score is the same and the mean B score associated with any A score is the same. When this is so, the covariance of the distribution is zero. Statistical independence and zero covariance alike therefore signify that no correspondence of the sort implicit in the customary use of the term correlation holds good ; but it is most important to realize that statistical independence implies far more than this. For instance, it implies that the variance of the distribution of B scores is the same for all values of A scores and *vice versa*.

The sort of correspondence which is our concern in this chapter takes no cognisance of this peculiarity of statistical independence, and it is easy to make up examples which illustrate the fact that covariance may be zero, when it does not hold good. In a grid lay-out which exhibits no trend of mean B scores as A scores increase or decrease and no trend of mean A scores as B scores increase, it may happen that the range of A scores tapers out at both ends and that the range of B scores tapers out at both ends, the variance of B (column) scores and A (row) scores being greatest in the middle of the grid. If so, equality of the means and zero covariance are alike consistent with a situation in which the cell entries do not obey the product rule, as the following set-up suffices to illustrate :

A Scores

		0	1	2	Total Frequency of B Scores	Mean
	0	0	$\frac{1}{8}$	0	$\frac{1}{8}$	1
B Scores	1	$\frac{1}{8}$	$\frac{1}{2}$	$\frac{1}{8}$	$\frac{3}{4}$	1
	2	0	$\frac{1}{8}$	0	$\frac{1}{8}$	1
Total Frequency of of A Scores		$\frac{1}{8}$	$\frac{3}{4}$	$\frac{1}{8}$	—	—
Mean		1	1	1	—	—

EXERCISE 8.06

1. An umpire tosses two tetrahedral dice of the sort shown in Fig. 70. Player A records as his score (x_a) the sum of the single-toss scores $(x_1$ and $x_2)$ of the umpire. Player B records as his score (x_b) the difference. Set out the grid exhibiting the relative frequencies of A and B scores at a single trial (double toss) and investigate its properties w.r.t. (*a*) statistical independence, (*b*) the numerical value of the covariance of the players' scores.

2. A roulette wheel of 16 equal segments carries scores 1, 2, 3, 4, 5 respectively allocated to 1, 4, 6, 4 and 1 segments. It rotates twice. Player A records as his own the score sum and Player B the score difference. Investigate this result from the same viewpoint.

3. Four dominoes respectively have pips on their left hand and right hand halves as follows :

	Left Hand	Right Hand
(i)	1	2
(ii)	2	1
(iii)	2	3
(iv)	3	2

At each trial the umpire lays them face down, and the players rearrange them in an order unknown to him. They do the same after he has taken one, recording the score before replacing it. At each 2-fold trial player A records as his score the sum of the number of pips on the left-hand side of the two dominoes chosen by the umpire. Player B records as his score the sum of the number of pips on the right-hand side of the same two dominoes. Set out the grid for the relative frequencies of A and B scores, and investigate its properties as above.

4. Investigate the effect of replacing the dominoes of Exercise 3 by the following set :

	Left Hand	Right Hand
(i)	1	1
(ii)	1	2
(iii)	2	1
(iv)	2	2

8.07 THE UNIQUENESS OF THE RANK SCORE

To say that (iv) in 8.03 and (vi) above have corresponding numerical values when correspondence is either perfect or absent does not in fact signify that the two formulæ must otherwise yield the same numerical value when applied to the same set of data, unless the method of scoring is the same. In general, a value of ρ is lower if based on a count, measurement or arbitrary number than a value of ρ based on a rank score.

Let us examine an example of 8.02 from this point of view. The original data from which we compiled two sets of rank scores in perfect reverse order were as follows :

A (Examination mark) .	. 55	40	60	95	45	80
B (Weekly allowance in pence) .	9	30	6	2	12	3

In virtue of the identity established in 8.06, both the Spearman formula and the covariance (*product-moment*) formula yield a coefficient of -1 for the correlation of rank scores ; but we do not obtain this result if we apply the latter to the *crude* scores. We then proceed as follows :

	A.	B.	A².	B².	AB.
	55	9	3025	81	495
	40	30	1600	900	1200
	60	6	3600	36	360
	95	2	9025	4	190
	45	12	2025	144	540
	80	3	6400	9	240
Totals	375	62	25675	1174	3025

Whence we have

$$M_a = 375 \div 6 = 62\cdot5,$$
$$M_b = 62 \div 6 = 10\cdot3$$
$$n \cdot M_a M_b = 6(10\cdot3)(62\cdot5) = 3875$$
$$n \cdot M_a^2 = 6(62\cdot5)(62\cdot5) = 23437\cdot5$$
$$n \cdot M_b^2 = 6(10\cdot3)(10\cdot3) = 640\cdot6$$
$$\therefore \rho = \frac{3025 - 3875}{\sqrt{(25675 - 23437\cdot5)(1174 - 640\cdot)6}} = \frac{-850}{1092\cdot3} = -0\cdot78.$$

The gross numerical divergence between the value of ρ ($= -0\cdot78$) calculated from the covariance formula by recourse to the raw data of the last example and the value of $\rho(= -1\cdot0)$ computed by recourse to the rank scores calls for clarification. With that end in view we should first recall the form of the question stated in 8.02 above.

If our aim is to assess what measure of correspondence exists between two sets of data, appropriate methods of assessing such correspondence should lead to the same result ; but such is not the task which we have undertaken in this chapter. What we set out to seek was an answer to the question : does any correspondence exist ? If that is our aim, the numerical value of the *summarising index* we employ to describe such correspondence as seemingly exists is merely a measure of our confidence in rejecting the null hypothesis. A particular value is of interest only in so far as we can specify how often a value at least as great would turn up, if the two sets of scores were independent.

To answer this question we must know the law of the distribution of the index within the framework of the assumption last stated, and we have investigated such a law in 8.04–8.05 above. We did so on the understanding that the scores we employ for the computation of our index in either form are *rank* scores, taking as the model of our null hypothesis two packs of cards numbered consecutively. On the assumption that choice of a card from one pack does not affect choice of a card from the other, the elementary rules of independent choice provide all we need to answer the question : what is the distribution of values of the sum of the square difference between corresponding paired scores ? That we are able to answer this question arises from the circumstance that we know the unit sampling distribution of the model universe, i.e. the frequency of each face score on the individual cards. That the question itself is relevant to practical statistics arises from the circumstance that we assign to every member of a group a rank in virtue of the ordinal status of its raw score, thus imposing on the universe of choice a law of distribution analogous to that of our model.

By using rank scores we thus prescribe what model is appropriate to an investigation of the distribution of the index on the assumption inherent in the null hypothesis, i.e. that there is no true correspondence. While this peculiarity of rank scoring gives it a great advantage in that it makes it easier than otherwise to grasp the logical implications of the use of our summarising index, it carries with it a penalty. *Ipso facto*, individuals with the same raw score (e.g. examination marks or pocket money) have the same rank ; and if individual scores are *ties* in this sense, the universe is not rectangular. Consequently, the law of distribution inferred on the assumption that all scores have equal frequency is no longer applicable. Strictly speaking, we are on safe ground only if our sample contains no *ties*. Otherwise, we cannot legitimately rely on the method of 8.03–8.05.

If we do *not* employ the rank method of scoring, we are not entitled to assume that the distribution of scores of either series in a trial is rectangular, and indeed it will scarcely ever happen that it will be. Consequently, we have no reason to suppose that a particular numerical value of the index ρ calculated by recourse to the rank score would have the same frequency in the absence of true correspondence as the same numerical value of ρ computed by recourse to the crude score on the same assumption. By the same token, the fact that recourse to different methods of scoring leads to different numerical values of ρ signifies no inconsistency inherent in our demonstration of the formal equivalence of the Spearman equation in 8.02 and the co-variance (*product-moment*) equation of 8.06. For samples of paired scores of a given size, the frequency with which the summarising index will assume a particular numerical value will in general depend on the method of scoring we adopt ; and it is sometimes convenient to use separate symbols ρ_{ab} and r_{ab} for the summarising index we compute respectively by recourse to

(*a*) either the Spearman or the product-moment formula when the score is ordinal ;
(*b*) the product-moment formula when the score is a measurement, count or grade other than rank.

From the point of view stated in the last few paragraphs, the dilemma arising from the numerical inconsistency under discussion is less formidable than it appears to be at first sight ; but it would be incorrect to infer that the foregoing considerations suffice to resolve it. By methods which we shall not explore in this volume it is possible to show that the sampling variance of the distribution of $r_{ab} = 0$ as defined above is equivalent to that of $\rho = 0$, being in fact $(n - 1)^{-1}$. Since the two distributions become approximately normal when the sample is *very large*, they would then be approximately identical ; and it is therefore necessary to seek for another explanation of the fact that the numerical value of r_{ab} is not necessarily identical with that of ρ for the same set of paired scores.

A reason is not far to seek. The examples which prompted us to explore this inconsistency suffice to emphasise an inescapable consequence of substituting rank scores for raw scores. By so doing we force our scores into a mould which restricts their range of independent variation. In this way, we may impose on a system of paired values a closer correspondence than they would otherwise have. In virtue of this fact we should expect, as we find, that r_{ab} would commonly have a lower numerical value than ρ. Commonly, also, grouping of raw scores for convenience of computation has the same effect, as when we score measurements or counts by a mid-value for each of a sequence of equally spaced intervals. We do so for instance if we specify age by years, designating a woman as $41\frac{1}{2}$ years at any time inclusive between her 41st birthday and the day before her 42nd, or in triennial age groups when we designate as $41\frac{1}{2}$ years of age any man who has attained his 40th but not as yet his 43rd birthday. When our pool of observations is very large, such coarsening of the scale of our scoring system reduces arithmetical labour without much sacrifice of information.

8.08 COMPUTATION OF THE PRODUCT MOMENT COEFFICIENT

As we have seen, Table 2 in 8.06 sets out a procedure for computing the correlation coefficient (r_{ab}) equally appropriate for scoring by rank or by crude scores such as counts or measurements specified by a mid-interval value. When a computing machine is not to hand it is useful to simplify the work involved by change of scale and/or origin.

An important property of r_{ab}, the covariance (product-moment) coefficient of correlation w.r.t. two sets of scores A and B resides in its identity with r_{pq} the corresponding coefficient of two sets of scores P and Q if the latter are respectively linear functions of A and B. In other words, changes of scale and/or origin of the correlated scores do not affect the numerical value of r. To demonstrate this we specify any such change of scale and origin by arbitrary constants in the equation of the line, viz. :

$$P = hA + g ; \quad Q = kB + c.$$

If M_p and M_q are the means and V_p, V_q the variances of the P and Q distributions

$$r_{pq} = \frac{\Sigma y . PQ - M_p M_q}{\sqrt{V_p . V_q}} \quad . \quad . \quad . \quad . \quad \text{(i)}$$

If M_a and M_b are respectively the means of the A and B scores, the variances of the score distributions being V_a and V_b

$$M_p = \Sigma yP = \Sigma y(hA + g) = h\Sigma yA + g\Sigma y$$
$$= hM_a + g.$$

Similarly

$$M_q = kM_b + c,$$
$$\therefore M_p . M_q = hk\, M_a M_b + ch\, M_a + gk\, M_b + cg,$$
$$\therefore ch\, M_a + gk\, M_b + cg = M_p M_q - hk\, M_a M_b \quad . \quad . \quad . \quad \text{(ii)}$$

By definition

$$\Sigma yPQ - M_p M_q = \Sigma y(hA + g)(kB + c) - M_p M_q$$
$$= hk\,\Sigma yAB + ch\,\Sigma yA + gk\,\Sigma yB + cg\,\Sigma y - M_p M_q$$
$$= hk\,\Sigma yAB + ch\, M_a + gk\, M_b + cg - M_p M_q.$$

Hence from (ii) above

$$\Sigma yPQ - M_p M_q = hk\,\Sigma yAB - hk\, M_a M_b. \quad . \quad . \quad . \quad \text{(iii)}$$

From (ii) in 8.05, we have

$$V_p = h^2 V_a \quad and \quad V_q = k^2 V_b,$$
$$\therefore \sqrt{V_p . V_q} = hk\,\sqrt{V_a . V_b}. \quad . \quad . \quad . \quad . \quad \text{(iv)}$$

From (iii) and (iv) by substitution in (i)

$$r_{pq} = \frac{\Sigma y . AB - M_a M_b}{\sqrt{V_a V_b}} = r_{ab}. \quad . \quad . \quad . \quad . \quad \text{(v)}$$

23

Subject to one restriction, we can therefore label numerically our scores A_1, A_2, A_3, etc., by a_1, a_2, a_3, etc., and B_1, B_2, B_3, etc., by b_1, b_2, b_3, etc., in any way most convenient for the parent scores respectively. All that matters is that the new sets are linear functions of the scores they replace. In practice, it is an economy to label them 0, 1, 2, 3 . . . n.

Example. The following (Table 3) fictitious data, set out first for computation by (x) of 8.06, refer to the heights of and mean weekly consumption of beer by twenty regular members of a men's social club.

TABLE 3.

1.	2.	3.	4.	5.
A.	B.			
Height in Inches.	Weekly Consumption in Pints of Beer.	A^2.	B^2.	AB.
53	21	2809	441	1113
56	12	3136	144	672
56	8	3136	64	448
57	0	3249	0	0
60	2	3600	4	120
63	0	3969	0	0
64	30	4096	900	1920
65	14	4225	196	910
65	7	4225	49	455
66	7	4356	49	462
67	8	4489	64	536
68	21	4624	441	1428
68	25	4624	625	1700
69	10	4761	100	690
69	10	4761	100	690
70	0	4900	0	0
70	5	4900	25	350
70	6	4900	36	420
71	12	5041	144	852
73	2	5329	4	146
1300	200	85130	3386	12912
ΣA	ΣB	ΣA^2	ΣB^2	ΣAB

For computation of r_{ab} by recourse to (x) in 8.06 we have

$$M_a = \tfrac{1300}{20} = 65. \qquad M_a^2 = 4225.$$
$$M_b = \tfrac{200}{20} = 10. \qquad M_b^2 = 100.$$

$$r_{ab} = \frac{12912 - 20 . 65 . 10}{\sqrt{(85130 - 20 . 4225)(3386 - 20 . 100)}}$$

$$= \frac{12912 - 13000}{\sqrt{(85130 - 84500)(3386 - 2000)}}$$

$$= - \frac{88}{\sqrt{630 \times 1386}}.$$

$$r_{ab} \simeq - 0{\cdot}0942.$$

Change of scale is often useful, when we group data to make mid-range scores increase by unity. The data of Table 3 are not sufficiently numerous to call for grouping of data ; but we may use them to illustrate the economy of effort involved in a change of origin by taking the lowest A score (i.e. 53) as our base line and setting the origin of the B scores as 15 without change of scale, as in Table 4. The student who has then memorised the squares of the first 25 integers —a useful accomplishment in work of this sort—will be able to perform all the necessary operations by mental arithmetic ; and the result as shown below will tally exactly with the more laborious procedure involved in use of the raw A scores.

TABLE 4

A. Height on Arbitrary Scale.	B. Pints.	A^2.	B^2.	AB.
0	6	0	36	0
3	− 3	9	9	− 9
3	− 7	9	49	− 21
4	-15	16	225	− 60
7	−13	49	169	− 91
10	−15	100	225	−150
11	15	121	225	165
12	− 1	144	1	− 12
12	− 8	144	64	− 96
13	− 8	169	64	−104
14	− 7	196	49	− 98
15	6	225	36	90
15	10	225	100	150
16	− 5	256	25	− 80
16	− 5	256	25	− 80
17	− 15	289	225	−255
17	−10	289	100	−170
17	− 9	289	81	−153
18	− 3	324	9	− 54
20	−13	400	169	−260
240 = ΣA	− 100 = ΣB	3510 = ΣA^2	1886 = ΣB^2	−1288 = ΣAB

In this case we have :

$$M_a = 12; \; M_b = -5; \; V_a = \frac{3510}{20} - 144 = 31\cdot5; \; V_b = \frac{1886}{20} - 25 = 69\cdot3.$$

$$Cov\,(AB) = \frac{-1288}{20} - 12(-5) = -4\cdot4.$$

$$r_{ab} = \frac{-4\cdot4}{\sqrt{(31\cdot5)(69\cdot3)}} \simeq -0\cdot0942.$$

EXERCISE 8.08

1. Calculate for each pair of the following sets of figures :

    (a) Spearman's rank coefficient based on the sum of the square differences.

    (b) The product moment coefficient for (i) the raw scores, (ii) the scoring by rank.

Distribution of professorial posts in Arts, Pure Science and Medical Faculties of British Universities, 1938.

(J. D. Bernal—*The Social Function of Science*, 1939.)

University.	Professors and/or Heads of Departments of :—		
	Arts Subjects.	Science.	Medicine.
Birmingham . .	23	6	5
Bristol . . .	10	9	6
Cambridge . .	46	23	2
Durham . .	20	16	9
Exeter . . .	6	5	0
Leeds . . .	19	9	7
Liverpool . .	22	10	13
London . .	100	68	78
Manchester . .	25	9	8
Nottingham . .	9	10	0
Oxford . . .	79	27	12
Reading . .	15	7	0
Sheffield . .	14	7	7
Southampton . .	9	6	0
Aberystwyth . .	15	8	0
Bangor . . .	13	5	0
Cardiff . . .	13	6	2
Swansea . .	8	5	0
Aberdeen . .	24	5	10
Edinburgh . .	40	5	12
Glasgow . .	22	5	9
St. Andrews . .	23	11	7

2. Calculate for each pair of the following sets of examination results the value of

    (a) Spearman's rank coefficient based on the sum of the square differences.

    (b) The product-moment coefficient for the raw scores.

In each case show that the product-moment formula gives the same result as (a) above, if the scoring is by rank.

Pupil.	Classics.	Mathematics.	History.	Divinity.	General Science.
A	95	62	87	42	75
B	90	74	89	65	20
C	88	61	64	38	55
D	87	54	61	35	84
E	86	78	43	80	32
F	80	96	83	32	92
G	72	32	52	75	67
H	65	45	44	87	53
I	62	86	25	23	63
J	55	28	76	94	47
K	48	70	38	53	41
L	47	47	72	60	28
M	45	42	30	44	37
N	44	50	57	90	19
O	39	98	63	48	49
P	36	3	2	54	17
Q	31	65	24	27	45
R	26	17	13	30	38
S	18	25	19	14	12
T	2	14	17	24	9

3. Investigate correlations between the male mortality for (a) Tuberculosis and respiratory diseases, (b) Tuberculosis and influenza, and (c) Respiratory diseases and influenza, by finding Spearman's rank coefficient and the product-moment coefficient in each case.

Male mortality from certain causes, England, 1933 (according to life table).

(Kuczynski—*Measurement of Population Growth*, 1935.)

Successive Age Groups by rank.	Deaths per thousand from :		
	Tuberculosis (all forms).	Disease of Respiratory System (excluding Tuberculosis).	Influenza.
1	0·94	13·66	1·01
2	2·31	7·89	1·13
3	1·06	1·33	0·27
4	0·84	0·58	0·20
5	2·97	0·88	0·53
6	5·17	1·08	0·60
7	4·90	1·15	0·74
8	4·86	1·69	1·07
9	5·15	2·78	1·72
10	5·46	3·60	2·35
11	5·81	5·28	3·09
12	5·59	6·56	3·17
13	4·80	7·51	3·21
14	3·46	8·12	3·28
15	2·20	9·95	3·40
16	1·18	12·27	4·29
17	0·41	13·95	5·20
18	0·22	19·45	7·02

8.09   THE CORRELATION GRID

The schema embodied in Table 2 sets out the quickest way of arriving at a numerical value for r_{ab}; but it is not the best way of exhibiting the data for preliminary inspection. It is then more instructive to lay out the crude data as a joint frequency distribution or correlation grid (Table 5). Each cell of such a grid shows the number (N_{ij}) of items to which we assign a particular A-score (A_i) and a particular B-score (B_j) in an assemblage of N paired score values. At the foot of the columns and right-hand margin of the rows we then record as N_{ai} the total number of items with a score A_i and the total number of items with a score B_j as N_{bj}. By definition

$$\Sigma N_{bj} = N = \Sigma N_{ai}.$$

The student should be able to insert in the above, as also below, the appropriate limits of each summation. If y_{ij} is the frequency of the joint score A_i, B_j, in the cell of column i and row j, by definition :

$$y_{ij} = N_{ij} \div N.$$

For the means (M_a, M_b) and variances (V_a, V_b) of the two score distributions, we may then write

$$M_a = \frac{1}{N}\Sigma N_{ai} \cdot A_i \quad and \quad M_b = \frac{1}{N}\Sigma N_{bj} \cdot B_j.$$

$$V_a = \frac{1}{N}\Sigma N_{ai} \cdot A_i^2 - M_a^2 \quad and \quad V_b = \frac{1}{N}\Sigma N_{bj} \cdot B_j^2 - M_b^2.$$

TABLE 5

B Scores.	A Scores.					Total No. of Entries.	Mean A-score (M_{ab})
	0	1	2	3	4		
0	N_{00}	N_{10}	N_{20}	N_{30}	N_{40}	N_{a0}	M_{a0}
1	N_{01}	N_{11}	N_{21}	N_{31}	N_{41}	N_{a1}	M_{a1}
2	N_{02}	N_{12}	N_{22}	N_{32}	N_{42}	N_{a2}	M_{a2}
3	N_{03}	N_{13}	N_{23}	N_{33}	N_{43}	N_{a3}	M_{a3}
4	N_{04}	N_{11}	N_{24}	N_{34}	N_{44}	N_{a4}	M_{a4}
5	N_{05}	N_{15}	N_{25}	N_{35}	N_{45}	N_{a5}	M_{a5}
Total No. of Entries.	N_{b0}	N_{b1}	N_{b2}	N_{b3}	N_{b4}	N	
Mean B-score (M_{ba}).	M_{b0}	M_{b1}	M_{b2}	M_{b3}	M_{b4}		

If there is monotonic correspondence between the two sets of scores, we may usually recognise it by a *diagonal* concentration of high values of N_{ij} from left to right *downwards* (positive correlation) or *upwards* (negative correlation).   In any case it will be apparent, if we lay out the mean values (M_{ba}) of B-scores associated with each column (A) border score as in Table 5 and the mean values (M_{ab}) of A-scores associated with each row (B) border score.   If we denote as M_{bi} the mean B score associated with an A score A_i and by M_{aj} the mean A score associated with a B score B_j the computation follows from definition, *viz.* :

$$M_{bi} = \frac{1}{N_{bj}}\sum N_{ij} . B_j \quad and \quad M_{aj} = \frac{1}{N_{ai}}\sum N_{ij} . A_i.$$

When we first set out our data in this way, the quickest way to compute the covariance is by using either of the following relations to find the mean value $M(A_i . B_j)$ of the product $A_i . B_j$.

$$\frac{1}{N}\sum N_{ai}(B_j M_{ai}) = M(A_i B_j) = \frac{1}{N}\sum N_{bj}(A_i M_{bj}).$$

These identities follow from the fact that A_i is a constant factor of the product $A_i . B_j$ within any column and B_j is a constant factor within any row, so that

$$M(A_i . B_j) = \frac{1}{N}\sum\sum N_{ij}A_iB_j$$

$$= \frac{1}{N}\sum A_i \sum N_{ij}B_j$$

$$= \frac{1}{N}\sum N_{bj} A_i \frac{1}{N_b,}\sum N_{ij}B_j$$

$$= \frac{1}{N}\sum N_{bj}(A_i M_{bj}).$$

The covariance of the joint distribution is, of course, $M(A_i . B_j) - M_a M_b$.

CHAPTER 9

THE NATURE OF CONCOMITANT VARIATION

9.01 THE UMPIRE-BONUS MODEL

THE treatment of correlation in our last chapter focused attention on the possibility of validifying the existence of correspondence between paired observations with a view to establishing a true concurrence as defined in 8.01. From that point of view the magnitude of the Spearman co-efficient is merely an indication of the unlikelihood that the paired observations are independent in the statistical sense of the term ; and we have examined the means of assessing the likelihood assigned to a particular range of values by the appropriate null hypothesis when the method of scoring is by rank. The null hypothesis relevant to the issue so stated postulates that the mean value of the summarising index is zero. If the observed value is significantly greater or less than zero, nothing we have hitherto discussed confers an ulterior meaning on its numerical magnitude.

Educational psychology provides many examples of enquiries in which repeated trials of pairs of tests applied to different groups of individuals yield high, and in the sense implied above, significant values of ρ * in fairly close agreement. In such circumstances, we are entitled to speak of a representative value of the summarising index relative to the situation, and to regard it as a measure of the concurrence of two tests, itself subject to sampling error. If so, we are also entitled to ask whether two particular numerical values of ρ each significantly different from zero are significantly different from one another. In this volume we shall not attempt to find a complete answer to this question ; but it is profitable to make a preliminary examination of some of its implications, if only to bring into focus two totally different issues which arise in connexion with the use of correlation indices and in connexion with their sampling distributions.

Hitherto, we have discussed sampling from different universes on the assumption that the act of choosing an item from one universe has no influence on the act of choosing an item from a second universe or *vice versa*. To give the problem stated in the last paragraph precision, we need a statistical model which places a constraint on the second act of choice, limiting its range in a particular way. It is in fact possible to devise many lottery models which fulfil this requirement ; and the one which will occupy our attention in this chapter is of special interest, because it permits us to get at the same time a close-up view of correlation both in the realm of *concurrence*, and that of *consequence*, as defined in 8.01. In doing so, we shall consistently com-pute the correlation coefficient (ρ) by recourse to the product moment formula of (vi) in 8.06.

If each of two players tosses the same penny three times, the *head*-scores of either may be 0, 1, 2, 3 at any trial ; and any of the four possible scores of one player may turn up with any one of the four possible scores of the other. Let us now suppose that : (*a*) the two players each toss once at a trial ; (*b*) an umpire tosses twice ; (*c*) the total score of each player at the 3-fold trial so conducted is the sum of his own single-toss score and the umpire's double-toss score. The player's score has therefore two components, one the result of his own luck, one common to that of his opponent ; and this imposes a constraint on the range of scores which his opponent can have at the same trial as himself. For instance, we may suppose that his total score is 2. This means one of two things—the score he gets in virtue of the umpire's toss is either 2 or 1.

---

* In this chapter we shall use ρ for the product moment index regardless of whether we score by rank or otherwise, in order to avoid confusion with r_a and r_b used elsewhere in this book for sample size.

If the first is true, his opponent's score can be 2 or 3, but cannot be 0 or 1. If he gets only one point as umpire's bonus, his opponent's score can be 1 or 2, but cannot be 0 or 3. At any single toss, there is equal chance that either player A or player B will score 0 or 1, and we can set out the ratios of their joint scores as a chessboard diagram. For economy of computation, we can use integers in the same ratio as the frequencies of the events concerned, if we weight our figures appropriately throughout :

	A 0	A 1
B 0	1	1
B 1	1	1

This result may go hand in hand with an umpire bonus score of 0, 1 or 2 in the ratio 1, 2, 1. This means that the total score of either A or B may be 1 or 2 twice as often as it may be : (a) 0 or 1; (b) 2 or 3. We may therefore set out all the possibilities with due weight to their relative frequencies as follows :

	A 0	A 1
B 0	1	1
B 1	1	1

	A 1	A 2
B 1	2	2
B 2	2	2

	A 2	A 3
B 2	1	1
B 3	1	1

We can thus build up the composite table :

		A 0	1	2	3	Total
	0	1	1	0	0	2
	1	1	3	2	0	6
B	2	0	2	3	1	6
	3	0	0	1	1	2
Total		2	6	6	2	16

We have now all the data for computing a product-moment coefficient w.r.t. the scores of the two players. The total number (n) of products is 16 and the sum of the products is

$$1(0)(0) + 1(0)(1) + 1(1)(0) + 3(1)(1) + 2(1)(2) + 2(2)(1)$$
$$+ 3(2)(2) + 1(2)(3) + 1(3)(2) + 1(3)(3)$$
$$= 0 + 0 + 0 + 3 + 4 + 4 + 12 + 6 + 6 + 9 = 44,$$
$$\therefore \frac{\Sigma AB}{n} = \frac{44}{16} = \frac{11}{4}.$$

Since it is immaterial whether A or B each tosses three times or each allows the umpire to toss twice out of three times for him, the mean score of either A or B and the variance of the score distributions w.r.t. either set is simply that of a 3-fold toss, so that $M_b = \frac{3}{2} = M_a$ and $V_a = \frac{3}{4} = V_b$

$$\therefore \quad \frac{\Sigma AB}{n} - M_a M_b = \frac{11}{4} - \frac{9}{4} = \frac{1}{2} = Cov\,(A, B),$$

$$V_a \cdot V_b = \frac{9}{16} \quad and \quad \sigma_a \cdot \sigma_b = \frac{3}{4},$$

$$\therefore \quad \frac{\dfrac{\Sigma AB}{n} - M_a \cdot M_b}{\sigma_a \cdot \sigma_b} = \frac{\frac{1}{2}}{\frac{3}{4}}.$$

$$\therefore \quad \rho_{ab} = \frac{2}{3} = 0 \cdot 6.$$

The mean score (M_u) of the umpire is $2(\frac{1}{2}) = 1$. Thus the numerical value of ρ is exactly equal to the proportionate contribution of that of the umpire to the player's mean score; and this is always true if: (a) the players toss the same die as the umpire; (b) each player records the same number of tosses. If the trial consists of ten tosses, and the umpire's bonus is the result of eight of them, the value of ρ is therefore 0·8. Table 1 exhibits the basic calculations for the build-up of the correlation table. The marginal totals of the latter for the 10-fold toss are in the same ratio as the coefficients of $(\frac{1}{2} + \frac{1}{2})^{10}$. This is, of course, in conformity with the product and addition rules of electivity, as a single item will suffice to illustrate. If the player's total score is 1, he may obtain it in two ways: (i) an umpire's bonus of 1 with an electivity of $8(\frac{1}{2})^8$ and a personal score of zero with an independent electivity of $(\frac{1}{2})^2$, the electivity of the composite event being $8(\frac{1}{2})^8 \cdot (\frac{1}{2})^2 = 8(\frac{1}{2})^{10}$; (ii) an umpire's bonus of zero with an electivity of $(\frac{1}{2})^8$ and a personal score of 1 with an independent electivity of $2(\frac{1}{2})^2$, so that the electivity of the joint event is $2(\frac{1}{2})^{10}$. The two possibilities (i) and (ii) are exclusive, and the electivity of a total score of 1 is therefore the sum $8(\frac{1}{2})^{10} + 2(\frac{1}{2})^{10} = 10(\frac{1}{2})^{10}$.

The computation of ρ in this case is as follows:

$$\begin{aligned}
\Sigma AB = \ & 7(0) + 12(1) + 36(2) + 61(4) + 144(6) + 176(9) + 56(8) \\
& + 336(12) + 322(16) + 112(15) + 504(20) + 392(25) \\
& + 140(24) + 504(30) + 322(36) + 112(35) + 336(42) \\
& + 176(49) + 56(48) + 144(56) + 61(64) + 16(63) + 36(72) \\
& + 12(81) + 2(80) + 4(90) + 1(100) \\
= \ & 110592
\end{aligned}$$

$$\frac{\Sigma AB}{4096} = 27 \qquad M_a = \frac{10}{2} = M_b.$$

$$\frac{\Sigma AB}{4096} - M_a M_b = 27 - 25 = 2.$$

$$V_a = \frac{10}{4} = V_b. \qquad \therefore \quad \sigma_a \sigma_b = 2 \cdot 5.$$

$$\therefore \quad \rho_{ab} = \frac{2}{2 \cdot 5} = 0 \cdot 8.$$

Instead of supposing that the die is a coin ($p = \frac{1}{2} = q$), let us now suppose that it is tetrahedral like those of Figs. 70–73; and three faces carry 1 pip, the other being blank. Scores of 0 and 1 at a single toss therefore occur with frequencies $\frac{1}{4}$ and $\frac{3}{4}$, i.e. in the ratio 1 : 3. For a 3-fold toss the mean score $M_a = M_b = 3(\frac{3}{4})$ and the variance of the distribution $\sigma_a^2 = \frac{9}{16} = \sigma_b^2$. If the umpire's bonus is the total score of two tosses out of the three, as in the foregoing example,

TABLE 1

Contingent Weighted Totals for Double Toss of A and B corresponding to Possible Results of the Umpire's 8-fold Toss of a coin with score (u), 0, 1, 2, . . . 8 Frequencies in the Ratio 1 : 8 : 28 : 56 : 70 : 56 : 28 : 8 : 1.

$u = 0$

B	0	1	2
0	1	2	1
1	2	4	2
2	1	2	1

$u = 3$

	3	4	5
3	56	112	56
4	112	224	112
5	56	112	56

$u = 6$

	6	7	8
6	28	56	28
7	56	112	56
8	28	56	28

$u = 1$

B	1	2	3
1	8	16	8
2	16	32	16
3	8	16	8

$u = 4$

	4	5	6
4	70	140	70
5	140	280	140
6	70	140	70

$u = 7$

	7	8	9
7	8	16	8
8	16	32	16
9	8	16	8

$u = 2$

B	2	3	4
2	28	56	28
3	56	112	56
4	28	56	28

$u = 5$

	5	6	7
5	56	112	56
6	112	224	112
7	56	112	56

$u = 8$

	8	9	10
8	1	2	1
9	2	4	2
10	1	2	1

the bonus score will be 0, 1 and 2 with frequencies in the ratio 1 : 6 : 9. We may therefore set out the three possible results of the 3-fold toss as follows :

B	A 0	A 1
0	1	3
1	3	9

B	A 1	A 2
1	6	18
2	18	54

B	A 2	A 3
2	9	27
3	27	81

The composite correlation table with 256 entries is

		A score				
		0	1	2	3	Total
B score	0	1	3	—	—	4
	1	3	15	18	—	36
	2	—	18	63	27	108
	3	—	—	27	81	108
	Total	4	36	108	108	256

From these figures we have

$$\Sigma AB = 7(0) + 15(1) + 36(2) + 63(4) + 54(6) + 81(9) = 1392,$$

$$\therefore \frac{\Sigma AB}{256} = \frac{87}{16}. \quad \therefore \frac{\Sigma AB}{256} - M_a M_b = \frac{87}{16} - \frac{81}{16} = \frac{6}{16}.$$

Since $\sigma_a \cdot \sigma_b = \frac{9}{16}$

$$\rho_{ab} = \frac{16}{9} \cdot \frac{6}{16} = \frac{2}{3} = 0\cdot\dot{6}.$$

Thus the foregoing result does not depend on whether p is equal to q.

    We can investigate correlation in a system involving more than two score classes at a single trial, if our tetrahedral die has one face with 1 pip, two faces each with 2 and the fourth face with 3 pips, as in Fig. 70. Thus all possible scores at a single trial are 1, 2 and 3 in the ratio $1 : 2 : 1$. We shall give each player one toss in a 3-fold trial, supplementing his score with an umpire's bonus based on the result (2, 3, 4, 5, 6) of a double toss. The bonus score distribution then accords with the binomial coefficients $1 : 4 : 6 : 4 : 1$. The total score set up is therefore as in Table 2:

TABLE 2

	3	4	5
3	1	2	1
4	2	4	2
5	1	2	1

	4	5	6
4	4	8	4
5	8	16	8
6	4	8	4

	5	6	7
5	6	12	6
6	12	24	12
7	6	12	6

	6	7	8
6	4	8	4
7	8	16	8
8	4	8	4

	7	8	9
7	1	2	1
8	2	4	2
9	1	2	1

Table 3 exhibits the composite grid:

TABLE 3

A score

		3	4	5	6	7	8	9	Total
	3	1	2	1	—	—	—	—	4
	4	2	8	10	4	—	—	—	24
	5	1	10	23	20	6	—	—	60
B	6	—	4	20	32	20	4	—	80
score	7	—	—	6	20	23	10	1	60
	8	—	—	—	4	10	8	2	24
	9	—	—	—	—	1	2	1	4
Total		4	24	60	80	60	24	4	256

The sum of the products is

$$9 + 48 + 128 + 30 + 400 + 475 + 192 + 1200 + 1152 + 420$$
$$+ 1680 + 1127 + 384 + 1120 + 512 + 126 + 288 + 81 = 9472.$$

$$\therefore \frac{\Sigma AB}{n} = \frac{9472}{256} = 37.$$

The mean score of the players is 6, so that $M_a \cdot M_b = 36$, and

$$\frac{\Sigma AB}{n} - M_a M_b = 1.$$

The variance $V_a = V_b$ of the 3-fold toss is given by

$$\frac{(-3)^2 + 6(-2)^2 + 15(-1)^2 + 20(0)^2 + 15(1)^2 + 6(2)^2 + (3)^2}{64}$$

$$= \tfrac{96}{64} = \tfrac{3}{2}.$$

Hence the value of $\sigma_a \cdot \sigma_b$ is 1·5 and

$$\rho_{ab} = \frac{1}{1\cdot5} = \frac{2}{3} = 0\cdot6.$$

Again, the value of ρ is the proportionate contribution of the umpire's bonus to the total mean score of either player.

Let us now examine how the mean score of one player varies w.r.t. a fixed value of the score of the other, and still within the framework of the restrictions that: (a) both players use the same die as the umpire; (b) each player has the same number of individual and of total tosses. The total number of tosses each player records will be 6; and we shall explore the consequences of varying the umpire's proportionate contribution from unity (six tosses) to zero (no tosses). Table 4 sets out separately correlation tables for six paired tosses of a coin on the assumption that the umpire tosses 6, 5, 4, 3, 2, 1 and 0 times, i.e. for values of ρ respectively equal to

$$1; \tfrac{5}{6}; \tfrac{4}{6}; \tfrac{3}{6}; \tfrac{2}{6}; \tfrac{1}{6}; 0.$$

TABLE 4

Result of 6-fold toss of Coin : Contingent A and B relative score frequencies with Mean (M_{ba}) of each Column and corresponding Relative Frequency of Column Total Score at Foot of Table

Umpire tosses six times ($\rho = 1$)

	0	1	2	3	4	5	6
0	1						
1		6					
2			15				
3				20			
4					15		
5						6	
6							1
M_{ba}	0	1	2	3	4	5	6
Total	1	6	15	20	15	6	1

Umpire tosses five times ($\rho = 0\cdot83$)

	0	1	2	3	4	5	6
0	1	1					
1	1	6	5				
2		5	15	10			
3			10	20	10		
4				10	15	5	
5					5	6	1
6						1	1
M_{ba}	$\frac{3}{6}$	$\frac{8}{6}$	$\frac{13}{6}$	$\frac{18}{6}$	$\frac{23}{6}$	$\frac{28}{6}$	$\frac{33}{6}$
Total	1	6	15	20	15	6	1

Umpire tosses four times ($\rho = 0\cdot6$)

	0	1	2	3	4	5	6
0	1	2	1				
1	2	8	10	4			
2	1	10	23	20	6		
3		4	20	32	20	4	
4			6	20	23	10	1
5				4	10	8	2
6					1	2	1
M_{ba}	$\frac{3}{3}$	$\frac{5}{3}$	$\frac{7}{3}$	$\frac{9}{3}$	$\frac{11}{3}$	$\frac{13}{3}$	$\frac{15}{3}$
Total	1	6	15	20	15	6	1

Umpire tosses three times ($\rho = 0\cdot5$)

	0	1	2	3	4	5	6
0	1	3	3	1			
1	3	12	18	12	3		
2	3	18	39	39	18	3	
3	1	12	39	56	39	12	1
4		3	18	39	39	18	3
5			3	12	18	12	3
6				1	3	3	1
M_{ba}	$\frac{3}{2}$	$\frac{4}{2}$	$\frac{5}{2}$	$\frac{6}{2}$	$\frac{7}{2}$	$\frac{8}{2}$	$\frac{9}{2}$
Total	1	6	15	20	15	6	1

Umpire tosses twice ($\rho = 0\cdot3$)

	0	1	2	3	4	5	6
0	1	4	6	4	1		
1	4	18	32	28	12	2	
2	6	32	69	76	44	12	1
3	4	28	76	104	76	28	4
4	1	12	44	76	69	32	6
5		2	12	28	32	18	4
6			1	3	6	4	1
M_{ba}	$\frac{6}{3}$	$\frac{7}{3}$	$\frac{8}{3}$	$\frac{9}{3}$	$\frac{10}{3}$	$\frac{11}{3}$	$\frac{12}{3}$
Total	1	6	15	20	15	6	1

Umpire tosses once ($\rho = 0\cdot16$)

	0	1	2	3	4	5	6
0	1	5	10	10	5	1	
1	5	26	55	60	35	10	1
2	10	55	125	150	100	35	5
3	10	60	150	200	150	60	10
4	5	35	100	150	125	55	10
5	1	10	35	60	55	26	5
6		1	5	10	10	5	1
M_{ba}	$\frac{5}{2}$	$\frac{8}{3}$	$\frac{17}{6}$	3	$\frac{19}{6}$	$\frac{10}{3}$	$\frac{7}{2}$
Total	1	6	15	20	15	4	1

Umpire does not toss ($\rho = 0$)

	0	1	2	3	4	5	6
0	1	6	15	20	15	6	1
1	6	36	90	120	90	36	6
2	15	90	225	300	225	90	15
3	20	120	300	400	300	120	20
4	15	90	225	300	225	90	15
5	6	36	90	120	90	36	6
6	1	6	15	20	15	6	1
M_{ba}	3	3	3	3	3	3	3
Total	1	6	15	20	15	6	1

If the umpire tosses six times, so that neither A nor B tosses at all, their scores must be identical. The relative frequencies of either set, represented by the figures along the diagonal, then correspond to the frequencies of successive scores of a single 6-fold toss as shown above the upper and outside the left-hand margin of the table. This is equivalent to stating that the mean B score for any column is equal to the A score associated with it, and has the same frequency. Hence the distribution of the mean B score for the six columns is the same as the distribution of the B scores as a whole ; and the variance of the distribution of the column means is equal to the variance of the distribution of individual B scores. Similarly, the variance of the distribution of the mean (A) scores of the six rows is the same as the variance of the distributions of the individual A scores.

When $\rho = 0$, A and B each toss six times independently. The means of the columns are then equal, as are also the means of the rows. Hence the variance of the distributions of either the column means or the row means is zero. If we label the variance of the distribution of the column means as $V(M_{ba})$, that of the row means as $V(M_{ab})$ and that of the individual B or A scores as $V_b = V_a$, we may thus define

(a) our criterion of *perfect correspondence* by the relation $V(M_{ba}) \div V_b = 1 = V(M_{ab}) \div V_a$;
(b) a necessary condition of *independence* by $V(M_{ba}) \div V_b = 0 = V(M_{ab}) \div V_a$.

When, as here, the table is symmetrical it is more convenient to write $V(M_{ba}) = V_m = V(M_{ab})$ and $V_a = V = V_b$. In this notation the criterion of perfect correspondence and a necessary condition of independence are respectively

(a) $\rho = 1 = V_m \div V$.
(b) $\rho = 0 = V_m \div V$.

These considerations suggest an examination of the ratio $(V_m \div V)$ when correspondence is imperfect. By reference to Table 4, which shows the mean column scores at the foot with the corresponding relative frequencies of column totals below, the reader will be able to compute the variance of the distribution of *mean B* scores as follows in accordance with the familiar formula in which M_b ($= 3$), being the mean of all B scores, is also the weighted mean of the column means (M_{ba}) :

$$V_m = \sum_{x=0}^{x=6} y\, M_{ba}^2 - M_b^2.$$

Umpire tosses six times

$$V_m = \frac{1(0)^2 + 6(1)^2 + 15(2)^2 + 20(3)^2 + 15(4)^2 + 6(5)^2 + 1(6)^2}{64} - 9 = \tfrac{21}{2} - 9 = \tfrac{3}{2}.$$

Umpire tosses five times

$$V_m = \frac{1(0\cdot5)^2 + 6(1\cdot3)^2 + 15(2\cdot16)^2 + 20(3)^2 + 15(3\cdot83)^2 + 6(4\cdot6)^2 + 1(5\cdot5)^2}{64} - 9 = \tfrac{241}{24} - 9 = \tfrac{25}{24}.$$

Umpire tosses four times

$$V_m = \frac{1(1)^2 + 6(1\cdot6)^2 + 15(2\cdot3)^2 + 20(3)^2 + 15(3\cdot6)^2 + 6(4\cdot3)^2 + 1(5)^2}{64} - 9 = \tfrac{29}{3} - 9 = \tfrac{2}{3}.$$

Umpire tosses three times

$$V_m = \frac{1(1\cdot5)^2 + 6(2)^2 + 15(2\cdot5)^2 + 20(3)^2 + 15(3\cdot5)^2 + 6(4)^2 + 1(4\cdot5)^2}{64} - 9 = \tfrac{75}{8} - 9 = \tfrac{3}{8}.$$

Umpire tosses twice

$$V_m = \frac{1(2)^2 + 6(2\cdot3)^2 + 15(2\cdot6)^2 + 20(3)^2 + 15(3\cdot3)^2 + 6(3\cdot6)^2 + 1(4)^2}{64} - 9 = \tfrac{55}{6} - 9 = \tfrac{1}{6}.$$

Umpire tosses once

$$V_m = \frac{1(2\cdot5)^2 + 6(2\cdot6)^2 + 15(2\cdot83)^2 + 20(3)^2 + 15(3\cdot16)^2 + 6(3\cdot3)^2 + 1(3\cdot5)^2}{64} - 9 = \frac{217}{24} - 9 = \frac{1}{24}.$$

Umpire does not toss

$$V_m = 0.$$

We now proceed to tabulate side by side with corresponding values of ρ and of u (the number of tosses the umpire performs) the ratio $V_m \div V$ in which $V = 6pq = \frac{3}{2}$:

u.	V_m.	$V_m \div V$.	ρ.
0	0	$0 \div \frac{3}{2} = 0$	0
1	$\frac{1}{24}$	$\frac{1}{24} \div \frac{3}{2} = \frac{1}{36}$	$\frac{1}{6}$
2	$\frac{1}{6}$	$\frac{1}{6} \div \frac{3}{2} = \frac{1}{9}$	$\frac{1}{3}$
3	$\frac{3}{8}$	$\frac{3}{8} \div \frac{3}{2} = \frac{1}{4}$	$\frac{1}{2}$
4	$\frac{2}{3}$	$\frac{2}{3} \div \frac{3}{2} = \frac{4}{9}$	$\frac{2}{3}$
5	$\frac{25}{24}$	$\frac{25}{24} \div \frac{3}{2} = \frac{25}{36}$	$\frac{5}{6}$
6	$\frac{3}{2}$	$\frac{3}{2} \div \frac{3}{2} = 1$	1

From the table above the following relation emerges consistently :

$$\rho^2 = V_m \div V \qquad . \qquad . \qquad . \qquad . \qquad . \qquad . \quad \text{(i)}$$

In (i) V is the variance of the distribution of individual B scores, and V_m is that of the distribution of the column means, i.e. mean values of B associated with successive values of A. The numerator varies between the limits : (a) zero when there is no tie-up between the two scores ; (b) *unity* when there is one to one correspondence between them. *At the two limits*, we may therefore regard the quantity on the right of the equation as the fraction of the total variance attributable to the association between the two sets of scores ; but nothing we have so far established entitles us to regard it as such except in the sense indicated by the use of italics. We may write its complement in the form

$$1 - \rho^2 = (V - V_m) \div V \qquad . \qquad . \qquad . \qquad . \quad \text{(ii)}$$

This leads us to seek a meaning for the residual component $(V - V_m)$. When $u = 4$ and $\rho = 0\cdot6$ in Table 4, the column mean B for an A score of 1 is $\frac{5}{8}$ and the variance of the B score distribution of the column as a whole is given by

$$\frac{2(0)^2 + 8(1)^2 + 10(2)^2 + 4(3)^2}{24} - \left(\frac{5}{3}\right)^2 = \frac{13}{18}.$$

In the same way we derive the variances of the B score distributions of other columns as at the foot of the ensuing table :

			A score					
	0	1	2	3	4	5	6]	Total
B score 0	1	2	1	—	---	—	—	4
1	2	8	10	4	..	..	----	24
2	1	10	23	20	6	..		60
3	—	4	20	32	20	4	--	80
4	--	-	6	20	23	10	1	60
5	.	—	-	4	10	8	2	24
6	—	—	-		1	2	1	4
TOTAL	4	24	60	80	6)	24	4	256
Mean (M_{ba})	$\frac{3}{3}$	$\frac{5}{3}$	$\frac{7}{3}$	$\frac{9}{3}$	$\frac{11}{3}$	$\frac{13}{3}$	$\frac{15}{3}$	3
Variance (V_{ba})	$\frac{1}{2}$	$\frac{13}{18}$	$\frac{77}{90}$	$\frac{9}{10}$	$\frac{77}{90}$	$\frac{13}{18}$	$\frac{1}{2}$	$\frac{3}{2}$

The weighted mean value (M_v) of the variances of the column score distributions is in this case

$$\frac{4(\tfrac{1}{2}) + 24(\tfrac{13}{18}) + 60(\tfrac{77}{90}) + 80(\tfrac{9}{10}) + 60(\tfrac{77}{90}) + 24(\tfrac{13}{18}) + 4(\tfrac{1}{2})}{256} = \frac{5}{6}.$$

Now the value of V_m for this case $(u = 4)$ is $\frac{2}{3}$ and V is in any case $\frac{3}{2}$. Hence we have

$$V_m + M_v = \frac{2}{3} + \frac{5}{6} = \frac{3}{2} = V.$$

For any set-up in Table 4 the same relation holds good, i.e.

$$V = V_m + M_v \qquad\qquad\qquad \text{(iii)}$$
$$\therefore M_v = V - V_m.$$

Hence from (ii) above

$$1 - \rho^2 = \frac{M_v}{V} \qquad\qquad\qquad \text{(iv)}$$

Evidently $M_v = V$ when A and B scores are completely independent so that $\rho = 0$; and $M_v = 0$ when correspondence is perfect, i.e. when $\rho = 1$. *At the two limits* the fraction on the right of (iv) is therefore that part of the total variance of either set of score distributions attributable to their independent behaviour.

Without prejudice to what ulterior interpretation we may appropriately confer on V_m and M_v, we shall henceforth speak of the first as the mean *inter*-class, and of the second as the mean *intra*-class, variance. Some statisticians respectively refer to these two components of a score distribution as *explained* and *unexplained* variation. This usage is exceptional for a reason which we shall examine at a later stage (p. 389). Meanwhile, it is well to remember that variance is not a unique measure of variation. As an index of the latter, it owes its popularity to its mathematical properties rather than to any justifiable claim to semantic identification with variation in general.

What we have hitherto discussed is the *concurrent* relation between the scores of the players arising from the common source of variation, *viz.* that of the umpire's contribution. The same set-up permits us to examine the correlation between the score of the player and that of the umpire, a relation which is purely *consequential*. For the case last discussed, *viz.* the toss of a coin, when the umpire tosses four out of six times and the player twice, the lay-out is as follows :

Umpire's score	.	.	0			1			2			3			4		
Player's score	.	.	0	1	2	1	2	3	2	3	4	3	4	5	4	5	6
Relative frequency	.	.	1	2	1	4	8	4	6	12	6	4	8	4	1	2	1

Whence we obtain the correlation grid :

Player's Score

		0	1	2	3	4	5	6	Total	Mean	Variance
	0	1	2	1	—	—	—	—	4	1	$\frac{1}{2}$
Umpire's	1	—	4	8	4	—	—	—	16	2	$\frac{1}{2}$
Score	2	—	—	6	12	6	—	—	24	3	$\frac{1}{2}$
	3	—	—	—	4	8	4	—	16	4	$\frac{1}{2}$
	4	—	—	—	—	1	2	1	4	5	$\frac{1}{2}$
	Total	1	6	15	20	15	6	1	64	3	—
	Mean	0	$\frac{2}{3}$	$\frac{4}{3}$	$\frac{6}{3}$	$\frac{8}{3}$	$\frac{10}{3}$	$\frac{12}{3}$	2	—	—

For this set-up we need to distinguish the umpire's mean (M_u) score and the variance (V_u) of the umpire's score distribution. We then obtain in the usual way :

$$M_u = 4(\tfrac{1}{2}) = 2 ; \quad M_a = 6(\tfrac{1}{2}) = 3 = M_b.$$

$$V_u = 4(\tfrac{1}{2})(\tfrac{1}{2}) = 1 ; \quad V_a = 6(\tfrac{1}{2})(\tfrac{1}{2}) = \tfrac{3}{2} = V_b.$$

$$\frac{\Sigma AU}{n} = \frac{\Sigma BU}{n} = \tfrac{1}{64}\{4(1) + 8(2) + 4(3) + 6(4) + 12(6) + 6(8) + 4(9)$$

$$+ 8(12) + 4(15) + 1(16) + 2(20) + 1(24)\} = 7.$$

$$\therefore \ Cov\,(AU) = \frac{\Sigma AU}{n} - M_a M_u = 7 - 6 = 1 = Cov\,(BU).$$

$$\rho_{au} = Cov\,(AU) \div \sqrt{V_u . V_a} = 1 \div \sqrt{1 . (\tfrac{3}{2})} \quad and \quad \rho^2_{au} = \tfrac{2}{3} = \rho^2_{bu}.$$

In this case the *square* of the product moment index is therefore the proportionate contribution of the umpire's mean score to that of the player ; and it is also equal to the proportionate contribution the variance of the row means $V(M_{au})$ makes to the total variance (V_a) of the player's score, since

$$V(M_{au}) = \frac{4(1^2) + 16(2^2) + 24(3^2) + 16(4^2) + 4(5^2)}{64} - 3^2 = 1.$$

$$\therefore \ V(M_{au}) \div V_a = 1 \div (\tfrac{3}{2}) = \tfrac{2}{3}.$$

If the reader will make similar calculations for the 6-fold toss to which the umpire contributes 1, 2, 3 or 5 of the total, he or she will find that the relation here stated also holds good, *viz.* in each case ρ^2, i.e. the square of the product moment index w.r.t. the score of the umpire and that of one or other player, is the proportionate contribution of the umpire's mean score to that of the player, as likewise the proportionate contribution of the variance of the player's mean scores to the player's total variance. In this context, we are therefore entitled to speak of $\rho^2 = V(M_{au}) \div V_a$ as the *explained* variation. Since the row variances (V_{au}) are identical, being 0·5, we have for their mean value $M(V_{au}) = \tfrac{1}{2}$.

$$\therefore \ V(M_{au}) + M(V_{au}) = 1 + \tfrac{1}{2} = \tfrac{3}{2} = V_a.$$

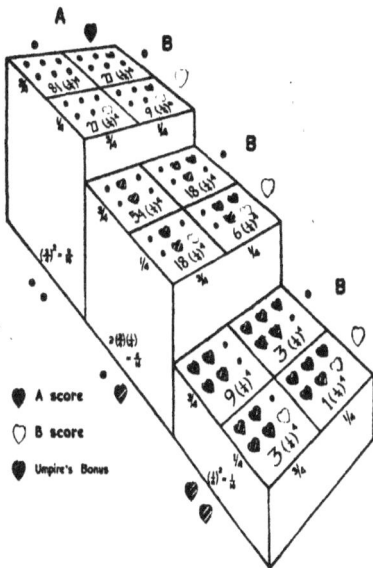

A Score

	0×0 81	0×1 27	0×2 •••	0×3 •••	TOTAL 108
	1×0 27	1×1 (9·54·63)	1×2 18	1×3 •••	108
B Score	2×0 •••	2×1 18	2×2 (6·9·15)	2×3 3	36
	3×0 •••	3×1 •••	3×2 3	3×3 1	4
TOTAL	108	108	36	4	256

FIG. 79. The Umpire Bonus Model. The umpire and the two players (A and B) draw from the same pack. The umpire picks two cards, replacing each card taken before drawing another. A picks one card and replaces it before B also picks one card. The total score of either A or B is the sum of that of the umpire and his own.

EXERCISE 9.01

For each example determine ρ w.r.t. the concurrent relation (ρ_{ab}) of the two players and the consequential relation (ρ_{au} or ρ_{bu}) of the score of the umpire with that of one or other player. Show that the covariance is the variance of the umpire's score (u) distribution and test the identities

$$(a)\ \rho = \frac{u}{n}; \quad (b)\ \rho^2 = V_m \div V.$$

1. The umpire tosses a penny three times. The two players each toss three times, scoring a head as success.

2. The umpire tosses the flat circular die of Fig. 67 four times, the players each toss the same die four times.

3. The umpire tosses the tetrahedral die of Fig. 70 twice and the players each toss it twice.

4. The umpire tosses an ordinary cubical die three times, each of the players tossing three times.

5. The umpire tosses twice the tetrahedral die of Fig. 73, each of the players also tossing twice.

6. The umpire draws two cards with replacement from a full pack, scoring spades as success. Each player then draws two cards with replacement.

7. The umpire draws three cards from a full pack with replacement scoring picture cards as a success. The players each draw three cards, replacing each card before another draw.

8. The umpire and each player draws singly three balls from an urn containing four black and one red ball, replacing each ball before drawing another and scoring red as success.

9. The umpire tosses a penny twice, each of the players tossing three times counting heads as successes.

10. The umpire tosses twice the tetrahedral die of Fig. 70, one player tossing three times the circular die of Fig. 67 and the other tossing once the tetrahedral die of Fig. 73.

9.02 THE UNRESTRICTED BONUS MODEL

We have so far explored the properties of the umpire bonus model subject to two restrictions : (a) that the two players record the same number of independent tosses ; (b) that the umpire and both players toss the same die or draw cards with replacement from the same pack. Within this framework, the long run expectation of winning the contest is the same for each player. The mean score of one player is the same as that of the other, and the variances of the two score distributions are the same. By recourse to numerical examples, we shall now explore the consequences of removing these restrictions.

Let us first suppose that the two players (A and B) and the umpire use the same die, a coin or flat disc with 1 pip on one face and no pip on the other. If the umpire tosses twice, A once and B twice, the set-up in conformity with our previous procedure is as follows :

	(a) Umpire scores zero				(b) Umpire scores 1				(c) Umpire scores 2	
		A				A				A
	0	1			1	2			2	3
0	1	1		1	2	2		2	1	1
B 1	2	2		2	4	4		3	2	2
2	1	1		3	2	2		4	1	1

The composite correlation table for the 3-fold toss A records and the 4-fold toss B records is therefore as below :

		A					
	0	1	2	3	4	Total	Mean
0	1	1	—	—	—	2	0·5
1	2	4	2	—	—	8	1·0
2	1	5	5	1	—	12	1·5
B 3	—	2	4	2	··	8	2·0
4	—	—	1	1	··	2	2·5
Total	4	12	12	4	0	32	1·5
Mean	$\frac{3}{8}$	$\frac{4}{3}$	$\frac{7}{6}$	$\frac{9}{4}$	0	2·0	—

From the foregoing table we obtain

$$M_a = 1\cdot5; \quad M_b = 2; \quad M_a M_b = 3;$$
$$V_a = 3(\tfrac{1}{2})^2 = 0\cdot75; \quad V_b = 4(\tfrac{1}{2})^2 = 1; \quad V_a \cdot V_b = 0\cdot75.$$
$$\frac{\Sigma AB}{n} = \frac{5(0) + 4(1) + 7(2) + 5(4) + 2(3) + 5(6) + 2(9) + 1(8) + 1(12)}{32} = \tfrac{7}{2}.$$
$$Cov\,(AB) = \tfrac{7}{2} - 3 = 0\cdot5.$$
$$\therefore\ \rho_{ab}^2 = \frac{(0\cdot5)^2}{0\cdot75} = \tfrac{1}{3} \quad . \qquad . \qquad . \qquad . \qquad . \qquad (i)$$

Let us now compare this value of ρ^2 with the two ratios $V(M_{ab}) \div V_a = 4V(M_{ab}) \div 3$, since $V_a = \tfrac{3}{4}$, and $V(M_{ba}) \div V_b = V(M_{ba})$, since $V_b = 1$:

$$V(M_{ab}) = \tfrac{1}{16}\{(0\cdot5)^2 + 4(1)^2 + 6(1\cdot5)^2 + 4(2)^2 + (2\cdot5)^2\} - (1\cdot5)^2 = 0\cdot25,$$
$$\therefore\ V(M_{ab}) \div V_a = 4V(M_{ab}) \div 3 = 4(0\cdot25) \div 3 = \tfrac{1}{3} = \rho_{ab}^2.$$
$$V(M_{ba}) = \tfrac{1}{8}\{1^2 + 3(1\cdot6)^2 + 3(2\cdot3)^2 + 3^2\} - 2^2 = \tfrac{1}{3},$$
$$\therefore\ V(M_{ba}) \div V_b = V(M_{ba}) = \tfrac{1}{3} = \rho_{ab}^2.$$

Once more we see that the identity of ρ^2 with the proportionate inter-class variance is equally applicable to either the B scores alone or the A scores alone. Since the umpire's mean score $M_u = 1$, his proportionate contribution (c_a) to the mean score of A is $\tfrac{2}{3}$ and his proportionate contribution (c_b) to the mean score of B is $\tfrac{1}{2}$, so that

$$c_a \cdot c_b = \tfrac{2}{3} \cdot \tfrac{1}{2} = \tfrac{1}{3} = \rho_{ab}^2 \quad . \qquad . \qquad . \qquad . \qquad . \qquad (ii)$$

Thus ρ is in this case the *geometric mean* of the respective proportionate contributions of the umpire to the two players' mean scores. A second example reinforces both conclusions suggested by this one. We next suppose that the umpire tosses the same die four times, A twice and B three times, so that $c_a = \tfrac{2}{3}$ and $c_b = \tfrac{3}{4}$. The set-up is as follows :

Umpire scores zero				Umpire scores 1				Umpire scores 2		
	0	1	2		1	2	3		2	3	4
0	1	2	1	1	4	8	4	2	6	12	6
1	3	6	3	2	12	24	12	3	18	36	18
2	3	6	3	3	12	24	12	4	18	36	18
3	1	2	1	4	4	8	4	5	6	12	6

Umpire scores 3				Umpire scores 4		
	3	4	5		4	5	6
3	4	8	4	4	1	2	1
4	12	24	12	5	3	6	3
5	12	24	12	6	3	6	3
6	4	8	4	7	1	2	1

The composite table is as below :

A score

	0	1	2	3	4	5	6	7	Total	Mean
0	1	2	1	—	—	—	—	—	4	$\frac{1}{7}$
1	3	10	11	4	—	—	—	—	28	$1\frac{1}{7}$
2	3	18	33	24	6	—	—	—	84	$1\frac{5}{7}$
3	1	14	43	52	26	4	—	—	140	$1\frac{6}{7}$
B score 4	—	4	26	52	43	14	1	—	140	$1\frac{1}{7}$
5	—	—	6	24	33	18	3	—	84	$4\frac{2}{7}$
6	—	—	—	4	11	10	3	—	28	$3\frac{1}{7}$
7	—	—	—	—	1	2	1	—	4	$4\frac{4}{7}$
Total	8	48	120	160	120	48	8	0	512	3
Mean	$\frac{9}{6}$	$1\frac{3}{6}$	$1\frac{7}{6}$	$2\frac{1}{6}$	$2\frac{5}{6}$	$3\frac{2}{6}$	$3\frac{3}{6}$	0	3·5	—

The student should check the details of the calculation, of which the following figures will therefore suffice

$$M_a M_b = 3(3{\cdot}5) = 10{\cdot}5 ; \quad V_a = 1{\cdot}5 ; \quad V_b = 1{\cdot}75 ; \quad V_a V_b = 2{\cdot}625.$$

$$\frac{\Sigma AB}{n} = \frac{23}{2} ; \quad Cov\,(AB) = 1.$$

$$\rho_{ab}^2 = \frac{1}{2{\cdot}625} = \frac{8}{21}.$$

$$V(M_{ab}) = \tfrac{4}{7} ; \quad V(M_{ba}) = \tfrac{2}{3}.$$

$$V(M_{ab}) \div V_a = \tfrac{4}{7} \cdot \tfrac{2}{3} = \tfrac{8}{21} = \rho_{ab}^2 = \tfrac{8}{21} = \tfrac{2}{3} \cdot \tfrac{4}{7} = V(M_{ba}) \div V_b.$$

In this case $c_a c_b = (\tfrac{2}{3})(\tfrac{4}{7}) = \tfrac{8}{21} = \rho^2$. Hence again ρ^2 is the geometric mean of the proportionate contributions of the umpire to the mean scores of the two players, and its square is in fact the proportionate interclass variation of either the A scores or of the B scores.

The identification of ρ_{ab} as a geometric mean when the common die is, as heretofore, one of the *binomial* type exhibited in Figs. 67, 70 and 73, including the coin as a special case of Fig. 67, discloses a clue to a more general interpretation of *covariance*, when we remove this restriction. Let us suppose that the face scores of the die are : 0, 1, 2, etc., wit frequencies referable to the definitive binomial $(q + p)^k$, denoting by u, a and b the number of individual tosses permitted to the umpire and to the two players respectively. Thus the total number of tosses assigned to the players are $(u + a)$ and $(u + b)$, and we may write, as in (ii) (p. 300) :

$$V_u = ukpq ; \quad V_a = (u + a)kpq ; \quad V_b = (u + b)kpq.$$

In the foregoing symbolism

$$c_a = u \div (u + a) \quad \text{and} \quad c_b = u \div (u + b).$$

Our numerical example illustrates the relation :

$$c_a \cdot c_b = \rho_{ab}^2 = \frac{u^2}{(u+a)(u+b)}.$$

If this is so :

$$\rho_{ab}^2 = \frac{ukpq}{(u+a)\,kpq} \cdot \frac{ukpq}{(u+b)\,kpq} = \frac{V_u^2}{V_a \cdot V_b}.$$

Hence also :

$$\frac{Cov\,(AB)}{\sqrt{V_a \cdot V_b}} = \frac{V_u}{\sqrt{V_a \cdot V_b}}.$$

Whence we derive :

$$Cov\,(AB) = V_u \quad \cdot \quad \cdot \quad \cdot \quad \cdot \quad \cdot \quad (\text{iii})$$

Here then the covariance of the whole distribution is the variance of the umpire score distribution, being zero when the umpire does not toss ($V_u = 0$). In short, the covariance of the paired score distribution is the variance (V_u) of the *concomitant* score distribution.

We shall now give the umpire and each of the players a *different die* to toss, assuming that

(a) the umpire tosses twice a tetrahedral die with 1 pip on one face, 3 pips on another and 2 pips on each of the remaining pair, so that his total scores of 2, 3, 4, 5, 6 occur with relative frequencies 1 : 4 : 6 : 4 : 1 ;

(b) the player A tosses once a tetrahedral die of which one face carries 1 pip, the other three faces 2 pips, so that his own scores of 1 and 2 respectively occur with relative frequencies 1 : 3 ;

(c) the player B tosses twice a flat circular die with 1 pip on one face and 2 on the other, so that his own scores of 2, 3 and 4 respectively occur with relative frequencies 1 : 2 : 1.

The variance of the distribution of the umpire's score is deducible from the weighted mean of the square deviations, *viz*.

$$V_u = \frac{(-2)^2 + 4(-1)^2 + 6(0)^2 + 4(1)^2 + (2)^2}{16} = 1 \quad \cdot \quad \cdot \quad \cdot \quad (\text{iv})$$

In the usual way, we set up separately unit correlation tables for each value of the umpire's bonus, as below :

Umpire scores 2

	A	
	3	4
4	1	3
B 5	2	6
6	1	3

Umpire scores 3

	A	
	4	5
5	4	12
B 6	8	24
7	4	12

Umpire scores 4

	A	
	5	6
6	6	18
B 7	12	36
8	6	18

Umpire scores 5

	A	
	6	7
7	4	12
B 8	8	24
9	4	12

Umpire scores 6

	A	
	7	8
8	1	3
B 9	2	6
10	1	3

The composite correlation table for the trial is now as below :

		A score							
		3	4	5	6	7	8	Total	Mean
	4	1	3	—	—	—	—	4	$4\tfrac{1}{1}$
	5	2	10	12	—	—	—	24	$4\tfrac{1}{1}$
	6	1	11	30	18	—	—	60	$4\tfrac{1}{1}$
B	7	—	4	24	40	12	—	80	$4\tfrac{1}{1}$
score	8	—	—	6	26	25	3	60	$7\tfrac{1}{1}$
	9	—	—	—	4	14	6	24	$4\tfrac{1}{1}$
	10	—	—	—	—	1	3	4	$4\tfrac{1}{1}$
	Total	4	28	72	88	52	12	256	$\tfrac{11}{4}$
	Mean	5	$\tfrac{30}{7}$	$\tfrac{19}{3}$	$\tfrac{79}{11}$	$\tfrac{108}{13}$	9	7	—

From the foregoing table we obtain

$$\Sigma AB = 10560 ; \quad \frac{\Sigma AB}{n} = \tfrac{165}{4}.$$

$$M_a = \tfrac{23}{4} ; \quad M_b = 7 ; \quad M_a M_b = \tfrac{161}{4},$$

$$\therefore \; Cov(AB) = \frac{165 - 161}{4} = 1.$$

Whence by (iv) in agreement with (iii) we have

$$Cov(AB) = 1 = V_u.$$

The second moments about zero of the A and B score distributions are respectively obtainable by weighting the squares of the scores at the head and at the side of the chessboard by the column and row totals as fractions of the grand total (256), so that

$$V_a = \frac{4(3)^2 + 28(4)^2 + 72(5)^2 + 88(6)^2 + 52(7)^2 + 12(8)^2}{256} - (\tfrac{23}{4})^2 = \tfrac{19}{16}.$$

$$V_b = \frac{4(4)^2 + 24(5)^2 + 60(6)^2 + 80(7)^2 + 60(8)^2 + 24(9)^2 + 4(10)^2}{256} - 7^2 = \tfrac{3}{2}.$$

In the usual way we obtain

$$V(M_{ab}) = \tfrac{2}{3} ; \quad V(M_{ba}) = \tfrac{846}{1001}.$$

Hence we have

$$V(M_{ab}) \div V_a = \tfrac{2}{3} \div \tfrac{19}{16} = \tfrac{32}{57}.$$

$$V(M_{ba}) \div V_b = \tfrac{846}{1001} \div \tfrac{3}{2} = \tfrac{1692}{3003}.$$

$$\rho_{ab}^2 = 1 \div (\tfrac{19}{16})(\tfrac{3}{2}) = \tfrac{32}{57}.$$

The identity of ρ^2 with the proportionate inter-class variance here holds good w.r.t. *only one* set of scores—the A scores. If we seek a clue to this anomaly, inspection of the correlation cable discloses an essential difference between it and those of any models discussed in what has gone before. In other examples of the Umpire Bonus model dealt with so far, the means of the columns and the means of the rows alike increase by a fixed increment. In other words the mean B score (M_{ba}) associated with an A score (x_a) is a linear function of x_a ; and the mean A score (M_{ab}) associated with a B score of x_b is a linear function of x_b. In this case, only the means (M_{ab}) of the rows, i.e. the mean A scores paired off with particular B scores increase by equal increments, and the identity of ρ^2 with the proportionate interclass variance holds good only for the ratio of the variance of the distribution of the A score means to the variance of the distribution of the A scores themselves. It is easy to remember the meaning of the symbols M_{ba} and M_{ab} (hence by analogy V_{ba}, V_{ab}) if one repeats aloud :

(i) M_{ba} is the mean B score associated with a particular A score.

(ii) M_{ab} is the mean A score associated with a particular B score.

If the column border (A) scores and column means (M_{ba}) each increase by equal steps, it is customary to say that there is *linear regression of the B score on the A score*. If the row border (B) scores and row means (M_{ab}) each increase by equal steps, it is likewise customary to say that there is *linear regression of the A score on the B score*. Our last example thus illustrates linear regression of the A score on the B score and *non-linear* regression of the B score on the A score. We shall now examine a situation in which neither the means of the columns nor the means of the rows are linear functions of the corresponding column and row scores. For the sake of variety our model will refer to three card packs consisting of : (*a*) equal numbers of clubs and hearts, being the pack from which the umpire draws one ; (*b*) equal numbers of clubs, spades and hearts, being the pack from which player A draws one ; (*c*) equal numbers of all four suits, being the pack from which B draws one. We shall count a heart as a success and a card of any other suit as a failure, assuming that there is replacement of each card chosen before drawing another. Thus the umpire may score either 0 or 1 with the following results :

	A	
	0	1
B 0	1	2
B 1	3	6

	A	
	1	2
B 1	1	2
B 2	3	6

From the above we derive the following composite table in which it is evident that : (*a*) the row mean is *not* a linear function of the row score ; (*b*) the column mean is *not* a linear function of the column score.

		A score				
		0	1	2	Total	Mean
	0	1	2	—	3	$\frac{8}{12}$
B score	1	3	7	2	12	$\frac{11}{12}$
	2	—	3	6	9	$\frac{20}{12}$
Total		4	12	8	24	$\frac{14}{12}$
Mean		$\frac{9}{12}$	$\frac{13}{12}$	$\frac{21}{12}$	$\frac{15}{12}$	—

In this case the umpire's score distribution is defined by the terms of the binomial $(\frac{1}{2} + \frac{1}{2})^1$, so that its variance (V_u) is 0·25. From the above table we obtain

$$\frac{\Sigma AB}{n} = \frac{41}{24}; \quad M_a = \frac{14}{12}; \quad M_b = \frac{15}{12},$$

$$\therefore \text{Cov}(AB) = \frac{41}{24} - \frac{15.14}{144} = \frac{6}{24} = 0\cdot25 = V_u.$$

Again the covariance is the variance of the distribution of the common score component. The values of V_a and V_b are respectively given by

$$V_a = \frac{4(0)^2 + 12(1)^2 + 8(2)^2}{24} - (\tfrac{14}{12})^2 = \tfrac{17}{36}$$

$$V_b = \frac{3(0)^2 + 12(1)^2 + 9(2)^2}{24} - (\tfrac{15}{12})^2 = \tfrac{7}{16}.$$

In the usual way, we obtain for the variance of the distributions of the means of the columns and rows

$$V(M_{ab}) = \tfrac{5}{36} \quad and \quad V(M_{ba}) = \tfrac{5}{32}$$

From these results we have

$$\rho_{ab}^2 = (\tfrac{1}{4})^2 \div \tfrac{17}{36} \cdot \tfrac{7}{16} = \tfrac{36}{119}.$$

$$V(M_{ab}) \div V_a = \tfrac{5}{36} \div \tfrac{17}{36} = \tfrac{5}{17}.$$

$$V(M_{ba}) \div V_b = \tfrac{5}{32} \div \tfrac{7}{16} = \tfrac{5}{14}.$$

As we anticipated, the ρ^2 criterion now holds good for *neither* variance ratio defined by (i) in 9.01. Thus the arithmetical results of removing all restrictions we imposed on our model in 9.01 suggest the two following conclusions :

(i) the covariance of the joint distribution is equivalent to the variance of the distribution of the common score and in that sense is an exact measure of the concomitant, as opposed to the residual, variation of the A and B scores ;

(ii) for the distribution of either set of scores the relation defined by (i) in 9.01 holds good if, and *only* if, their means with respect to particular values of the alternate set are linear functions of the latter.

In Vol. II we shall see that the identity of the ratio $(V_m \div V)$ with ρ^2 computed by the product-moment formula is a necessary and sufficient condition of *linear regression*, i.e. that the mean values of one set associated with successive equally spaced values of the other increase by equal increments. The insight we have now obtained w.r.t. the meaning of a correlation table by varying the proportionate contribution the umpire can make to the mean total score of either player clarifies another issue, *viz.* in what terms it is appropriate to describe a law of concomitant variation, when each player, as heretofore, adds to his individual score that of the umpire. At one limit, only the umpire tosses. The individual score of either player is necessarily zero ; and there is then no residual variation arising from the independent contributions of the players themselves. Accordingly, their joint total scores then tally with those definitive of the diagonal cells of the table, and the means of either columns or rows necessarily increase by equal increments being identical with the successive scores of either player. In that sense, the law of concomitant variation descriptive of our Umpire Bonus Model is strictly *linear*. On the other hand, numerical examples which illustrate the effect of allocating different dice to the umpire and to the two players suffice to show that such a linear law of joint variation does not necessarily signify a linear relation between the mean of one set of scores associated with a particular value of the alternate score, when correlation is imperfect. What we here call a linear law of concomitant variation is, in fact, consistent with three possibilities :

(a) there is linear regression of *both* sets of scores w.r.t. the alternate set ;

(b) there is linear regression of *only one* set of scores w.r.t. the alternate set ;

(c) there is linear regression of *neither* set of scores w.r.t. the alternate set.

When reciprocal linear regression exists the so-called *correlation ratio* $V_m \div V$ computed w.r.t. either set of scores is numerically equivalent to ρ^2. If only the column means increase by equal steps, this relation holds good only w.r.t. B scores definitive of the rows, and it holds good only w.r.t. A scores definitive of the columns if only the row means increase by equal steps. The identity of the product-moment coefficient with the square root of the *correlation ratio* is therefore a *criterion of linear regression ;* but we should here be alert to a verbal pitfall. While all statisticians agree that regression does not necessarily, or commonly, signify a relation of *consequence* as defined in 8.01, the laws of experimental physics cast a long shadow over the mathematical formulation of scientific laws in general ; and we too lightly yield to the temptation of identifying linear regression with linearity in the domain of physical law. In the latter, a linear or other relation between B and A embodies the result we expect if the variation of B is subject to no agency other than the variation of A. To say that there is *non-linear* regression of one set of scores, e.g. A scores, with respect to the alternate, e.g. B scores, in the domain of statistics does *not* necessarily mean that the A scores would prove to be a non-linear function of the B scores in the absence of agencies which confer on them their freedom to vary independently.

We can, of course, modify the foregoing procedure in order to impose on the universe of the bonus model a non-linear law of concomitant variation. We shall suppose that

(a) the umpire and players use the same die, a coin with equally likely scores of 0 or 1 at a single toss ;

(b) The umpire tosses three times, each of the players twice ;

(c) A adds to his individual score $(x_{a.o})$ the score x_u of the umpire, and B adds x_u^2 to his individual score $(x_{b.o})$.

In this instance, the law of concomitant variation is quadratic in the sense that the B score becomes an exact quadratic function of the A score when we diminish the freedom of the two sets to vary independently by indefinitely increasing the proportionate number of tosses of the umpire. The unit correlation tables consonant with the foregoing prescription are

	A				A				A				A		
	0	1	2		1	2	3		2	3	4		3	4	5
B 0	1	2	1	1	3	6	3	4	3	6	3	9	1	2	1
1	2	4	2	2	6	12	6	5	6	12	6	10	2	4	2
2	1	2	1	3	3	6	3	6	3	6	3	11	1	2	1

The composite table is

A score

	0	1	2	3	4	5	Total	Mean
0	1	2	1	—	—	—	4	1
1	2	7	8	3	—	—	20	$\frac{8}{5}$
2	1	8	13	6	—	—	28	$\frac{11}{7}$
3	—	3	6	3	—	—	12	2
4	—	—	3	6	3	—	12	3
5	—	—	6	12	6	—	24	3
6	—	—	3	6	3	—	12	3
7	—	—	—	—	—	—	0	0
8	—	—	—	—	—	—	0	0
9	—	—	—	1	2	1	4	4
10	—	—	—	2	4	2	8	4
11	—	—	—	1	2	1	4	4
Total	4	20	40	40	20	4	128	2·5
Mean	$\frac{8}{5}$	$\frac{8}{5}$	$\frac{14}{5}$	$\frac{23}{5}$	$\frac{32}{5}$	$\frac{30}{5}$	4·0	—

(*B score* labels the rows.)

From this table we obtain

$$\frac{\Sigma AB}{n} = \frac{1568}{128} = \frac{49}{4}.$$

$$Cov\,(AB) = \frac{49}{4} - 4(2\cdot5) = \frac{9}{4}.$$

In this case, the covariance of the player's joint total score distribution is not the same as the variance of the umpire score distribution, since

$$V_u = 3(\tfrac{1}{2})(\tfrac{1}{2}) = \tfrac{3}{4}.$$

As stated above, the peculiarity of this set-up is that the mean score of B approaches an exact quadratic function of A as we diminish the freedom of B and A to vary independently by increasing the proportionate number of tosses the umpire performs. It will bring into focus the distinction it helps to elucidate if we use the symbols x_u, $x_{a\cdot o}$ and $x_{b\cdot o}$ for the scores referable to the *individual* and independent tosses of the umpire and players to set out as below the build-up of the *total* scores of the players when the law of concomitant variation is quadratic and, as in all examples cited other than the last, when it is linear as we here use the term

Linear Law of Concomitant Variation	Quadratic Law of Concomitant Variation
$x_a = x_{a\cdot o} + x_u$ $x_b = x_{b\cdot o} + x_u$	$x_a = x_{a\cdot o} + x_u$ $x_b = x_{b\cdot o} + x_u^2$

All the examples we have discussed so far refer to situations in which the correlation between the total score of one player and that of the other is *positive*. We shall now vary our procedure to accommodate negative correlation as follows :

(i) the umpire tosses twice the tetrahedral die of Fig. 70 with 1, 2, 3 pips in the ratio 1 : 2 : 1 ;

(ii) player A tosses three times the flat circular die of Fig. 67 with 1 and 2 pips respectively on its two faces, *adding* to his individual score the score of the umpire ;

(iii) player B tosses once the customary cubical die and *deducts* from his individual score that of the umpire.

Thus the *individual* score distributions are

Scores	1	2	3	4	5	6
(i) Umpire	0	1	4	6	4	1
(ii) Player A	0	0	1	3	3	1
(iii) Player B	1	1	1	1	1	1

The unit correlation tables for total scores of player A and player B w.r.t. the several possible scores of the umpire are as shown in Table 5.

TABLE 5

Umpire scores 2

A

	5	6	7	8
−1	1	3	3	1
0	1	3	3	1
1	1	3	3	1
B 2	1	3	3	1
3	1	3	3	1
4	1	3	3	1

Umpire scores 3

A

	6	7	8	9
−2	4	12	12	4
−1	4	12	12	4
0	4	12	12	4
B 1	4	12	12	4
2	4	12	12	4
3	4	12	12	4

Umpire scores 4

A

	7	8	9	10
−3	6	18	18	6
−2	6	18	18	6
−1	6	18	18	6
B 0	6	18	18	6
1	6	18	18	6
2	6	18	18	6

Umpire scores 5

A

	8	9	10	11
−4	4	12	12	4
−3	4	12	12	4
−2	4	12	12	4
B −1	4	12	12	4
0	4	12	12	4
1	4	12	12	4

Umpire scores 6

A

	9	10	11	12
−5	1	3	3	1
−4	1	3	3	1
−3	1	3	3	1
B −2	1	3	3	1
−1	1	3	3	1
0	1	3	3	1

The complete table for the players' total scores is as below (Table 6):

TABLE 6

		A									
		5	6	7	8	9	10	11	12	Total	Mean
	−5	—	—	—	—	1	3	3	1	8	10·5
	−4	—	—	—	4	13	15	7	1	40	9·7
	−3	—	—	6	22	31	21	7	1	88	9·05
	−2	—	4	18	34	35	21	7	1	120	8·63
	−1	1	7	21	35	35	21	7	1	128	8·5
B	0	1	7	21	35	35	21	7	1	128	8·5
	1	1	7	21	35	34	18	4	—	120	8·22
	2	1	7	21	31	22	6	—	—	88	7·95
	3	1	7	15	13	4	—	—	—	40	7·3
	4	1	3	3	1	—	—	—	—	8	6·5
	Total	6	42	126	210	210	126	42	6	768	8·5
	Mean	1·5	0·93	0·36	−0·21	−0·79	−1·35	−1·92	−2·5	−0·5	—

Since the umpire's sampling distribution is $(\frac{1}{2} + \frac{1}{2})^2$, the variance of the score sum distribution w.r.t. 2-fold samples is $2.2(\frac{1}{2})(\frac{1}{2})$, i.e.

$$V_u = 1 \quad . \quad . \quad . \quad . \quad . \quad . \quad . \quad \text{(viii)}$$

To determine V_a and V_b, the variances of the total score sum distribution of the players, we recall that each is the sum or difference of two independent scores. Hence the variance of the total score sum distribution of A is the sum of the variance of the umpire's score sum distribution and the variance of A's *individual* score sum distribution. Likewise that of B is the sum of the variance of the umpire's score sum, and the variance of B's individual score sum distribution. We shall here denote the variances of the individual score sum distributions respectively by $V_{a \cdot o}$ and $V_{b \cdot o}$. The former is that of a distribution defined by the binomial $(\frac{1}{2} + \frac{1}{2})^3$; and the latter that of the rectangular distribution $(\frac{1}{6} + \frac{1}{6} + \frac{1}{6} + \frac{1}{6} + \frac{1}{6} + \frac{1}{6})$. From results cited in 7.05 and 7.03, we therefore have

$$V_{a.o} = \tfrac{3}{4} \quad \textit{and} \quad V_{b.o} = \frac{6^2 - 1}{12} = \frac{35}{12}.$$

$$V_a = V_u + V_{a.o} = 1 + \tfrac{3}{4} = \tfrac{7}{4},$$

$$V_b = V_u + V_{b.o} = 1 + \tfrac{35}{12} = \tfrac{47}{12},$$

$$\therefore\ V_a V_b = \tfrac{329}{48}.$$

The reader may check these figures by reference to the column scores and totals and to the row scores and totals of the foregoing table from which we obtain

$$M_a = 8.5; \quad M_b = -0.5; \quad M_a M_b = -4.25,$$

$$\frac{\Sigma AB}{n} = \frac{-21}{4} = -5.25.$$

$$\therefore \ Cov\,(AB) = -5.25 + 4.25 = -1.$$

Hence from (viii) above

$$V_u = -Cov\,(AB).$$

Thus the covariance is again numerically equivalent to the variance of the distribution of the umpire's bonus, and its sign indicates a reverse-order law of concomitant variation. Evidently, the law of concomitant variation is linear in the sense defined above, though regression is linear in only one dimension of the table.

From the above values of $Cov\,(AB)$, V_a and V_b, we obtain

$$\rho_{ab} = -\sqrt{\tfrac{48}{329}} = -0.382.$$

In the computation of this example, we have made explicit a balance sheet of variation consonant with preceding results. With respect to each player we may split the variance of the total score distribution into two additive components, which it is appropriate to designate as *concomitant* (explained) and *unconstrained* (unexplained) :

$$V_a = V_u + V_{a.o}; \quad V_b = V_u + V_{b.o}.$$

Thus we may define a concomitant fraction (E_a) of the total variance of the A score distribution by

$$E_a = \frac{V_u}{V_a} = \frac{Cov\,(AB)}{V_a} = \rho_{ab}\sqrt{\frac{V_b}{V_a}}. \qquad \qquad \text{(ix)}$$

Likewise, for that of the B scores

$$E_b = \frac{V_u}{V_b} = \rho_{ab}\sqrt{\frac{V_a}{V_b}}. \qquad \qquad \qquad \text{(x)}$$

Accordingly, the unconstrained fractions $(U_a = 1 - E_a$ and $U_b = 1 - E_b)$ are definable thus

$$U_a = 1 - \rho_{ab}\sqrt{\frac{V_b}{V_a}}; \quad U_b = 1 - \rho_{ab}\sqrt{\frac{V_a}{V_b}} \qquad \qquad \text{(xi)}$$

When regression is linear, as we shall later see, ρ_{ab} is thus the geometric mean of the concomitant fractions, commonly called *regression coefficients*, i.e.

$$\rho_{ab}^2 = E_a . E_b.$$

The reader will recall that the last example in 9.01 suggests the possibility of a different causal partition of variation in the domain of the *consequential* relation between the score of the player and that of the umpire. Let us therefore examine such a situation without imposing the restriction that umpire and player toss the same die, *viz.* the umpire tosses twice the tetrahedral die of Fig. 70 and the player (A) tosses three times the flat circular die of Fig. 67. The correlation table is one which the reader should be able to construct as below without difficulty ; and we obtain therefrom in the usual way

$$M_u = 4; \quad M_a = 8.5; \quad V_u = 1; \quad V_a = \tfrac{7}{4}$$
$$Cov(AU) = 1 = V_u$$
$$\rho_{au}^2 = \tfrac{4}{7}.$$

In this set-up the *square* of the product moment index is the true measure of the proportionate contribution of the variance attributable to the source of concomitant variation, i.e. that of the score distribution of the umpire. In the usual way we also find :

$$V(M_{au}) = 1 \quad and \quad V(M_{au}) \div V_a = \tfrac{4}{7} = \rho_{au}^2.$$

Thus the meaning we attach to the numerical value of ρ and the sense in which we can partition the variance of a set of scores with reference to a particular source of variation in the model universe of this chapter depends on whether the relation is *concurrent* or *consequential*. An interesting feature of the latter relationship is that the row variances (V_{nu}) are equal as also in the last example of 9.01. When this is so, we say that the score distribution is *homoscedastic* in the appropriate dimension of the grid.

A's Total Score

		5	6	7	8	9	10	11	12	Total	Mean	Variance
	2	1	3	3	1	—	—	—	—	8	6·5	0·75
	3	—	4	12	12	4	—	—	—	32	7·5	0·75
Umpire's	4	—		6	18	18	6	—	—	48	8·5	0·75
Score	5	—	—	—	4	12	12	4	—	32	9·5	0·75
	6	—	—	—	—	1	3	3	1	8	10·5	0·75
Total		1	7	21	35	35	21	7	1	128	8·5	1·75
Mean		2	2·57	3·14	3·71	4·29	4·86	5·43	6·0	4	—	1·0
Variance		0	0·245	0·408	0·490	0·490	0·408	0·245	0	1·0	0·75	—

EXERCISE 9.02

Set out the table and evaluate M_v, V_m, V and ρ for the following score situations. In each case A and B each adds the umpire's score to his independent score.

1. Umpire tosses once ; A tosses three times ; B tosses twice with the circular die of Fig. 67.

2. Umpire tosses twice ; A tosses five times ; B tosses four times with the die of Fig. 73.

3. Umpire tosses four times ; A tosses three times ; B tosses twice with the die of Fig. 70.

4. Umpire tosses twice with the tetrahedral die of Fig. 70 ; A tosses twice with the circular die of Fig. 67 ; B tosses twice with the tetrahedral die of Fig. 73.

5. Umpire throws twice an ordinary cubical die, scoring 1 to 6. A has two throws with the tetrahedral die of Fig. 70 ; B has two throws with that of Fig. 73.

6. Umpire tosses the circular die of Fig. 67 four times ; A draws four cards, with replacement, from a standard 52-card pack, scoring 1 for black cards and 3 for red cards ; B throws four times with the tetrahedral die of Fig. 70.

7. Umpire throws an ordinary cubical die, scoring 1 to 6, once ; A throws the die of Fig. 70 twice ; B throws the die of Fig. 73 three times.

8. Umpire draws four cards with replacement from a standard 52-card pack, scoring 2 for clubs, 3 for diamonds, 4 for hearts and 5 for spades ; A tosses the circular die of Fig. 67 five times ; B throws a cubical die twice.

9. Umpire draws five times, with replacement, from an urn containing seven black and three red balls, scoring one for each black ball drawn ; one black ball is removed from the urn and replaced by a red one, A then draws four times, with replacement ; ten black balls are now added to the urn and B draws three times, also with replacement, adding to their own scores that of the umpire.

10. Umpire throws the die of Fig. 70 four times ; A throws the die of Fig. 73 five times ; B throws six times with two conventional cubical dice, each marked 1 to 6, scoring the difference between the scores on the two dice.

9.03 THE ALGEBRAIC PROPERTIES OF THE UMPIRE-BONUS MODEL

We shall now examine formally the validity of some results illustrated numerically in 9.01 and 9.02. Let the umpire have u tosses, player A have a independent tosses, and player B have b. We denote the raw score of the umpire by x_u and the individual raw scores of the players respectively by $x_{a.o}$ and $x_{b.o}$. If the total scores of the players are respectively x_a and x_b

$$x_a = x_{a.o} + x_u \quad and \quad x_b = x_{b.o} + x_u \qquad . \qquad . \qquad . \qquad (i)$$

The mean individual score of A based on a tosses and the mean individual score of B based on b tosses, we denote respectively by $M_{a.o}$, $M_{b.o}$, in contradistinction to M_a, M_b the means of the players' total scores. More fully we may write

$$M_{a.o} = M(x_{a.o}); \quad M_a = M(x_{a.o} + x_u).$$
$$M_{b.o} = M(x_{b.o}); \quad M_b = M(x_{b.o} + x_u).$$

The relations of M_a to $M_{a.o}$ and M_b to $M_{b.o}$ are as defined by the chessboard for the distribution of a score sum in Fig. 46 of 4.04

$$M_a = M(x_{a.o} + x_u) = M(x_{a.o}) + M(x_u) = M_{a.o} + M_u \qquad . \qquad . \qquad (ii)$$
$$M_b = M(x_{b.o} + x_u) = M(x_{b.o}) + M(x_u) = M_{b.o} + M_u \qquad . \qquad . \qquad (iii)$$

By V_u we denote the variance of the distribution of x_u. Similarly

$$V_{a.o} = V(x_{a.o}); \quad V_a = V(x_{a.o} + x_u).$$
$$V_{b.o} = V(x_{b.o}); \quad V_b = V(x_{b.o} + x_u).$$

In virtue of the independence of the umpire's score and the individual score of either player :

$$V_a = V(x_{a.o} + x_u) = V(x_{a.o}) + V(x_u) = V_{a.o} + V_u \qquad . \qquad . \qquad (iv)$$
$$V_b = V(x_{b.o} + x_u) = V(x_{b.o}) + V(x_u) = V_{b.o} + V_u \qquad . \qquad . \qquad (v)$$

It will here be convenient to write the customary formula for the *mean* square deviation, i.e. the variance (V) of the distribution of a score x in a form more economical than (iii) of 3.06 :

$$V = M(x - M_x)^2 = M(x^2) - M_x^2.$$

In this notation

$$V_u = M(x_u^2) - M_u^2; \quad V_{a.o} = M(x_{a.o}^2) - M_{a.o}^2; \quad V_{b.o} = M(x_{b.o}^2) - M_{b.o}^2. \qquad . \qquad (vi)$$

25

In the same notation we may write

$$Cov\,(AB) = M(x_a \cdot x_b) \quad - M_a M_b$$
$$Cov\,(x_{a.o}, x_u) = M(x_{a.o} \cdot x_u) - M_u \cdot M_{a.o.}$$
$$Cov\,(x_{b.o}, x_u) = M(x_{b.o} \cdot x_u) - M_u \cdot M_{b.o.}$$
$$Cov\,(x_{a.o}, x_{b.o}) = M(x_{a.o} \cdot x_{b.o}) - M_{a.o} \cdot M_{b.o.}$$

In 8.06, we have seen that it is a fundamental property of the chessboard for equipartition of opportunity, i.e. statistical independence, that the covariance of the joint border score distribution is zero. Since the players' individual scores are independent of one another and of that of the umpire, we may therefore write

$$Cov\,(x_{a.o},\ x_{b.o}) = Cov\,(x_{a.o},\ x_u) = Cov\,(x_{b.o},\ x_u) = 0. \qquad . \qquad . \qquad \text{(vii)}$$

By recourse to these symbols we may now proceed to establish the general conclusion that the *variance of the umpire's score distribution is the covariance of the players' joint total score distribution* :

$$x_a\ x_b = (x_{a.o} + x_u)(x_{b.o} + x_u) = x_u^2 + x_{a.o} \cdot x_u + x_{b.o} \cdot x_u + x_{a.o} \cdot x_{b.o}$$
$$\therefore\ M(x_a \cdot x_b) = M(x_u^2) + M(x_{a.o} \cdot x_u) + M(x_{b.o} \cdot x_u) + M(x_{a.o} \cdot x_{b.o}),$$
$$M_a M_b = (M_{a.o} + M_u)(M_{b.o} + M_u) = M_u^2 + M_{a.o} M_u + M_{b.o} M_u + M_{a.o} M_{b.o}.$$
$$Cov\,(AB) = M(x_a \cdot x_b) - M_a M_b$$
$$= M(x_u^2) - M_u^2 + M(x_{a.o} \cdot x_u) - M_{a.o} M_u + M(x_{b.o} \cdot x_u) - M_{b.o} M_u$$
$$+ M(x_{a.o} x_{b.o}) - M_{a.o}\ M_{b.o}$$
$$= V_u + Cov\,(x_{a.o}, x_u) + Cov\,(x_{b.o}, x_u) + Cov\,(x_{a.o}, x_{b.o}).$$

Whence by (vii) above

$$Cov\,(AB) = V_u \qquad . \qquad . \qquad . \qquad . \qquad . \qquad \text{(viii)}$$

This theorem is of unrestricted applicability to the model, its truth being independent of any assumption w.r.t. the number of tosses the players make or whether they use the same die. If they do use one and the same die, we may denote the variance of a unit sample (single toss) by V_1. That of the score sum of an r-fold sample will be as in (iii) (p. 289) rV_1, so that :

$$V_u = uV_1 ; \quad V_a = (u + a)V_1 ; \quad V_b = (u + b)V_1$$
$$\therefore\ \rho_{ab}^2 = \frac{u^2 V_1^2}{(u + a)(u + b)V_1^2} = \frac{u^2}{(u + a)(u + b)} . \qquad . \qquad . \qquad . \qquad \text{(ix)}$$

That is to say, the correlation coefficient is then the ratio of the number of tosses of the umpire to the geometric mean of the total number of tosses which the players respectively score, or the proportionate contribution of the mean score of the umpire to the *g.m.* of the players' total mean scores. If the players perform the same number of individual tosses, as in 9.01, so that $a = b$ and $(u + a) = n = (u + b)$, the preceding formula becomes

$$\rho_{ab} = \frac{u}{n} \qquad . \qquad . \qquad . \qquad . \qquad . \qquad . \qquad \text{(x)}$$

If we put $x_{b.o} = 0$ in (i), that of B becomes identical with the umpire's score so that $V_b = V_u$ and $\rho_{ab} = \rho_{au}$, the coefficient of correlation between A's total score and that of the umpire. We thus have

$$\rho_{au} = \frac{V_u}{\sqrt{V_a \cdot V_u}} = \sqrt{\frac{V_u}{V_a}} \qquad . \qquad . \qquad . \qquad . \qquad \text{(xi)}$$

By the same token

$$\rho_{bu} = \sqrt{\frac{V_u}{V_b}} \qquad . \qquad . \qquad . \qquad . \qquad . \qquad . \qquad (xii)$$

From the above it follows that

$$\rho_{au} \cdot \rho_{bu} = \frac{V_u}{\sqrt{V_a \cdot V_b}} = \rho_{ab} \qquad . \qquad . \qquad . \qquad . \qquad (xiii)$$

Let us now examine the implications of linear regression of one player's score w.r.t. that of the other player. To employ the symbols of preceding sections consistently, we shall assume that we lay out the scores of player A on the top margin of the table *horizontally* and those of player B *vertically* on the left of it. We then say that there is linear regression of the score B w.r.t. the score A if the mean values of B scores (i.e. the column means) corresponding to successive values of the score A increase by equal steps. With appropriate constants k_{ba} and C_{ba} respectively definitive of the scale and origin of the graphical representation of this relation, we may therefore write

$$M_{ba} = k_{ba} \cdot x_a + C_{ba} \qquad . \qquad . \qquad . \qquad . \qquad . \qquad (xiv)$$

Similarly linear regression of A on B implies that the row means increase by equal steps, so that

$$M_{ab} = k_{ab} \cdot x_b + C_{ab} \qquad . \qquad . \qquad . \qquad . \qquad . \qquad (xv)$$

To obtain k_{ab} or k_{bu} in terms of ρ_{ab} or of expressions related to ρ_{ab}, it is convenient to define the constants specifying the origin in terms of the mean values of the two scores, i.e.

$$C_{ba} = (M_b - k_{ba} \cdot M_a) \quad and \quad C_{ab} = (M_a - k_{ab} \cdot M_b).$$

We may now write (xiv) in the form commonly referred to as the equation of linear regression of B on A :

$$M_{ba} - M_b = k_{ba}(x_a - M_a) \qquad . \qquad . \qquad . \qquad . \qquad (xvi)$$
$$\therefore (M_{ba} - M_b)^2 = k_{ba}^2 (x_a - M_a)^2.$$

By definition the mean value of $(x_a - M_a)^2$ is V_a, and that of the expression on the left is $V(M_{ba})$ because M_b is the mean of the column means. We may therefore write

$$V(M_{ba}) = k_{ba}^2 V_a \qquad . \qquad . \qquad . \qquad . \qquad . \qquad (xvii)$$

To evaluate k_{ba} in an alternative form, we can multiply both sides of (xvi) by x_a, so that

$$x_a M_{ba} - x_a \cdot M_b = k_{ba} x_a^2 - k_{ba} x_a \cdot M_a. \qquad . \qquad . \qquad . \qquad (xviii)$$

Again we take the mean value of the expressions on either side of the equation. For the expression on the right we then have

$$k_{ba} \cdot M(x_a^2) - k_{ba} \cdot M_a \cdot M(x_a) = k_{ba} \cdot M(x_a^2) - k_{ba} \cdot M_a^2 = k_{ba} \cdot V_a \qquad . \qquad . \qquad (xix)$$

Also we may write the expression on the left as

$$M(x_a \cdot M_{ba}) - M_b \cdot M(x_a) = M(x_a \cdot M_{ba}) - M_a \cdot M_b.$$

By scrutiny of Fig. 78, the reader will see that we may obtain the mean value of the products AB either by summing the cell products of the correlation table weighted by their appropriate cell frequencies, or by summing the products of the column means and corresponding border (A) scores weighted by the frequenices of the latter. That is to say

$$M(x_a \cdot M_{ba}) = M(x_a \cdot x_b),$$
$$\therefore M(x_a \cdot M_{ba}) - M_a \cdot M_b = Cov\ (AB).$$

By substituting the last expression together with (xix) in (xviii) we thus obtain

$$Cov\,(AB) = k_{ba}\,.\,V_a,$$
$$\therefore\ k_{ba} = V_u \div V_a = \rho_{au}^2 \qquad\qquad \text{(xx)}$$

If we now substitute this value of k_{ba} in (xvii), we obtain

$$V(M_{ba}) = V_u^2 \div V_a,$$
$$\therefore\ V(M_{ba}) \div V_b = V_u^2 \div V_a\,.\,V_b = \rho_{ab}^2 \qquad\qquad \text{(xxi)}$$

This signifies that linear regression of B on A implies the identity

$$V(M_{ba}) \div V_b = \rho_{ab}^2.$$

Similarly we may show that linear regression of A on B implies the identity

$$V(M_{ab}) \div V_a = \rho_{ab}^2.$$

The corresponding linear equation which denotes this relationship is

$$M_{ab} - M_a = k_{ab}(x_b - M_b) \qquad\qquad \text{(xxii)}$$

It is customary to call k_{ba} the *regression coefficient of B on A* and k_{ab} the *regression coefficient of A on B*. The same reasoning which led us to (xx) shows that

$$k_{ab} = V_u \div V_b = \rho_{bu}^2,$$
$$\therefore\ k_{ab}\,.\,k_{ba} = \rho_{ab}^2 \qquad\qquad \text{(xxiii)}$$

It is possible to express k_{ab} and k_{ba} in a form which does not involve explicit information about the umpire's score distribution :

$$k_{ba}^2 = \frac{V_u^2}{V_a^2} = \frac{V_u^2}{V_a\,.\,V_b}\,.\,\frac{V_b}{V_a} = \rho_{ab}^2\,.\,\frac{V_b}{V_a},$$
$$\therefore\ k_{ba} = \rho_{ab}\,.\,\frac{\sigma_b}{\sigma_a} \qquad\qquad \text{(xxiv)}$$

Similarly we may write

$$k_{ab} = \rho_{ab}\,.\,\frac{\sigma_a}{\sigma_b}. \qquad\qquad \text{(xxv)}$$

We are now in a position to appreciate more clearly in what circumstances we can make a balance sheet of variation resolving the variance of a score distribution into additive components appropriately described as *explained* or otherwise. With respect to each player of the umpire-bonus set-up, we may split the variance of the total score distribution into additive components, as follows

$$V_a = V_u + V_{a,o} \quad and \quad V_b = V_u + V_{b,o} \qquad\qquad \text{(xxvi)}$$

In either case, the same component is the variance of the distribution of the concomitant umpire's score, the residual component being the variance of the player's independent score distribution. Any meaningful sense of the epithets explained and unexplained w.r.t. a common source of variation, therefore justifies us in applying them respectively to V_u and $V_{a,o}$ or $V_{b,o}$; and we may accordingly express the proportionate contributions of two components so described by writing (xxvi) in the form

$$\frac{V_u}{V_a} + \frac{V_{a,o}}{V_a} = 1 = \frac{V_u}{V_b} + \frac{V_{b,o}}{V_b}. \qquad\qquad \text{(xxvii)}$$

As in (ix) and (x) of 9.02, we denote by E_a and E_b respectively the *explained* components, and in virtue of (xi)-(xii).

$$E_a = V_u \div V_a = \rho_{au}^2,$$
$$E_b = V_u \div V_b = \rho_{bu}^2.$$

Except when ρ_{ab} $(= \rho_{au} \cdot \rho_{bu})$ is zero or unity, neither E_a nor E_b is therefore identical with ρ_{ab}^2, nor identical with the correlation ratios $(V_m \div V)$ equivalent to ρ_{ab}^2 when regression is linear. It is therefore pertinent to re-examine the relation exhibited at the end of 9.01, where we first saw that the chessboard set-up for the umpire-bonus model admits a break-down of variance into components defined by the relation

$$\frac{V(M_{ab})}{V_a} + \frac{M(V_{ab})}{V_a} = 1 = \frac{V(M_{ba})}{V_b} + \frac{M(V_{ba})}{V_b}. \qquad \text{(xxviii)}$$

We shall later see (10.04) that this identity has nothing to do with correlation as such. It is, in fact, a *numerical* property of any chessboard set-up, depending on no distinctively statistical postulates, though the foregoing analysis has shown that the first component of the above relations has a special statistical significance when regression is linear. If regression is linear in both dimensions

$$\frac{V(M_{ab})}{V_a} = \rho_{ab}^2 = \frac{V(M_{ba})}{V_b} \qquad \text{(xxix)}$$

The last equation refers to a correlation table setting out the association of the total scores of the two players, and the coefficient ρ_{ab} is then a summarising index of their *concurrent* correspondence in the sense defined in 8.01 ; but we may also compute the corresponding ratio from the data of a correlation table setting out either the *consequent* variation of the total score of player A on that of the umpire, or the *consequent* variation of the total score of player B on that of the umpire. In that case, the reader should be able to show that regression of the player's score on that of the umpire is necessarily linear, and

$$\frac{V(M_{au})}{V_a} = \rho_{au}^2 = E_a \quad and \quad \frac{V(M_{bu})}{V_b} = \rho_{bu}^2 = E_b \qquad \text{(xxx)}$$

It would therefore seem that the use of the terms *explained* and *unexplained* for the components of the relation exhibited in (xxviii) is justifiable only in special circumstances. It is a true bill for the *consequent* relation of the total score of the players to that of the umpire, but it is not a true bill for the *concurrent* relation of the total scores of the two players. Thus the ratio $V_m \div V$ has no unique title to rank as the fraction of explained variance, and the ratio $M_v \div V$ has no unique claim to identification with what fraction of total variance is residual or unexplained in this sense. What we are entitled to say is that

(a) the limiting value of the ratio $V_m \div V$ is unity when there is perfect concomitant variation, in which case $\rho = 1 = \rho^2$, and is zero when there is no common source of variation, in which case $\rho = 0 = \rho^2$;

(b) the limiting value of the ratio $M_v \div V$ is unity when there is no common source of variation, in which case $(1 - \rho) = 1 = (1 - \rho^2)$, and is zero when there is perfect concomitant variation, in which case $(1 - \rho) = 0 = (1 - \rho^2)$.

Between these limits, we can justify the use of the terms *explained* and *unexplained* respectively for the inter-class and mean intra-class variances, only if we have other reasons for identifying them as such.

9.04 PARTIAL CORRELATION

In the use and design of performance tests we often meet the following type of situation. Paired scores of one and the same individual w.r.t. two test batteries A and B constructed with a view to assessing a particular ability (U) yield a high product moment index ρ_{ab} ; but one has reason to suspect that this concurrence arises partly from another source (W) for which we have a standard test deemed satisfactory as such. If so, we may determine correlation coefficients (ρ_{aw} or ρ_{bw}) based on paired scores of individuals assessed respectively by the A and the W test or the B and the W test. Having done so, we may ask what numerical value the residual correlation coefficient ($\rho_{ab \cdot w}$) w.r.t. tests A and B would have in the absence of the extraneous component assessed by the W test itself. If we can answer this question we can modify the original design of A and B to assure a higher measure of concurrence attributable to U alone, and so define more precisely which of their ingredients are most diagnostic of the ability we aim to assess thereby.

Formally the problem is to define $\rho_{ab \cdot w}$, which is a more satisfactory measure than ρ_{ab} of the concurrence arising from U, in terms of observed values of $\rho_{ab}, \rho_{aw}, \rho_{bw}$. Within the framework of our present model, we may regard A and B as player's total scores each with a component attributable to individual variation and each with components referable to a bonus from each of two umpires U and W. We shall assume that we know the value of : (a) the correlation coefficients (ρ_{aw}, ρ_{bw}) for the score of the player and that of one umpire W ; (b) that of the total A and B scores (ρ_{ab}) inter se. Our problem is to extract from this information what the product moment coefficient ($\rho_{ab \cdot w}$) of the A-B scores would be, if umpire W withheld his bonus or—what comes to the same thing—donated one and the same fixed bonus at each trial. Any relationship we seek will conform to the condition that $\rho_{ab \cdot w}$ must be zero if all correlation between A and B scores arises exclusively from the bonus of W, there being only one umpire bonus. If so, by virtue of (xiii) in 9.02

$$\rho_{ab} - \rho_{aw} \cdot \rho_{bw} = 0.$$

Since $\rho_{ab \cdot w} = 0$ when the contribution of the U bonus to the recorded total of A and B scores is zero, we therefore infer that the numerator of a relation of the type we seek may have as a factor some power of ($\rho_{ab} - \rho_{aw} \cdot \rho_{bw}$). Tentatively, we may explore the simplest assumption, that it involves a first power. If we then write D for an undetermined denominator :

$$\rho_{ab \cdot w} = \frac{\rho_{ab} - \rho_{aw} \cdot \rho_{bw}}{D}.$$

Now $\rho_{ab \cdot w}$ must be equivalent to ρ_{ab}, if umpire W contributes nothing to the total score of A or B. If so, $\rho_{aw} = 0 = \rho_{bw}$ and $D = 1$. A condition which satisfies this relation is that D is a function of a product of the form $(1 - \rho_{aw}^x)(1 - \rho_{bw}^x)$. With these clues to the relation we seek, we may now get down to cases. We shall suppose that :

(i) A and B each toss a coin twice ;

(ii) Umpire W tosses a coin once and umpire U three times ;

(iii) A and B record as their total scores the results of their respective tosses supplemented by both the U and W score at the same trial.

Since the umpires toss independently, we may regard the distribution of their joint toss as equivalent to that of a third umpire Z who tosses $(1 + 3) = 4$ times, so that we may write $V_z = 4(\frac{1}{2})(\frac{1}{2}) = 1 = Cov\ (AB)$. Likewise, $V_w = \frac{1}{4}$ and $V_u = \frac{3}{4}$. When both umpires toss, A and B each record the outcome of six independent tosses, so that

$$V_a = 6(\tfrac{1}{2})(\tfrac{1}{2}) = \tfrac{3}{2} = V_b$$

and

$$\rho_{ab} = \frac{V_z}{\sqrt{V_a \cdot V_b}} = \frac{1}{\sqrt{(\tfrac{3}{2})^2}} = \tfrac{2}{3} \quad \ldots \quad \text{(i)}$$

Likewise, we derive the correlation coefficient for the player's total and that of umpire W

$$\rho_{aw}^2 = \frac{V_w}{V_a} = \tfrac{1}{6} = \rho_{bw}^2 \quad \ldots \quad \text{(ii)}$$

$$\therefore \ \rho_{ab} - \rho_{aw} \cdot \rho_{bw} = \tfrac{2}{3} - \tfrac{1}{6} = \tfrac{1}{2} \quad \ldots \quad \text{(iii)}$$

When the umpire W withholds his bonus, A and B record the result of only five independent tosses. If we write the variances of the final score distribution respectively as $V_{a \cdot w}$ and $V_{b \cdot w}$, we have

$$V_{a \cdot w} = \tfrac{5}{4} = V_{b \cdot w}.$$

If we denote by $Cov\,(AB)_w$ the covariance of the final scores when umpire W withholds his bonus, and $\rho_{ab \cdot w}$ as the corresponding product-moment index, we have

$$Cov\,(AB)_w = \tfrac{3}{4} = V_u.$$

$$\rho_{ab \cdot w} = \frac{V_u}{\sqrt{V_{a \cdot w} \cdot V_{b \cdot w}}} = \tfrac{3}{4} \div \tfrac{5}{4} = \tfrac{3}{5} \quad \ldots \quad \text{(iv)}$$

Let us now examine our previous proposal, viz.

$$\rho_{ab \cdot w} = \frac{\rho_{ab} - \rho_{aw} \cdot \rho_{bw}}{D}.$$

From (iii) and (iv) we then get

$$D = \tfrac{1}{2} \div \tfrac{3}{5} = \tfrac{5}{6}.$$

From (ii) we have

$$1 - \rho_{aw}^2 = \tfrac{5}{6} = 1 - \rho_{bw}^2.$$

$$\therefore \ D = \sqrt{(1 - \rho_{aw}^2)(1 - \rho_{bw}^2)}.$$

Our numerical investigation thus suggests as a basis for further enquiry the relation

$$\rho_{ab \cdot w} = \frac{\rho_{ab} - \rho_{aw} \cdot \rho_{bw}}{\sqrt{(1 - \rho_{aw}^2)(1 - \rho_{bw}^2)}}.$$

This relation is easy to establish as a general property of our model without restriction w.r.t. the identity of the dice involved or the numbers of individual tosses allocated to the players. It is, in fact, valid for the more general case of linear concomitant variation, when the players receive different multiples (l and m) of each umpire bonus in conformity with the equations

$$x_a = x_{a \cdot o} + l_u \cdot x_u + l_w \cdot x_w.$$
$$x_b = x_{b \cdot o} + m_u \cdot x_u + m_w \cdot x_w$$

If we here derive it for the simpler case which arises when l_u, l_w, m_u and m_w are each equal to unity, the student should be able to adapt the reasoning with a view to removal of this restriction. For what follows, we are thus concerned with a system of joint scores

$$x_a = x_{a \cdot o} + x_u + x_w$$
$$x_b = x_{b \mid o} + x_u + x_w$$

It will be useful to set out the relevant symbols as below :

(a) $\rho_{ab.w}$ w.r.t. the players' total scores in the absence of a bonus from an umpire W ;

(b) $\rho_{ab.u}$ w.r.t. the players' total scores in the absence of a bonus from an umpire U.

In conformity with this symbolism

$V_{a.w}$ is the variance of the distribution of the joint individual score of the umpire U and player A ;

$V_{a.u}$ is the variance of the distribution of the joint individual score of the umpire W and player A ;

$Cov\,(AU)_w$ is the covariance of the score of umpire U and the joint score of U and A ;

$Cov\,(AW)_u$ is the covariance of the score of umpire W and the joint score of W and A ;

$Cov\,(AB)_w$ is the covariance of the total scores of A and B in the absence of the contribution of umpire W ;

$Cov\,(AB)_u$ is the covariance of the total scores of A and B in the absence of the contribution of umpire U.

Symbols for corresponding parameters of the B score distribution in the absence of a contribution from one or other umpire are analogous to the above. Without the additional subscript $Cov\,(AB)$, ρ_{ab}, V_a, V_b refer to total score distributions involving the double bonus. The symbols $V_{a.o}$ and $V_{b.o}$ refer, as elsewhere, to the individual score distributions when the players receive neither bonus.

The following relations must exist between the variances in virtue of the fact that the individual player's own score is independent of that of either umpire and the score of one umpire is independent of that of the other :

$$V_a = V_{a.o} + V_u + V_w; \quad V_{a.w} = V_{a.o} + V_u,$$
$$\therefore \; V_a - V_w = V_{a.o} + V_u = V_{a.w} \qquad . \qquad . \qquad . \qquad . \qquad (\text{v})$$

Similarly,

$$V_a - V_u = V_{a.u}; \quad V_b - V_u = V_{b.u}; \quad V_b - V_w = V_{b.w} \qquad . \qquad . \qquad (\text{vi})$$

If the distributions of the umpire scores x_u and x_w are referable to a binomial, the theorem of 7.05 shows that the distribution of the total score $x_{uw} = (x_u + x_w)$ is likewise a binomial with variance $V_{uw} = V_u + V_w$; and we can regard the distribution of x_{uw} as the source of total concomitant variation, i.e.

$$Cov\,(AB) = V_{uw} = V_u + V_w$$

The above relation does not, however, presume that the dice tossed by the umpires are of the binomial specification exhibited in Figs. 67, 70 and 73. It is deducible in the same way as the corresponding relation, when there is only one umpire, by the method used to derive (viii) in 9.03. The reader should be able to take this hurdle as an exercise. We then have

$$\rho_{ab} = \frac{(V_u + V_w)}{\sqrt{V_a \cdot V_b}} \qquad . \qquad . \qquad . \qquad . \qquad . \qquad (\text{vii})$$

Similarly we may put

$$\rho_{ab.w} = \frac{V_u}{\sqrt{V_{a.w} V_{b.w}}} \qquad . \qquad . \qquad . \qquad . \qquad (\text{viii})$$

$$\rho_{aw} = \frac{V_w}{\sqrt{V_a \cdot V_w}} = \sqrt{\frac{V_w}{V_a}} \qquad . \qquad . \qquad . \qquad (\text{ix})$$

$$\rho_{bw} = \sqrt{\frac{V_w}{V_b}} \qquad \qquad \text{(x)}$$

$$\rho_{aw} \cdot \rho_{bw} = \frac{V_w}{\sqrt{V_a \cdot V_b}} \qquad \qquad \text{(xi)}$$

Hence from (vii) above

$$\rho_{ab} - \rho_{aw} \cdot \rho_{bw} = \frac{(V_u + V_w)}{\sqrt{V_a \cdot V_b}} - \frac{V_w}{\sqrt{V_a \cdot V_b}} = \frac{V_u}{\sqrt{V_a \cdot V_b}} \qquad \text{(xii)}$$

Also from (ix) and (x)

$$1 - \rho_{aw}^2 = \frac{V_a - V_w}{V_a}; \quad 1 - \rho_{bw}^2 = \frac{V_b - V_w}{V_b}.$$

Hence from (v) and (vi)

$$\sqrt{(1 - \rho_{aw}^2)(1 - \rho_{bw}^2)} = \frac{\sqrt{V_{a,w} \cdot V_{b,w}}}{\sqrt{V_a \cdot V_b}} \qquad \qquad \text{(xiii)}$$

By combining (xii) and (xiii) we get

$$\frac{\rho_{ab} - \rho_{aw}\rho_{bw}}{\sqrt{(1 - \rho_{aw}^2)(1 - \rho_{bw}^2)}} = \frac{V_u}{\sqrt{V_{a,w}V_{b,w}}}.$$

Whence by (viii) above

$$\rho_{ab.w} = \frac{\rho_{ab} - \rho_{aw}\rho_{bw}}{\sqrt{(1 - \rho_{aw}^2)(1 - \rho_{bw}^2)}} \qquad \qquad \text{(xiv)}$$

In the same way, we obtain

$$\rho_{ab.u} = \frac{\rho_{ab} - \rho_{au}\rho_{bu}}{\sqrt{(1 - \rho_{au}^2)(1 - \rho_{bu}^2)}}.$$

EXERCISE 9.04

By the methods of this section, investigate a set-up in which the following are the rules of the game :

(i) The umpire tosses the circular die of Fig. 67 three times.

(ii) A tosses the die of Fig. 73 twice and adds to his individual score that of the umpire.

(iii) B tosses the die of Fig. 73 three times, adds to his score that of the umpire, and deducts that of A.

(iv) C tosses the die of Fig. 70 once and adds to his score B's final score.

CHAPTER 10

PREVIEW OF SAMPLING SYSTEMS

10.01 THE LEXIS MODEL

IN Chapters 4 and 7 our theme was the distribution of differences and sums of scores from one and the same universe on the assumption that choice is independent. In the last chapter our concern was with two-way systems of scoring subject to some constraint or source of concomitant variation, so that there is not equipartition of opportunity for association between one set (A) of scores and the other (B). For any such set-up, we have had occasion to notice a common property by recourse to arithmetical examples. The meaning of this we shall now recall to explore more fully. If V_a stands for the total variance of the distribution of A scores, $V(M_{ab})$ for the variance of the distribution of the mean value of the A-score associated with a particular value of the B score, and $M(V_{ab})$ for the mean value of the variance of the A score distribution corresponding to a particular value of the B score

$$V_a = V(M_{ab}) + M(V_{ab}).$$

Mutatis mutandis the same relation holds good for the B score distribution, i.e.

$$V_b = V(M_{ba}) + M(V_{ba}).$$

When the context makes clear which sets of scores are our concern, we may write more succinctly

$$V = V_m + M_v \qquad \qquad \text{(i)}$$

If the mean A scores constitute an arithmetic series, when we lay out the B border scores consecutively in equal steps, as in all the examples of Chapter 9, our arithmetical illustrations have suggested a second relation a full consideration of which we shall defer to Part II, *viz.* :

$$\rho^2 = V(M_{ab}) \div V_a.$$

Subject to a like restriction, i.e. linear regression of the B scores on the A scores

$$\rho^2 = V(M_{ba}) \div V_b ;$$

and if regression is linear in both dimensions of the grid, we may write this relation in the more general form

$$\rho^2 = V_m \div V \qquad \qquad \text{(ii)}$$

It is important to notice that the numerical value of the B border scores of an A-B correlation grid do not enter into the computation of either $V(M_{ab})$ or $M(V_{ab})$. For that purpose the B border scores are merely labels to distinguish one column from another and as such we might replace them by any distinctive symbols. Similarly the A border scores of the grid do not enter into the computation of $V(M_{ba})$ or $M(V_{ba})$, in that context, the A scores being merely classificatory categories with no numerical significance. This circumstance gives (ii) a special interest *vis-à-vis* an alternative formulation of the class of problems dealt with in our last chapter. If there were no tie-up of any sort between our A-scores and our B-scores in the model situations of Chapter 9, the grid lay-out would be the same as if we extracted unit samples from each of two identical universes. If so, the score distribution of each column would be the same, and the score distribution of each row would be the same. These considerations invite an exploration of the properties of a model for sampling from separate universes with a view to discovering whether they are of different composition. The universes of Fig. 80 are four urns

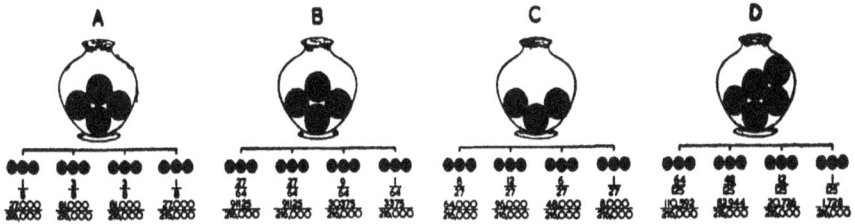

		A	B	C	D	TOTAL
O	●●●	27,000	9,125	64,000	110,592	292,717
I	●●●	81,000	9,125	96,000	82,944	351,069
2	●●●	81,000	30,375	48,000	20,736	180,111
3	●●●	27,000	3,375	8,000	1,728	40,103
TOTAL		216,000	216,000	216,000	216,000	864,000
MEAN		$\frac{3}{2}$	$\frac{3}{4}$	I	$\frac{3}{5}$	$\frac{77}{80}$
VARIANCE		$\frac{3}{4}$	$\frac{9}{16}$	$\frac{2}{3}$	$\frac{12}{25}$	$\frac{2809}{3840}$

FIG. 80. A Lexis Model. Each column exhibits the theoretical sampling distribution w.r.t. repeated 3-fold trials from one and the same urn. The player replaces each ball before drawing another. For convenience of computation the cell entries definitive of the frequency of the corresponding border score in samples from the appropriate urn are whole numbers referable to a common denominator, *viz.* 216,000.

from each of which we take repeatedly the same number (3) of balls, replacing each ball taken before taking another. With respect to each urn, the number of 3-fold trials is identical and *indefinitely large*. From the entries in Fig. 80, recording the long-run result of such sampling, we obtain the following figures for the grand mean (M) and the total variance ($V = V_0 - M^2$):

$$M = \frac{292,717(0) + 351,069(1) + 180,111(2) + 40,103(3)}{864,000} = \frac{77}{80},$$

$$\therefore M^2 = \frac{5929}{6400} = \frac{800,415}{864,000}.$$

$$V_0 = \frac{292,717(0) + 351,069(1) + 180,111(4) + 40,103(9)}{864,000} = \frac{1,432,440}{864,000}.$$

$$\therefore V = \frac{1432,440}{864,000} - \frac{800,415}{864,000} = \frac{632,025}{864,000} \quad . \quad . \quad . \quad . \quad . \quad . \quad \text{(iii)}$$

For the *inter*-class variance (V_m) we have

$$V_m = \frac{(\frac{3}{2})^2 + (\frac{3}{4})^2 + 1^2 + (\frac{3}{5})^2}{4} - \frac{5929}{6400} = \frac{747}{6400} = \frac{100,845}{864,000}.$$

For the *intra*-class variance (M_v) we have

$$M_v = \frac{\frac{3}{4} + \frac{9}{16} + \frac{2}{3} + \frac{12}{25}}{4} = \frac{2951}{4800} = \frac{531,180}{864,000}.$$

Whence we obtain:

$$V_m + M_v = \frac{100,845}{864,000} + \frac{531,180}{864,000} = \frac{632,025}{864,000}.$$

Hence from (iii) in agreement with (i) above:

$$V = V_m + M_v.$$

If we seek to generalise the results exhibited in Fig. 80, we should remind ourselves : (a) that it refers to the extraction of equal numbers of 3-fold samples from each urn ; (b) that the cell entries of one and the same column of the Lexis model record a long-run result.   In fact, they are merely frequencies as we have used the term elsewhere in this book, when divided by the common denominator 216,000 shown as each column total.   In other words, each cell entry divided by the column total is a term of the binomial expansion definitive of the column score distribution. The sum of all such quotients in one and the same column is therefore unity, and the grand total for a grid of c columns is c.   If the proportion of balls in the urn of the jth column is p_j, the mean of the score distribution whose relative frequencies successive cell entries of the jth column define is rp_j and the corresponding variance is rp_jq_j.   If M is the grand mean, being the mean value of rp_j, we may therefore write

$$M_v = \frac{1}{c}\sum_{j=1}^{j=c} rp_jq_j ; \quad V_m = \frac{1}{c}\sum_{j=1}^{j=c} r^2p_j^2 - M^2,$$

$$\therefore M_v + V_m = \frac{1}{c}\sum_{j=1}^{j=c} (rp_jq_j + r^2p_j^2) - M^2 \quad . \qquad . \qquad . \qquad . \qquad \text{(iv)}$$

By a now familiar property of the binomial distribution the mean square score of the jth column is $(rp_jq_j + r^2p_j^2)$.   Hence the mean square score (V_0) of the entire grid is

$$\frac{1}{c}\sum_{j=1}^{j=c} (rp_jq_j + r^2p_j^2).$$

Since $V = V_0 - M^2$

$$V = \frac{1}{c}\sum_{j=1}^{j=c} (rp_jq_j + r^2p_j^2) - M^2 \quad . \qquad . \qquad . \qquad . \qquad \text{(v)}$$

Hence from (iv) above, we obtain (i), i.e.

$$V = V_m + M_v.$$

It is tempting to regard the relation last stated as a sort of balance sheet exhibiting the indebtedness of the total variance of the system to the circumstance that the urns are different. In one sense this is permissible, in so far as the inter-class variance (V_m) becomes zero when all the urns of Fig. 80 are identical.   In any other sense, it is false to identify the ratio $V_m \div V$ with the fraction of total variance attributable to the differential composition of the urns themselves. To say so would signify that the inter-class variance represents *how much of the total variance would disappear if, in fact, the urns were all alike ;* and any such assertion is meaningful only if we define how we propose to eliminate differences w.r.t. urn composition.   An example will make this clear.   Let us consider a 12-fold replacement trial from each of three urns constituted thus :

Urn	Numbers of		Total.	Proportion of Red Balls.
	Red Balls.	Black Balls.		
A	3	9	12	$\frac{1}{4}$
B	6	6	12	$\frac{1}{2}$
C	4	8	12	$\frac{1}{3}$
Total	13	23	36	$\frac{13}{36}$

The definitive distributions are then $(0.75 + 0.25)^{12}$, $(0.5 + 0.5)^{12}$, $(0.6 + 0.8)^{12}$ with column (*red*-score) means and variances as follows :

	A	B	C
Mean	3	6	4
Variance	$\frac{9}{4}$	3	$\frac{8}{3}$

We derive in the usual way

$$V_m = \tfrac{14}{9} ; \quad M_v = \tfrac{95}{36} ; \quad V = \tfrac{151}{36} = \tfrac{453}{108} \qquad \cdots \qquad \text{(vi)}$$

Inter alia we may eliminate all variation attributable to the circumstance that the urns are different in three ways : (i) we replace B and C by A ; (ii) we replace A and C by B ; (iii) we replace A and B by C. If we establish uniformity of column structure in accordance with (i) the total variance is $12(\tfrac{1}{4})(\tfrac{3}{4}) = \tfrac{81}{36}$, the reduction effected being $\tfrac{70}{36}$. If we do so in accordance with (ii) the total variance is $12(\tfrac{1}{2})(\tfrac{1}{2}) = \tfrac{108}{36}$, the reduction being $\tfrac{43}{36}$. If we do so in accordance with (iii) the total variance is $12(\tfrac{1}{3})(\tfrac{2}{3}) = \tfrac{96}{36}$, the reduction effected being then $\tfrac{55}{36}$.

Any of the foregoing procedures is arbitrary ; but there is one which is less so. We might, in fact, mix the contents of all three urns and sample from the composite urn, or what comes to the same thing—extract three samples from each of three urns, each containing the same proportion of red balls as the composite urn. The result of mixing the contents of the urns will in general depend on whether the urns A, B and C contain, as above, the *same number* of balls. It is therefore important to recognise a restriction inherent in the numerical example under discussion, namely that the total number of balls in each urn is the same. To construct the standard urn we thus impose the condition of *equiproportionate contribution of each original urn*. Having done so, we have reconstructed our frequency grid with identical columns whose definitive binomial is

$$\left(\frac{13 + 23}{36}\right)^{12}.$$

The total variance (V_s) of this *standard grid* is, of course, the same as its intra-class variance, since the variance of the distributions of the columns are identical, i.e.

$$V_s = 12(\tfrac{13}{36})(\tfrac{23}{36}) = \tfrac{299}{108}.$$

This is not in fact equivalent to what the total variance of the original grid would be, if diminished by a figure equivalent to the *inter-class variance*, the residual variation being then as defined above, i.e.

$$\tfrac{95}{36} = \tfrac{285}{108}.$$

It is thus clear that the inter-class variance of the model exhibited in Fig. 80 is *not* the amount by which we have to diminish the total variance in order to obtain the variance of a standard grid referable to identical urns containing the same proportion of red balls as an urn made up of equal contributions from each of the original ones.

Within the framework of our last convention it is, however, possible to exhibit a meaningful balance sheet of variance, and a clue to its components emerges from the last example cited. In the light of the results cited as (vi) above, let us examine how much effect on the total variance of the grid equalisation of the sort last dealt with entails :

$$V - V_s = \frac{453}{108} - \frac{299}{108} = \frac{154}{108} = \frac{(11)(14)}{(12)(9)} = \frac{(12 - 1)}{12} \cdot V_m.$$

This suggests a new balance sheet of variance for r-fold trials :

$$V = V_s + \frac{r-1}{r} V_m \qquad . \qquad . \qquad . \qquad . \qquad . \text{ (vii)}$$

We may proceed to establish the truth of (vii) as follows in accordance with our definition of the standard grid, $viz.$:

(a) we take an *equal number* N of balls from each of c urns, the proportions (p_a, p_b, etc.) of red balls in each N-fold set being as in the original urns A, B, etc. ;

(b) we make up a standard urn by mixing the sets, and replace the urns A, B, etc., by others each containing red and black balls in the same proportion as the standard urn and therefore each identical with every other one.

For the proportion of red and black balls in the standard urn we shall respectively use p and q, the symbols V_m and V for the Lexis model having the same meaning as heretofore ; and we extract repeated r-fold samples of equal size from each standardized urn. Since the urns are now identical, the inter-class variance V_m is zero and in virtue of (i) the mean intra-class variance of the standard grid is the same as its total variance : *

$$V_s = rpq.$$

If the number of red balls in the original urn of column j is x_j, the proportion of red in a total of n balls is $(x_j \div n) = p_j$. Our assumption is that the number (n) of balls in each of the c original urns is the same, and on mixing their contents we obtain the standard urn in which the total number of balls is nc, and the number of red balls

$$\sum_{j=1}^{j=c} x_j = \sum_{j=1}^{j=c} np_j.$$

If we use p and q to signify respectively the proportion of red and black balls in the standard urn, we thus obtain

$$p = \frac{1}{c}\sum_{j=1}^{j=c} p_j; \quad q = \frac{1}{c}\sum_{j=1}^{j=c} q_j = \frac{1}{c}\sum_{j=1}^{j=c} (1 - p_j), \qquad \text{(viii)}$$

$$\therefore \ V_s = \frac{r}{c^2}\sum_{1}^{c} p_j \sum_{1}^{c} (1 - p_j)$$

$$= \frac{r}{c^2}\sum_{1}^{c} p_j (c - \sum_{1}^{c} p_j)$$

$$= \frac{r}{c}\sum_{1}^{c} p_j - \frac{r}{c^2}\sum_{1}^{c} p_j \sum_{1}^{c} p_j$$

$$= \frac{1}{c}\sum_{1}^{c} rp_j - \frac{1}{r}\left(\frac{1}{c}\sum_{1}^{c} rp_j . \frac{1}{c}\sum_{1}^{c} rp_j \right)$$

$$= M - \frac{1}{r}M^2 \qquad . \qquad . \qquad . \qquad . \qquad . \text{ (ix)}$$

* This consideration sidesteps a limitation in (b) of our prescription for imposing the condition of homogeneity on the c urns, $viz.$ we can do so only if the total number of red balls is consistent with the relation $x_j = Np_j$ is an integer. If each of the c urns of the Lexis model has the same number (N) of balls, our prescription implies that we pool all their contents to make our standard urn containing cN balls. The total variance (V_s) of the score distribution of an r-fold sample from such a standard urn is the same as the total variance of an indefinitely large number of c-fold sets of trials from c identical urns each containing N balls, when it is in fact possible to partition the standard urn in this way, i.e. if Σx_j is an exact multiple of c.

We have seen from (v) above that

$$V = \frac{1}{c}\sum_1^c (rp_jq_j + r^2p_j^2) - M^2$$

$$= \frac{1}{c}\sum_1^c rp_j(1 - p_j) + \frac{1}{c}\sum_1^c r^2p_j^2 - M^2$$

$$= \frac{1}{c}\sum_1^c rp_j - \frac{1}{rc}\sum_1^c r^2p_j^2 + \frac{1}{c}\sum_1^c r^2p_j^2 - M^2$$

$$= M - M^2 + \frac{r-1}{rc}\sum_1^c r^2p_j^2 \quad . \quad . \quad . \quad . \quad . \quad (x)$$

Since the square of the jth column mean is $r^2p_j^2$, we may write

$$V_m + M^2 = \frac{1}{c}\sum_1^c r^2p_j^2.$$

Thus (x) becomes

$$V = M - M^2 + \frac{r-1}{r}(V_m + M^2)$$

$$= M - \frac{1}{r}.M^2 + \frac{r-1}{r}.V_m.$$

Hence from (ix) above

$$V = V_s + \frac{r-1}{r}V_m \quad . \quad . \quad . \quad . \quad . \quad . \quad (xi)$$

This is a true bill in the sense that the second component on the right-hand side *exactly* represents the difference between : (*a*) the total variance of the grid when the urns are not all alike, (*b*) the total variance the grid would have if we eliminated all variance attributable to differences between the urns by replacing each urn by an urn of composition identical with that of a standard urn having an equal contribution of balls from each column urn of the original grid, the ratio of red to black in each contribution being the same as in the contributory urn. We may express this alternatively by regarding the standard urn as a parent universe, and the set of different urns as a *stratified* universe therefrom by redistributing its contents in such a way that the total of items in each stratum from which we sample is equal ; but we are entitled to look on it in this way only if the condition last stated holds good.

We may write (xi) alternatively in a form recalling a result established in Chapter 7 (p. 304)

$$V_m = \frac{r}{r-1}.V - \frac{r}{r-1}V_s \quad . \quad . \quad . \quad . \quad (xii)$$

It is, however, important to notice that neither V nor V_s in this context is an *estimate* (s^2) of the form (vi) in 7.06 in which σ_o^2 is the sample variance :

$$s^2 = \frac{r}{r-1}\sigma_o^2.$$

Both V and V_s in this context signify the true variances of the corresponding *theoretical* sampling distribution. Thus V_m itself is a true measure of variance attributable to the stratification of the parent (standard urn) universe only when r is very large, so that $r \div (r-1) \simeq 1$.

A casual reading of some text-books in common use might give the beginner the wrong impression that (i) above has a unique causal significance,* such as that which we have clarified above. Needless to say, this is not so. Our examination of the umpire-bonus model has shown us that the meaning we attach to the two components on the right-hand side of the equation depends on whether our concern is with the *consequent* relation of the score of the player with that of the umpire or the *concurrent* relation of the scores of the two players. If the latter is our concern, (i) above contributes nothing to a meaningful interpretation of *explained* variation ; but if V_m is the variance of the player's mean score associated with a particular value of the umpire's score, it represents that part of the variance of the player's score distribution attributable to the umpire's contribution. On the other hand, it would *not* be true to say that V_m represents the part of the variance of the score distribution in our Lexian universe attributable to the stratification of the universe. If $r = 2$ in (vii) above, the part of the variance attributable to the stratification of the universe is in fact $\frac{1}{2}V_m$.

When we turn our attention to a related model illustrated in Fig. 81 we shall see that V_m in (i) above does in fact have a causal meaning in terms of a standard universe, but not *en rapport* with (xi) ; and it will surprise us less to find that the meaning we can rightly attach to V_m in this equation depends on the statistical set-up under consideration, if we first take cognisance of the fact that the additive relation it exhibits has nothing to do with the theory of probability. It is in fact an arithmetical property of any grid set-up like the grids of Chapter 9 with two sets of border scores or the grid of Fig. 80 with its single set of border scores, and we might well have sidestepped the derivation of (v) above by exhibiting it as a particular case of a general rule.

Let us suppose that a grid exhibits a set of B scores laid out at the margin of its R rows. The C columns of the grid we shall label 1 to C. We shall denote the B score of a cell in the rth row and cth column as b_{rc} with frequency y_{rc}. By definition, the sum of all such frequency cell entries in the grid is unity, i.e.

$$\sum_{r=1}^{r=R} \sum_{c=1}^{c=C} y_{rc} = 1.$$

We shall also use y_c and y_r respectively to denote the column and row frequency totals, so that

$$\sum_{r=1}^{r=R} y_{rc} = y_c \quad and \quad \sum_{c=1}^{c=C} y_{rc} = y_r.$$

If M_b is the mean value of all the B-scores and V_b is the variance of the B-score distribution for the grid as a whole

$$M_b = \sum_{r=1}^{r=R} \sum_{c=1}^{c=C} y_{rc} b_{rc} \quad and \quad V_b = \sum_{r=1}^{r=R} \sum_{c=1}^{c=C} y_{rc} \cdot b_{rc}^2 - M_b^2 \qquad . \qquad . \quad \text{(xiii)}$$

Within the cth column we may write for the mean B-scores whose total frequency is y_c

$$M_{bc} = \frac{1}{y_c} \sum_{r=1}^{r=R} y_{rc} \cdot b_{rc} . \qquad . \qquad . \qquad . \qquad . \quad \text{(xiv)}$$

The mean value of M_{bc} is, of course, M_b, as is evident from the fact that we may write it in the form

$$\sum_{c=1}^{c=C} y_c \cdot M_{bc}.$$

---

* Churchill Eisenhart, *Biometrics* (1947), Vol. 3 (pp. 1-21) is noteworthy as one of the few writers who have undertaken a *logical* analysis of the assumptions inherent in current procedures invoking the partition of variance.

But from (xiv) and (xiii)

$$\sum_{c=1}^{c=C} y_c M_{ba} = \sum_{c=1}^{c=C} \sum_{r=1}^{r=R} y_{rc} b_{rc} = M_b.$$

For the mean value of the square B-score in the C columns, from definition we thus obtain as the variance of the distribution of the B-score column means

$$V(M_{ba}) = \sum_{c=1}^{c=C} y_c . M_{ba}^2 - M_b^2 \quad . \quad . \quad . \quad . \quad . \quad \text{(xv)}$$

Within a column the variance of the B-scores is

$$V_{ba} = \frac{1}{y_c} \sum_{r=1}^{r=R} y_{rc} b_{rc}^2 - M_{ba}^2.$$

The mean value of the variance of the B score distribution within a column is therefore

$$M(V_{ba}) = \sum_{c=1}^{c=C} y_c . V_{ba}$$

$$= \sum_{c=1}^{c=C} \sum_{r=1}^{r=R} y_{rc} . b_{rc}^2 - \sum_{c=1}^{c=C} y_c M_{ba}^2 \quad . \quad . \quad . \quad . \quad \text{(xvi)}$$

By addition from (xv) and (xvi)

$$V(M_{ba}) + M(V_{ba}) = \sum_{r=1}^{r=R} \sum_{c=1}^{c=C} y_{rc} b_{rc}^2 - M_b^2.$$

Hence from (xiii)

$$V_b = V(M_{ba}) + M(V_{ba}) \quad . \quad . \quad . \quad . \quad . \quad \text{(xvii)}$$

The foregoing argument is evidently valid for both dimensions of a grid with border scores and border frequencies in both dimensions, as in Chapter 9. If, as in Fig. 80, we lay out border scores in only one dimension, the foregoing derivation is equally applicable, since an alternative set of A scores does not come into the picture. If we allocate one urn to each column, $y_c = \frac{1}{C}$, but if we care to include more than one urn of identical composition in a set-up of u urns of which we allocate k of identical composition to the cth column, $y_c = k \div u$.

Thus the relation exhibited by (i) above is one which merely summarises the lay-out of any grid whose cell entries are frequencies referable to border scores laid out in one or other dimension. As such there is nothing sacrosanct about it from a statistical viewpoint, still less in terms of the semantics of causation, except in so far as it is true that

(a) Statistical independence implies that $V_m = 0$, though the converse is not necessarily true ;

(b) One-to-one correspondence of A with B scores in a grid with both column and row border scores signifies that $M_v = 0$ in both dimensions.

Against this background let us now examine the properties of the model illustrated by Fig. 81. Instead of taking samples of equal size from urns of different composition, as in the Lexian model of Fig. 80, we now take samples of different size from one and the same urn. The assumption is that the number of samples of each size is equal and indefinitely large. The chart exhibits an arithmetical example which we may generalise without difficulty by recourse to (xvii) above. Our problem is the long-run distribution of scores in a set-up which involves extracting simultaneously a sample of 1, 2, 3, etc., from a single universe in which the fixed expectation of success at a single trial is p, that of failure being $q = (1 - p)$. We lay out scores

26

EXPECTATION-ACCOMPLISHMENT MODEL FOR SAMPLES OF DIFFERENT SIZE

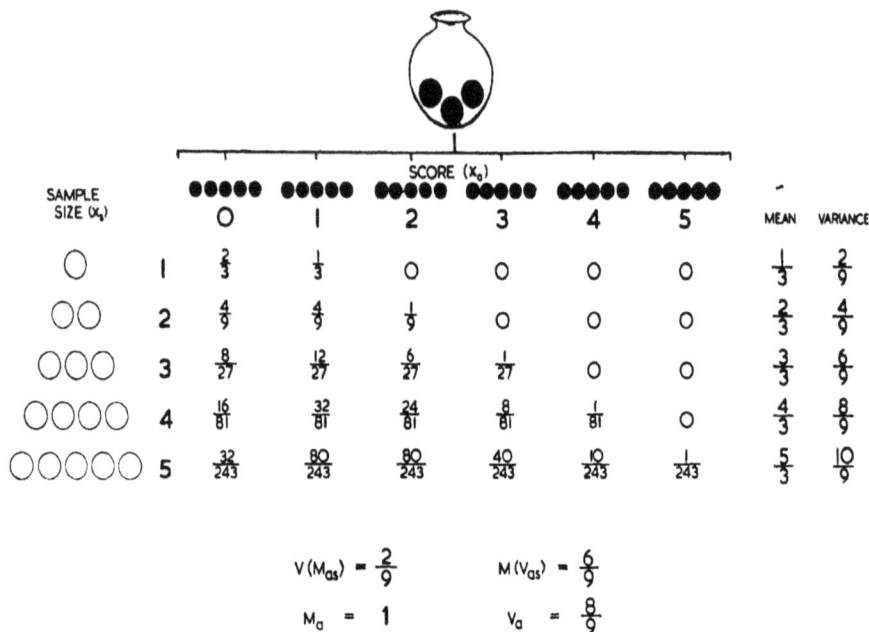

SAMPLE SIZE (x_s)		SCORE (x_a) 0	1	2	3	4	5	MEAN	VARIANCE
○	1	$\frac{2}{3}$	$\frac{1}{3}$	0	0	0	0	$\frac{1}{3}$	$\frac{2}{9}$
○○	2	$\frac{4}{9}$	$\frac{4}{9}$	$\frac{1}{9}$	0	0	0	$\frac{2}{3}$	$\frac{4}{9}$
○○○	3	$\frac{8}{27}$	$\frac{12}{27}$	$\frac{6}{27}$	$\frac{1}{27}$	0	0	$\frac{3}{3}$	$\frac{6}{9}$
○○○○	4	$\frac{16}{81}$	$\frac{32}{81}$	$\frac{24}{81}$	$\frac{8}{81}$	$\frac{1}{81}$	0	$\frac{4}{3}$	$\frac{8}{9}$
○○○○○	5	$\frac{32}{243}$	$\frac{80}{243}$	$\frac{80}{243}$	$\frac{40}{243}$	$\frac{10}{243}$	$\frac{1}{243}$	$\frac{5}{3}$	$\frac{10}{9}$

$$V(M_{as}) = \frac{2}{9} \qquad M(V_{as}) = \frac{6}{9}$$

$$M_a = 1 \qquad V_a = \frac{8}{9}$$

FIG. 81. Expectation-Accomplishment Model for Samples of Different Size. In this set-up the player extracts *with* replacement samples of 1, 2, 3, 4 or 5 balls repeatedly from one and the same urn. The columns exhibit the theoretical sampling distribution w.r.t. a sample of a particular size. The cell entries exhibit the frequencies in fractional form. To compute the variance of the total score distribution directly, the reader will find it convenient to express them, as in Fig. 80, by recourse to whole numbers referable to the common denominator 243.

(x_a) at the head of the columns, and each row represents the score distribution for a sample of a given size. If x_s is the size of the sample definitive of a row score distribution, M_{as} the corresponding mean score, V_{as} the corresponding intra-row variance, and r_m the mean sample size, we derive from the familiar properties of the binomial distribution

$$M(V_{as}) = M(x_s pq) = pqM(x_s) = r_m pq.$$

If V_a is the variance of the distribution of successes (x_a), we obtain by recourse to (xvii):

$$V_a - V(M_{as}) = r_m pq \qquad . \qquad . \qquad . \qquad . \qquad . \qquad \text{(xviii)}$$

We can eliminate in many ways all variance attributable to the circumstance peculiar to the model situation, i.e. that the sample size is itself variable, but the least arbitrary choice is to make the size of each sample the same as the sample size mean (r_m), as is possible if r_m is an integer. If so, the variance (V_s) of the sample distribution so standardised would be

$$V_s = r_m pq.$$

With this interpretation of what we mean by eliminating the source of variation peculiar to the model situation we therefore obtain the identity

$$V_a - V_s = V_m.$$

This recalls our partition of variance w.r.t. the correlation of the player's score with that of the umpire in 9.03; but is evidently *not* equivalent to (xi) above.

EXERCISE 10.01

A number of black and red balls are placed in four urns so that the proportion of black balls is as follows :

Urn	. . . . . A	B	C	D
Proportion of black balls .	. $\frac{1}{10}$	$\frac{1}{5}$	$\frac{1}{2}$	$\frac{1}{4}$

1. Four balls are drawn with replacement from each urn. Set up the Lexian grid for black ball scores and evaluate V_m, M_r and V w.r.t. the B scores (x_b) set out at the margin of the rows in Fig. 80.

2. One ball is drawn from urn A ; it is replaced in the urn and two balls are drawn simultaneously one from urn A and one from B. After replacing the balls, three are drawn simultaneously, one each from A, B and C. Finally, after replacing the balls, four are drawn one each from A, B, C and D. Set up a grid for repeated trials of this experiment scoring the black balls. Work out V_m, M_r and V.

3. Insert the expected frequencies (x_s) at the head of the columns of the Lexis grid obtained in 1. From the *expectation and accomplishment* grid thus set up, evaluate $Cov\ (EB)$, and verify that

$$\rho_{eb}^2 = \frac{V_e}{V_b}.$$

4. The Danish actuary Arne Fisher set up 20 urns each containing 40 balls, black and white in different proportions. From each urn he drew 500 balls with replacement and obtained the following results :

No. of black balls in urn	.	20	21	22	23	24	25	26	27	28	29	30	31	32	33	34	35	36	37	38	39
No. of white balls drawn	.	251	246	222	216	193	176	183	173	156	135	140	127	115	96	78	69	55	43	29	19

Work out the mean and variance of this distribution and compare them with the theoretically expected values.

10.02 THE POISSON MODEL

One consequence of the foregoing partition of variance in the Lexian system of sampling from urns containing equal numbers of balls has a special interest in connection with a seemingly paradoxical consequence of taking unit samples from different universes. If V_e is the variance of the score distribution w.r.t. extraction of samples from the standardised urn (Bernoullian universe), and V_L is the variance of the score distribution for extraction of samples of equal size from urns of different composition (stratified universe), we have seen that

$$V_L = V_e + \frac{r-1}{r} V_m.$$

If the sample size (r) is unity, this means that

$$V_L = V_e.$$

Hence in virtue of the partition (xvii) of 10.01, which we have also established as a property of any score-frequency grid

$$V_e = M_v + V_m.$$

To get the meaning of this breakdown of variance into focus, let us consider a simple numerical illustration. We suppose that we may take *one* ball from each of three urns constituted as follows :

Urn A	4 Red	8 Black
Urn B	6 Red	6 Black
Urn C	3 Red	9 Black

If we score the extraction of a red ball as a success, we may set out a frequency distribution of the result as follows :

Score	*Urn A*	*Urn B*	*Urn C*
0	$\frac{2}{3}$	$\frac{1}{2}$	$\frac{3}{4}$
1	$\frac{1}{3}$	$\frac{1}{2}$	$\frac{1}{4}$
Mean	$\frac{1}{3}$	$\frac{1}{2}$	$\frac{1}{4}$
Variance	$\frac{2}{9}$	$\frac{1}{4}$	$\frac{3}{16}$

From the above we obtain

$$V_m = \tfrac{1}{3}(\tfrac{1}{9} + \tfrac{1}{4} + \tfrac{1}{16}) - \tfrac{1}{9}(\tfrac{1}{3} + \tfrac{1}{2} + \tfrac{1}{4})^2 = \tfrac{7}{648},$$
$$M_v = \tfrac{1}{3}(\tfrac{2}{9} + \tfrac{1}{4} + \tfrac{3}{16}) \qquad\qquad = \tfrac{95}{432},$$
$$\therefore M_v + V_m = \tfrac{299}{1296}.$$

Now the composition of our standard urn will be

Red	4	+	6	+	3	=	13
Black	8	+	6	+	9	=	23
					Total		36

Thus the probabilities of drawing, or of failing to draw, a red ball from the standard urn in a single trial will be

$$p = \tfrac{13}{36} \quad and \quad q = \tfrac{23}{36}$$

If V_s is the variance of the unit sample distribution, we therefore obtain

$$V_s = \tfrac{13}{36} \cdot \tfrac{23}{36} = \tfrac{299}{1296}.$$

Hence, as above, we obtain for the unit sample

$$V_s = M_v + V_m = V_L.$$

Let us now suppose that we take a unit sample from each of the same three urns, and score the *sum* as the result of the 3-fold trial. We may set out the frequency distribution of the total score as a staircase diagram (Fig. 82), from which we get the following frequency distribution for the 3-fold trial involving the extraction of a unit sample from each of the three urns.

Score	0	1	2	3
Frequency	$\frac{6}{24}$	$\frac{11}{24}$	$\frac{6}{24}$	$\frac{1}{24}$

If we denote the variance of this distribution by V_p

$$V_p = \frac{11(1) + 6(4) + 1(9)}{24} - \left[\frac{11(1) + 6(2) + 1(3)}{24}\right]^2 = \frac{95}{144} = \frac{285}{432}.$$

POISSON TRIAL OF THREE

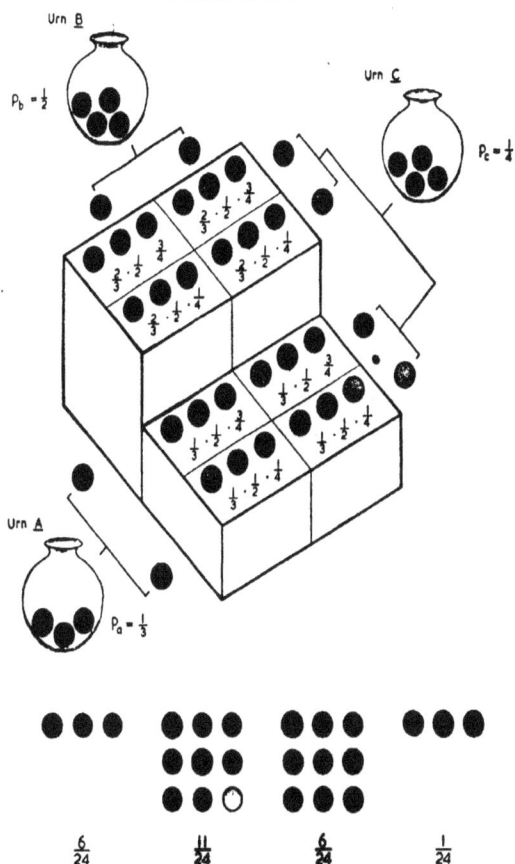

FIG. 82. The Poisson Model. This staircase diagram exhibits the derivation of theoretical sampling dis tribution of an indefinitely large number of 3-fold trials, each involving extraction of a *unit sample* from on of three different urns.

The reader will note that :

(a) The last figure is three times the mean variance $(M_v = \frac{95}{432})$ for the unit sample distribution in the stratified (*Lexian*) universe ;

(b) It is less than the variance of a 3-fold trial from the standard urn, since the latter is

$$3\,p\,.\,q = \frac{3\,.\,13\,.\,23}{36^2} = \frac{299}{432}.$$

Here then is a sort of stratified sampling which admits of less variation than sampling from the homogeneous parent universe of the standard urn. It is common to refer to it as the *Poisson* system, and we may set out the relations illustrated above more formally as follows. We shall denote the probabilities of drawing a red ball from the three urns A, B, C as p_a, p_b and p_c respectively, corresponding variances for the unit sample distributions being then $p_a q_a = p_a(1 - p_a)$,

$p_b q_b = p_b(1 - p_b)$ and $p_c q_c = p_c(1 - p_c)$. Since the trials are independent, the variance (V_p) of the score sum distribution is given by

$$V_p = p_a(1 - p_a) + p_b(1 - p_b) + p_c(1 - p_c).$$

But the mean variance of the system of *unit* samples in the 3-urn Lexian universe, denoted above by M_v is

$$\frac{p_a(1 - p_a) + p_b(1 - p_b) + p_c(1 - p_c)}{3},$$

$$\therefore V_p = 3M_v.$$

For the variance of the score distribution of 1-fold samples from the 3-urn universe, we obtained

$$V_L = pq = M_v + V_m.$$

In the above V_m is the variance of the distribution of unit sample means, i.e. the distribution of p_a, p_b, p_c, and we may therefore write more explicitly $V_m = V_p$, so that

$$3pq = 3M_v + 3V_p$$
$$= V_p + 3V_p$$
$$\therefore V_p = 3pq - 3V_p.$$

More generally if the number of urns from which we draw unit samples is r, and V_p as before is the variance of the distribution of the score total :

$$V_p = \sum_{s=1}^{s=r} p_s(1 - p_s) = \sum_{s=1}^{s=r} (p_s - p_s^2)$$

$$= \sum_{s=1}^{s=r} p_s - \sum_{s=1}^{s=r} p_s^2 \qquad \qquad \qquad \text{(i)}$$

To construct our standard urn in which the proportion of red balls is p, we postulate that the r urns contain the same number of balls, in which case p is the mean value of p_s, and

$$V_p = \frac{1}{r}\sum_{s=1}^{s=r} p_s^2 - p^2$$

$$\therefore r \cdot V_p = \sum_{s=1}^{s=r} p_s^2 - r \cdot p^2 \qquad \qquad \qquad \text{(ii)}$$

For the same reason

$$\sum_{s=1}^{s=r} p_s = r \cdot p \qquad \qquad \qquad \text{(iii)}$$

Hence by substitution of (ii) and (iii) in (i) we get

$$V_p = rp - (r \cdot V_p + rp^2)$$
$$= rp - rp^2 - r \cdot V_p$$
$$= rp(1 - p) - r \cdot V_p$$
$$= rpq - r \cdot V_p \qquad \qquad \qquad \text{(iv)}$$

This result exhibits the variance (V_p) of the distribution of the score sum of r unit samples from different universes in terms of the variance (rpq) w.r.t. r-fold sampling from the standard universe and the variance of the mean unit sample score within the stratified universe.

SCORE GRID AND FREQUENCY GRID

	P	Q	R	Score Total	Row Mean	Row Variance
I	3	5	2	10	10/3	14/9
II	1	4	4	9	3	18/9
III	3	1	2	6	2	6/9
Score Total	7	10	8	25	M_{ab}	V_{ab}
Column Mean	7/3	10/3	8/3	M_{ba}		
Column Variance	8/9	26/9	8/9	V_{ba}		

Grand Mean $(M_a = M_b)$ 25/9

$V(M_{ba})$ Variance of Column Means 14/81 $V(M_{ab})$ Variance of Row Means 26/81

$M(V_{ba})$ Mean Variance of Columns 126/81 $M(V_{ab})$ Mean Variance of Row 114/81

Total Variance 140/81 = Total Variance 140/81

B-dimensional Border Scores

	P	Q	R	
1	1/9	1/9	0	
2	0	0	2/9	
3	2/9	0	0	
4	0	1/9	1/9	
5	0	1/9	0	Grand Total
Total Frequency	1/3	1/3	1/3	1
M_{ba}	7/3	10/3	8/3	
V_{ba}	8/9	26/9	8/9	

Grand Mean (M_b) = 25/9

Total Variance (V_b) = 140/81

$V(M_{ba})$ = 14/81; $M(V_{ba})$ = 126/81

A-dimensional Border Scores

	1	2	3	4	5	Total Frequency	M_{ab}	V_{ab}
I	0	1/9	1/9	0	1/9	1/3	10/3	14/9
II	1/9	0	0	2/9	0	1/3	3	18/9
III	1/9	1/9	1/9	0	0	1/3	2	6/9
					Grand Total	1		

Grand Mean (M_a) = 25/9; Total Variance (V_a) = 140/81

$V(M_{ab})$ = 26/81; $M(V_{ab})$ = 114/81

FIG. 83. Score Grid and Frequency Grid. A grid exhibiting cell entries as scores in columns and rows definitive of a qualitative two-way classification without reference to border frequencies or border scores is reducible to a frequency-score grid in either dimension, though not simultaneously in both.

10.03 The Book-Keeping of a Score Grid

Our Lexis model has shown us that it is possible to partition the variance of a distribution in such a way as to separate components attributable to different sources of variation, if we agree to adopt a particular criterion of homogeneity, *viz.* a standard urn of a particular composition. The student may surmise with good reason that a corresponding yardstick of homogeneity is by no means easy to formulate in connexion with statistical problems of everyday life; and we shall now examine a model of which it is possible to envisage more clearly the causal interpretation of a breakdown of variance, and to do so in a less arbitrary sense.

Our new model, the handicap score-grid model, like those which went before is the model of a *universe.* It is necessary to state this because the score-board of Fig. 84 recalls the lay-out of sample scores for the statistical technique known as *analysis of variance;* and its design is to clarify what assumptions the use of the latter invokes with special reference to a class of situations involving two putative sources of variation superimposed upon a third. The following is a hypothetical illustration. A laboratory worker records a single determination of the blood calcium level of 5 rabbits at 6 four-hourly intervals from noon to 8 a.m. inclusive. In such a situation, there is a basic substratum of scatter due to instrumental errors inherent in the method employed in the determination. Superimposed on this score of variation are : (*a*) intraspecific variation at one and the same time ; (*b*) possible variations of the blood calcium level of the same animal customarily included under the term diurnal rhythm. In such a situation we can lay out our data on a 5 × 6 chessboard assigning all figures referable to the same rabbit to one of the five columns, and all figures referable to the same time of day to one of the six rows.

When we do this, it is important to be clear about : (*a*) what arithmetical properties of such a score-grid are inherent in any such lay-out regardless of our statistical preoccupations ; (*b*) what sort of statistical questions we may ask about it and what they severally imply concerning the structure of the universe from which we extract our sample. At the outset, we should also be clear that the several statistical issues to which we customarily apply the term *analysis of variance* invoke identities which have no necessary connexion with statistics. When we arrange data as indicated above, our score-grid of R rows and C columns exhibits a single score in each cell with frequency $\frac{1}{RC}$; but we can lay it out (*see* Fig. 83) in either dimension as a frequency grid with appropriate border scores, though not simultaneously in both. Consequently the relation exhibited in (xvii) in 10.01 is valid w.r.t. both columns and rows. In either dimension, the total variation of the score distribution is necessarily the same, since we are dealing with only one set of scores, and the frequency of any cell entry of a grid with R rows and C columns, as in the schema below, is then the reciprocal of $RC = N$ the total of cells.

Column

	1	2	. . .	c	. . .	C	No. of Cells
1	x_{11}	x_{12}	. . .	. . .	. . .	x_{1C}	C
2	x_{21}	x_{22}	. . .	. . .	. . .	x_{2C}	C
Row . . .	. . .	. . .	. . .	. . .	. . .	. . .	C
r	. . .	. . .	. . .	x_{rc}	. . .	x_{rC}	C
. . .	. . .	. . .	. . .	. . .	. . .	. . .	C
R	x_{R1}	x_{R2}	. . .	x_{Rc}	. . .	x_{RC}	C
No. of cells	R	R	R	R	R	R	RC N

Without guidance from what follows, the student may indeed find it helpful to repeat as an exercise the derivation of (xvii) in 10.01 as it applies to such a set-up, making use of the fact that each column contains R cells and each row contains C cells, so that

$$y_{rc} = \frac{1}{RC}; \quad y_c = \frac{R}{RC} = \frac{1}{C}; \quad y_r = \frac{C}{RC} = \frac{1}{R}.$$

If x_{rc} is a score in a cell referable to the rth row and cth column, we thus have for the grand mean (M) and total variance (V)

$$M = \frac{1}{RC}\sum_{c=1}^{c=C}\sum_{r=1}^{r=R} x_{rc} \quad and \quad V = \frac{1}{RC}\sum_{c=1}^{c=C}\sum_{r=1}^{r=R} x_{rc}^2 - M^2. \qquad . \quad . \text{ (i)}$$

For the score distribution within the rth row the means and variances will be

$$M_r = \frac{1}{C}\sum_{c=1}^{c=C} x_{rc} \quad and \quad V_r = \frac{1}{C}\sum_{c=1}^{c=C} x_{rc}^2 - M_r^2.$$

Also for the grand mean of the row means we have

$$\frac{1}{R}\sum_{r=1}^{r=R} M_r = \frac{1}{RC}\sum_{r=1}^{r=R}\sum_{c=1}^{c=C} x_{rc} = M.$$

Whence we get

$$V(M_r) = \frac{1}{R}\sum_{r=1}^{r=R}(M_r^2 - M^2) \quad and \quad M(V_r) = \frac{1}{RC}\sum_{r=1}^{r=R}\sum_{c=1}^{c=C} x_{rc}^2 - \frac{1}{R}\sum_{r=1}^{r=R} M_r^2. \qquad . \quad \text{(ii)}$$

By combining (ii) in virtue of (i) we thus have

$$V = V(M_r) + M(V_r) \qquad . \qquad . \qquad . \qquad . \qquad . \quad \text{(iii)}$$

Likewise, for the intra-column means (M_c) and variances (V_c)

$$V = V(M_c) + M(V_c) \qquad . \qquad . \qquad . \qquad . \qquad . \quad \text{(iv)}$$

In 10.04 we shall examine what circumstances endow another statistic V_e with a special meaning. Its definition is

$$V_e = V - V(M_r) - V(M_c) \qquad . \qquad . \qquad . \qquad . \qquad . \quad \text{(v)}$$

Hence by (iii) and (iv) we can write

$$V_e = M(V_c) + M(V_r) - V \qquad . \qquad . \qquad . \qquad . \qquad . \quad \text{(vi)}$$

To dissect the expression on the right, it is convenient to write the intra-class variances in the form

$$V_c = \frac{1}{R}\sum_{r=1}^{r=R}(x_{rc} - M_c)^2 \quad and \quad V_r = \frac{1}{C}\sum_{c=1}^{c=C}(x_{rc} - M_r)^2.$$

$$\therefore M(V_c) = \frac{1}{RC}\sum_{c=1}^{c=C}\sum_{r=1}^{r=R}(x_{rc} - M_c)^2 ; \quad M(V_r) = \frac{1}{RC}\sum_{r=1}^{r=R}\sum_{c=1}^{c=C}(x_{rc} - M_r)^2.$$

Likewise we may write

$$V = \frac{1}{RC}\sum_{r=1}^{r=R}\sum_{c=1}^{c=C}(x_{rc} - M)^2.$$

Since $RC = N$, the total number of cells in the grid, (vi) becomes

$$NV_e = \sum_{r=1}^{r=R}\sum_{c=1}^{c=C}\{(x_{rc} - M_c)^2 + (x_{rc} - M_r)^2 - (x_{rc} - M)^2\}$$

$$= \sum_{r=1}^{r=R}\sum_{c=1}^{c=C}\{x_{rc}^2 - 2M_c x_{rc} - 2M_r x_{rc} + 2Mx_{rc} + M_c^2 + M_r^2 - M^2\},$$

$$= \sum_{r=1}^{r=R}\sum_{c=1}^{c=C}[(x_{rc} - M_c - M_r + M)^2 - 2(M^2 + M_c M_r - M_c M - M_r M)]$$

$$= \sum_{r=1}^{r=R}\sum_{c=1}^{c=C}(x_{rc} - M_c - M_r + M)^2 - 2\sum_{r=1}^{r=R}\sum_{c=1}^{c=C}(M - M_c)(M - M_r).$$

The sum on the extreme right is

$$2 \sum_{r=1}^{r=R} (M - M_r) \sum_{c=1}^{c=C} (M - M_c) = 0.$$

$$\therefore V_e = \frac{1}{N} \sum_{r=1}^{r=R} \sum_{c=1}^{c=C} (x_{rc} - M_c - M_r + M)^2 \qquad . \qquad . \qquad . \qquad . \qquad . \qquad \text{(vii)}$$

We shall later find useful the following relation which is likewise an arithmetical property of any score-grid. If M_u is any number

$$(M_c - M) = (M_c - M_u) - (M - M_u) \quad and \quad (M_c - M_u) = (M_c - M) + (M - M_u),$$

$$\therefore \quad (M_c - M)^2 = (M_c - M_u)^2 - 2(M_c - M_u)(M - M_u) + (M - M_u)^2.$$

Since $$\qquad\qquad\qquad (M_c - M_u) = (M_c - M) + (M - M_u),$$

$$2(M_c - M_u)(M - M_u) = 2(M - M_u)(M_c - M) + 2(M - M_u)^2,$$

$$\therefore \quad (M_c - M)^2 = (M_c - M_u)^2 - 2(M - M_u)(M_c - M) - (M - M_u)^2.$$

The sum of the deviations $(M_c - M)$ of the column means M_c from the grand mean, which is also the weighted mean of the column means, is zero. In this expression both M and M_u, hence $(M - M_u)$ are constants, so that

$$2 \sum_{1}^{C} (M - M_u)(M_c - M) = 2(M - M_u) \sum_{1}^{C} (M_c - M).$$

The expression under the summation sign on the right is the mean of the deviations of the column means, hence by definition zero, so that

$$2 \sum_{1}^{C} (M - M_u)(M_c - M) = 0.$$

If therefore we sum the terms in the above over the C column values, we obtain

$$\sum_{1}^{C} (M_c - M)^2 = \sum_{1}^{C} (M_c - M_u)^2 - \sum_{1}^{C} (M - M_u)^2$$

$$= \sum_{1}^{C} (M_c - M_u)^2 - C(M - M_u)^2,$$

$$\therefore V(M_c) = \frac{1}{C} \sum_{1}^{C} (M_c - M)^2 = \frac{1}{C} \sum_{1}^{C} (M_c - M_u)^2 - (M - M_u)^2 \qquad . \qquad \text{(viii)}$$

In the same way we may obtain

$$V(M_r) = \frac{1}{R} \sum_{1}^{R} (M_r - M_u)^2 - (M - M_u)^2 \qquad . \qquad . \qquad . \qquad . \qquad \text{(ix)}$$

10.04 THE HANDICAP SCORE-GRID MODEL

In contradistinction to the arithmetical tautologies set forth in the preceding section, the score-grid lay-out of a sample has statistical properties which depend on the universe from which the sample comes. What we customarily call analysis of variance covers statistical questions of three sorts, and it will be easier to distinguish them by reference to our rabbit parable at the beginning of 10.03 than by stating them in formal terms. We have there specified three putative sources of scatter within such a table :

(a) common to all cells, "residual" errors of measurement inherent in the method of determination ;

(b) a between-row component ascribable to diurnal rhythm ;

(c) a between-column component ascribable to intra-specific variation.

With respect to such a situation the three sorts of questions we may ask are as follows :

(i) is any one source of variation negligible ?

(ii) what estimates of the several components are unbiassed in the sense defined in 7.06 ?

(iii) to what limits of errors are such estimates subject ?

We may speak of (i) as testing the null hypothesis that the universe is *homogeneous* in one or other of three dimensions, of the second and third as that of constructing a *balance sheet of variation* and of validifying the entries. As Churchill Eisenhart has emphasised, few among the many research workers who use the analysis of variance sufficiently recognise that the construction of any meaningful balance sheet of variation invokes assumptions irrelevant to the rejection of a null hypothesis, and that its validification invokes additional assumptions irrelevant to adequate reasons for conceiving the possibility of constructing one. The problem of significance *vis-à-vis* testing the null hypothesis or assessing the error inherent in the construction of the balance sheet will be the subject of separate treatment in Vol. II. Here our concern is :

(a) to exhibit what properties of a *universe* justify a true bill in the sense specified by (ii) above ;

(b) to indicate that criteria of homogeneity can stand on their own feet without recourse to the assumption that the universe may have such properties.

The handicap-score grid model of Fig. 84 exhibits the build-up of a universe whose structure permits us to entertain the possibility of making a balance sheet of variation. What we there call the players' score-board exhibits in each major-cell a complete distribution of the red-score of a single individual in a 4-fold trial with replacement of each ball chosen before drawing another from an urn containing 2 red and 2 black balls. Each such major cell has therefore 16 sub-cells to accommodate scores of 0, 1, 2, 3, 4 in the ratio 1 : 4 : 6 : 4 : 1, so that a score of 0 or 4 each occur in 1 sub-cell only, a score of 1 or 3 in 4 sub-cells and a score of 2 in 6 sub-cells. To suggest the idea that the scores of different players at successive trials are independent we make the arrangement of the scores in the sub-cells different, though the distribution and mean values of scores in each major cell of 16 sub-cells is the same.

We can now assume that each player at each trial receives : (a) one and the same bonus (*handicap*) in virtue of his membership of a " column team " ; (b) one and the same bonus in virtue of his membership of a row team. The score-boards on the left and right respectively exhibit the result of allocating the column bonus and the row bonus singly. The *final* score-board assigns to each player at each trial his total score in virtue of the addition to his own of the *combined* bonus. In this universe of scores we therefore have : (a) a *residual* component of variation attributable to the individual score distribution common to all the players ; (b) a *between-row* component of variation arising from the fact that different row teams receive a different bonus ; (c) a *between-column* component of variation arising from the fact that different column teams receive a different bonus. The lay-out implies that the distributions of the two bonus systems are *independent* of one another and of the player's luck.

The beginner will find it helpful to check over the marginal figures for row and column means and the variances of the row and column distributions. Each score board illustrates the two numerical tautologies :

$$M(V_r) + V(M_r) = V = M(V_c) + V(M_c) \qquad . \qquad . \qquad . \qquad . \text{ (i)}$$

If we define V_e as in (vii) of 10.03, the four grids also illustrate the numerical tautologies specified by (iv) and (v) in the same context, *viz.*:

$$M(V_r) + M(V_c) - V = V_e = V - V(M_r) - V(M_c) \qquad . \qquad . \qquad . \quad \text{(ii)}$$

Either of the foregoing is, in fact, a property of any grid-wise lay-out of numbers, regardless of their statistical terms of reference, if any. To appreciate what the grid teaches we must take stock of the variance of each bonus system. For the column-bonus distribution, the mean is $(1 + 2 + 0) \div 3 = 1$ and the variance $V(b_c)$ is:

$$\tfrac{1}{3}(0^2 + 1^2 + 2^2) - 1^2 = \tfrac{2}{3}.$$

That of the row-bonus distribution $V(b_r)$ is:

$$\tfrac{1}{4}(0^2 + 1^2 + 2^2 + 4^2) - (\tfrac{7}{4})^2 = \tfrac{35}{16}.$$

The specifically statistical properties which the model brings into focus are:

(*a*) as defined by (ii) above and (viii) of 10.03, the statistic V_e of the final score-board is also the total variance of the original homogeneous (players') score-board, and as such is an exact measure of *residual* variation;

(*b*) the mean value of the within-column variance in the final score-board is:

$$M(V_c) = \tfrac{51}{16} = 1 + \tfrac{35}{16} = V_e + V(b_r);$$

(*c*) the mean value of the within-row variance of the final score-board is:

$$M(V_r) = \tfrac{5}{3} = 1 + \tfrac{2}{3} = V_e + V(b_c);$$

(*d*) the total variance of the final score-board is

$$V_4 = \tfrac{185}{48} = \tfrac{35}{16} + \tfrac{2}{3} + 1 = V(b_r) + V(b_c) + V_e.$$

These relations—which are statements about the structure of a *universe* in contradistinction to statements about a sample—are easy to derive from the properties of the model in virtue of the assumptions that

(*a*) the three score distributions are *independent*;

(*b*) the three score components are strictly *additive*.

If we denote by x the player's own score, by b_c the column-bonus and b_r the row-bonus, we may write the total variance of the final score-board as

$$V = V(x + b_c + b_r).$$

In virtue of independence, we therefore have

$$V = V(x) + V(b_c) + V(b_r).$$

In this expression $V(x)$ is the variance of the distribution of scores in the homogeneous universe of the players' score-board, and there denoted V_1, so that

$$V = V_1 + V(b_c) + V(b_r) \qquad . \qquad . \qquad . \qquad . \qquad . \quad \text{(iii)}$$

This equation exhibits the total variance as the sum of three components, each being the variance of the distribution definitive of one of three sources of variation. Within a row of major cells, we may write

$$V_r = V(x + b_c + b_r).$$

Since b_r is constant within the row, it merely changes the origin without affecting the scale of the distribution and therefore makes no contribution to the variance, i.e.

$$V_r = V(x) + V(b_c).$$

Now the distribution of the x-score component in each major cell is the same in each row and the same as for the original players' score-board, i.e.

$$V_r = V_1 + V(b_c).$$

Since each term in this expression is a constant of the set-up

$$M(V_r) = V_1 + V(b_c).$$

Likewise

$$M(V_c) = V_1 + V(b_r).$$

By definition

$$V_e = M(V_r) + M(V_c) - V.$$

Whence we derive for the final score-board value of V_e

$$V_e = V_1.$$

Thus the universe values of V_e, $M(V_r)$ and $M(V_c)$, hence alternatively of V_e, $V(M_r)$ and $V(M_c)$ in virtue of (i) and (ii) are all expressible in terms of its three components of variance. Hence also it is possible, as we shall see in Vol. II, to use the *sample* values of these three statistics to estimate the components of variation, if the universe has the structure implicit in the model. To say that it has such a structure is to say that the sources of variation are strictly additive, and it is important to recognise that this condition is essential, if a balance sheet of components of variation attributable to one or other sources is in fact a true bill. If we examine a sample with a view to making such a balance sheet it is therefore important to be quite clear about whether the additive postulate is applicable to the situation. The reader, if also a biologist, may therefore reflect with profit on the following remarks of Churchill Eisenhart :

" . . . Hence, when additivity does not prevail, we say that there are *interactions* between row factors and column factors. Thus, in the case of varieties and treatments considered above, additivity implies that, under the general experimental conditions of the test, the true mean yield of one variety is greater (or less) than the true mean yield of another variety by an amount—an additive constant, not a multiplier—that is the same for each of the treatments concerned, and, conversely, the true mean yield with one treatment is greater (or less) than the true mean yield with another treatment by an amount that does not depend upon the variety concerned ; which is exactly what is meant when we say that there are no " interactions " between varietal and treatment effects. . . ."

Though it is necessary to postulate the additive principle as a basis for estimating components of variation, and to make additional assumptions about the nature of the component score distributions if our aim is to assess the error to which such estimates are subject, we shall now see that the additive postulate is irrelevant to the definition of criteria of homogeneity. It is therefore somewhat unfortunate that the single expression analysis of variance has come into use to describe each or all of the procedures distinguished as (i)-(iii) at the beginning of 10.03, the more so because the most widely acceptable procedure included as such does not strictly entail a *break-down*, and hence the possibility of a balance sheet.

10.05 CRITERIA OF A HOMOGENEOUS UNIVERSE

In the jargon of statistics, we distinguish the four grids of Fig. 84 as being respectively : (a) homogeneous in both dimensions, as we assume to be true of the original players' score-board ; (b) homogeneous in neither dimension, as is true of the final score-board in virtue of the two-way bonus allocation; (c) homogeneous in one dimension only, as is true of the remaining pair. Commonly, the term *homogeneous* implies sampling from the same universe ; but this is not inconsistent with our use of the term in this context. If two players (A and B) who respectively receive a bonus of 2 and 3 each draw a 6-fold sample from the same universe, in this case an urn containing two red balls and one black one, recording as their original scores the number of red balls in the sample, we may regard their final score as respectively recording the result of drawing 8-fold and 9-fold samples from a stratified universe of which one stratum from which each player extracts six balls is the same, the other being an urn consisting of red balls only. Thus we may put our question in the form : do all the players draw the same number of items (balls) from the same universe (urn) ?

Let us first examine the consequences of the null hypothesis that a grid such as the final score-board of Fig. 84 is homogeneous w.r.t. the vertical arrangement of score entries by columns, i.e. that players in each column team receive the same bonus—or none at all. To do justice to the issue we must retrace our steps to the argument of 7.06. Our null hypothesis predicates that all our column samples are samples from a single universe, the variance of the unit sampling distribution of which we denote as usual by V_u. If R, the number of entries per column were very large, we could assume that $V_c \simeq V_u$, and hence that the *mean* value $M(V_c)$ of V_c is a good estimate of V_u. Now the variance of the column means is itself an estimate of the mean score for R-fold samples from the same universe. In accordance with (vii) of 7.03, we may therefore write $V(M_c) \simeq V_u \div R$. We can thus arrive at two independent estimates of V_u, and they must agree within whatever limits of confidence we care to impose on permissible sampling error. This gives us the expected value of the ratio $M(V_c) \div R \cdot V(M_c) \simeq 1$ when R is large. In fact, R may be small. To apply this criterion, we must then define a statistic whose long-run mean value is V_u in terms of $M(V_c)$ empirically defined as in 10.04, and a statistic whose long-run value is V_u in terms of $V(M_c)$ defined likewise.

In accordance with (iii) of 7.06, we can define one statistic whose long-run mean value is V_u by $R \cdot V_c \div (R - 1)$. It will be convenient to sidestep use of unwieldy symbols for summation of all possible samples weighted by their appropriate frequencies, by recourse to a more economical notation for the expected (i.e. long-run *mean*) value of a statistic (v) *, viz. :

$$E(v) = \sum v_t \cdot v.$$

So we now write

$$E\left(\frac{R}{R-1} \cdot V_c \right) = V_u = \frac{R}{R-1} \cdot E(V_c).$$

The C columns provide us with C such estimates, the sum of whose expected values is therefore $C \cdot V_u$ so that we may put

$$\frac{R}{R-1} E\left(\sum_1^C V_c \right) = C \cdot V_u,$$

$$\therefore E\left(\sum_1^C V_c \right) = \frac{CR - C}{R} \cdot V_u,$$

$$\therefore E\left(\frac{1}{C} \sum_1^C V_c \right) = \frac{CR - C}{CR} \cdot V_u,$$

$$\therefore E[M(V_c)] = \frac{N - C}{N} \cdot V_u \quad . \quad . \quad . \quad . \quad . \quad (i)$$

* *Vide* note in 7.06.

FIG. 84.

HANDICAP SCORE - GRID MODEL

(a) COLUMN TEAMS

(b) ROW TEAMS

Players' Red-Score Board for a 4-fold Trial with Replacement

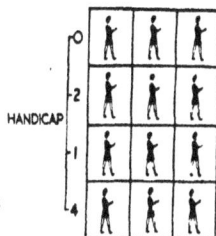

$M(V_r) = 1 = M(V_c); \quad V(M_r) = 0 = V(M_c)$

Total Variance $(V_c) = 1$

Column Team Score Board

$V(M_r) = 0; \quad M(V_r) = \frac{5}{3}$

$V(M_c) = \frac{2}{3}; \quad M(V_c) = 1$

Total Variance $(V_2) = \frac{5}{3}$

Row Team Score Board

$V(M_r) = \frac{35}{16}; \quad M(V_r) = 1$

$V(M_c) = 0; \quad M(V_c) = \frac{51}{16}$

Total Variance $(V_3) = \frac{51}{16}$

Final Score Board

$V(M_r) = \frac{35}{16}; \quad M(V_r) = \frac{5}{3} \qquad V(M_c) = \frac{2}{3}; \quad M(V_c) = \frac{51}{16}$

$V_c = 1$

Total Variance $(V_4) = \frac{185}{48}$

FIG. 84. For explanation see text 10.04.

From the same point of view, let us now consider the relation of $V(M_c)$ to V_u, denoting the true mean of the putative common universe of our null hypothesis by M_u, the true variance of the sampling distributions of the means of R-fold and N-fold samples therefrom respectively by $V_u(M_c)$ and $V_u(M)$. As in 10.04, we denote the grand mean of the sample as M of which M_u is the expected value. In accordance with (vii) of 7.03, we then have

$$E(M_c - M_u)^2 = V_u(M_c) = V_u \div R = C \cdot V_u \div N,$$

$$E(M - M_u)^2 = V_u(M) = V_u \div N,$$

$$\therefore \; E(M_c - M_u)^2 - E(M - M_u)^2 = (C - 1) \cdot V_u \div N \quad . \quad . \quad . \quad \text{(ii)}$$

By definition, our empirical statistic $V(M_c)$ satisfies the relation

$$V(M_c) = \frac{1}{C} \sum_1^C (M_c - M)^2.$$

In accordance with (viii) of 10.04 above, we may therefore put

$$E[V(M_c)] = \frac{1}{C} E \sum_1^C (M_c - M_u)^2 - E(M - M_u)^2.$$

The long-run mean value of the first expression in parenthesis on the right-hand side is the long-run mean value of a sum of C such terms, so that

$$\frac{1}{C} E \sum_1^C (M_c - M_u)^2 = E(M_c - M_u)^2,$$

$$\therefore \; E[V(M_c)] = E(M_c - M_u)^2 - E(M - M_u)^2.$$

Whence, from (ii)

$$E[V(M_c)] = \frac{(C - 1)V_u}{N} \quad . \quad . \quad . \quad . \quad . \quad \text{(iii)}$$

By combining (i) with (iii), we have

$$\frac{N}{C - 1} E[V(M_c)] = V_u = \frac{N}{N - C} E[M(V_c)].$$

Thus V_u is the mean value of either of two statistics which respectively refer to variation *within* and *between* the columns, viz. :

$$\frac{N}{N - C} \cdot M(V_c) \quad and \quad \frac{N}{C - 1} \cdot V(M_c).$$

The expected value of the ratio of the two statistics is therefore unity, if the null hypothesis is correct. Thus we can test the null hypothesis, if we can determine the sampling distribution of this ratio in accordance with the assumption stated w.r.t. its mean value, i.e.

$$\frac{(C - 1)M(V_c)}{(N - C)V(M_c)} \simeq 1 \quad . \quad . \quad . \quad . \quad . \quad \text{(iv)}$$

In accordance with (iii) of 7.03 our estimate of V_u based on V, the total variance of the N-fold sample distribution, itself would be

$$V \simeq \frac{(N - 1)}{N} V_u \quad . \quad . \quad . \quad . \quad . \quad \text{(v)}$$

Another criterion of homogeneity of the grid is admissible if we formulate the problem with due regard to the possibility that the allocation of the bonus to both teams of the final scoreboard of the model of Fig. 84 may be heterogeneous by recourse to the relation between : (a) the total observed variance (V); (b) the variances of both sets of means. That is to say

$$V = V(M_r) + V(M_c) + V_c \qquad . \qquad . \qquad . \qquad . \qquad \text{(vi)}$$

In accordance with (vii) of 10.04

$$V_e = \frac{\Sigma\Sigma(x_{rc} - M_r - M_c + M)^2}{N}.$$

The expected value is deducible from (v) and (iii) above, *viz.* :

$$E(V_e) \simeq E(V) - E \cdot V(M_r) - E \cdot V(M_c)$$

$$= \frac{N-1}{N} \cdot V_u - \frac{R-1}{N} \cdot V_u - \frac{C-1}{N} \cdot V_u$$

$$= \frac{(N - R - C + 1)}{N} V_u.$$

Since $N = RC$,

$$N - R - C + 1 = (R - 1)(C - 1),$$

$$\therefore E \cdot (V_e) \simeq \frac{(R-1)(C-1)}{N} V_u,$$

We thus have : (a) two estimates of V_u respectively based on $V(M_r)$ and $V(M_c)$; (b) a third estimate based on V_e defined by (viii) and (ix) in 10.03 being the difference between their sum and the total variance. Accordingly, a test of homogeneity which takes stock of the variance of the means in both dimensions is

$$\text{(a)} \quad \frac{N \cdot V_e}{(R-1)(C-1)} \simeq V_u \simeq \frac{N}{(R-1)} V(M_r).$$

$$\text{(b)} \quad \frac{N \cdot V_e}{(R-1)(C-1)} \simeq V_u \simeq \frac{N}{(C-1)} V(M_c),$$

$$\therefore \quad \frac{V_e}{(C-1)V(M_r)} \simeq 1 \simeq \frac{V_e}{(R-1)V(M_c)} \qquad . \qquad . \qquad . \qquad \text{(vii)}$$

The credentials of tests devised by R. A. Fisher and by Snedecor for assessing the significance of ratios defined by (iv) and (vii) rest on a theoretical foundation we shall explore in Vol. II. What calls for emphasis in this context is that the several criteria of homogeneity applicable to a grid set-up of experimental data can stand up to criticism without the testimonial of a balance sheet purporting to set forth additive components of variation, each uniquely attributable to a particular score.

The criteria themselves are deducible without any assumptions concerning the nature of the score distribution, normal or otherwise. On the other hand, significance tests based on them necessarily invoke such assumptions. Similarly, the validification of estimates of a balance sheet, if the construction of a balance sheet is indeed a legitimate aim, calls for special postulates about the distribution of the row and column score-increments ; and this itself imposes certain restrictions on the sampling process.